DISCARDED

Perspectives in American History

No. 24
PLOUGHS AND POLITICKS

PLOUGHS AND POLITICKS

Charles Read of New Jersey And His Notes on Agriculture

1715–1774

BY CARL RAYMOND WOODWARD

PORCUPINE PRESS
Philadelphia 1974

Library of Congress Cataloging in Publication Data

Woodward, Carl Raymond, 1890-
  Ploughs and politicks.

  (Perspectives in American history, no. 24)
  Reprint of the 1941 ed. published by Rutgers University Press, New Brunswick, N. J., which was issued as no. 2 of Rutgers University studies in history.
  Bibliography: p.
  1. Read, Charles, 1715-1774. 2. Agriculture--United States. I. Read, Charles, 1715-1774. Notes on agriculture. 1974. II. Title. III. Series: Perspectives in American history (Philadelphia) no. 24. IV. Rutgers University, New Brunswick, N. J. Rutgers University studies in history, no. 2.
F137.R3W6 1974   974.9'02'0924 [B]   73-16351
ISBN 0-87991-338-X

First edition 1941
(New Brunswick: Rutgers University Press, 1941)

Reprinted 1974 by
PORCUPINE PRESS, INC.
1317 Filbert St.
Philadelphia, Pennsylvania 19107

Manufactured in the United States of America

TO

JAMES CHARLES READ

WITH AFFECTION AND RESPECT

Contents

Foreword   xi

BOOK I

Charles Read of New Jersey

One: The Man and His Times	3
Two: Youth	22
Three: New Jerseyman	39
Four: Customs Collector	54
Five: Land Speculator	64
Six: Countryman	70
Seven: Ironmaster	86
Eight: Secretary	97
Nine: Legislator	121
Ten: Councillor	145
Eleven: Colonel	164
Twelve: Indian Commissioner	179
Thirteen: Jurist	195
Fourteen: Exile	212

BOOK II

Read's Notes on Agriculture

Introduction	229
One: The Husbandry of the Soil	235
Two: The Husbandry of Plants	254

Three: The Husbandry of Animals	322
Four: The Husbandry of Bees	366
Five: Farm Structures and Farm Implements	368
Six: The Husbandry of the Household	385
Seven: Fisheries	399
Appendix:	
A. Sketch of Charles Read	404
B. Inventory of the Personal Estate of Charles Read IV	407
Bibliography	413
Glossary	443
Index	451

Illustrations

Title Page of Worlidge's Systema Agriculturae	xv
Page of Letter from Charles Read to Jared Eliot	xx
Specimens of Charles Read's and Benjamin Franklin's Handwriting	xxi
The London Coffee House, Philadelphia, Birthplace of Charles Read	25
Locale of Charles Read's Burlington Residence; Smith House and Site of Noble House	42
High Street, Old Burlington	59
Log Stable near Lumberton and Ruins of Old House at Sharon	66
Map of Sharon, 1750	72
Map of Breezy Ridge, 1761	77
Sites of Taunton Furnace and Etna Furnace	88
Old Store at Batsto and Old Hotel at Atsion	94
Old Barracks at Trenton and at Burlington	176
Council with the Indians at Easton, 1758	187
Charles Read's Letter on Indian Policy	191
Page on Red Clover Grass from Read's Manuscript	256
Relics of Charles Read's Time	266
Page of Worlidge's Text showing Read's Marginal Notes	282
Vegetable Gardening Calendar from Read's Notes	294
Diagram of Oak Leaf	321
Charles Read's Sharon as it appears today	324
Hickory Grove and first Library Building at Burlington	338
Plan of Five-Post Barracks	370
Diagram of Bucket Engine	382
Pattern for a Cap	397
Diagram of Fishing Weir	400

Foreword

THIS book is the by-product of a search for a farm. When in 1840 Jared Sparks's edition of *The Works of Benjamin Franklin* appeared, it included a portion of a letter addressed to Jared Eliot—clergyman, physician, and farmer of Connecticut—presumably written by Franklin about 1749.[1] Although the original of this letter was incomplete, the signature and the date missing, it was with a collection of letters, later acquired by Yale University, which bore Franklin's signature, and was therefore accepted as his work. It told of a farm of about three hundred acres which the author had purchased "near Burlington," and described his efforts to bring it into profitable cultivation. For the hundred years since Sparks's collection was published, the belief has prevailed that Benjamin Franklin, in addition to his many other interests, was for a time a New Jersey farmer. The letter was reprinted in subsequent editions of Franklin's works and the farm was repeatedly mentioned by Franklin's biographers. Strangely, no one knew where the farm was situated. It was believed by some to have been the country estate subsequently acquired by William Franklin, the philosopher's son, who was the last royal governor of New Jersey, but a tracing of the title showed that this place was never owned by Benjamin. Searches of the records at the State Capitol and at the county seat yielded no clue of Benjamin Franklin ever having owned any land in New Jersey, nor was any specific reference to the New Jersey farm found in Franklin's voluminous papers.

About fifteen years ago, in making a study of the history of agriculture in New Jersey, I became interested in the mysterious Franklin farm. Burlington County, I found, was a veritable treasure ground for the agricultural historian, and I was fortunate in enlisting the aid of Charles A. Thompson,

[1] Jared Sparks, *The Works of Benjamin Franklin* (Boston, 1840), VI, 83-86.

Foreword

then county agricultural agent, and of his assistant, Leonard R. Smith. On one of my visits to the county agent's office in 1929, Mr. Thompson placed an ancient volume in my hands. It was coverless, and browned with age, the edges worn away, and the title page gone. It proved to be a copy of *Systema Agriculturae* by John Worlidge, the now rare London edition of 1681, but more important than the printed text were the copious notes and records with which it was interleaved. The text was freely underlined, and the margins bore many notations. Scattered dates indicated that the entries were made between 1746 and 1777. It was a find to delight the heart of an agricultural historian. I recognized at once that here was valuable source material, as rare of its kind as it was rich in the facts of colonial agriculture, but I did not then foresee that the browned and coverless book would also furnish the key to the mystery of the Franklin farm.

Mr. Thompson explained that the volume had been given to him by the county librarian, who had received it in a lot of books once owned by the family of the late Joseph J. White. Miss Elizabeth White, daughter of Joseph, however, was unable to say where her father had acquired it. By arrangement with Mr. Thompson the book was placed in the Rutgers University Library, and a careful study began.

The notes covered a wide range of agricultural topics: soil management, field crops, horticulture, animal husbandry, home management, apiculture, even fish culture. Although the name of the author was nowhere in evidence, clues appeared here and there in references to persons and to places. Altogether over one hundred persons were mentioned. These were listed, and the following in particular were selected for further study: "My farm in Springfield"; J. Hough, "a neighbor"; Edward Tonkin, "an honest neighbor"; a reference to "B. Franklin, Esqr."; "Oliphant's Mill"; "Joseph Moore's Land is very high Dry & poor opposite my Pl[an]t[atio]n on Ancocas"; "James Logan Read"; and "William Logan."

Was the author's name "James Logan Read"? A thorough

Foreword   xiii

search of *New Jersey Archives* and Burlington County records failed to reveal this name. But William Logan was a prominent Philadelphian, and "Read" was a common family name, associated with Benjamin Franklin. "Springfield" was obviously Springfield Township in Burlington County. If J. Hough, Edward Tonkin, and Joseph Moore were the author's neighbors, it appeared that a running down of the county records might disclose the author's name.

During my pursuit of these clues, several of which proved disappointing, Mr. George M. Sleeper, late editor of the *Mt. Holly Herald*, became interested and suggested that Mr. George DeCou, of Moorestown, could help me. On learning of my search Mr. DeCou promptly directed me to the Joseph Moore farm on the southern bank of the Rancocas Creek, between Hainesport and Lumberton. I visited the place on January 16, 1933, and found it answered the description in the notes.

Here, indeed, was a promising lead! The author had written, apparently in 1757 or 1758, that Moore's plantation was opposite his on the Rancocas—that is, on the north bank. Two farmhouses now stood across the river from the old Moore farm. Within an hour I had visited these places, ascertained their present ownership, and something of their earlier history. At the county clerk's office I examined the record of their ownership, and at last came across the name "Read." On the northern bank of the Rancocas not far from the Moore place, a parcel of land had been acquired by one Charles Read in 1747, about ten years prior to the date of the entry in the notes.

Following this clue to the State Capitol, I found in the index of deeds more than one hundred land transfers in the name of Charles Read. There was the plantation on the Rancocas; and there was the farm in Springfield Township. Jonathan Hough and Edward Tonkin had owned neighboring farms, while Read had owned another tract near Oliphant's

Mill. This was strong evidence that Charles Read was the author, but positive proof was still lacking.

A search of various documents for biographical facts disclosed that Charles Read had been an outstanding figure in pre-Revolutionary New Jersey. Agriculture had been one of his many interests. Point after point of his life story was found to check with the agricultural notes—dates, names, facts, while conflicting evidence was negligible.

The final test was the handwriting. Among the Pemberton Papers belonging to the Historical Society of Pennsylvania, I found many of Read's signed letters. When compared with the manuscript of the notes on agriculture, the handwriting was found without question to be the same. In this, my own conclusion was confirmed by Miss Mary Townsend, specialist in the manuscript department of the society's library.

The handwriting, which alone might be adequate proof that Read was the author of the notes, was supported by numerous other points, most of which by themselves would prove little, but taken in the aggregate leave no room for doubt. It is upon this evidence, here summarized, that the notes are presented as the work of Charles Read:

1. The notes deal mainly with Burlington County, the home of Charles Read.

2. Evidently they were written by a man of education and culture. Read studied under private tutors in Philadelphia, and continued his education in England.

3. The author refers to many English books on agriculture. Read had access to the English works in his father's library and in the Logan library and to the collections of the library companies of Philadelphia and of Burlington.

4. The notes mention the island of Antigua, and contain several other references to the West Indies, including observations made there. Read visited the West Indies in his early manhood, and married the daughter of an Antigua planter.

5. In the notes, the author refers to an acquaintance in

St. Croix. Read visited St. Croix the year referred to in the notes.

6. The author mentions over one hundred persons, most of them from Burlington County, some of them, such as Governor Belcher and Robert Ogden, prominent in the life of the colony. Read, as a provincial official, had an intimate personal acquaintance with these leaders of the colony.

7. The notes indicate an interest in shipping and commerce, containing computations of profit, and other accounting data. Read, the son of a merchant, at one time had been a sailor, had engaged in trade, and later became collector of the port of Burlington.

8. The author apparently corresponded with Benjamin Franklin. There is evidence that Charles Read was a relative of Dr. Franklin's wife, Deborah Read.

9. The author gives a "Mr. Waddingham" as the source of certain facts about corn culture in "Carolina." Samuel Waddingham, a planter of South Carolina, married one of Charles Read's nieces.

Title page of Worlidge's Systema Agriculturae

10. The author refers to "my farm in Springfield." During the years referred to in the notes, Read owned Sharon, a farm of over two hundred acres in Springfield Township.

11. The author mentions Edward Tonkin, J. Hough, and others as his neighbors. These men owned farms near Read's Sharon.

12. The author refers to Joseph Moore's farm as "opposite my Pl[an]t[atio]n on Ancocas." Read owned several tracts on the north side of the south branch of the Rancocas near Hainesport, one of which was across the creek from the Joseph Moore farm.

13. The book contains the names Alice T., Elizabeth, Lucy Ann, and Jane Chambers, under the date 1851. Anne Troth, a great-granddaughter of Charles Read, married Charles Chambers in 1816. This may account for these names.

14. The name "James Logan Read" is written boldly across a page in the body of the book. One of Charles Read's great-grandsons was James Logan Read, a Methodist clergyman of Ohio and western Pennsylvania.

15. The author of the notes discusses fowl meadow grass and frequently refers to Jared Eliot's *Essays*. William Logan, of Philadelphia, in writing to Jared Eliot, mentions "my Cosn Cha. Read of Burlington" as testing some fowl meadow grass seed which Eliot had sent, indicating Read's interest in agriculture.

16. The author devotes a section to fishing, and presents a diagram of a fish weir. Charles Read in 1763 leased a fishery on the Delaware River below Trenton.

17. The handwriting of a large collection of Charles Read's signed letters corresponds minutely with the handwriting of most of the agricultural notes. A few entries in a different hand correspond with the handwriting of his son Charles. Handwriting experts concur in this opinion.

WITH the authorship established beyond reasonable question, the notes were edited for publication and a biography of the author was prepared to accompany them. Meanwhile, the search for the Franklin farm went forward intermittently for several years, without success. Then, unexpectedly, came word from Mr. DeCou. He had examined the Eliot letter in the Yale University library, he said, and had some doubt that it was written in Franklin's hand. I was reminded at once of the account of an experiment with red clover in Read's notebook which was strikingly similar to the description of an experiment in the Eliot letter, and I compared the text of the two. So closely did the dates, measurements, and other details correspond, that I suspected they related to the same experiment. I visited Yale University, and with Mr. Bernhard Knollenberg, the librarian, examined the Eliot letter. It was carefully compared with several signed letters from Franklin to Eliot and also with the entries in Read's notebook. We were convinced that Franklin was not the author of the letter. We concluded also that the experiment described in the notebook was the same as the one in the letter, but we were not sure that the handwriting in both cases was identical. The penmanship of the Eliot letter did not seem to be in Read's customary style, but it did correspond closely with certain pages of the notebook, which differed somewhat from the typical script, and which we thought might have been copied by another person. The question now before us was— Who was the author of the letter to Eliot? Could it have been Charles Read?

In the company of Dr. Linwood Lee, of the United States Soil Conservation Service, I visited the farm which Read had owned in Springfield Township to see if it answered the description in the letter. The features we sought were eighty acres of "deep meadow," the soil "Fine and black" underlain with "a fatt Bluish Clay," also a "Round Pond of Twelve Acres" drained for planting, all of which were mentioned in the letter. This time we were not disappointed. The eighty-acre meadow, the deep black muck, the layer of blue-green marl, the twelve-acre depression that was the pond —here they all were. This, then, in all probability was the long sought "Franklin farm"; only it had belonged to Charles Read instead of Benjamin Franklin!

For further advice, I consulted Dr. Samuel Tannenbaum, hand-writing specialist, of New York, who on comparing the letter with signed Read papers, and without consulting its contents, declared it was in Read's hand. So, in his opinion, was the main body of the agricultural notes, including many pages whose variation had puzzled us. The significance of Read's notebook, therefore, as a contribution to agricultural science was substantially augmented by the Eliot letter. By the same token, a share of the reputation heretofore accorded to Franklin as an agricultural scientist must now be accredited to Charles Read.

To sum up, we have the following evidence that Franklin was not the author of the letter to Eliot:

1. Since the last page of the letter is missing, no signature is available for identification.

2. In 1751, Franklin wrote Eliot "for want of Skill in Agriculture I cannot converse with you on that valuable Subject."[2] It seems improbable that he would have written this if two years earlier he had corresponded about farm experiments.

3. Aside from this letter, no indication has been found that Franklin ever owned a farm in New Jersey.

4. The handwriting is not Franklin's.

Supporting the above on the positive side is the following evidence that the author of the letter was Charles Read:

1. The letter begins with the saluation "Sr." This was characteristic of Read's letters.

2. The address and date line are not on the first page. In Read's letters these were customarily noted on the last page with his signature.

3. The author of the letter mentions his good fortune and ability to indulge himself. Read's wife was wealthy, and he had a good income.

4. The author of the letter mentions Eliot's *Essays upon Field Husbandry*.[3] These are frequently quoted in Read's agricultural notes.

5. The handwriting is identical with that of several pages

[2] Albert H. Smyth, *The Writings of Benjamin Franklin* (New York, 1905), III, 61.
[3] Jared Eliot, *Essays upon Field Husbandry in New England* (1748-1760).

in the notebook. A handwriting expert declares the letter is in Read's hand.

6. The author of the letter, which appears to have been written in 1749, states that eighteen months previously he had bought "about three hundred Acres of Land near Burlington." In 1747, Read purchased a 212-acre farm in Springfield Township, and two other tracts aggregating 61 acres in Northampton Township, Burlington County.

7. Charles Read's Sharon is similar in soil and topography to the farm "near Burlington" described in the letter.

8. The account of an experiment with red clover in the letter corresponds in minute detail with certain entries in the notebook. Note the following comparison:[4]

From the letter:	From the notebook:
"of this *deep Meadow* I have about Eighty acres"	"The Ground was ye *deep Meadow* & very light"
"on *the twenty third of August* I sowed near thirty acres with red Clover and Herd grass, allowing Six quarts of Herd grass and four pound of red clover to an acre in most parts of it, in other parts four quarts Herd grass and three pound red clover"	"I sowed *ye 23d of August* it did well" "in Aug. 1749 I Sowed abt 10 lb [red clover] to an Acre"
"the red Clover came up in *four days* and the Herd grass in Six days"	"in *4 days* it appeared with 2 leaves"
"where I allowed *the most Seed* it protects itself the better against the *Frost*"	"where it was *thickest* it stood the *winter* best"
"the Red Clover *heaves out* much for want of being thicker" "the Herd grass" was "less affected by the Frost than the red Clover whose roots I measured in the *last of October*, and found that many of their *Tap*	"in ye End of October ye tap root was abt 5 inches long & from it went near 30 horrizontal roots of wch some were 6 inches long, & branched yet it hove out much where thin sown"

[4] The italics do not appear in the original, but are given here to aid in the comparison.

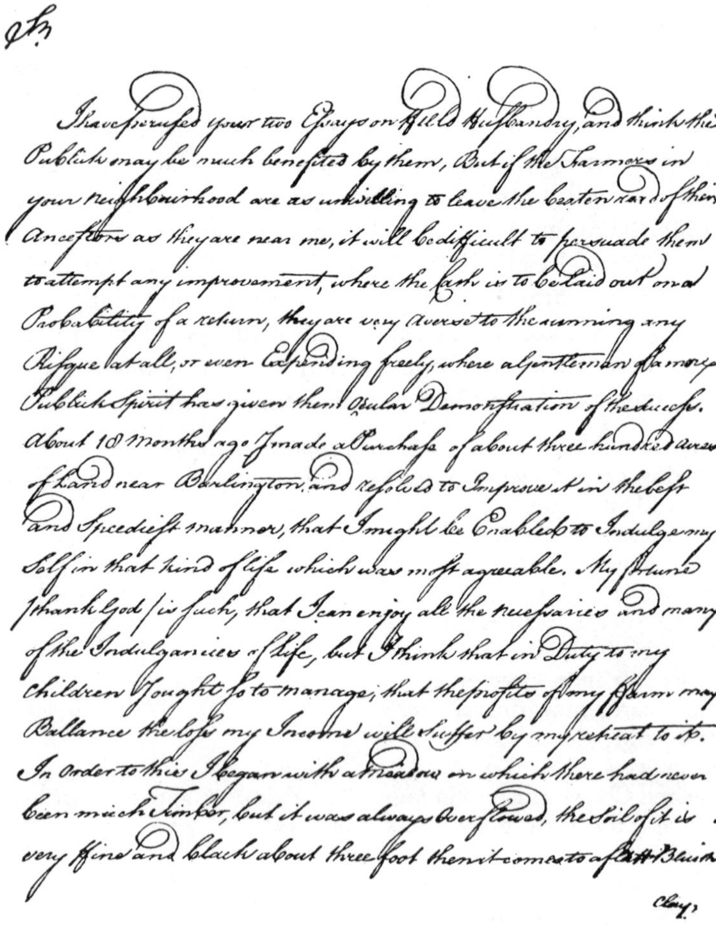
Sir

I have perused your two Essays on Field Husbandry, and think the Publick may be much benefited by them, But if the Farmers in your neighbourhood are as unwilling to leave the beaten road of their Ancestors as they are near me, it will be difficult to persuade them to attempt any improvement, where the Cash is to be laid out on a Probability of a return, they are very averse to the running any Risque at all, or even Expending freely, where a gentleman of a more Publick Spirit has given them Ocular Demonstration of the success. About 18 Months ago I made a Purchase of about three hundred acres of Land near Burlington, and resolved to Improve it in the best and Speediest manner, that I might be Enabled to Indulge my Self in that kind of life which was most agreeable. My Fortune (thank God) is such, that I can enjoy all the necessaries and many of the Indulgancies of life, but I think that in Duty to my Children I ought so to manage, that the profits of my Farm may Ballance the loss my Income will suffer by my retreat to it. In Order to this I began with a Meadow in which there had never been much Timber, but it was always overflowed, the Soil of it is very Fine and black about three foot then it comes to a Stiff Bluish Clay,

FIGURE I.

FIGURE I. *First page of letter to Jared Eliot, in which Charles Read describes Sharon;* FIGURE II. *Entry in Read's notebook showing comparison of handwriting with that of letter to Eliot;* FIGURE III. *Portion of letter from Benjamin Franklin to Jared Eliot, July 16, 1747, showing the contrast with Read's handwriting.*

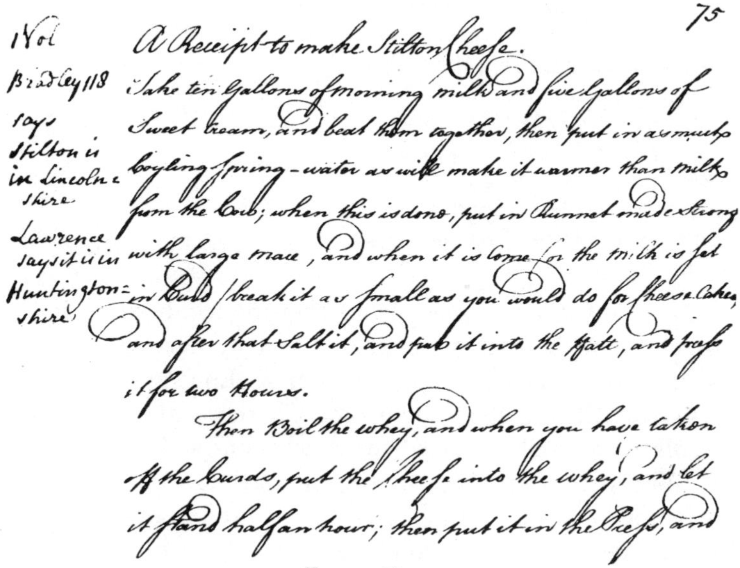

Foreword

1 Vol
Bradley 118
says
Stilton is
in Lincoln-
shire

Lawrence
says it is in
Huntington-
shire

75

A Receipt to make Stilton Cheese.

Take ten Gallons of morning milk and five gallons of Sweet Cream, and beat them together, then put in as much boyling Spring-water as will make it warmer than milk from the Cow; when this is done, put in Runnet made strong with large mace, and when it is come, for the milk is set in Curd, break it as small as you would do for Cheese Cakes, and after that Salt it, and put it into the Fatt, and press it for two Hours.

Then Boil the whey, and when you have taken off the Curds, put the Cheese into the whey, and let it stand half an hour; then put it in the Press, and

FIGURE II.

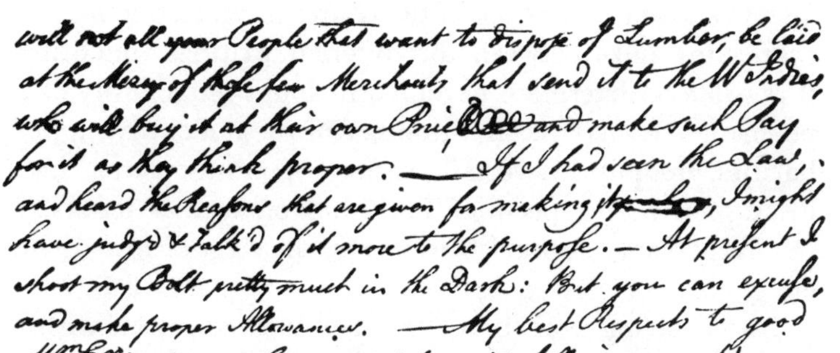

will not all your People that want to dispose of Lumber, be laid at the Mercy of those few Merchants that send it to the W. Indies, who will buy it at their own Price, and make such Pay for it as they think proper. — If I had seen the Law, and heard the Reasons that are given for making it so, I might have judg'd & talk'd of it more to the purpose. — At present I shoot my Bolt pretty much in the Dark: But you can excuse, and make proper Allowances. — My best Respects to good Mr Elliot & your Sons; and if it falls in your Way, my Service to the kind hospitable People near the River, whose Name I am sorry I've forgot. I am, &c. with the utmost Regard,

Your obliged humb Servt.

B. Franklin

FIGURE III.

roots penetrated five Inches, and from it's Sides threw out near thirty Horizontal roots, some of which were Six Inches long and branched . . . wherever it is thin Sown it is generally hove so far out, as that but a few of the Horizontal and a Small part of the Tap root remain covered"

Dr. Tannenbaum, after comparing the handwriting in the notebook with original letters written by Charles Read, as well as with photostats of the letter to Eliot and of several letters by Charles Read, by his son Charles, and by Benjamin Franklin, as well, reported his conclusions as follows:

"The specimens of handwriting of Charles Read, Sr., covered a period of about thirty years and showed many variations in the style and formation of his letters. Notwithstanding all these factors and the natural evolution which the writing of a fluent penman undergoes in the course of years, one may find connecting links between the different writings which make it possible to determine beyond any doubt the identity of the penman. You may recall a number of these identifying peculiarities, notably his certain variety of minuscular *r*, his *ff*, his capital *P*, his peculiar flourishes on capital letters, his relatively scanty abbreviation marks, the slant of his writings, etc.

"After having thus thoroughly familiarized myself with his writing habits, I examined the photostats of the four pages of the alleged Benjamin Franklin letter at Yale. In them I discovered so many of Charles Read's writing peculiarities, I could not help reaching the conclusion that these four pages are in the handwriting of Charles Read and not of Benjamin Franklin. To put this matter beyond a doubt, I made a careful analysis of the handwriting of Benjamin Franklin in a two-page letter written by him on July 16, 1747. In the Franklin letter we have none of Read's identifying peculiarities. On

Foreword

the other hand, in the Charles Read papers and in the four questioned pages there are none of Franklin's characteristic marks.

"In the printed book, with two or three exceptions, all the memoranda are in the handwriting of Charles Read, Sr."[5]

READ'S NOTES, of course, were not written for publication, and are too sketchy to give a complete picture of the agriculture of colonial New Jersey. However, they set forth a host of facts and comments which add substantially to our understanding of the farm practices of his time. As presented in Book II of this volume the notes are rearranged, with annotations and analytical comment, but otherwise they have not been changed from the original form.

The biography which comprises Book I has been built in part around the framework happily left us in the diary of a contemporary. Aaron Leaming, shrewd and successful Quaker, was associated with Read for many years in the provincial assembly and in numerous public enterprises. Upon learning of Read's death, Leaming entered in his diary a pungent summary of his friend's career. The diarist's critical comments make possible an appraisal of Read's character as well as an interpretation of his activities which could not be drawn from any other source.

Such a kaleidoscopic character as Read's is a problem for the biographer. It would be next to impossible, at best it would be disconcerting to the reader, to treat all the episodes of his life in strict chronology. The facts assembled fall naturally into two groups: those of a personal nature, together with matters of private enterprise; and those dealing with his public career. In order to provide a narrative with a minimum of interruption, his entire life is sketched in broad outline in the opening chapter. The six chapters which follow present in approximate chronology matters which deal mainly with his private interests. Then we turn back to the begin-

[5] Samuel A. Tannenbaum to Carl R. Woodward, Sept. 5, 1940.

ning of his public career, and in the remaining chapters follow his fortunes as they were interlaced with the affairs of the colony. The several chapters are shaped around the principal activities of his life, both private and public, and for this reason they represent a compromise betwen chronology and topical treatment.

WITHOUT the generous cooperation of a host of friends, this book would never have been written. My indebtedness to them is greater than I can express. I have already mentioned the assistance of Messrs. DeCou, Thompson, Smith, Sleeper and Knollenberg, Dr. Lee, Dr. Tannenbaum and Miss Townsend. To them must go a large measure of credit for solving the authorship of the notebook and of the alleged Franklin letter. Also, I am especially indebted to Mr. Franklin Bache, of West Chester, and to Mr. George Simpson Eddy, of New York City. By suggestion and counsel, and by giving me access to their personal collections, they contributed substantially to clearing up the mystery of the Franklin farm.

To mention all the persons who have assisted me in my library searches would require more space than is at my disposal. I am under a lasting obligation to Mr. George A. Osborn, librarian of Rutgers University, and to the members of his staff who cheerfully and efficiently have performed countless favors. I am likewise indebted to Miss Katharine E. Rogers, of the New Jersey State Library; to Mrs. Maud H. Greene, librarian of the New Jersey Historical Society; to Mr. Charles S. Aitkin, archivist of the State of New Jersey; to Dr. Victor Hugo Paltsits, of the New York Public Library; to Dr. Julian Boyd, formerly librarian of the Historical Society of Pennsylvania, now of Princeton University; to Dr. Austin Gray and Mr. Barney Chesnick, of the Library Company of Philadelphia; to Miss Claribel R. Barnett, librarian of the United States Department of Agriculture; to Miss Belle daCosta Greene, of the Pierpont Morgan Library; to Miss E. Marie Becker, of the Monmouth County Historical

Foreword

Society; to Mr. Edward Post, of the Cape May County Historical and Genealogical Society; to Miss Mary S. Aslin, librarian of the Rothamsted Experimental Station, at Harpenden, England; to all these and to the members of their staffs who assisted me in one way or another. I gratefully acknowledge courtesies also received while working at the Library of Congress and the libraries at Harvard University, Columbia University, the Massachusetts Historical Society, and the New York Historical Society, and through correspondence with the North Carolina State Library, the Carnegie Library of Pittsburgh, and the libraries of Stevens Institute of Technology, the University of Pennsylvania, the University of Pittsburgh, Princeton University, and the American Antiquarian Society.

Many others helped by correspondence, by investigating clues, or by sending me memoranda and original material. Among these are Miss M. Atherton Leach, Mr. Joseph B. VanSciver, Mr. Bertram Lippincott, Mr. Barclay White, and Mr. Charles Montgomery, all of Philadelphia; Mr. Henry C. Beck, Haddonfield; the Reverend John Talbot Ward and the Misses Margaret and Caroline Haines, of Burlington; Mr. Herbert N. Moffett, of the Historic American Buildings Survey; Mr. Jonathan H. Kelsey and Mr. Horace Newbold, of Mt. Holly; Miss Gertrude Newbold, Bordentown; Miss Caroline Allinson, Yardville; Judge Harold E. Pickersgill, Perth Amboy; Mr. Thomas C. Summerill, Salem; Mr. Walter R. White, Westtown, Pa.; Dr. Roscoe West, of the Trenton State Teachers College; Mr. Hilary Tilghman, owner of Pine Lane Farm, at Jobstown; and Dr. Sydney L. Wright, of the Franklin Institute. Especially valuable has been the assistance given by Mr. John M. Okie, of Lansdowne; by Mr. H. Clifford Campion, of Swarthmore; by Mr. Nathanial R. Ewan, of Moorestown; by Dr. J. Bennett Nolan, of Reading; and by Dr. Harry J. Carman, of Columbia University.

I have been greatly encouraged in preparing the work by the generous interest of Mr. James Charles Read, of Greenwich, Conn., a descendant of Charles Read, who placed the family papers at my disposal. In the revision of the manuscript, I have been immeasurably helped by my colleagues at Rutgers University, whom I am restrained from naming, and by other friends. Dr. William T. Hutchinson, of the University of Chicago, Dr. Herbert A. Kellar, of the McCormick Historical Association, and Dr. Robert T. Thompson, of the Rutgers faculty, each read the manuscript and made valuable suggestions. I am indebted also to Mr. Charles A. Philhower, of Westfield, for criticizing Chapter XII, and to Mr. George Miller, of Perth Amboy, for examining Chapter XIII. Finally, I wish to thank the members of my office staff and of my family who have shared the physical heat and burden of getting the book into print.

Such merit as this book may possess is due in large measure to these co-workers, whose sympathetic and understanding interest I value even more than their material aid.

CARL R. WOODWARD

New Brunswick,
New Jersey
May 1, 1941

BOOK I

Charles Read of New Jersey

One: The Man and His Times

A MAN of many affairs was Charles Read, and a dominant figure in pre-Revolutionary New Jersey. Scarcely an event of major importance transpired in the colony during his time in which he did not play some part. Born and bred in Philadelphia, schooled in London, trained in the British navy, and married in the West Indies, in 1739 he came to New Jersey to begin a career which in a unique way was identified with the pioneering activities of the colony. Merchant, planter, ironmaster, soldier, jurist, statesman, Charles Read was a man whose life throbbed to the gripping interplay of forces—social, economic, political, and military—which were forging a new nation. In and about his public services and his private enterprises is woven the history of New Jersey through the three decades which preceded the Revolution. In his life is dramatized the story of a commonwealth in the making.

This Charles Read was the third of a primogenitive line of five sires and sons, all bearing the identical name, which in three generations reached the heights of distinction, and in two more touched the depths of disgrace. His grandfather, Charles Read I, was a Quaker who at the age of twenty-eight left the ancestral estate Trevascon,[1] in Cornwall; joining a group of emigrants to the New World, he arrived at Burlington about 1679, and subsequently settled in Philadelphia.[2] His father, Charles II, had a mercantile business in Philadelphia, and was for a time mayor of the city. Charles III, the subject of this study, inherited his father's commercial tastes

[1] Benjamin and Jane Harvey to Israel (?) Pemberton, Apr. 16, 1755. Clement Papers, Liber J, 28.
[2] Samuel Smith, *The History of the Colony of Nova Caeserea, or New Jersey* (2nd ed., Trenton, 1877), p. 109. Although positive documentary proof is lacking, some records indicate that Charles Read I was a passenger on *The Shield*, which arrived at Burlington in December, 1678, the first ship to sail that far up the Delaware.

and interest in public affairs, but instead of exercising them in his native city, he chose New Jersey as the field of his career.

When Charles III established his home at Burlington, fifty years had passed since his grandfather had arrived at this same place, and it was a hundred years since the Swedes, together with a few Finns, had founded their modest colony on the banks of the lower Delaware. During the century the population of New Jersey had grown to 50,000, and at the time was increasing at the rate of more than 1,500 a year.[3] Eleven counties had been formed—Bergen, Essex, Morris, Middlesex, Somerset, and Monmouth in the Eastern Division; Hunterdon, Burlington, Gloucester, Salem, and Cape May in the Western Division. Centrally situated among the English colonies that bordered the Atlantic, midway between commercial New England and the agricultural South, New Jersey, more than any other colony, had become a melting pot of European peoples who differed widely in character, language, and habits. In the Middle Colonies, with their mixed populations, was fused a culture more typically American than that of New England or of the South. In the middle member of these Middle Colonies, this was particularly so. Though New Jersey was small in area, its long coastline and navigable waters offered easy access to the immigrant. Of the thirteen colonies it was the only one whose political boundaries almost wholly followed natural frontiers. To the east were the Hudson River, the Raritan Bay, and the Atlantic Ocean; to the west and south the Delaware River and Bay.

Before and after the Swedes, and greatly outnumbering them, came the Dutch, who tarried along the borders of the Hudson, the Passaic, the Hackensack, and the Raritan. Rural-minded, they shrewdly selected for their homes the fertile valleys akin to their own native lowlands. To both the Eastern

---

[3] The total population of New Jersey in 1737 was 47,402, of whom 3,981 were slaves (Eastern Division 25,437; Western Division 21,965); in 1745 it was 61,383, of whom 4,605 were slaves (Eastern Division 29,472; Western Division 31,911). (Hugh MacD. Clokie, "New Jersey as a Separate Royal Province," *New Jersey—A History*, I, 207-227).

The Man and His Times

and the Western Divisions came the English, attracted by enthusiastic reports of a rich soil, abundant vegetation, and pleasant climate. Large numbers of these English were of the Quaker faith, especially those who settled in the Western Division.[4] They were on the whole a homogeneous group—middle-class farmers, craftsmen, and tradesmen. To the Eastern Division, too, especially to Monmouth County came Quakers and Baptists from New England and Long Island to escape harsh treatment at the hands of the Puritans and the Dutch. Lured also, perhaps, as much by the spirit of adventure as by the quest for religious freedom, they and other New England groups, notably the Congregationalists, were in the forefront of the great westward movement. Smaller numbers of Scotch and Irish were among the early immigrants—the former to the eastern counties, the latter to Gloucester. In the Eighteenth Century groups of Scotch-Irish penetrated into the western counties. Likewise, modest groups of Germans and Moravians settled in Somerset, Morris, and Hunterdon Counties, and some French Huguenots came to Bergen County.

Though of diverse origins, these peoples were bound together by certain common aims—to escape the oppression of Europe, to enjoy liberty of religion and of conscience, to leave behind personal economic troubles and get a new start, to wrest a more satisfactory living from the soil of the new land. What they sought was a commonwealth built by axe and plow, not by sword and gun, and New Jersey was born of this dream. Except for some discrimination against the few Catholics in the colony, in spite of strongly opposing points of view the spirit of tolerance generally prevailed. All sects were united in opposition to ecclesiastical despotism. Episcopalians, Presbyterians, Dutch Reformed, Congregationalists, Quakers, Lutherans, and Baptists severally established their own congregations. Sectional feeling, jealousies, and rivalries were present, but in the main the different groups lived in

[4] In 1745 there were 3,557 Quakers in the Eastern Division, 6,079 in the Western Division. (J. R. Sypher and E. A. Apgar, *History of New Jersey*, . . . Philadelphia, 1871, p. 264).

harmony and worked cooperatively. Houses of worship reared their spires in the midst of the growing communities. During the Great Awakening, the ecclesiastical giants, Whitefield, Tennent, and Frelinghuysen, found the harvest of human souls in central New Jersey as rich as the produce of its fertile soil. Under the stimulus of their vigorous preaching, thousands gained a new perspective of life.

The settlements in the main followed the courses of the navigable streams, and pushed inland along the crude roads which were emerging from the old Indian trails. Most of the early farms fronting on river or creek had their own landings and boats. The principal towns of the province were growing up in a belt which stretched across the colony from the mouth of the Hudson southwesterly to the mouth of the Delaware: Bergen, settled by the Dutch; Newark, founded in 1666 by a group of Congregationalists from New England; Elizabethtown, the first English settlement (1665); Woodbridge, a few miles to the south; Perth Amboy, the capital of the Eastern Division; New Brunswick, at the ford of the Raritan; Trenton, at "the Falls of the Delaware"; Bordentown, a few miles down the river; Burlington, the capital of the Western Division; Gloucester across the Delaware from Philadelphia; and the Quaker town of Salem, the first English settlement (1675) in the Western Division.

This belt, following the natural channels of trade, joined the larger centers of New York and Philadelphia, and embraced the main routes of travel between them. The principal highway, the first of major import in the province, extended from Elizabethtown to New Brunswick; thence one branch continued to Trenton, soon to become the most popular of all routes, and another swung southerly to Burlington. The passenger bound from Philadelphia to New York could either go by land to the ferries on the Delaware at Burlington or at Trenton, or by boat up the river to these towns, thence overland by stage wagon to New Brunswick, where he had the option of continuing on to Elizabethtown and ferrying to New

The Man and His Times 7

York, or of transferring at once to boat for the trip down the Raritan and up New York Bay. Another route frequently used was overland from Burlington to Perth Amboy, thence by boat to New York. Over jolting roads ungraded and crude, and on frail craft subject to wind and tide, the journey was fraught with discomfort and at times with danger. The ninety miles between Philadelphia and New York usually took the greater part of a week, but with good luck might be covered in three days. The postal service was irregular at best. Post riders carried letters for their own profit, but would tarry on their way to transact business for their customers. Competing with them were stage drivers and travelers who welcomed the chance to carry letters for small fees. Travel was improving, however, as new highways were opened and new stage lines established.[5]

The prevailing industry and thrift of the Dutch, the Scotch, the Quakers, and the Puritan English, were gaining for New Jersey a prosperity which invited a large immigration. But at the time Read moved to New Jersey, aside from these centers and the farming communities which were pushing out from them, together with a few more scattered settlements, New Jersey was still largely a wilderness—of wooded hills to the northwest, of pine forests and cedar swamps to the southeast. It offered a virgin field for the enterprise of a pioneering spirit like Charles Read's.

In sharp contrast with the religious harmony enjoyed by the New Jersey colonists, the province from the beginning had been a political battleground, and to a lesser degree it also suffered from economic strife, which lasted well into the national period. Within a single decade three flags flew successively over New Jersey soil, as the Dutch took possession of New Sweden in 1654, only to be conquered by the British in 1664. When soon thereafter the Duke of York granted to Sir George Carteret and to Lord Berkeley as proprietors the lands which subsequently were the provinces of East New

---

[5] Wheaton J. Lane, *From Indian Trail to Iron Horse* . . . (Princeton, 1939), pp. 104-105.

Jersey and West New Jersey, the stage was set for a long series of political controversies. The clash between the proprietary and the anti-proprietary parties, which represented the conflicting interests of proprietors and freeholders, found its counterpart years later in the struggle between loyalist and patriot.

The plan of government, designed to encourage planters to migrate to the new country, was set forth in the "Concessions and Agreements,"[6] in effect a constitution dealing with matters of land settlement, private rights, and political organization. Philip Carteret, the first governor of East New Jersey, who was a cousin of Sir George, on arriving at Elizabethtown "went up from the Place of his Landing . . . carrying a Hoe on his Shoulder, thereby intimating his Intention of becoming a Planter . . ."[7] The subsequent sale of proprietary rights by Berkeley and Carteret to other proprietors, prominent among whom were William Penn and his Quaker associates, stimulated the English settlements, but raised further questions of land rights and governmental authority.

The first settlers of Elizabethtown and of Monmouth County, who purchased their lands from the Indians under an earlier (the Nicolls) patent, protested against the yearly quit-rents laid by Governor Carteret and, in defiance of English law, claimed that the proprietors had no right to tax. Likewise in other communities, freeholders refused to pay their quit-rents; simple deeds of violence led to organized revolt; rioters were arrested, but encouraged by overwhelming popular support, they defied the court and refused to pay their fines. Clouded titles, obscure boundaries and careless surveying were further causes of controversy over lands. Popular resentment was stirred to a high pitch, passions were inflamed, courts were closed, and lawless disregard for official action resulted in indescribable confusion. Weary of this discord which had retarded the normal development of the

---

[6] *The Grants, Concessions, and Original Constitutions of the Province of New Jersey*, compiled by Aaron Leaming and Jacob Spicer (Philadelphia, 1755).

[7] *An Answer to a Bill in the Chancery of New Jersey* . . . (New York, 1752), p. 20.

The Man and His Times

province, unable to maintain order, and discouraged with a venture which had yielded them little or no profit, the proprietors proposed a surrender of the government to the Crown, but they, unlike the proprietors in other colonies, were to retain title to the land. Thus in 1702, soon after the accession of Queen Anne, the separate proprietary colonies of East New Jersey and West New Jersey became the united royal province of New Jersey. Practically the whole of New Jersey at this time was under Quaker control. Men of conviction, self-reliant and determined, the Quakers were to play a part in New Jersey history quite out of proportion to their numbers. Their mild and tolerant spirit leavened the whole course of the colony's growth.

Although the proprietors were now relieved of the trials and tribulations of government, the sources of controversy had not been removed. The new constitution was not so liberal as the old "Concessions and Agreements," and furthermore, there were now two authorities in the province, one of government and one of property. The governor, council, and assembly were responsible for the ordinary functions of government, while on the other hand, all affairs relating to the landed interests of the proprietors were entrusted to a council of proprietors in each of the old divisions. This body managed the sale and transfer of lands, purchased lands from the natives, ordered surveys, granted warrants, inspected the rights of claimants, and when they could, collected quit-rents.

Inevitably conflict followed—conflict between the proprietors and the Crown; conflict between the assembly, representing the common people, and the governor and his council, who represented the royal authority. When personal rights were challenged, quiet Quakers and peaceful Puritans could display a spirit which would do credit to the most militant partisans. For seventy years and more, until the final break with the mother government, the struggle continued, now violent, now subdued, under a succession of royal governors who displayed varied degrees of executive ability and statesmanship.

For thirty-six years after becoming a royal province New Jersey shared her governor with New York. Lord Cornbury, Lord Lovelace, Richard Ingoldsby, Robert Hunter, William Burnet, John Montgomerie, and William Cosby in turn each presided jointly over the two provinces. The rights of the proprietors, the powers of the assembly as compared with the royal prerogative, the support of military forces and the issuing of paper money, were matters of constant wrangling. Indifference, and ignorance of conditions in America on the part of the British Lords of Trade who were in charge of colonial affairs, did not improve the situation. Meanwhile the assembly petitioned that a separate governor be assigned to New Jersey, and the request was granted in 1738 when Lewis Morris was appointed the first full-time governor, independent of the authorities of New York.

Such turbulent events as these shaped the political scene for Charles Read's public career, which began with the administration of Governor Morris and came to fruition during the subsequent administrations of Governor Belcher and Governor Franklin.

In 1739 this 24-year-old Philadelphian purchased the clerk's office of Burlington, and there established his home. The political and the commercial center of West New Jersey, Burlington was a perfect locale for a colonial man of affairs. Opportunities beckoned, and Read was not slow to seize them. Shortly afterward he procured appointment as collector of customs at Burlington, and in a few years had added other offices in quick succession: clerk of the circuits, deputy-secretary of the province, and surrogate of the prerogative court. His progress up the official ladder was swift and determined. With Governor Lewis Morris he readily found favor, and he became the confidant and counsellor of Governor Jonathan Belcher who succeeded Morris. During Belcher's administration he was a commanding figure in the official family. He

The Man and His Times

served two terms in the assembly, during the first of which he was speaker. In 1749 he accepted appointment as associate justice of the supreme court, but five years later resigned to give more time to his personal affairs. Apparently without formal study in law, he was admitted to the bar and was said to have had the best run of practice of any attorney of his time. Later, under Governor Franklin he served as a member of the council, and also for a time as chief justice. "No man knew so well as he how to riggle himself into office nor keep it so long nor make so much of it."[8]

Read's mercantile heritage, his boyhood on the Philadelphia waterfront, his experience abroad and at sea, his close personal relations with the Philadelphia merchants—altogether were such as eminently to fit him for the office of customs collector. Once he had been appointed, it seems, this post was good for life. Even before the Eighteenth Century, New Jersey had a considerable export trade in the products of farm and of forest, shipped mainly from the ports of Burlington, Salem, and Perth Amboy. By the time Read became collector, however, New York and Philadelphia had outstripped the New Jersey ports, and had absorbed much of the colony's trade. During Read's regime the exports from Burlington continued to dwindle. By 1749 Governor Belcher reported that about twenty New Jersey vessels of 1,500 tons carrying 160 men were trading in "Provisions and Lumber" exported to Europe and the West Indies. The staple provisions besides timber were flour, beef, and pork, valued at £30,000 a year. In exchange, the ships brought imports of "Cloths, Hats, Cuttlery & other Smiths Work, most sorts of British Wollen Manufacture, and some East India Goods to the Value of £20,000 Sterling p Annum."[9] In contrast with these figures, within a few years after the French and Indian

[8] Aaron Leaming's Diary, Nov. 14, 1775 (see Appendix A of this book); also J. Granville Leach, "Colonel Charles Read," *Pennsylvania Magazine of History and Biography*, XVII (1893), 190-194.

[9] Jonathan Belcher to the Secretary of the Lords of Trade, Apr. 21, 1749; *Archives of the State of New Jersey*, 1st ser., VII, 244-245 (hereafter referred to as *N. J. Arch.*).

War, the exports from New York amounted to £526,000 annually, while Pennsylvania's exports were valued at £705,500.[10]

As collector of the port of Burlington, Charles Read was in the midst of the commercial activity of the colony. He saw the rise of New Jersey industries, and their inevitable focus upon Philadelphia and New York. He saw a declining export trade from New Jersey give way to a flourishing domestic commerce. He served as watchdog against the illicit trading which was common on New Jersey's exposed shores. He sought to enforce the embargoes against exports to enemy and neutral ports during the colonial wars. He dreamed of industrial progress and prosperity, and sought by legislation and by personal promotion to preserve and protect the infant industries of the colony from disintegrating forces.

Midst all his public duties, Read somehow found time for an extraordinary variety of private business ventures. On coming to New Jersey, he did not wholly relinquish his mercantile interests. In this pre-banking age, however, land represented the principal form of investment, and it was in the undeveloped lands of New Jersey that he saw his greatest business opportunity.

It was wholly natural that he should become a land promoter and speculator of the first magnitude. From 1740 to 1767, scarcely a year passed that he did not acquire some new tracts, by purchase or by proprietary grant, until altogether he made at least seventy-five separate purchases of lands. These were mainly in Burlington County, embracing more than 35,000 acres,[11] and for them he paid approximately £7,000. Likewise, the sale of at least 26,000 acres of these lands in fifty or more separate deals is recorded. He appears to have made no great profit in these transactions as a whole, although there were gains in some individual cases. Many of

[10] *American Husbandry* (London, 1775), I, 124, 181.
[11] From deeds recorded in the Office of the Secretary of State, Trenton. It is possible that some of the transactions noted were for another person. There is some indication, not well established, that there was another Charles Read in Burlington County during this period. However, collateral evidence provides ample proof that most of these land dealings, if not all, were by the subject of this study.

his holdings came from the location of original surveys on large proprietary grants. As a high official of the province, he was in a strategic position to receive special consideration from the proprietors, and he likely took full advantage of it. The largest of his holdings came from the purchase of proprietary rights to unlocated lands, and also from grants which the proprietors made directly to him.

Next to easy money from the speculative sale of his lands, the timber it bore held promise of substantial profits. So Read soon found himself a lumber merchant. When he came to the province, the heartless conquest of New Jersey's forests had been in progress for four-score years. Gradually, relentlessly, the steel implements of the settlers were gnawing away nature's gift of the ages. On rolling upland, the "brave oaks" which had awed the explorers, the ancient walnuts, chestnuts, hickory, and ash, fell before the onslaught of axe and saw. Native stands of juniper, poplar, fir, beech, birch, gum, pine, and maple—all were now bending to the will of man. Thousands of acres were slashed and burned, to lay bare to sun and rain the fertile loam. From out the forest rose the farmsteads—first the crude log structures, later the low-lying houses and broad-eaved barns, framed in massive whiteoak, and enclosed in durable cedar, forerunners of the more lasting brick houses. Posts of locust and rails of cedar fenced the open fields which were the pride of the hopeful husbandman.

Such an opportunity for exploitation could not be resisted by a person of Charles Read's temper. The spirit of the pioneer was abroad. Land and timber in abundance waited on the kind of human enterprise Read so abundantly possessed. Here and there on his large timber tracts, he dammed the streams and set up his sawmills. Here he turned out timber for farmsteads, shipyards, and city homes, which joined the procession of rafts and flatboats down the Rancocas and the Delaware, carrying the material that went into the building of Philadelphia, the best of timber markets. Governor Lewis Morris had facetiously told the Board of Trade in 1742 that without New Jersey timber, "Pensilvania cannot build a ship, or even a tolerable House, nor ship off a Hogshead or a pine stave."[12]

But not for timber alone did Read utilize his lands. With Governor Belcher and other country gentlemen of his day, he became deeply interested in agriculture. Farming was the basic industry of the colonies, but with rare exception it was practiced in a crude and wasteful manner. Read saw the important role that agriculture must play in the prosperity and development of the colony, and strove to advance its interest as a matter of public policy. Furthermore, he was a student of the science of agriculture. He had read books on the subject by English authors, and had corresponded with the handful of Americans who had given attention to agricultural science. He foresaw the limitless possibilities of improvement in agricultural practices. Hence he purchased several tracts of land adaptable to cultivation, and set about to build them into profitable plantations.

Although he retained an interest in agriculture all his life, his principal agricultural activity occurred between 1747 and 1760. During this period, he invested in two farms which claimed his close attention—Sharon in Springfield Township, Burlington County, and Breezy Ridge between the forks of the Rancocas near Mt. Holly. Whether or not he made his home on either of these estates is not clear, but he left record of operations on "his farm in Springfield" and his "plantation on Ancocas." He sought to improve these lands and make them as profitable as possible, he acquired new varieties of plants and observed their growth, he noted the effects of manures, and he studied different methods of soil management. With a wide circle of acquaintances he exchanged ideas and experiences on agricultural subjects. Frequently he asked friends and neighbors to try out new methods for him, and the results he entered in a voluminous notebook, along with

[12] *The Papers of Lewis Morris* . . . , *New Jersey Historical Society Collections,* IV (Newark, 1852), 156.

his own observations. This notebook comprises one of the richest sources of information now available on the agriculture of colonial America.

Closely allied with agriculture in colonial times was the fishing industry. The shellfish in New Jersey's shallow bays, the vast runs of shad in the Delaware and the Raritan, and the abundance of sturgeon in the Delaware were an important source of food supply. Likewise, salt fish, fish oil, and the by-product glue, figured prominently in colonial trade. True to form, Read invested in a fishery on the Delaware near Trenton, but like other of his "schemes for the improvements of the Country," it was a financial failure.

Of all Read's commercial activities, his adventures in iron manufacturing were the most ambitious and worked the greatest influence upon the progress of colonial industry. By virtue of four separate enterprises, developed between 1765 and 1770, he became "the most noted ironmaster in West Jersey prior to the Revolution."[13] Though not the first in the field, his works furnished the real impetus to the bog iron industry which for many years prospered in southern New Jersey.

Read's four iron furnaces were all situated in Burlington County, at Taunton, at Etna, at Batsto, and at Atsion. These sites were chosen because they provided the three essentials for iron manufacture—deposits of iron ore, abundant wood for fuel, and water power. Each iron manor was a substantial community, largely self-sufficing. In addition to the furnaces and forges, it comprised farm and timber lands, saw and grist mills, manor house, commissary, and cottages and shacks to house the laborers, both slave and free. An abundance of capital was required. Read invested heavily of his own savings and enlisted funds from others in erecting in quick succession these four manufacturing plants. But they were scarcely started before the economic depression followed in

[13] *The Bulletin*, Philadelphia, XX, no. 33, Sept. 29, 1886; Clement Papers, Liber K, 73.

the wake of the French and Indian War, and Read was obliged to dispose of a large share of his interests.

Although the iron ventures resulted in Read's financial undoing, from the standpoint of public welfare and industrial progress they were quite the opposite to failure. His four ironworks, by manufacturing munitions during the Revolutionary War, contributed substantially to the success of American arms. Furthermore, he had set in motion an industry which extended over a period of eighty years, and transformed an almost untrodden forest into a hive of activity.

The failure of Read's private business ventures is not surprising in view of the heavy demands which his several public offices exacted of him. Although for posterity his notes on agriculture represent his greatest contribution, during his time it was in the councils of government that his talents found their highest expression. At one time or another he held almost every important office in the colony—legislative, executive, and judicial—save that of sheriff and the governorship itself. His initial years as secretary of the province provided the best possible schooling in practical politics. The secretary was second only to the governor in executive authority—the governor's proclamations and other important papers appeared over the secretary's signature. In this position Read became the guiding genius of the machinery of state, a veritable "prime minister," who exercised a powerful political influence through the dispensing of patronage. According to his colleague Aaron Leaming, he "took the whole disposal of all offices," and "little consulted the merits of the person he preferred," so long as it suited his party principles.[14]

A born political leader, from 1747 to 1771 Read had "the almost absolute rule of Governor, Council & Assembly," except during the brief term of Governor Boone.[15] Sometimes the governor and council would do things against their in-

---

[14] A. Leaming's Diary, Nov. 14, 1775, *loc. cit.*
[15] *Ibid.*

The Man and His Times

clinations to please him, and he often persuaded the assembly to do so.

During the Eighteenth Assembly, over which he presided as speaker from 1751 to 1754, Read displayed the skill of a master politician which approached high statesmanship. In the three previous years, Governor Belcher had been at odds with the assembly. The old controversy over the taxing of proprietary lands and the collection of quit-rents from freeholders, was still raging. The assembly, composed of delegates elected in the several counties and municipalities, represented the interests of the common people. The members of the council, on the other hand, were appointed by the Crown, and in the main favored the proprietors' interests. The democratic assembly, where most legislation originated, had its counterpart in the British House of Commons, and the aristocratic council in the House of Lords. For the passage of a law, adoption by both assembly and council was necessary, as well as approval by the governor and the Crown. As during the preceding administrations, the governor and assembly found themselves at an impasse, as the assembly refused to vote funds for the support of the government unless the governor acceded to their demands. The assembly jealously guarded its control of the purse, for this was the heart of colonial self-government. But within a few weeks after Read became speaker, the long-standing controversy was quickly settled; the governor adopted a more agreeable attitude, the council and assembly compromised on the tax issue, and the assembly voted support for the government. Counseled by Read, Belcher found that peace and harmony were most easily obtained by siding with the assembly. Read had proved himself more than a legislator. He had brought about an orderly functioning of the governmental machinery, and had encouraged Belcher in a mild and conciliatory policy.

Thenceforth Belcher's administration was one of constructive action. The principal public issues in the years which followed were concerned with the settlement of differences with the Indians, with raising military forces for the French and Indian War, and with financing the military program of the province. In all of these Charles Read played a leading role. Although in 1758 he received appointment to the council, he elected to remain in the Nineteenth Assembly, until it was dissolved in 1760, where he could be of greater service. Particularly in sponsoring issues of currency from year to year to raise the necessary funds for the prosecution of the war, and in winning approval in the face of opposition in London, Read rendered outstanding service. He was regarded as the chief authority on public finance in the colonial government.

In other ways, also, he gave war service. Besides commanding the Burlington militia with rank of colonel, he was a member of a commission which purchased supplies for the troops, and he was joint author of the plan for the erection of military barracks in five of the colonial towns. Subsequently, he was one of the committee which supervised the building of the old barracks in Trenton, the only one of these structures which has survived.

While a member of the assembly, Charles Read gave earnest attention to the problem of the Indians. Except in the early Dutch settlements between 1630 and 1660, when both sides were guilty of atrocities, the relations between the New Jersey colonists and the natives were generally friendly. Notably the Quakers were just and considerate in their dealings with the Indians. Thanks to the goodwill thus engendered, for many years after the province came under British control, it was spared the shadow of dread and terror which hovered over the colonies exposed to Indian attack. In sharp contrast with the unhappy experiences of other colonies, the New Jersey officials in their dealings with the Indians displayed a high order of statesmanship. Charles Read took the lead in developing an equitable and sound Indian policy. He shared in marked degree the Quaker spirit of justice and fairness toward the Indians. None of his public achievements was

The Man and His Times

more distinguished, none more genuinely devoted to the public welfare, than his service to the Indians.

In spite of generally peaceful relations with the settlers, the New Jersey Indians claimed that some of their lands had been taken unfairly, and their complaints came to a head with the French and Indian War. Read was appointed a commissioner to examine the Indian claims, and to effect adjustment. Two conferences at Crosswicks, one at Burlington, and the establishment in 1758 of an Indian reservation at Brotherton in Burlington County, led to a satisfactory settlement with the Indians south of the Raritan. An agreement with the more quarrelsome tribes of northern New Jersey was reached the same year at the famous treaty at Easton, attended by the governors of Pennsylvania and of New Jersey. The success of this conference was an enduring tribute to Read's policy of conciliation, and to his superior diplomacy.

For ten years of William Franklin's administration, Read was officially in an advisory position as a member of the council. Political and economic stress followed the French and Indian War and the storm was brewing which broke in 1776. Franklin was an able governor, and he remained loyal to the Crown, but the policy of the home government made it increasingly difficult for him. Trade restrictions, special taxes, and the Stamp Act widened the rift which separated the assembly from the governor and council. During these years Read rendered valuable service as liaison officer between the two houses.

At the beginning of Belcher's administration there had been four political factions: reactionaries who controlled the council; liberals who controlled the assembly; radicals, mainly small farmers who acquired their lands through Indian titles; and the center, or Quaker, party, to which Read belonged. Although not of the Quaker faith, he was associated with the Friends in the public mind, and they contributed largely to his political success. Although he and the Quakers parted ways on military policy, according to Leaming he usually supported them, but "with a high & prominent hand, taking to himself the mastery."[16]

In the controversy with the Crown, Read stuck to a middle course. He believed a union of the colonies was inevitable, and he urged the London agent to impress upon the home authorities the mistaken judgment of their colonial policy. But he gave no evidence of violent opposition to the Crown. During this period the council came off winner in the majority of its clashes with the assembly, reversing its position of the previous decade, when Read was a member of the assembly.

Read's multifold public activities, of course, brought him associations which were a decided asset in his law practice. As a member of the assembly, as councillor, as secretary, as surrogate, as court clerk and as justice, he was concerned with law from all sides—its making, its administration, and its interpretation. Possessing a fine memory, he had a thorough understanding of the law, and "spoke very well off hand but short and to the purpose." Nevertheless, he was not so capable "of arranging and delivering a long train of Ideas; nor of replying and mending his first essay, either in speech or writings." In Leaming's opinion he was "upright as a Judge" and "a better judge than Lawyer."[17]

When Chief Justice Morris died in 1764, Read seemed to be the best qualified man available as his successor, and on the urging of Lord Stirling, Governor Franklin named him for the place. The temporary appointment, however, was not confirmed, and a few months later Read was again sworn in as associate justice. He continued on the bench for nine years, sitting at Perth Amboy and at Burlington, and holding circuit courts in the several counties.

Read's life was not all business and politics. He had a decided cultural bent, and was identified with important educational projects of his time. As a youth he was a charter member of Benjamin Franklin's Library Company in Phila-

[16] *Ibid.* [17] *Ibid.*

delphia; after coming to New Jersey he became a founder of the Burlington library. He subscribed to funds for the building of Nassau Hall, at Princeton, and was a member of the council when the charter was granted to Queen's College, forerunner of Rutgers University, in 1766. In recognition of his scientific interests, he was elected to membership in the American Philosophical Society.

The late 1760's witnessed the decline of the Read household. Read's own illness and the death of his wife followed hard upon financial difficulties. Read's interests were so diversified as to extend beyond the capacity of any one person. His four iron manufactories alone would have required executive skill of the highest order to operate successfully. But added to these were his land and his agricultural interests, his fishery, his legal practice, his duties in the governor's council and in the supreme court, and the result was too much of a strain for one mortal. It is not surprising that he cracked financially, physically, and mentally.

With his creditors pressing, in 1773 he took passage to St. Croix, apparently to salvage some financial interests in the West Indies. Thenceforth his life is shrouded in mystery. Apparently he never returned to New Jersey. Two years later came word that he had died in an obscure spot in North Carolina where, mentally deranged, he had reverted to his mercantile tastes, and had spent his closing days keeping a small store—an ending both strange and tragic for "so large a character."[18] So closed an extraordinary career —a career of contradictions and extremes; many-sided, yet intensive; prosperous at its peak, poverty-ridden in the end; once rich in achievement, but barren and helpless at its close. To his children he left an estate hopelessly depleted. Among his personal effects was an old notebook, seemingly of little value, which miraculously has survived, and now becomes his greatest contribution to posterity.

[18] *Ibid.*

Two: Youth

THE first Charles Read, like the other early settlers of Burlington, took up lands in West New Jersey, but unlike most of them, he did not remain in the province. When a few years later Philadelphia was founded, he joined in the new enterprise. Here he engaged in mercantile pursuits and soon became one of the leading citizens of the community. In the city charter issued by William Penn in 1701 he was named alderman,[1] and he also served as a member of the colonial assembly. In 1701 he purchased from Laetitia Penn, daughter of the founder, a lot at the southwest corner of Front and High (Market) Streets, and on this site the following year he built a three-story gable-roofed house.

Charles Read was twice married. By his first wife he had at least one son, Charles II, born about 1686. His second wife was Amy Child. From this union were born two daughters, through whom by marriage the Reads enhanced their social position in the province. Rachel (1691-1765) became the wife of Israel Pemberton (1685-1754), prosperous merchant, and Sarah (1699-1754) married James Logan (1674-1751), secretary of the province under William Penn. Originally a Quaker, Read followed the fortunes of George Keith, and transferred his allegiance to the Church of England, in which he brought up his son. Amy Child, however, was a devout Quaker, and the two girls chose the faith of their mother, a connection strengthened by their marriage to Quaker leaders. Read seems to have been on the friendliest terms with the Quakers, commanding their confidence and respect. James Logan, writing to William Penn in 1702,

[1] Robert Proud, *The History of Pennsylvania in North America* (Philadelphia, 1797-98), II, Appendix, p. 46.

called him "a truly honest man."[2] He was not permitted long to enjoy his new house, for his death occurred in 1705.[3]

His will, drawn the year before he died, was that of a substantial public spirited citizen in comfortable circumstances. His daughter Rachel received an engraved silver tankard and a sealskin trunk; to Sarah went a small silver tankard, "a French Louis d'Ore in gold," and two large silver spoons. To his wife he bequeathed a silver cup, the best bed, and the house furnishings. His "Dwelling house and Grainery," he provided, was to become the property of his son Charles when he reached the age of twenty-one. Meanwhile, the young man was to have an allowance of £10 a year for his clothes. Bequests of £5 each were made to the minister of the church and to the poor of Philadelphia. The residuary estate was to be divided among widow, son and daughters, with reasonable allowances for the girls' maintenance and education. In the event the heirs should not survive, he provided contingently that his estate should go in equal shares to "the poor of the freedmen of Philadelphia" and to "the maintenance of a Church of England Minister in Philadelphia."[4]

Amy Read did not long survive her husband. On October 4, 1705, "being sick and very weak of body," she made her will, which was probated eight days later. She bequeathed Charles II £5 "current money of Pennsylvania," and provided that her residuary estate be divided equally between the two daughters.[5]

The second Charles followed his father's career as a Philadelphia merchant and public official. Coming into possession of the house at Front and High Streets, he made this his home. His first wife, Rebecca Freeland, died in 1712, and a year later he married Anne Bond. The first child of this

---
[2] J. Thomas Scharf and Thompson Westcott, *History of Philadelphia, 1609-1884* (Philadelphia, 1884), II, 855.
[3] John Meredith Read, "Charles Read," *Penn. Mag. Hist. and Biog.*, IX (1885), 339-343; J. Granville Leach, "Colonel Charles Read," *ibid.*, XVII (1893), 194; *N. J. Arch.*, 1st ser., X, 426.
[4] Philadelphia County (Penn.) Wills, Liber C, 4-7.
[5] *Ibid.*, 13-14.

union was Charles Read III, the subject of this book. He was born February first, 1715, and when twenty days old was baptized in Christ Church, where for many years his father served as vestryman. At least five other children were born to Charles II and Anne Bond. A son Thomas died in 1725, at the age of eight. A daughter Mary, born in 1720, died in her second year, and a son Robert, born in 1721, died before reaching the age of seven. Two other children survived: James, born in 1718, and Sarah, born in 1723. There appears also to have been a son Andrew, and possibly another daughter. When Charles was sixteen years old, his mother died. Two years later his father made his third marriage, to Sarah, widow of Joseph Harwood, a saddler of Philadelphia.[6]

Of the third Charles Read's boyhood and youth there is but meagre record, yet circumstances of the city's life and of the family's activities suggest something of the pattern it probably followed. He was to grow up in the heart of the sprouting metropolis. Nowhere could his father's home have been better situated to expose him to the milling forces of the youthful town. The Front Street windows commanded a wide prospect of the river, and down the slope at the foot of High Street lay the wharf. Boats crossing from the New Jersey shore with passengers and produce tied up here under the eye of the inquisitive boy. Ships arriving from distant ports and discharging their cargoes along the waterfront, touched the young mind with romance and fired the imagination. Month after month, year after year, the growing lad saw shiploads of settlers debark. No sooner could he walk than he must have accompanied his father to the docks when a ship came in. Later, like all normal boys, he would play along the river front and acquire at first hand, naturally and subconsciously, a concept of life at sea. He saw cargoes of America's raw produce dispatched to foreign lands; he saw friends

---

[6] Charles P. Keith, *The Provincial Councillors of Pennsylvania* (Philadelphia, 1883), pp. 186, 187; Charles R. Hildeburn, *Records of Christ Church, Philadelphia; Baptisms, 1709-1760* (Philadelphia, 1893), p. 75.

The London Coffee House, Philadelphia, Birthplace of Charles Read. An etching by E. T. Scowcroft

and relatives set out for missions abroad. The clatter about the docks, the hammering in the shipyards, the lingo of the sailors, filled the atmosphere he daily breathed.

Likewise, past his home on High Street coursed the main channel of the city's life. Here was the tie between town and country; between inland settlement and waterfront. Here passed rich and poor; merchant and mechanic. Tradesman met with farmer and landsman with sailor. Provincial officials passed in review while shipmasters bargained for the services of the bond-slaves they had brought ashore. Now and then appeared straggling Indians, increasingly bewildered by the invasion of Western culture. Here daily, young Read saw Quaker deal with churchman; here he heard the *Hoch Deutsch* of the Palatine immigrants mingle with the Cockney of the London slums. Here was a *pot pourri* of activity which was a liberal education for anyone. Nowhere else in America could a growing boy better gain a speaking acquaintance with the forces that were building a new world.

His father's store fitted into the mercantile atmosphere. Doubtless he spent some of his time here, and when old enough took his turn at clerking. It was a varied business that his father conducted, suited to the divers needs of the rising community. The Philadelphia ladies of 1720 had their choice at Mr. Read's counters of "Flower'd Callaminco," "Figur'd linnen" and "Red and blew Strouds."[7] The appearance of a newspaper in the youthful city offered the enterprising merchant an opportunity to advertise his wares, and accordingly he announced that at his store "any Person may have Cocoa Ground, or be supply'd with right good Chocolate Cheap." At the same time he let it be known that he was experimenting on the side with "Very good Season'd Pine boards and Cedar shingles."[8] Also, with Andrew Bradford the printer, he was engaged in the sale of lampblack.

---

[7] William M. Hornor, Jr., *Blue Book [of] Philadelphia Furniture* . . . (Philadelphia, 1935), p. 27.
[8] Carl Bridenbaugh, *Cities in the Wilderness* (New York, 1938), p. 189; *American Weekly Mercury*, no. 19, p. 41, Apr. 28, 1720.

Charles was a boy of twelve when his father ventured upon another important business enterprise. With James Logan and others, the elder Charles formed a company which in 1727 took up 6,000 acres of land at Durham, on the Delaware River, rich in deposits of limestone and iron. Here they erected extensive works for the manufacture of iron, the first of the kind in Bucks County, subsequently the Cooper-Hewitt works.[9] This venture probably served to focus young Charles' attention upon iron manufacture, and to implant an interest which years later found expression in a large way.

But more than by commercial pursuits, young Charles appears to have been influenced by his father's example as a public official. During the plastic years he had ample opportunity to observe his father as a community leader, now in one capacity, now in another. When the boy was but two years old, Charles the elder became a member of the common council of Philadelphia. Five years later he began a long period of service as alderman, broken only during 1726-1727 when he was elevated to the office of mayor. Young Charles also saw his father in turn serve as sheriff, justice of the peace, commissioner of the loan office, collector of the excise, judge of the admiralty, and clerk of the orphan's court. By the time the boy was eighteen (in 1733), Mr. Read Senior was honored with membership on the provincial council. Such was the heritage of political aptitude and of public service which fell upon young Charles Read and his brother James.

From his father, too, he seems to have inherited a taste for books and an interest in gardening. Charles II counted among his friends both John Bartram, the Quaker botanist of Philadelphia, and the latter's London correspondent, Peter Collinson. Their letters indicate that the elder Read was party to the introduction of some new varieties of plants into America. In 1737, for example, Collinson wrote to Bartram, "Pray does the marsh Trefoil, or Buck-bean increase, that

[9] *American Weekly Mercury*, no. 1085, Oct. 16, 1740; *Penn. Mag. Hist. and Biog.*, VII (1883), 75.

was sent to our friend Charles Reed? It grows wonderfully, in very moist, shallow watery places."[10] Young Charles could observe what was going on in his own backyard, and thus gain an introduction to the mysteries of plant growth.

For his academic training, he enjoyed the dual advantage of his father's library and of private tutors. The elder Charles, like his brother-in-law, James Logan, was a lover of books and owned many fine volumes, which he lent with generous frequency.[11] The young man's formal education was in keeping with his father's social standing. For some years he attended the private school of Alexander Annard, scholarly pedagogue who taught the sons of well-to-do Philadelphia families of the time. Here he got a fair training in Latin which was to stand him in good stead when years later he entered the profession of law.

In school and out, he probably saw a great deal of his cousins, destined like himself to prominent careers. William Logan was three years his junior, the same age as his brother James. The three boys must have been much together. Charles was about thirteen when old Jamie Logan moved to his new country estate Stenton, near Germantown.[12] About this time, too, William was sent to London for his schooling and Charles was thus deprived temporarily of his companionship. Stenton was a good six miles from Front and Market Streets, too far for neighborhood calls, but not for occasional visits. Moreover, it housed an excellent library, probably the largest in Philadelphia at that time. Perhaps now and then Uncle James invited the boy into the library and showed him some of the treasures stored in books. Then there were William's two charming sisters, whose company would delight the heart of any youth—even a first cousin. Sarah, only a few months younger than Charles, in due time married the merchant

---

[10] William Darlington, *Memorials of John Bartram and Humphry Marshall* (Philadelphia, 1849), p. 95.
[11] C. Bridenbaugh, *op. cit.*, p. 457; *Pennsylvania Gazette*, no. 452, p. 4, Aug. 4-11, 1737.
[12] Sarah Logan Wistar Starr, *History of Stenton* (Philadelphia, 1938), p. 7.

Isaac Norris, who became the leader of the Quaker party in Philadelphia and served for many years as the speaker of the provincial assembly. Hannah, four years younger than her sister, after a courtship beautifully narrated in her suitor's diary, married the gentlemanly John Smith of Burlington.[13]

Scarcely less important than the Logans were the Pembertons. Israel and Rachel Read Pemberton had three sons, the oldest of whom, Israel Junior, was about Charles Read's age. James and John were, respectively, eight and twelve years younger. These three brothers were to carry on the wealthy mercantile house founded by their father, and the younger Israel, because of his leadership among the Friends, was to become known as "the King of the Quakers."[14] Charles Read remained closely associated with these cousins throughout his life.

In a modest house on High Street between Third and Fourth lived another relative, John Read, a carpenter. His daughter Deborah was some years older than Charles III. One Sunday morning in October 1723, there landed at the Market Street wharf a 17-year-old journeyman printer from Boston who was destined to play a vital role in the fortunes of the Read family. As Benjamin Franklin moved awkwardly up Market Street, he must have passed the big house on the corner of Front Street before reaching the carpenter's residence where he presented a crude figure before the amused Deborah. By the time young Charles was fifteen, Benjamin took Deborah to wife, and Mr. Franklin the printer became "Cousin Benjamin." Deborah's precise relationship to the Charles Reads is not clear, as no documentary evidence of her father's parentage has been found. Perhaps he was a brother or a first cousin of Charles II. Benjamin, in his letters to James Read, occasionally signed himself "your very affectionate friend and cousin" or "your affectionate kinsman."[15] Furthermore, James's son Collinson customarily referred to

[13] Albert Cook Myers, *Hannah Logan's Courtship* (Philadelphia, 1903).
[14] *Ibid.*, p. 94.
[15] A. H. Smyth, *op. cit.*, II, 290; III, 296.

the philosopher as "Cousin Benny." There seems, therefore, to be no reasonable doubt of a close kinship.

Those were impressionable years in Charles Read's life, while Benjamin Franklin was becoming established in the Quaker City. The youthful Charles must have seen much of the thrifty printer, attacking his work with a diligence and an intelligence which were rapidly winning him widespread respect. Franklin had loosed new cultural forces in the conservative community. The Junto Club, which he initiated, combined the functions of a modern luncheon club, a scientific society, and a reading circle. Since the club was made up mainly of inconspicuous middle-class tradesmen and mechanics, Charles Read and others of his social standing were not included. However, when Franklin in 1731 propounded the idea of a public library, "Charles Read, Jr." was among the first to pay the fee of 40 shillings, and became one of the original thirty members of the Philadelphia Library Company. Since young Charles was then but sixteen, it is a safe inference that his father was interested in the enterprise, and probably set up the boy to a share in the company to encourage him in the use of books.[16]

When Franklin opened his own printshop and bookstore, Charles II, being in a position of political influence, probably helped him by throwing some work his way. As a merchant, also, it seems that he supplied Franklin with some of the paper he needed for printing. Franklin's ledger for 1730 shows the purchase from Read, in two lots, of 17 reams of paper at a cost of £12-0-0.

From a long list of entries in Franklin's ledger, under an open account that ran for at least seven years, we find the printer furnishing Charles Read II a miscellany of services and products. Here the merchant and public official procured his stationery and office supplies, purchased his books, contracted for printing of official forms, and arranged for his

---

[16] Library Company of Philadelphia, Minutes, 1731, I, 3; also Austin K. Gray, *Benjamin Franklin's Library* (New York, 1937), p. 8.

advertising. We can visualize the errand boy Charles stopping at the shop for a shilling bottle of ink, or for a "Quire of Post Paper" and lingering long enough to glimpse the printer set a stick of type or pull from the press a sheet of the *Pennsylvania Gazette*. Now it was an order for legal forms—"300 Summons," £1-5-0, or "150 Deputations," £0-12-6; or "3 Doz. penal Bills," £0-2-3; or "a Book [of] Bills of Lading," £0-8-0, that went into the ledger. At times Charles II replenished his library at the bookshop, and brought to the printer books to be repaired. The ledger shows such items as: "An Herbal," £2-15-0; "Ward's Introduction," £0-11-0; "A Laws," £0-2-0; "binding a Common Prayer" £0-2-0; "binding a Bible and two other Books," £0-11-6; and "binding a Dictionary." In 1735 an "Acc[oun]t Book" cost £0-12-6. The Reads, of course, subscribed to the *Pennsylvania Gazette*; under the date of June 12, 1731, is an entry for "Gazette 7 Quarters" £0-17-6. And for an advertisement in the *Gazette* (no. 66), posted June 16, 1730, a charge of 3s. was entered.[17]

In a different class from these printshop items, was an entry in December, 1736, for "1 Doz. Bottles Claret, £2-0-0." One would naturally wonder how the printer came to be selling this brand of Christmas cheer to Charles Read. It was not simply a chance incident. For Franklin was not averse, in the manner of many modern drug stores, to carrying some miscellaneous merchandise as a sideline to his book and printing business. There were times when a bottle of wine made a good companion with a book. This purchase, however, appears to have been the last one charged to the account of the elder Read before his death occurred the following year.

As Charles Read neared the age of eighteen, he outgrew Master Annard's private school. Inasmuch as provincial schools were considered inadequate for youth of his birth and breeding, it was decided he should finish his education abroad. For help in carrying out the plan, his father naturally turned to Sir Charles Wager, privy councillor and first Lord of the

[17] Benjamin Franklin's Ledger AB, pp. 178, 179.

Admiralty, who was a relative of the young man's mother. Sir Charles, whose father had escorted King Charles II in his triumphant return for the Restoration, and who himself had fought the French and Spaniards in West Indian waters, had once visited the Read family in Philadelphia.[18] Sometime in 1733, it appears, young Read sailed for London,[19] where he came under the benevolent eye and the patronage of Sir Charles.

While in London he doubtless made the acquaintance of other friends of the family, among them the genial Peter Collinson. In a letter to John Bartram, March 14, 1737, Collinson expressed regret that a trustworthy ship was not available for sending some American birds to England and some squirrels to be installed in Wager Park, commenting, "If our friend Charles Reed's son Charles should have a ship, we might have some hopes." Several years later, Collinson referred to the younger Charles as "my friend Charles Read."[20] Just how long Read remained in England is not clear, likely a couple of years. He probably stayed for a time with the Wager family at their seat in Surrey. The admiral interested himself actively in the lad's behalf and secured for him a berth as midshipman in the British navy, while his father forwarded remittances sufficient to support him in that rank. In due time he was assigned to the *Penzance*, man-of-war of twenty guns, which sailed for the West Indies.

But Charles was not born for the sea. Having been bred mid Philadelphia luxuries and having tasted London pleasures, for him the rough life of the sailor had little appeal. Moreover, a war was approaching, and he was not inspired by "that Boisterous, romantic honor that characterizes the Seaman."[21]

Sometime in 1736 or early in 1737, Charles took advantage of his presence in the West Indies to visit the Thibou family

[18] J. Bennett Nolan, *Printer Strahan's Book Account* . . . (Reading, Pa., 1939), p. 7.
[19] Library Company of Philadelphia, Minutes, December, 1733; I, 36-37.
[20] W. Darlington, *op. cit.*, pp. 93, 143; also J. B. Nolan, *op. cit.*, p. 8.
[21] A. Leaming's Diary, Nov. 14, 1775, *loc. cit.*

on the Island of Antigua. Jacob Thibou, now past fifty, was well established as one of the wealthy planters of this prosperous little island. His father Lewis Thibou, native of the province of Orleans, had been among the early settlers of Charleston, South Carolina. Here Jacob was born. Subsequently, the elder Thibou acquired lands in Antigua, and took his family there in 1699. Like many planters of his day, he also had commercial interests, and a few years later appeared in Philadelphia in the role of merchant. While there in 1704 he issued a letter of attorney to his son Gabriel, then bound for Antigua. How many years he remained in Philadelphia does not appear, but it must have been long enough to become acquainted with the Reads. When the *Penzance* put in at Antigua, it was quite the natural thing for the young midshipman, hungry for a touch of home, at the first chance to look up these friends of his family. Charles was captivated with the charms of the place and of its people. He found Jacob Thibou not only the lord of many acres and of many slaves, but a commanding figure on the island. For a time Thibou served as chief justice of the insular court. Furthermore, in 1705, in Antigua, he had married Dorothy Blissard, and he now presided over a household of two sons and ten daughters. The sixth of these children was Alice, born in 1719. She was described as "not hansom, nor gentele, but talked after the Creole accent."[22] Notwithstanding any deficiency in physical beauty, Charles promptly fell in love with her. As to a choice between bachelorhood in the navy and a home career with Alice, he quickly made up his mind. He sold his commission, and they were married at St. John on April 11, 1737.[23]

Meanwhile, bad news had come from home. The health of young Read's father, now fifty years of age, was failing. Furthermore, while devoting attention to his public duties, the elder Charles had sacrificed his private interests and had

[22] *Ibid.*
[23] Vere Langford Oliver, *The History of the Island of Antigua* (London, 1899), III, 124-125.

become deeply involved in debt. Possibly it was in an effort to save some of his property for the son, then absent with the British navy but soon to turn twenty-one, that Charles and Sarah Read, on December 11, 1736, conveyed to "Charles Read, Jr.," a tract of 258 acres in Philadelphia on the east side of the Schuylkill River. This was part of a tract of 1,258 acres they had acquired just three days earlier, and was deeded to the new owner in prosaic legal phrase, "for and in Consideration of the Natural Love and Affection which they have and do bear toward their . . . Son and for his better preferment in the World."[24]

Charles Read II was not spared to witness his son's return home. His death occurred January 6, 1737. The passing of so prominent a character received more than casual notice. Andrew Bradford, mentioning his death in the *American Weekly Mercury*,[25] eulogized him as "a sincere Christian, tender Husband, indulgent Father, kind Master, faithful Friend, good Neighbor and agreeable Companion." Likewise complimentary was Franklin's *Pennsylvania Gazette*,[26] which referred to him as "always Discharging his respective Trusts with Applause, and has left behind him an excellent Character."

As we have seen, his financial status was not so fortunate. The diarist Aaron Leaming records that "his estate was 7000 lb. worse than even with the world."[27] Whether or not young Charles had realized the extent of his father's embarrassment, it is probable that he had in a measure shared the secret of the family's financial situation. For this reason, it has been intimated that his matrimonial venture may have been inspired by more than love. At any rate, he passed for a rising young gentleman, the son of a rich merchant and eminent grandee from Philadelphia, which doubtless carried weight in the Thibou household, and he chose not to reveal the real state of his finances until after his marriage.

[24] Philadelphia County (Penn.) Deeds, Liber F9, 187.
[25] No. 889, p. 3, Jan. 13, 1737.
[26] No. 422, p. 3, Jan. 6-13, 1737.
[27] *Loc. cit.*

So he soon set sail with his bride for Philadelphia to take possession of his supposed estate. When he came away his father-in-law ordered the negroes to "rool" him out thirty-seven hogsheads of rum and had many more consigned to him. We may be sure that on his arrival at Philadelphia he made the most of the situation, posing as a wealthy merchant. The homecoming must have created something of a stir. Here was the son of a former mayor, who had served in the British navy, returning from abroad with a West Indian wife and a cargo of West Indian rum to boot. He had married a fortune, but what manner of woman was this foreign wife? Quaker bonnets doubtless bobbed with animated gossip, accompanied by a plentiful lifting of Quaker eyebrows and the shaking of Quaker heads.

Of Charles Read's next two years in Philadelphia, we know but little. He probably made his home with his step-mother and his brother James in the Front Street house, while endeavoring to carry on his father's mercantile enterprises. One of his first responsibilities was to help in settling the estate of his father, who had not left a will.[28] Presumably, it was in this connection that on August 12, 1737, he took title to a tract of 778 acres in Makefield Township, Bucks County, Pennsylvania, conveyed to him by the trustees of the General Loan Office of Pennsylvania. Six days later he deeded to his brother James 261 acres of this tract.[29]

In disposing of his father's effects, we find him also endeavoring to reassemble the family library. As we have seen, the elder Read had a considerable library from which he was wont to lend books widely, but not always discreetly. To Charles II fell the task of rounding up these loans. Apparently it was he who inserted the following advertisement in the *Pennsylvania Gazette*:[30]

Whereas the Library late of Charles Read, Esq., is very much

---

[28] Letters of Administration were granted to Sarah Read Jan. 26, 1736. Philadelphia County (Penn.) Book of Administration C, 352.
[29] Phila. Co. (Penn.) Deeds, Liber F9, 414.
[30] No. 452, Aug. 4-11, 1737, p. 4.

despers'd, and many Sets of Books broken, particularly the Spectator, Tatler, Guardian, Conquest of Mexico, Athenian Oracle, &c. and some whole Sets lent out, together with several valuable Treatises.

These are therefore to desire those who have any Books lately belonging to the said Charles Read, that (in order to prevent the Expence of repairing the Library aforesaid) they would generously and gratefully return them.

How successful young Read was in recovering his father's books does not appear. The experience should have convinced him that the public library offered a better way to give book service to the community than did loans from a personal library.

Even though Charles Read had not long been away from home, he found marked changes when he returned. Adolescent Philadelphia had passed the awkward stage, and was rapidly coming of age. With a population of 12,000 it was by now no mean city. In a few years it would outstrip Boston and become the first in the American colonies. Real estate was booming. New streets lined with red-brick houses were pushing northward and westward, while local brick and timber yards worked overtime to keep up with the procession. The provincial assembly was meeting in the new State House on Chestnut Street. The first residences, small and plain, were giving way to more elaborate three-story structures. Thrifty Quakers were laying the foundations of substantial fortunes. Their aristocratic homes adorned Front Street and they were dominating the economic and social life of the city. The rapid growth of Philadelphia was "the most of any settlement in the World for its time."[31]

Situated in the heart of the finest agricultural region of the colonies, Philadelphia was a natural center of commerce. New shops and warehouses were rising to accommodate the growing trade. The life of the city converged upon High Street, with its market sheds stretching down the middle of the thoroughfare, flanked with batteries of public pumps. On

[31] C. Bridenbaugh, *op. cit.*, p. 329.

market days twice a week farmers, tradesmen, and townspeople flocked to this busy place. Heavy wagons drawn by four-horse teams brought the produce from farms ten to one hundred miles distant. New Jersey farmers, who had to ferry their produce across the Delaware, competed with Pennsylvanians for the market. Intruding New Jerseymen and peddlers so roused the ire of Philadelphia tradesmen that they were branded "public nuisances." Philadelphia butchers complained about the large amounts of meat brought into town from New Jersey.

The waterfront teemed with activity. From the shipyards new shallops and sloops, and larger sailing vessels of two hundred to three hundred tons, periodically swelled the city's mercantile fleet. From day to day the arrival and the departure of vessels from Europe, from the West Indies and from other American ports, added to the excitement. Each sailing took the produce of forest and farm from both sides of the Delaware to distant shores; each arrival from Europe brought the goods of civilization and new quotas of settlers, prospective tradesmen, and planters, both bond and free. Along the row of public wharves shouting draymen guided their heavy loads through the rutted streets. Noisy sailors crowded the taverns. Drunkenness and vice, with their consequent disorder, were keeping pace with the growth of population and commerce. Hard by the market, the stone jail enjoyed an active patronage. As if an antidote were needed, a wave of religious enthusiasm was about to grip the community, inspired by the dynamic preaching of Whitefield and the Tennents.

Opportunities were not lacking for skilled labor and the manual trades. Sons of parents of all degrees of ambition were serving their apprenticeships, bound out to occupations as varied as merchant, goldsmith, limner, distiller, wagon maker, printer, bookseller, weaver, hatter, cordwainer, and cooper. Benjamin Franklin's bookstore-printshop was yielding a comfortable profit, and Philadelphians had come to look to its proprietor for leadership in all community enterprises.

Although at first the population was predominantly Quaker, a non-Quaker element was gaining increasing power. Churchmen clashed with Quakers in the effort to control the common council and the assembly. The proprietors' interests did not always coincide with those of the citizens, at least in the opinion of the latter. The secretary's post was no sinecure, but James Logan, representing the proprietors, discharged the duties of the office diplomatically and with credit, and "made his influence felt in every sphere of Pennsylvania life."[32] It was a fortuitous period in the life of the Quaker city.

But with all the promise that Philadelphia offered the ambitious young man, Charles Read saw greener pastures across the Delaware. Older than Philadelphia, but closely allied with it economically and socially, was the Western Division of New Jersey. The settlements on the eastern side of the Delaware in effect, were a part of the larger Quaker community. It was to this land of opportunity that Charles Read now turned his attention. His grandfather had settled here; his father, though residing in Philadelphia, had held the post of collector of the port of Burlington. Not satisfied, apparently, with his mercantile ventures, and following family precedent, young Charles now sought public office. In 1739 he bought of Peter Bard the office of court clerk of Burlington County and received his commission on August tenth of that year.[33] Soon afterward, through the influence of Sir Charles Wager, he was appointed to the office of collector of the port of Burlington at an annual salary of £60. Here, in this thriving little city of fewer than two hundred houses, he now established his home. Burlington was not only the political capital and principal port of the Western Division of New Jersey, but about it revolved the agricultural and commercial activity of the counties which touched the Delaware on the New Jersey side

[32] *Ibid.*, p. 253.
[33] New Jersey Commissons, Liber AAA, 348.

above Philadelphia. Perhaps more than any other settlement it reflected the economic and the social life which was unfolding throughout the province, particularly in the Western Division. Situated midway between Trenton and Philadelphia, it was the natural center for a public career which for a third of a century was to be intimately identified with the life of the province.

Three: New Jerseyman

THE year 1739 was one of new experiences for Charles and Alice Read, new work, a new home, and a new son. The record of their first years in Burlington, however, is not as complete as we would wish. The precise place of their residence is not known. As yet, apparently, they did not own any property in the town; probably at first they rented a house.

To the young couple in the succeeding years came the normal joys and cares that attend the founding of a home and the raising of a family. Their first-born was more than two years old before they took him on April 19, 1742, to St. Mary's Church and had him christened "Charles," with benefit of formal Episcopalian rites.[1] The birth of two other sons—Jacob in 1742, and James, brought increased home cares. The infant James survived the critical second summer, then fell "very ill with his teeth," and died, after suffering "several fitts."[2]

After coming to Burlington, Read lost no time in broadening his interests, both public and private. Not content with his dual office of court clerk and collector of customs, he began to deal in lands and in lumber, and as we have seen, was soon seeking other public offices. Some idea of life in the Read household through these years is suggested by the scraps of information which have survived. Quite naturally, old ties with friends and relatives in Philadelphia were closely maintained. Read continued to purchase books and stationery at Benjamin Franklin's shop, where, like his father, he kept an open account.[3] On July 2, 1745, he stood in debt to Franklin

[1] Parish Register of St. Mary's Church, Burlington, I, 19.
[2] Charles Read to James Pemberton, Nov. 7, 1749. Pemberton Papers, V, 170.
[3] Benjamin Franklin's Ledger D, p. 97. Soon after Read's removal to New Jersey, Franklin entered in the ledger under "Chs. Read Esqr. Burln" the following items:
    2 Qr. of brown paper & gilt Paper   -1-2
    1 Qr. of Mourning Paper   -1-6
    Art of Preaching   -6

in the amount of £6-11-10, when according to an appended note, the balance was carried to another ledger. When the account was paid does not appear. About this time, too, Read ordered for his wife a piece of "good Irish Linnen" from a shipment John Smith had received in Philadelphia.[4]

Mid strict Quaker surroundings Alice Read doubtless had her homesick days when she yearned for the easygoing plantation life of the West Indies. It was hardly to be expected that the Creole wife of the rising young official would always be acceptable to conservative Quaker society. After spending close to ten years in the Philadelphia-Burlington vicinity, she decided to visit her native Antigua. It was hoped that the voyage would improve her health, which had been impaired of late, and in addition, there were matters of business that needed her attention. About Christmas time, 1748, she arrived at Antigua for a stay of several months. Letters the following spring reported her health "much restored" by the voyage. Thinking to please her on her return home, Charles dispatched a letter to James Pemberton, then in London, ordering from the English shops a liberal supply of merchandise. And since he was now a member of the supreme court, he asked Pemberton to procure some needed law books, and likewise some personal embellishments, such as a cane, in keeping with the dignity of the office. Apparently, too, he had joined Governor Belcher in the distinction of owning a carriage:

I have seen some French Plate at the Governors which is handsome enough and Comes very Cheap as it is iron plated wth Silver and will Serve me as well as any as it will be Seldom used. I should be glad you would buy for me two small Cans about half a pint or more each, a Tea Pot, a Sugar dish, a small bowl about a Quart and a Case with One dozn. Plain Ivory Knives and Forks & a

---

Other entries appeared at intervals, for example:
For 400 Blanks & an Alm[anac]k interl[eave]d	2-18-4
For blank Book	15s
Two Health Poems	4s
Parchm[en]t & Licen[c]e	51s-4

[4] Charles Read to John Smith, Jan. 31, 1744. Smith Papers, II, 72.

place for Spoons, a neat Small Turrene of good Pewter, a good half hunter horse whip, and a plain Strong Cane with a plain Cap of Cocoa nut on it, a Steel Seal, and 4 or 5 hundred Quills; these things I would have Put up in a Strong Port Manteau trunk with a suit of Cloaths Mr. Partridge is to send me.[5]

Then, too, he asked Pemberton to procure for him a personal seal. In 1747 Benjamin Franklin had ordered a large seal for Read from James Turner, silversmith of Boston. The price quoted was £12 but Turner subsequently wanted £15.[6] Perhaps Read refused to accept it at this figure. At any rate, he now wrote Pemberton: "I would have it of the larger size of seals wore to a Watch. The arms are A Griffin Seagrant Or in a Field Azure the Crest a Demi Griffin with a Truncheon in his paw The Motto—Nec Spe, Nec Metu."[7] Two days later he notified Pemberton that he had sent £454 with a Captain Arthur, presumably to cover the previous order.[8] About this time also he added to his household force a "Mulatto Woman named Dinah aged about twenty-five Years together with her Bed and Bedding," and "wearing Apparrel" at a cost of £48.[9]

Pemberton likewise was useful in keeping the Read wine cellar adequately stocked. In November, 1750, Read dispatched to Pemberton, then at Philadelphia, £20-8-8 with an order for "¼ cask of white wine; ¼ cask of good Port; & if you can a Doz of Ale" to be shipped by the Burlington packet.[10] The same year he wrote John Pemberton: "If you See Capt. Reeves tell him He need not buy me Razors to my Strop He is to gett me, as I have them already."[11]

With the possible exception of brief periods in the country, Read appears to have made his residence through these years

[5] Charles Read to James Pemberton, May 20, 1749. Pemberton Papers, V, 86.
[6] James Turner to Benjamin Franklin, July 6, 1747. American Philosophical Society, Franklin Papers, I, pt. 1, p. 5.
[7] "Neither hope nor fear." Charles Read to James Pemberton, May 20, 1749. Pemberton Papers, V, 86.
[8] Charles Read to James Pemberton, May 22, 1749. Pemberton Papers, V, 88.
[9] New Jersey Deeds, Liber IK, 61.
[10] Charles Read to James Pemberton, Nov. 24, 1750. Pemberton Papers, VI, 138.
[11] Charles Read to James Pemberton, July 31, 1750. Pemberton Papers, VI, 76.

in Burlington. Apparently for a time he lived in John Smith's house at 320 High Street, the mansion built by Richard Smith in 1720. In 1755 Read wrote his landlord:

> As the House I now live in is exceedingly out of repair & has been so ever since I lived in it, & the repairs so as to make it convenient to me would amount to more than I would desire you to lay out, I have rented your kinsman Samuel Noble's house, to go into it when I leave yours, & I shall do so at Expiration of this quarter wch will end abt. ye 15th of August, & that will give you time to do the repairs wch are absolutely necessary for the preservation of the House, & to take in a Tenant as I know of one who will offer.[12]

Samuel Noble lived in Philadelphia and his Burlington house was situated on High Street adjacent to the Smith mansion, on the site of the buildings at present numbered 308 to 314, the lot "Extending Back as farr as the back street Commonly Called the Church Lane."[13] Read took a seven-year lease on the residence at an annual rental of £24. Noble allowed Read to spend £50 on repairs, for which he reimbursed him and Read also spent money on the property on his own account. It was agreed that in the event of Read's death before the expiration of the lease, his executors would have the right to remove from the property his "Stable, Chimney backs, Extraordinary locks and other household furniture though affixed to the freehold."

Busy as he was, Read probably spent little time with the members of his father's family, although now and then he was in touch with them. His brother Andrew, it seems, was an ironworker, for in 1753 Charles wrote to James Pemberton that Andrew desired £10 "to lay out in Iron Mongery ware."[14] Beyond this, little is known of Andrew, but he was still living about 1775, for in a letter by the son of Charles III, there is mention of "my own poor Uncle Andrew."[15]

---

[12] Charles Read to John Smith, June 30, 1755. Smith Papers, IV, 233.
[13] N. J. Deeds, Liber M, 456. Church Lane was subsequently named Wood Street.
[14] Charles Read to James Pemberton, July 26, 1753. Pemberton Papers, IX, 53.
[15] Charles Read (IV) to Jacob Read, Dec. 16, 1775 (?). Elting Papers, Book of Jurists, p. 60.

Locale of Charles Read's Burlington Residence; Smith House erected 1720 in center; Site of Noble House at extreme right. Sketch by Henry S. Haines

West Side of High Street, Old Burlington, near Site of Charles Read's Residence. Sketch by Henry S. Haines

New Jerseyman 43

His sister Sarah married Thomas Shoemaker, and their daughter Rebecca became the wife of Samuel Waddingham, a planter of South Carolina. Apparently Read continued his acquaintance with his niece, for he referred to Mr. Waddingham in his agricultural notes.

Like him in many respects, and almost equally prominent in public affairs was his brother James, a man of scholarly bent, with broad and varied interests. As a young man James had spent some months in England, where he met Peter Collinson, for whom in time he developed so great an admiration as to name his only son in the botanist's honor. He also visited Sir Charles Wager, and became acquainted with Charles Wesley and the book dealer, William Strahan. Moved by a deeply religious nature, he formed a close attachment for George Whitefield, the English evangelist, which was strengthened a few years later when Whitefield visited America. He served as trustee of the charity school and house of worship where Whitefield preached, the structure which, as the subsequent home of Franklin's academy, is associated with the founding of the University of Pennsylvania. In 1752 James Read moved to Reading to begin a career as court official of Berks County, Pennsylvania. By appointment of the provincial government he filled the county offices of prothonotary, recorder, register, and clerk of the orphans' court, and of the court of quarter sessions continuously from the time of the organization of the county until the outbreak of the Revolutionary War. In fact, all the county offices except that of sheriff were centered in this inveterate office holder, who contrived to usurp most of the fees of the new district. He served also as a member of the Continental Congress, of the supreme executive council, and of the general assembly.[16] During the Revolution he returned to Philadelphia and in 1781 became register of the admiralty.

With all his intellectual brilliance, James Read seems to

[16] C. P. Keith, *op. cit.*, pp. 188-191; J. Bennett Nolan, *Neddie Burd's Reading Letters* (Reading, Pa., 1927), p. 36; Francis von A. Cabeen, "The Society of the Sons of St. Tamany of Philadelphia," *Penn. Mag. Hist. and Biog.*, XXVI (1902), 340.

have been quite eccentric, a tendency augmented by frequent over-indulgence in strong drink. In 1749 Benjamin Franklin wrote of him to William Strahan, who was endeavoring to collect from James a bill long overdue for an order of books, "He has a great many good Qualities for which I love him; but I believe he is, as you say, sometimes a little crazy."[17] With his brother Charles he had much in common—intellectual brilliance, a wide diversity of interests, a penchant for public office, a streak of eccentricity, a practical interest in gardening. Likewise, both in their time experienced financial reverses. He died during the yellow fever epidemic of 1793.

As to Sarah Harwood Read, widowed step-mother of Charles and James, misfortune came to cloud her declining years. In 1750 she was obliged to sell for £1,095 the lot on High Street she had inherited from her first husband.[18] The family mansion at High and Front Streets she sold to her brother-in-law, Israel Pemberton, who lived in it for a time. Meanwhile, Charles and James seemed quite indifferent to her needs. In 1753, Israel Pemberton wrote of the brothers to Peter Collinson:

> Cousin C. Read is now in a Station & circumstances to be of Service to the People among whom he lives & I think fills the sev'l posts he holds in that government with Reputation & in particular that of a Judge in their Supreme Court, for he hath acquired much more knowledge in the Law than his Friends ever expected . . . Their Mother (My uncle's widdow) by indiscreet generosity to them & want of Oconomy is reduc'd to poverty, so that our Family are now oblig'd to maintain her.[19]

In 1754 the Read mansion which was Charles's birthplace was turned to commercial use. In time it became known as

---

[17] Benjamin Franklin to William Strahan, Apr. 29, 1749. A. H. Smyth, *op. cit.*, II, 375. For a comprehensive and entertaining narrative of this controversy, as well as for other events in James Read's colorful life, see *Printer Strahan's Book Account*, by J. B. Nolan.

[18] *Archives of the Commonwealth of Pennsylvania*, 3rd ser., IX, 556-560 (hereafter referred to as *Penn. Arch.*).

[19] Israel Pemberton to Peter Collinson, July 12, 1750. Collinson's Letter Books; courtesy of Mr. Charles Montgomery and Dr. J. Bennett Nolan. See also *Printer Strahan's Book Account*, by J. B. Nolan, p. 71.

the London Coffee House, for years a picturesque landmark. Here occurred many of the leading events of Eighteenth-Century Philadelphia—real estate and slave auctions, conferences of the governor and other public leaders, and such political demonstrations as the burning of Stamp-Act papers and the burning in effigy of the enemies of liberty.[20]

Since Read was closely associated with the leading Quakers in Pennsylvania and New Jersey, and politically was a member of the "Quaker party," in the public mind he was at times identified with the Society of Friends; he admitted on occasion that he was disliked by "Eastern Men" (the people of East New Jersey) for being a Quaker.[21] Though inclined toward Quaker ideals, he was too liberal to be bound by their strict economic and social codes, and accordingly was not a member of the society. As we have seen, he was not averse to owning slaves. Doubtless, however, he was acquainted with the pioneer abolitionist John Woolman, and he must have felt the force of the anti-slavery sentiment which Woolman inspired among the Quakers.

Read's interest in books, clearly evident throughout his notes on agriculture, was not that of the introvert bookworm. He regarded books as tools indispensable to one's business or profession. But more than this, he had a high appreciation of the value of books as an educational agency making for social progress. The cultural riches of books were not to be selfishly hoarded, but to be shared with others.

As already noted, when scarcely more than a boy he joined the Library Company of Philadelphia. The first catalog of books belonging to the company printed in Franklin's shop in 1741 listed a four-volume set of *Cato's Letters*, given by Charles Read, Esq.[22] When on March 25, 1742, the company

[20] J. T. Scharf and T. Westcott, *op. cit.*, II, 855; *Penn. Mag. Hist. and Biog.*, LIII (1929), 6; John F. Watson, *Annals of Philadelphia and Pennsylvania in the Olden Time* (Philadelphia, 1898), I, 393-395; III, 203-204.
[21] Charles Read to James Pemberton, Nov. 17, 1748. Pemberton Papers, IV, 156.
[22] *A Catalogue of Books belonging to the Library Company of Philadelphia* (Philadelphia, 1741), pp. 4, 23, 25, 48.

received a charter at the hand of the proprietors, Charles Read was named as one of the charter members.[23]

Incidentally, among the books listed in the infant library were Miller's *The Gardener's Dictionary*, Bradley's *New Improvements of Planting and Gardening*, Ellis' *The Practical Farmer*, and other English works on agriculture which Read cited in his notes.[22] Perhaps he used these very volumes in his studies of scientific farming. Perhaps also he borrowed agricultural works from the splendid library of James Logan, which later, as the valued "Loganian" collection, was conveyed to the Library Company by William Logan.

It is not surprising, then, that some years after becoming a citizen of Burlington, Read should have been a leader in the movement to establish a library in that city. In November, 1757, a group of sixty "inhabitants of New Jersey, thinking that a Library Company in . . . Burlington would be of great benefit to the members, as well as the public in general, . . . formed themselves into a company. . . ."[24] Each one agreed to pay 10s. a year in support of the organization. The directors chosen to organize the library met for the first time in the parlor of Thomas Rodman's house on High Street. In 1758 the Library Company of Burlington was established by royal charter, one of the few royal charters granted to libraries in the American colonies. As secretary of the province Read was instrumental in having the charter granted through Governor Reading. He headed the list of the founders named in the charter, and likewise was the first-named of the directors, serving in this capacity four years.[25] Among the charter members were several of the men Read mentioned in his notebook: Daniel Doughty, William Foster, Hugh Hartshorne,

---

[23] *The Charter, Laws and Catalogue of Books of the Library Company of Philadelphia* (Philadelphia, 1764), p. 3.

[24] E. Morrison Woodward and John F. Hageman, *History of Burlington and Mercer Counties, New Jersey* (Philadelphia, 1883), p. 132.

[25] *Catalogue of Books belonging to the Burlington Library* (Burlington, 1876), p. 3.

New Jerseyman

William Masters, William Skeeles, Daniel Smith, James Smith, Samuel Smith, and William Smith.[26]

Within a year the library had acquired more than five hundred volumes—most of them donated by the founders. Charles Read contributed a collection of books which apparently had been accumulating for some years in the family library. They represented several fields of learning: history, law, literature, politics, and theology. Among his gifts to the new library was a large three-volume set of *A Collection of Voyages and Travels* (London, 1744), and three other works of folio size: *Tables of Chronology, containing a View of Universal History, from the Creation to the Year 1700; Historical Collections,* by Heywood Townshend, M.P. (London, 1680); and William Sheppard's *An Epitome of All the Common and Statute Laws now in Force* (London, 1656). Also there were smaller volumes: a set of *The Poetical Register; or the Lives and Characters of all the English Poets* (London, 1723); *A Conference, about the next Succession to the Crown of England* (1681); and *Maxims, Theological Ideas, and Sentences,* extracted by J. Gambold (London, 1751). Probably Read did not part with any of his favorite volumes for the benefit of the library. Some of the books were obviously out of date. No recent works on law and none on agriculture were among the gift volumes. Several works on agriculture, however, were donated by Read's friends. William Smith gave a copy of *The Husband-man's Guide, in 4 Parts,* second edition, printed in New York in 1712. John Smith gave a copy of *The New Art of Gardening,* by Leonard Meager (London, 1699), *A Treatise on Fruit Trees,* by Thomas Hitt (London, 1755), and *A Complete Body of Husbandry,* by Thomas Hale (London, 1756). Samuel Smith contributed *The Manner of raising, ordering and improving Forrest Trees,* by M. Cooke (London, 1676), and the first catalog of the library lists Philip Miller's *The*

---

[26] *The Charter, Laws and Catalogue of Books of the Library Company of Burlington* (Philadelphia, 1758), p. 7.

Gardener's Dictionary in three volumes (4th edition, London, 1754).[27]

It was to be expected that Read would be interested in the books and magazines which now and then came from the colonial printing presses. In 1757 he expressed pleasure in the proposed *The Jersey Magazine*, and procured a subscription from John Smith.[28] This seems to have been the magazine that appeared the following January under the title *The New American Magazine*, from the press of James Parker, proprietor of New Jersey's first printshop at Woodbridge, and "King's Printer."[29] The first three issues of the magazine carried a special department on agriculture under the title *The Country Farmer*. Again, Read joined Samuel and John Smith in endorsing a new book by Thomas Powell, master of the Boarding School at Burlington, entitled *The Writing Master's Assistant*. The testimonial was used for an advertisement which appeared in the *Pennsylvania Gazette* for August 2, 1764.[30]

Read, too, must have had more than a casual interest in Samuel Smith's *The History of the Colony of Nova-Caesarea, or New Jersey*, which in 1765 came off Parker's press. To print this work, which is still used as an authoritative record of colonial New Jersey, Parker moved his press from Woodbridge to Burlington, and maintained it there until the printing was completed. The book received the warm commendation of Francis Bernard, one-time governor of New Jersey. Writing to Read from Boston December 8, 1767, he remarked:

> I desire my Compliments to all my Friends & particularily to Mr. S. Smith with Thanks for his Ingenious Book, whose only Fault is that there is not more of it.[31]

Read's acknowledged interest in science and letters received

---

[27] *Ibid.* Read refers to the works of Hale, of Hitt, and of Miller in his notes on agriculture.
[28] Charles Read to James Pemberton, Dec. 3, 1757. Pemberton Papers, XII, 82.
[29] *N. J. Arch.*, 1st ser., XX, 177.
[30] *N. J. Arch.*, 1st ser., XXIV, 399.
[31] Francis Bernard to Charles Read, Dec. 8, 1767. Bernard Papers, V, 248-250.

appropriate recognition in 1768. On April 19, his name was proposed for membership in the American Philosophical Society, and on May 18 following he was accepted.[32] Here in Franklin's learned society he was in a stimulating company—with Lord Stirling, with John Bartram, William Logan, and Israel Pemberton, of Philadelphia; with Richard Stockton and Dr. John Witherspoon, of Princeton; with John and Samuel Smith, of Burlington; and with Dr. John A. DeNormandie, of Bristol.

To foregather with this brilliant group was a delightful experience, though for Read not an altogether novel one. For years he had been on intimate terms with many of these men, and had exchanged ideas with them on scientific topics. But the meetings of the society gave him opportunity to widen his circle of friends, and doubtless he felt a measure of pride and satisfaction in the simple fact that he now "belonged" to the intellectually elect.

Following the commercial bent of his father and his grandfather, after leaving Philadelphia Read occasionally engaged in private mercantile ventures, in addition to an extensive lumber trade. It would have been surprising if he did not have some interest in shipping. A letter to James Pemberton written in 1754 states, "Our Boats usually lye at Crooked Billet Wharf."[33] In 1760 he asked Pemberton to recommend some "Industrious Merchant at Hampton in Virginia . . . . who is safe & not above Business."[34]

Again in 1766, when his iron enterprises were under way, Read reported that John Brick, who had purchased one of his places, was then "in Cumberland County after fatt Cattle" for him,[35] perhaps to provision his forces at the works, or perhaps as a strictly speculative venture.

At another time the salt industry which dotted the shallow

---

[32] *Early Proceedings of the American Philosophical Society . . . from 1714 to 1838* (Philadelphia, 1884), p. 14.
[33] Charles Read to James Pemberton, Nov., 1754. Pemberton Papers, XVII, 112.
[34] Charles Read to James Pemberton, Sept. 7, 1760. Pemberton Papers, XIV, 59.
[35] Charles Read to James Pemberton, Oct. 15, 1766. Pemberton Papers, XIX, 8.

bays and thoroughfares along the seacoast captured Read's fancy. The bordering forests furnished an abundance of firewood for the evaporation of salt water. With his friend Jacob Spicer, of Cape May, who stopped over at Burlington in April, 1755, and with a "gentleman from Philadelphia," Read discussed the erection of a salt works at Cape May. Spicer owned a suitable site for the enterprise; he was interested and might go into it with them, he wrote in his diary, "provided I like the Scheme & am further enabled to Judge of the Expediency by an experiment made at an Easy expence," and provided also they could procure legislative authority or a charter of incorporation from the Crown.[36] In the absence of any further mention of the scheme, we infer that it never materialized.

When in 1748 Read's neighbor John Smith married Hannah Logan, the bonds of the Smith-Logan-Read circle were further strengthened. There was a steady coming and going among the families, and a frequent exchange of letters. Read's letter of congratulation to Smith on the birth of a daughter, written in November, 1749, reflects the intimate relationship. After apologizing for delay in paying a bill he owed Smith, explaining that he had been in Perth Amboy "farr from my Bank & in Expensive company," he continued:

> I am heartily glad of the Encrease of your Family & that Mrs. Smith is likely to do well   The text says the tree is known by its fruits wch implys that the fruit partakes also the Nature of the tree   if so I am assured that my Young kinswoman must (if her life be spared) be a Comfort to you & her mother & I hope to have the pleasure of seeing it so.[37]

Hannah Logan Smith was mentioned in another letter which also offers a glimpse of a friendly circle:

> My Wife tells me that my cousin Hannah wrote to her or me for Mistletoe of the Oak   I know nor can I hear of any   Mrs. Noble tells me that tis plenty in Maryland, if the mistletoe of the gum will do, on a line from you I will procure some &

[36] Jacob Spicer's Diary, Apr. 5, 1755.
[37] Charles Read to John Smith, Nov. 7, 1749. Smith Papers, III, 126.

convey it to you I suppose tis for Sally Loyd I wish it may do her good.

I send you two of our Treaties I wish I had it in my power to oblige yo wth more they come in a packett to yr Brother.[38]

Through association in the provincial council, Read became well acquainted with James Alexander, who, while residing in New York, was a member of the New Jersey council from 1723 to 1756; and with his son, William Alexander, Earl of Stirling, who served as councillor from 1761 to 1775 and was a General in Washington's army during the Revolution. For many years, too, Read lived in intimate touch with James Alexander's son-in-law, John Stevens, distinguished head of a distinguished family.

One of Read's letters to the elder Alexander, written in September, 1748, gives a passing glimpse of the Read home. After mentioning the dispatching of certain public documents, he wrote:

Mr. Sandin the Swedish Missionary at Racoon & his Wife lodged at my house this Spring & He had a printed Disertation on plants & a Letter to Mr. Colden wch He desir'd me to forward & I really thought I had done it. It was laid on a table in my wifes Chamber where we breakfasted & by her or ye Girl put into a Drawer & being out of my View I forgott it, till the rect. of your Letter & was then convinced of my great Neglect I have sent it to Mr. Nichols as you directed.[39]

Read's correspondence with Lord Stirling indicates a close personal relationship. In 1753 he wrote Stirling about delivering two pairs of gloves which apparently he had been instrumental in procuring for him. Incidentally, the letter reveals that Read had "very little acquaintance" at New York.[40]

---

[38] Charles Read to James Pemberton, Dec. 20, 1758. Pemberton Papers, XIII, 18. The "Treaties" here mentioned were probably copies of the pamphlet printed by Franklin and Hall, reporting the Indian conference held at Easton the previous October (*Indian Treaties Printed by Benjamin Franklin, 1736-1762*, ed. J. P. Boyd, Philadelphia, 1938, pp. 213-244).

[39] Charles Read to James Alexander, Sept. 16, 1748. Theodore Stevens Collection.

[40] Charles Read to William Alexander, Jan. 22, 1753. New York Historical Society MSS.

Occasionally, Charles Read saw members of Benjamin Franklin's family. While Benjamin was on his second mission to England, his daughter Sarah wrote him about an incident which disclosed Read's willingness to go out of his way to do a favor for a friend. The letter was written March 25, probably in 1766. It read in part:

I meet Mr. Read of Burlington last evening he told me he had been down to Capt. Egdon's Wreck and among the things he saw a parcel of nice Wax work fruit, which the Capt. told him was put on board by Dr. Franklin for his Daughter. he then had a box made for it (for the things had been strangely hauld about), packed it carefully, and it was coming round. I told him I was much obliged to him for his kindness, but did not think it belong'd to me, as I was sure you would have mentioned it, if you had sent it.[41]

Among Read's close associates was Aaron Leaming of Cape May, who was wont to tarry overnight at Read's home when in Burlington. Leaming survived Read and on learning of his death, wrote in his diary a graphic character sketch of the man.[42] After describing his public career, the diarist in cleverly turned phrase introduced a chapter of Read's personal history which has not appeared from any other source. The reason for his frank treatment of a subject which otherwise might be taboo, he explained, was that Read himself, as a point of honor and pride, would not have wanted it to go unmentioned:

He had more vices than virtues he had many of both and those of the high rank. He was intrieguing to the highest degree. . . .

His intrigues with women, tho' they employed a large share of his thought, were not worth naming; they were rather the foibles than the vices in so large a character yet because I know he would never have pardoned the man that should attempt his story without making honourable mention of them I draw them into his shade. He was so vain of them that if he had penned this character they would have filled many pages. . . .

[41] William Duane, *Letters to Benjamin Franklin from His Family and Friends* (New York, 1859), p. 28.
[42] A. Leaming's Diary, Nov. 14, 1775, *loc. cit.;* also his Diary for 1761 (Jan. 2).

The large-scale promoter whose schemes, though well planned, carried him beyond his personal and financial resources, also was pictured:

> His greatest virtues were found among his vices. His offices furnished him with a constant flow of Cash. This power & flow of Cash enlarged his mind above himself. Instead of founding a fortune to his two sons as he ought to have done in those prosperous times, he ran upon schemes for the improvements of the country, witness his Fishery at Lamberton, his Iron Works and many other schemes which tho' virtuous in a very high degree in a man of great fortune, it ought to be treated with distrust with men of little estates. He was industrious in the most unremitting degree. No man planned a scheme so well as he, nor executed them better. He loved the country better than his family. And knew no friend but the man that could serve him.

In recounting mannerisms not normally found in a man who for years demonstrated unquestioned powers of influence and leadership, Leaming listed certain points of weakness in character which might be expected to preclude a successful public career:

> His airs and action was much after the french manner, ever on the wing & fluttering never long fixed frequently courting, frequently whispering as if to make the person believe they were in his confidence a little too severe in enmity and not grateful for good offices, high strung and selfish unwilling to forgive an injury. . . . Timerous almost to cowardice, Whimsical to the borders of insanity, which he inherited maternally . . .

Not a very complimentary picture—this. We wonder how a person with such flagrant defects could have become the commanding figure that Read once was. Perhaps here Leaming was not describing the man in his prime. Read came to the close of his career broken in spirit, mind, and fortune. More than likely, at the time Leaming wrote this appraisal, his perspective was warped by the memory, still fresh, of Read's last unhappy years.

Four: Customs Collector

READ'S duties as collector of the port of Burlington probably did not consume a great deal of his time, but through this office he had his finger upon the pulse of New Jersey's commerce and industry. Burlington was one of the three authorized ports of entry of colonial New Jersey. Of the others, Perth Amboy was the focal point of shipping in East New Jersey; Salem the commercial center of the southern counties.

With the growth of the colony, the primary occupations of fishing and agriculture had gone forward, while side by side with them, and out of them, had risen home and village industries which now fed a constantly growing commerce. Bordered by a long coast line, indented by numerous bays and harbors, and pierced by navigable rivers and creeks, New Jersey offered exceptional facilities for water travel and trade.

Whalers and fishermen from New England had early visited New Jersey waters about Cape May, and an export trade in oil, whalebone, and fish, largely with Spain, Portugal, and the Canary Islands, had followed. As the settlers pushed inland, they had close at hand the natural resources for the beginnings of industry. Primeval forests stood ready for the cutting, and the network of small streams offered abundant water power. Every few miles down their winding courses wooden dams widened out the narrow creeks into placid millponds, and sawmills rose on the thickly wooded sites. From these mills came the harvest of timber, of shingles, of barrel staves, and of pipe staves that found their way to seaport and shipyard, to growing town and village. From the pine forests came quantities of pitch, tar, turpentine, and rosin.

The sawmill frequently heralded the beginning of a permanent settlement. As the forests gave way to cultivated fields, gristmills rose beside the sawmills, or replaced them. Around these picturesque structures buttressed with massive water wheels, clustered the rural villages which dotted the countryside. Often the gristmill was the heart of the rural community. It ground the grain for farm and household use; but more than this, it furnished a ready market for the produce of the farm and opened the door to the channels of trade. From the gristmills of colonial New Jersey came quantities of flour which were shipped to New York, to Philadelphia, and to foreign ports.

Each colonial farm was in large measure self-sufficient—a center of such varied household industries as the dipping of tallow candles, the spinning of flax and wool, the making of soap, and the weaving of homespun. These commodities in the main were for home consumption, but a portion of them reached the public market. Beyond the home unit, even in this early day, village and town industries began to absorb in ever increasing volume the products of the farm, and assumed an important place in colonial economy. Near the gristmill the tanner laid out his tanning vats and received the raw hides brought in by the farmer. The resulting leather he divided—half he returned to the farmer, the other half he kept as his compensation, and sold it, perhaps to the local shoemaker, the saddler, or the harness maker. Or perhaps he sent it to a merchant in Burlington or Salem, whence in all probability it was shipped to New York, or to Philadelphia, or to a foreign port. In the Eastern Division, Newark early took the lead in the manufacture of leather goods.

Every village of any consequence had its blacksmith shop or wheelwright shop, or a combination of the two, where the versatile smith was equally at home in shoeing a horse, making a plow, or repairing a cartwheel. In the larger villages and towns there came in time weaving mills, fulling mills, and paper mills. In Salem County, in 1738, the glass industry of New Jersey had its birth. In addition to the cider produced on the farms, large quantities were manufactured at cider mills set up in the towns and villages. The cider from Newark and vicinity was of such superior excellence as to be in great demand at home and abroad. At many of the mills, the cider was distilled into brandy, the popular high-powered "applejack," or "Jersey lightning," which became an important article of commerce. To meet the demand for milder liquors, breweries and malt houses took their places in the larger towns, and to supply the necessary barrels, the cooperage became an essential adjunct to the brewery and the distillery.

For building purposes, it did not take the settlers long to discover that excellent brick could be made from New Jersey clay. The brickyards set up at strategic points even before 1700 began to provide materials for the larger and more substantial buildings which replaced the first hastily raised structures.

These village industries throve in New Jersey, not because of encouragement from the mother country, but because of the abundance of raw materials and the growing local markets, and also because of the skilled craftsmen who manned them. Here opportunity beckoned to the carpenter, the bricklayer, the mason, the joiner, the ropemaker, the cabinet-maker, the tailor, the shoemaker, the tanner. Skilled New Jersey hat makers prospered so far that they interfered seriously with the sale of hats made in England, which was highly disturbing to the English manufacturers. Accordingly, in 1732, Parliament passed a law which prohibited the transportation of hats from one colony to another and limited the number of apprentices. Colonial manufacturers, however, successfully evaded the law. As early as 1699 a similar restriction had been placed on trade in woolen and linen cloth. The manufacture of cloth at Burlington appears to have been smothered by the selfish policy of the mother country.

Even before 1700, New Jersey had a considerable export trade. Among the exports to the West Indies were a "great plenty of Horses, and also Beef, Pork, Pipe staves, Bread,

Customs Collector

Flour, Wheat, Barley, Rye, Indian Corn, Butter and Cheese."[1] From Salem shipments of "Bread, Beer, Beef and Pork . . . also Butter and Cheese" were sent to Barbadoes; and skins, furs, rice, and pickled cranberries were exported to Europe.[2] Cedar posts, shingles, barrel staves, hemp, flaxseed, sheep, tallow, wax, and iron also were shipped to other ports. In exchange for this produce commodities of wide variety were brought into New Jersey ports: Muscovado sugar, rum, and molasses from the West Indies; linen and manufactured goods from England and Ireland; wine, salt, olive oil, silks, and spices from Lisbon; and naval stores from North Carolina.[3]

As New York's foreign trade developed and as Philadelphia outstripped Burlington, New Jersey's commerce to an increasing degree cleared through these neighboring cities, which possessed superior advantages as seaports. In spite of official efforts to promote them, the New Jersey ports never attracted a large volume of foreign commerce. Nevertheless, the produce shipped from its home ports was only a small portion of the colony's trade. Much was produced for home consumption—the markets and fairs of the earliest towns were important agencies for commercial exchange. Beyond the home towns, however, the principal markets were New York and Philadelphia. Trade with the former was mainly by way of the Raritan, the Passaic, the Hackensack, and the Hudson; and with the latter by way of the Delaware and its tributaries. From Bergen, Essex, Monmouth, Morris, Somerset, Middlesex, and Monmouth counties produce was shipped to New York; from the counties of Burlington, Gloucester, and Salem, shipments were to Philadelphia. The produce of Sussex and of Hunterdon counties went both to New York and to Phila-

[1] William A. Whitehead, *East Jersey under the Proprietary Governments*, N. J. Hist. Soc. Col., I (Newark, 1846), 267.

[2] Gabriel Thomas, *An Historical and Geographical Account of the Province and Country of Pensilvania, and of West-New-Jersey in America* (London, 1698), 2nd pt., pp. 15, 32-33.

[3] Israel Acrelius, *A History of New Sweden . . .* ; *Memoirs of the Historical Society of Pennsylvania*, XI (Philadelphia, 1874), 145-146; William Douglass, *A Summary, Historical and Political . . . of the British Settlements in North-America* (Boston, 1750), II, 294.

delphia. Part of this produce was consumed in these cities, much of it was reshipped to foreign lands, much more, in fact than was exported directly from Burlington, Salem, and Perth Amboy. New Jersey producers resented the monopoly of foreign trade held by the merchants of New York and Philadelphia, which doubtless prevented a larger development of the home ports—then and also in later years. The regular stage service between New York and Philadelphia after 1730 stimulated further the trade of New Jerseymen with these cities. Likewise it brought the merchants of these two provincial capitals into closer rivalry with each other, while intervening New Jersey, divided into two economic districts, served as a commercial battleground. A European traveler in 1760 commented that this condition had deprived New Jersey "of those riches and advantages, it would otherwise soon acquire." He remarked that "although it is extremely well cultivated, and thickly seated, . . . yet having no foreign trade it is kept under."[4]

Another English writer commented in 1765 that the people of New Jersey "having no considerable foreign trade of their own, they exchange their commodities at those two places [New York and Philadelphia] for foreign goods, and consequently leave a profit there, which otherwise they might have themselves."[5] By 1770 the New Jersey trade was reported to be "solely with and from New York and Pennsylvania."[6]

With his catholicity of interests and his zeal for the public welfare, Read must have reveled in the typical market-day scene which, about the middle of the century, surrounded him as he made his way to the waterfront. Here early in the morning he saw the farmers and their wives gathering from the countryside, in oxcart and on horseback, to barter the yield of their farms for goods from foreign lands and city

---

[4] Andrew Burnaby, *Travels through the Middle Settlements in North-America* . . . (London, 1775), pp. 59-60.
[5] Robert Rogers, *A Concise Account of North America* (London, 1765), p. 76.
[6] [Alexander Cluny], *The American Traveler* . . . (Philadelphia, 1770). Letter XV, pp. 59-61.

High Street, Old Burlington, nearly opposite the Site of Charles Read's Residence. Drawn on stone by John French and lithographed by T. Sinclair

shops. Butter and cheese, fruits and vegetables, poultry and eggs, bacon and ham, were exchanged for bonnets and calico, for coffee and tea. Cattle, sheep, and swine were herded in the paddocks—a noisy, odorous prelude to the butcher's shambles. Burlington tradesmen and buyers from the Philadelphia export houses and mercantile firms swaggered past the array of produce jealously guarded by its owners, who automatically viewed all city traders with suspicion, and were prepared to drive a sharp bargain before parting with their possessions. The occasional appearance of the governor's chaise was a signal for a brief but pleasant interlude in the day's business. Town and country folk alike stood aside, as intent upon a good view of the uncommon vehicle as of the great man it carried.

Close to the market stood the malthouses, the brewhouses, and the bakehouses. Nearby were the timber yards. Over the market was the hall where the assembly sometimes met. The waterfront was a busy spot—it was the commercial barometer of West New Jersey, for the prosperity of the towns on the Delaware depended mainly upon the river trade. At the wharf lay sailing vessels discharging their cargoes or preparing for the next sailing: A smart sloop ready to cast off for a quick trip to Philadelphia, weighted low in the water with sides of beef and of mutton, tubs of butter and of cheese, and barrels of salted Delaware river shad; another, Philadelphia-bound, taking aboard barrels of salt pork, hams, and bacon, for reshipment to the West Indies, since New Jersey pork had "deservedly gained reputation through all the islands."[7]

The arrival of a merchantman from Lisbon was an occasion now sufficiently rare to bring the collector to the waterfront, since Philadelphia had absorbed most of the foreign trade. He might find casks of wine, and barrels of salt piled on her deck, awaiting the permit to unload. Flour, ship bread, and other local products would be gathered on the dock to load for the return journey. Joining the clerk, who in the

[7] S. Smith, *op. cit.*, p. 496.

absence of the collector normally handled these matters, Read would seek out the captain of the merchantman, examine his papers, and collect the various fees fixed by law. The captain must pay 15s. for entering the vessel, and an equal amount in advance for the clearance certificate. For the permit to unload, the fee was 18d. After the cargo lists were checked and the valuations figured, the current duties would be levied and collected. Other fees were:

> Register, *ten shillings*
> Endorsement of a Register, *two shillings and six pence*
> A Cocket,[8] *three shillings*
> Bill of Store, *five shillings*
> Bill of Health, *seven shillings and six pence*
> Bond for enumerated goods, *two shillings and six pence*
> Certificate to cancel ditto, *one shilling and six pence*
> Permits for Boats or Shallops to and from New York or Pennsylvania for their whole load of Country produce, *two shillings and six pence*.[9]

A typical scene would find the customs inspector moving among barrels of pork and beef piled on the dock. Under a law enacted by the assembly in 1725 to avoid "Frauds and Deceits," each barrel exported must carry the official stamp *N.J.*, indicating that it met certain specifications as to size of container and quality of its contents.[10] Likewise, flour for export (except to Pennsylvania and New York) must be passed by the inspector, and each barrel must be stamped with his initials and the official mark *New Jersey*.[11] Even in these early days, standards of quality were deemed essential to the protection of trade, and governmental authority was invoked to prevent abuses.[12]

---

[8] A certificate that goods have been entered and the duty paid.

[9] *The Acts of the General Assembly of the Province of New Jersey;* compiled by Samuel Nevill, I (Philadelphia, 1752), 350, also II (Woodbridge, 1761) (hereafter referred to as *Nevill's Laws*).

[10] *Ibid.*, I, 139-142.

[11] *Ibid.*, 445-448.

[12] Joseph Borden, Jr., operator of stage lines across the province complained to the assembly in 1760 while Read was a member of this house, that non-residents of Burlington County were branding their pork barrels with the word "Burlington," hoping thereby to benefit from the reputation Burlington pork

Customs Collector

Floating down the river came flatboats from Sussex and Hunterdon counties, laden with wheat, flour, flaxseed, hams, butter, skins. Likewise, sixty-foot Durham boats carried iron from the mines in northern New Jersey and in Bucks County, Pennsylvania. Some of these craft might stop at Burlington, but more likely they would pass on to Philadelphia. Immense lumber rafts moved slowly by, also destined for Philadelphia.

Within easy view was the shipyard where towered the skeletons of new vessels soon to join the colonial merchant marine. At several points down the river, along the bayshore in bight and harbor, brigs and other sailing vessels were being built. Some of these were for British as well as for colonial owners. With an abundance of good oak and pine at hand, they could be built 30 per cent cheaper here than in England. Shipbuilding created a demand for cordage, and provided a lively market for hemp raised on New Jersey farms.

Mid such a scene, the duties of the collector of customs were not lacking in human interest. Here he could observe at first hand the play of social and economic forces which were shaping the course of American life.

In addition to collecting the customary import duties, as well as the taxes levied intermittently upon exports, it was Read's duty to submit quarterly to the governor, to be transmitted to the Lords of Trade, a list of vessels entered and cleared at Burlington.[13] It was also his responsibility to prevent the smuggling of foreign goods into the province. This was no easy task, for New Jersey was notorious as a center of illicit trading. The abundance of creeks and inlets along the shore invited wholesale smuggling, and government officials found it impossible to apprehend the host of clandestine trad-

---

enjoyed in the markets. He urged the assembly to enact a law which would require inhabitants to brand their pork with their proper names, together with the name of the county in which they dwelt, "by which every Man and County will have the Merit they are justly intituled to." (*N. J. Assembly Minutes*, Nov. 25, 1760, p. 41). The principle suggested by Mr. Borden was finally incorporated in a law passed in 1774, entitled "An Act to regulate the Packing of Beef and Pork, and to ascertain the Size of the Casks" (*Allinson's Laws*, pp. 450-453).

[13] Belcher Papers, VIII, 324.

ers who would avoid the payment of import duties at the established ports. In 1749 Governor Belcher reported to the Lords of Trade that the collectors at the three ports were "Vigilent and prevent any Clandestine Trade as much as in them lies," voicing eloquently in his qualifying clause the hopelessness of completely stamping out the practice.[14]

It was also Read's duty during the colonial wars to take precautions that the exports from Burlington should not reach enemy hands. In 1741 Governor Morris, fearing the French fleet in the West Indies would depend for supplies upon shipments from the colonies, directed him "not to allow any vessel laden with provisions to clear, unless sufficient security is given that her cargo is not intended to be landed in any Dutch, Danish, French or other foreign place in America."[15]

Similarly, in 1757, an embargo was placed on shipping to neutral ports, and in 1762 Governor Hardy placed an embargo on all ships sailing from New Jersey ports, except those in His Majesty's service, to prevent food getting to the enemy.[16] In 1763 Read issued notice of the governor's proclamation that shipmasters must report to the customhouse for officers to go on board for inspection, since "many Vessels, trading to Plantations not belonging to the King of Great Britain, returning with Cargoes of Rum, Sugar and Molasses, have found Means to smuggle the same into His Majesty's Plantations, without paying the King's Duty."[17]

Although, according to available records, Read remained customs collector for life, the position did not require much of his time, particularly in the later years when Burlington had lost the bulk of its foreign trade to Philadelphia. The customs post was not so significant in the duties involved or in the income it yielded, as in the relationship it represented with colonial commerce and industry. Trade was a vital link

---

[14] Jonathan Belcher to the Secretary of the Lords of Trade, *N. J. Arch.*, 1st ser., VII, 244-246.
[15] *The Papers of Lewis Morris, op. cit.*, p. 128.
[16] *N. J. Arch.*, 1st ser., XVII, 75-79, 314-315.
[17] *Ibid.*, XXIV, 290-291.

between the colonies and the mother country; upon it largely depended the prosperity both of agriculture and of industry; out of it during Read's lifetime arose one of the burning issues which led to the Revolution. The first-hand association with commerce which Read thus acquired was useful in other spheres of his career. As farmer and as ironmaster he could better understand the essential balance between production and marketing. As a legislator he was better able to advise a sound commercial policy, and to promote measures for the protection of the industrial interests of the colony. That he made full use of his understanding of trade problems is apparent in both his private enterprises and his public services.

Five: Land Speculator

CHARLES READ had been in New Jersey but a short time when he began to trade in lands. The large tract in Philadelphia County which he had acquired from his father, he sold in 1742 to Robert May, of Philadelphia County, for £100.[1] It appears, also, that he had inherited certain lands in New Jersey, for between 1740 and 1742 he sold a tract of land and a 250-acre "plantation" in Burlington County.[2] By this time he had begun to make new acquisitions of land. In 1741 he located in his name a survey of 8 acres of flats on the Delaware River in Willingboro Township, Burlington County.[3] Then in 1745 he purchased a 6-acre lot in Burlington,[4] and the following year an acre lot on the east side of High Street.[5] Although a person of his enterprise would be expected to build a residence of his own, we have no record of his having done so.

Read's principal real estate transactions, however, were not in city lots, but in large parcels of farm and timber lands. In 1744 he purchased for £50 from Thomas Gardiner, of Burlington, 1,725 acres of unappropriated land in West New Jersey, under which surveys were subsequently located on both uplands and swamps in Springfield and Evesham townships in Burlington County, and in a cedar swamp in Gloucester County.[6] By 1749 he could write to James Pemberton, "my Estate lays chiefly in Land."[7] Many of his acquisitions appear to have been primarily for speculative purposes. In 1750 he purchased for £212 rights to a grant of 5,000 acres

---

[1] Philadelphia County (Penn.) Deeds, Liber G4, 205.
[2] N. J. Deeds, Liber S, 153; Liber AK, 205.
[3] Basse's Surveys of New Jersey, p. 286.
[4] N. J. Deeds, Liber GH, 411.
[5] *Ibid.*, Liber X, 429.
[6] *Ibid.*, Liber GH, 409; New Jersey Surveys, Liber BB, 277, Liber S, 190, 258, 259.
[7] Charles Read to James Pemberton, April 28, 1749. Pemberton Papers, V, 71.

Land Speculator

from the executors of John Johnson in Perth Amboy, part of which was located in Gloucester County and part sold within a short time.[8] Several other purchases of proprietary rights followed soon; likewise some sales, among them 1,000 acres of unappropriated land to Joseph Burr, Jr., "yeoman" of Bridgetown (Mt. Holly), for £60 "proclamation" money.[9] In 1747 he sold John A. DeNormandie 21 acres in Northampton Township for the development of an iron works.[10] Then in 1751, Read received from the Council of Proprietors of West New Jersey, two large grants of unappropriated lands—one of 5,378 acres, the other of 3,750 acres.[11] Parts of these were sold promptly without being located; others were surveyed and recorded within the next six years, including tracts in Morris, Hunterdon, and Gloucester Counties, and several tracts of cedar swamp in Burlington County.

From the sale of several parcels from these grants, totaling over 5,000 acres, he received approximately £400. From 1753 to 1755 he acquired numerous other lands, by grant and by purchase from the Council of Proprietors, part of which were located in pine and cedar-swamp lands, and part sold. Another speculative venture which suggests modern real estate promotion was the purchase in 1757 of 42 acres in Bear Swamp, Burlington County, for £80.[12] Five years later the tract was broken up into small parcels of 2½ to 3 acres each, which he sold to various local residents for prices ranging from £24-12-0 to £30-0-0, realizing a return of over £200 altogether.[13]

Read acquired at least one tract of land—banked meadow on Timber Creek—by the foreclosure of a mortgage in the amount of £110.[14] Few other mortgages in his name are on record; he appears to have preferred actual ownership of land

[8] N. J. Deeds, Liber GG, 460; Liber IK, 140; N. J. Surveys, Liber E, 269.
[9] N. J. Deeds, Liber IK, 1.
[10] Ibid., Liber H, 127.
[11] Basse's Surveys of New Jersey, pp. 271, 272.
[12] N. J. Deeds, Liber O, 316.
[13] Ibid., Liber Q, 515, 517, 519; Liber R, 505, 512, 514, 516, 518.
[14] Ibid., Liber GH, 327.

and speculation in its sale, to lending money on real estate as collateral.

The large cedar-swamp and pine areas were acquired with the view of utilizing their timber. For example, in 1745 Read purchased for £500 a cedar-swamp sawmill tract of 2,000 acres on Cotoxing Creek, a branch of the Rancocas below Lumberton. In 1747 he leased the sawmill and tract for four years to Benjamin Moore, Jr., and four others, who agreed to cut at least 50,000 feet of merchantable pine boards annually, one-fifth of which Read was to receive for himself.[15]

The property in time became known as Read's Mill. The agreement of 1747 mentions, in addition to the sawmill and log dwelling house, a "Logg Stable and Hay house." The stable apparently was erected about 1740, or before there was a sawmill in the neighborhood. Battered and blackened by the storms of two centuries, it still stands on the property. Its dovetailed corners with the door jambs pinned to the ends of the hand-hewn logs of white cedar; its hand-split floor boards and roof of hand-split shingles—speak eloquently of an age of farm architecture that was crude but lasting. Inside is clear evidence of stall divisions and of a manger formed from three logs running the length of the building.

This tract in due time was divided; some of it, including the site of the stable, came into the hands of Benjamin Moore. The sawmill and adjacent lands remained in Read's possession until 1773 when these and other holdings were conveyed to his son Charles for £1,320.[16]

By lease or by purchase, over the years Read expanded his timber interests. In 1746 he leased for twenty years a 10-acre tract of swampland at an annual rental of £6. Under the terms of the lease, it was agreed that Read should "clear the said Swamp and make it meadow fit for the scythe," and it was agreed that he should "make and keep the fencing hogtight."[17]

[15] *Ibid.*, Liber GH, 225; Liber H, 43.
[16] *Ibid.*, Liber AF, 179.
[17] *Ibid.*, Liber GG, 105.

Log Stable, near Lumberton, on Farm purchased by Charles Read in 1745

Ruins of Old House, Sharon

Land Speculator

Additional purchases of timber lands alone or in partnership with others (among them his "honest neighbor," Edward Tonkin) followed in quick succession: a 20-acre tract of cedar swamp, designated as Unknown Swamp, for "10 Spanish pistoles," and a 50-acre tract near "Bever Pond";[18] also lands in Shrewsbury Township, Monmouth County; and tracts on Great Timber Creek and elsewhere in Gloucester County. Among the sawmill and cedar tracts acquired in Upper Freehold Township, in Monmouth, was Success Mills, which Read operated in partnership with Jonathan Thomas.[19]

In September, 1753, Read purchased a one-seventh interest in two parcels of pine land and cedar swamp, totaling 2,240 acres, in the vicinity of Cotoxing Pond. The same year he also acquired a large tract adjacent to Braddock's Bridge and Edge Pillock sawmill.[20] On some of these holdings, Read appears to have worked the stand of timber himself, others he leased out for cutting. On occasion he also handled the timber from the woodlands of others. In 1751 he contracted for a one-half share of the pine trees cut from a 664-acre tract adjoining his lands at Goshen Neck.[21] Probably for the purpose of cutting timber, and possibly also with the prospect of eventually utilizing the iron deposits, in 1755 he procured a 999-year lease of a tract of 1,128 acres on "Atsionk Creek."[22]

One of his sawmills with the adjoining tract in Evesham Township, which he called Penzance Mill, was leased for a seven-year term in 1757 to Robert Powell and Abraham Prickitt. The tenants agreed to cut at least 100,000 feet of lumber a year, and to pay Read one-half of all boards, planks, and other sawed stuff, to be chosen by Read semi-annually. He agreed to supply four steel saws and four dozen files, also

---

[18] *Ibid.,* Liber R, 520; Liber HH, 274.
[19] *Ibid.,* Liber H2, 2; also Charles Read to James Alexander, Feb. 25, 1750, New York Historical Society MSS.
[20] N. J. Deeds, Liber K, 379, 411, 460.
[21] *Ibid.,* Liber IK, 325.
[22] *Ibid.,* Liber Y, 113.

a pair of timber wheels and a wagon for the use of the lessees. Profits on sawing other people's timber were to be divided.[23]

That he gave close personal attention to the timber business appears in the letters to his merchant cousins, the Pembertons. His critical eye readily diagnosed the imperfections of the tools of his day. Writing in 1750, he remarked:

> There is one Article by which you may in the Course of your Trade make some profitt of Enquire for one Stedman or Steadman who Makes Saw Mill Saws, and direct him to Make you a Dozen Encourage him to Make them about Six foot or Six foot four inches Long He must Lett them be thicker & stronger in the Neck than usual & carry their thickness to the Back & if the Whole Saw was a little thicker it would be better He will understand these directions as He makes many to Send to Philadelphia but rather too slightly. We sell here by retail at three pounds & £3.5. but He will perhaps honestly tell you the Price & Saws made according to these directions would be worth 10s currency more than Common If you bring over a Dozn lett these direction[s] be sent him I will Engage you 120 pct. & immediately pay for the whole.[24]

Again, in 1753, he wrote about the sale of his lumber in Philadelphia, and expressed dissatisfaction with the method of grading:

> Pray be kind enough to dispose of the Boards on your wharf wch belong to me for the best price they will bring There are many good boards but perhaps faulty in a Spott or Stained. It does not Suit us Saw mill Men to have them so Close culled.[25]

Other woodland tracts were acquired between 1756 and 1761 in Bear Swamp, Burlington County; on a branch of the Little Egg Harbor River and on Little Timber Creek in Gloucester County. Portions of these were disposed of within a few years. In 1759 Read signed articles of agreement giving Henry Woodrow rights to lumber under the lease with Powell and Prickitt, by which Woodrow was to pay Read at the rate of £4-10-0 per 1,000 feet of "heart stuff," £2-12-6 per 1,000

---

[23] *Ibid.*, Liber N, 369.
[24] Charles Read to John Pemberton, July 31, 1750. Pemberton Papers, VI, 76.
[25] Charles Read to James Pemberton, Jan. 2, 1753. Pemberton Papers, VIII, 118.

Land Speculator

feet of "common stuff," and £5-10-0 per 1,000 feet of "2-inch plank."[26] It was also provided that any disagreement between Read and Woodrow would be settled by a board of arbitrators, among whom were William Foster and Joshua Bispham.

One of Read's last land ventures took him beyond the bounds of New Jersey. In 1770 he headed the list of sixty-nine persons, mainly from New Jersey and Pennsylvania, who received letters patent to a large tract in Otsego County, then a part of Albany County, New York. The tract, sometimes referred to as the "Otsego Tract," compromised approximately 100,000 acres, and lay between the Mohawk and the Susquehanna Rivers. It was proposed to divide the tract into parcels of 1,000 acres each and allot one to each of the sixty-nine patentees. An annual quit-rent of £0-2-6 per 100 acres was required. In order for the patentees to retain title, it was required that the tract be settled within three years to the extent of one family per 1,000 acres and the cultivation of three acres in every fifty.

Read seems to have been the leader in the enterprise. Among the patentees were many of his intimate friends—Jonathan Odell, William Lovett Smith, Samuel Allinson, Daniel Smith, James Kinsey, Daniel Ellis, James Verree, John Smith, Samuel Smith, Thomas Rodman, and John Lawrence, a number of them prominent figures in colonial New Jersey.[27] This project came too late in Read's life for its full consummation. Important as were his lumbering interests, his land investments were of significance mainly is so far as they led to his adventures in agriculture and in iron manufacture, as will presently appear.

[26] N. J. Deeds, Liber P, 111.
[27] New York State Book of Patents, XIV, 535-541.

Six: Countryman

AFTER dealing in lands and in timber for several years, Read turned his attention seriously to agriculture. Beyond a business interest in farming, he sought to indulge himself in the life of a country gentleman, which he regarded of all estates as "the most agreeable." To carry out this design, he purchased for £70 from John Arthur of North Wales a 212-acre tract in Springfield Township, in November, 1747. Shortly thereafter he purchased 6½ acres of meadow land from his neighbor Jonathan Hough, and in partnership with Edward Tonkin, a 50-acre tract on Beaver Pond. The Springfield holdings were enlarged in 1750 by the purchase from Samuel Black of an adjoining tract of 71 acres for £230. Thus expanded, this splendid plantation was designated Sharon, a name it carried for many years.[1]

It was Sharon that Read described in a letter to Jared Eliot, written about 1749.[2] After praising Eliot for his *Essays* and commenting upon the inertia of the rural population, Read confessed to feeling the call of the countryside and disclosed the eternal hope of the urbanite to make his country estate pay its way. His description of the farm, of his management of the soil, and of his crop experiments reveals the agricultural scientist well versed in the art of field culture:

I have perused your two Essays on Field Husbandry, and think the Publick may be much benefited by them, But if the Farmers in your neighbourhood are as unwilling to leave the beaten road of their Ancestors as they are near me, it will be difficult to persuade them to attempt any improvement; where the Cash is to be laid out on a Probability of a return, they are very Averse

[1] N. J. Deeds, Liber HH, 257, 266, 274; Liber GG, 429; Barclay White, "Early Settlements in Springfield Township, Burlington County, N. J.," *Proceedings, Constitution, By-Laws, List of Members, etc., of the Surveyors' Association of West New Jersey,* (Camden, 1880), pp. 62-68.
[2] This letter was long believed to have been written by Benjamin Franklin. For a discussion of its authorship, see pp. xvi-xxiii.

to the running any Risque at all, or even Expending freely, where a Gentleman of a more Publick Spirit has given them Ocular Demonstration of the Success. About 18 months ago I made a Purchase of about three hundred Acres of Land near Burlington, and resolved to Improve it in the best and Speediest Manner, that I might be Enabled to Indulge my Self in that kind of life which was most agreeable. My fortune (thank God) is such, that I can enjoy all the necessaries and many of the Indulgancies of Life, but I think that in Duty to my children I ought so to Manage, that the profits of my Farm may Ballance the loss my Income will Suffer by my retreat to it. In order to this, I began with a Meadow, on which there had never been much Timber, but it was always overflowed; the Soil of it is very Fine and black about three foot then it comes to a fatt Bluish Clay; of this deep Meadow I have about Eighty Acres, forty of which had been Ditched and mowed the Grass which comes in first after Ditching is Spear grass and white clover, but the weeds are to be mowed four or five years before they will be Subdued, as the Vegetation is very Luxuriant. This meadow had been ditched & planted with Indian Corn of which it produced about Sixty Bushells pr acre. I first Scoured up my Ditches and drains & took off all the Weeds, then I plowed it and Sowed it with Oats in the last of May, in July I Mowed them down, together with the Weeds, which grew plentifully among them, and they made good Fodder. I immediately Plowed it again, & kept harrowing 'till there was an appearance of Rain and on the twenty-third of August I sowed near thirty Acres with red Clover and Herd grass, allowing Six quarts of Herd grass and four pound of red clover to an Acre in most parts of it, in other parts four quarts Herd grass and three pound red clover, the red Clover came up in four days and the Herd grass in Six days, and I now find that where I allowed the most Seed, it protects itself the better against the Frost. I also Sowed an Acre with twelve pound of red clover it does well, I Sowed an Acre more with two bushells of Rye Grass Seed and five pound of Red Clover, the rye grass seed failed and the Red clover heaves out much for want of being thicker, however in March next I intend to throw in Six pound more of red Clover as the Ground is open and loose. As these Grasses are represented not durable, I have Sown two bushells of the Sweeping of Haylofts (where the best Hay was used) well Riddled, pr Acre supposing that the Spear Grass and white clover seed would be more equally Scattered when the other shall fail. What Surprized me was to

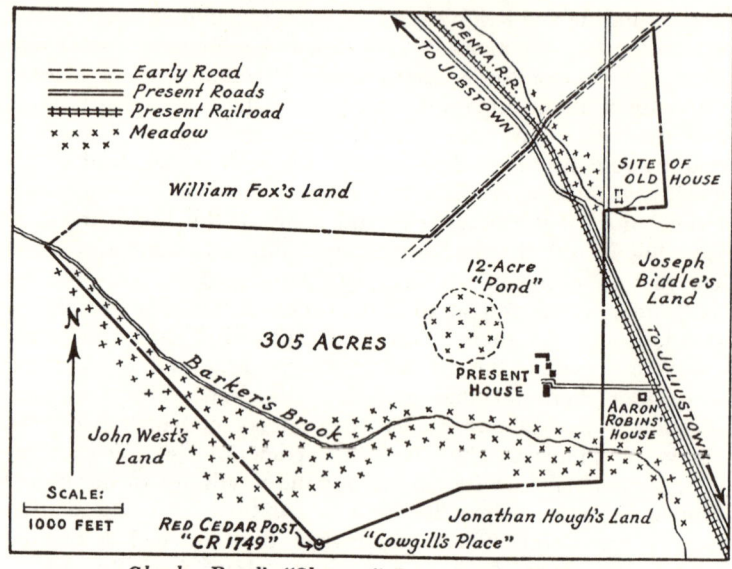

Charles Read's "Sharon," Boundaries as in 1750

find that the Herd grass, whose Roots are small and Spread near the Surface, should be less affected by the Frost than the red Clover whose Roots I measured in the last of October, and found that many of their Tap roots penetrated five Inches, and from it's Sides threw out near thirty Horizontal roots, some of which were Six Inches long and branched: From the figure of this root I flattered myself that it would endure the heaving of the Frost, but now See that wherever it is thin Sown it is generally hove so far out, as that but a few of the Horizontal and a Small part of the Tap root remains covered, and I fear will not recover. Take the whole together it is well matted & looks like a Green Corn field. I have about ten Acres more of this Ground ready for Seed in the Spring, but expect to Combat with the weeds a Year or two; That Sown in August I believe will rise so soon in the Spring as to Suppress them in a Great Measure. My next undertaking was a Round Pond of Twelve Acres; Ditching round it with a large Drain through the Middle and other smaller Drains laid it perfectly dry, this, having first taken up all the Rubbish, I plowed up and Harrowed it many times over till it was smooth, it's Soil is blackish but in about a foot or ten Inches you come to a Sand of the same colour with the upland: From the

Birch that grew upon it I took it to be of a Cold Nature, and therefore I procured a Grass which best Suit that kind of Ground, intermixt with many others, that I might thereby see which suited it best. On the eighth 7br, I laid it down with Rye, which being Harrowed in, I threw in the following Grass seed. A bushel of Salem Grass or feather Grass, half a bushell of Timothy or Herd Grass, half a bushell of rye Grass, a peck of Burden grass or blue bent, and two Pints of red Clover pr. Acre, all the Seed in the Chaff except the Clover, & bushed them in; I could wish they had been clean, as they would have come up Sooner, & been better grown before the Frost, and I have found by Experiment that a Bushell of clean Chaff of Timothy or Salem Grass will yield five quarts of Seed. The Rye looks well and there is abundance of Timothy or Salem Grass come up amongst it, but it is yet small, and in that State there is Scarce any knowing those Grasses apart. I expect from the Sands laying So near the Surface that it will Suffer much in dry Weather, but if it will produce good . . .[3]

This is the most complete account we have of Read's Sharon, limited though it is. Although the intention of residing in the country is clearly indicated, there is no evidence that Read ever actually lived on the place. He recorded in his notebook, in words almost identical with those of the letter, the above-described experiment with red clover. And from other entries we get a few odd glimpses of farm life at Sharon. For example, he noted that rolling protected young grass from the frost:

I saved my young grass at Springfield . . . by hand rollers 3 ft & ½ long and 7 or 9 inches over by wch a man could press down ye roots of grass & Earth so as to Cover what was raised by the Frost & do 3 acres and more in a day in The latter End of Feby & it raised no more that yr.

With his livestock he experienced considerable bad luck. A cow turned out to fresh pasture in 1749 so gorged herself that she died of "wind puff." The following year, he reported, "I had a Sow & piggs & the Sow killed herself by drinking too much new whey and some Beef Brine being put into some

[3] The sentence is not finished.

whey & given to my hoggs caused them to play Antick Tricks and foam at ye mouth & 5 Piggs out of 19 Dyed."

The number of laborers engaged, or the number of oxen and horses, is not suggested. We do know that Read owned slaves and that he made use of indentured servants. One of these, "a Spanish Mullatoe . . . named George," ran away and "went a privateering." He was caught and returned, and Read in 1748 got rid of the trouble-maker by selling his services for a period of 4½ years to one George Marpole. The price was £35.[4]

At Sharon Read was visited with typical line fence difficulties. In 1749 he brought suit for trespass against John West, whose lands were adjacent. The plea was heard before the May term of the supreme court, and the controversy was referred to a commission of seven. West was found guilty, a survey placed the boundary, and Read recovered the disputed lands.[5] It is not strange that, with the careless surveys of the time, such disputes between neighbors should arise. To what extent Read's political influence was a factor in the commission's decision—he was a member of the supreme court at the time—may only be conjectured. It is perhaps a safe inference that the cards were stacked against any man who had the temerity to oppose Read in court.

At any rate, Read took steps to remove all future doubt about the boundaries of Sharon. At a corner adjoining the property of Joseph Smith he set a red cedar post carved with the characters *C R 1749*, which was mentioned in subsequent deeds of conveyance, and stood a mute warning to trespassers for more than a century.[6]

Read seems to have recognized the superior quality of the Springfield Township lands for farming purposes, for in 1748 he paid £100 for a 10½-year lease of another farm, upon

---

[4] *N. J. Arch.*, 1st ser. XII, 549; N. J. Deeds, Liber H, 159.

[5] N. J. Deeds, Liber HH, 345.

[6] The post was subsequently removed, and is now a prized family relic in the possession of Barclay White, of Philadelphia, grandson of a one-time owner of Sharon.

which one Hugh Cowperthwaite dwelt, and leased from him certain other lands for seven years. The terms of the lease are of more than ordinary interest—Read was to pay further one shilling rent annually if legally demanded, and on the expiration of the lease was to leave a like quantity of land sowed with "the like Grain in good Season and tillage" as he found on taking possession, and was not to mow meadows more than once a year, nor to plow them more than twice during the term of the lease, nor take off more than two loads of hay per year, nor plow upland more than once in two years.[7] Such were the measures of soil conservation practiced two centuries ago.

However, Read did not long retain Sharon. Given an opportunity to turn it over at a good profit, he sold it to his Quaker friend Daniel Doughty in 1750. The price, including an interest in some small tracts of land, was £1350.[8]

After disposing of Sharon, Read turned his attention to a farming enterprise in Morris County. Acquiring a 180-acre tract on the Musconetcong, he rented it to John Mackibbinn, a weaver of Hunterdon County, for a period of seven years. The tenant was to pay only a nominal rental—one shilling a year, plus taxes, but in return for the use of the land, he was to make certain improvements according to a plan Read had designed. From all that he pledged to do, it appears that the tenant got no easy bargain. Under the terms of the lease, he agreed to erect a log house 16 by 24 feet, covered with oak shingles; within eighteen months to set 100 apple trees

[7] N. J. Deeds, Liber GG, 258.

[8] *Ibid.*, Liber HH, 345. This is the figure given in the deed. Aaron Leaming made the following entry in his diary Oct. 14, 1750: "Cha. Read has Sold his planta[tio]n for £1450 lb. proc." This splendid farm became in time a place of consequence; it was shown on both Lewis Evans' *Map of Pennsilvania, New Jersey, New York, and the Three Delaware Counties*, published in 1749, and William Faden's map of *The Province of New Jersey . . .* , drawn in 1769 and published in London in 1777; also on Suetter's Map, *Pennsylvania, Nova Jersey et Nova York . . .* (*ca.* 1748). In 1778 it passed by bequest to Doughty's grandson, Daniel Doughty Smith, and subsequently it came into the possession of Barclay White, father of the late Joseph J. White, once New Jersey's foremost cranberry grower. From the two rows of great pine trees bordering the lane, it lately has been known as Pine Lane Farm.

at 40 feet distance from each other, "and to support and defend them against the cattle also to plant out ten cherry trees and ten peach trees . . . and to trim and take care of them." Furthermore, he was to cut his rails "on the land he clears as far as the timber will allow" and to sell no timber off the land. He agreed not to take from the same ground more than one crop in two years, and at the end of seven years to deliver up possession of all and every part of the premises to Read.[9]

A third center of farm enterprises to which Read turned his attention lay in the triangle between the north and the south forks of the Rancocas, in the vicinity of Mt. Holly. Between 1747 and 1753 he held a 40-acre tract on the South Branch of the Rancocas.[10] It lay just across the creek from the sandy uplands of Joseph Moore, mentioned by Read in his agricultural notes. In this neighborhood, it seems, he continued his scientific observations. Three entries appear in his notes under the year 1753. He reported that he "stood by and Counted how many Fork fulls of old Dung went into a Waggon body with only a tail board." Moderately heaped, the wagon held 475 forkfuls. From this he computed that three loads of manure was an ample dressing for an acre. Also in October, 1753, he noted his experience in threshing clover seed. After grinding 16 bushels of the heads of red clover under a tanners' bark stone for four hours, he cleaned 10 quarts of seed. Four quarts of the seed, he found, weighed 6¾ pounds. Again, he reported seeding 3 quarts of timothy on an acre of loamy soil.

Between 1754 and 1758 he purchased a series of tracts in this area, aggregating over 600 acres, at a cost of at least £1,060. Two of these acquired in the spring of 1756 were major purchases: 256 acres of loamy upland and black meadow from Andrew Conarro for £300, and 234 adjacent acres from

[9] N. J. Deeds, Liber K, 450.
[10] *Ibid.*, Liber IK, 322; Liber T, 52.

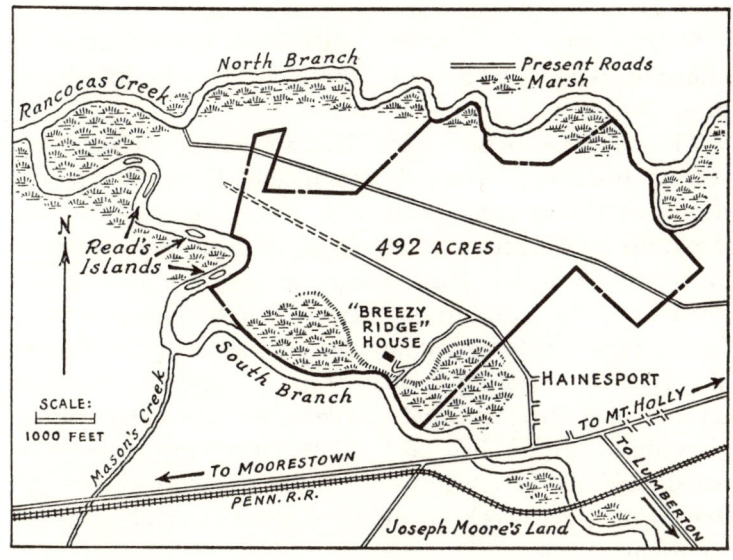

Charles Read's "Breezy Ridge," Boundaries as in 1761

John Erwin for £400.[11] Together these two places comprised the plantation he called Breezy Ridge.[12] From the limited evidence at hand it appears that Breezy Ridge was a pretentious establishment. The substantial hip-roofed house looking southward across the stream stood on an eminence exposed to the breezes that swept up the Rancocas Valley. Nearby stood a boathouse marked with the initials *C R*.[13] There is mention of "Read's Dam," also, on the South Branch of the Rancocas which may have been nearby.[14] Five islands in the Rancocas, totaling 6½ acres in area, also were acquired in 1756—one by purchase, the others by survey under one of Read's proprietary grants.[15]

Although the precise place of Read's residence through these years is not certain, there is evidence that for a time he

[11] *Ibid.*, Liber N, 132, Liber L, 525.
[12] Burlington County (N. J.) Deeds, Liber F, 401.
[13] N. J. Deeds, Liber L, 306.
[14] *Ibid.*, Liber L, 535.
[15] *Ibid.*, Liber N, 104; Liber L, 565.

lived at Breezy Ridge—he is referred to in deeds given in 1760 as "Charles Read of Breezy Ridge."[16] Perhaps he maintained his Burlington home and Breezy Ridge simultaneously, using the latter as a summer retreat.

The entries in Read's notes indicate unusual agricultural activity from 1756 to 1758. Presumably Breezy Ridge was the locale. Experiments as diverse as the seeding of millet, the grafting of fruit and the feeding of cattle were recorded during these years. That stumpy grafts were superior to slender ones he found to be "strickly true by Experim[en]t." On selling 20 cattle he had fattened on grass, he made a profit of £57. On harvesting his hay crop in 1756, he noted the capacity of his barracks, and also computed the weight of hay per cubical unit. When he butchered his hogs in 1757, he noted carefully the loss in weight of the dressed carcasses. No step in the day's work on the farm was too small to escape his attention.

Again at Breezy Ridge, Read experienced the common vexation of runaway servants. In the *Pennsylvania Journal* for August 17, 1758, appeared an advertisement offering 20s. reward for an Englishman named Joseph Dealy—who ran away June 7 "from the plantation of Charles Read, Esq., at Breesey-ridge in Burlington County."[17] The advertisement was signed by Hugh Dunn, who may have been Read's tenant, or the superintendent of the plantation.

It was probably Breezy Ridge also which harbored the stray cattle recorded, as was the custom of the times, in the minute book of Northampton Township in January, 1760. By "Order of Colo[ne]l Read," the township clerk entered a description of the cattle, with their earmarks, as follows:

At the plantation of Colol Read two streys one yearling heifer a brindle & white ½ crop in both ears one in the uper the other in the under side of ear the other a half crop in the uper Side of The Ear.[18]

[16] *Ibid.*, Liber X, 32; Liber Q, 9.
[17] *N. J. Arch.*, 1st ser., XX, 264-265.
[18] Minutes of Northampton Township, Burlington County, New Jersey, 1697-1803, Jan. 10, 1760, p. 368.

Much of the soil on Breezy Ridge was light and sandy, not so suitable for the growing of grain and grass crops as for sweet potatoes and other vegetables. Coveting for the cultivation of crops the rich but undrained marshlands that bordered the plantation, Read engaged one Thomas Rakestraw to ditch and bank the meadows on the north side of this branch of the Rancocas. The terms of the agreement reveal a creditable knowledge of agricultural engineering, which is evidenced also by the description of the water engine in Read's notes. The bank was the kind "Commonly Called twelve by Four" (presumably 12 feet wide at the base and 4 feet high), "the ditch to be Out Side and a five-feet Drain behind the Bank with the necessary Sluices, and always hereafter shall be repaired made and amended and Skoured & Other necessary work Done for keeping the Same Dry."[19] In 1761 Read sold Breezy Ridge to Thomas Bispham of the town of Gloucester.[20]

After 1760, the entries in Read's notes began to fall off, although they did not cease entirely. Then, too, he made numerous undated entries which recorded the results of his experiments or the experiences of others, most of which apparently occurred between 1750 and 1760. For example, he wrote "on a tryall I made" 60 bushels of ashes brought a piece of ground into clover sooner than any other manure. At times after describing a method—the eradication of weeds, the cleaning of timothy seed, or the growing of radishes—he would conclude with "I tryed this with success," or "I have proved this." He noted his success with grass seed which he imported from England and listed the prices of grain at Bear Key, London. In discussing fruit culture, he confessed "A peach

---

[19] N. J. Deeds, Liber M, 547.
[20] Burlington County (N. J.) Deeds, Liber F, 401. Years later this farm came into the possession of the VanSciver family, who found that the valuable underlying deposits of sand and gravel were veritable "acres of diamonds," of a quality highly suited for concrete construction. As a result, it was caught in the march of industrial progress, and under the onslaught of power shovel and truck, the ridge was leveled to the lowly estate of sand pit and mud flat. As if conspiring to complete the devastation, fire visited the sturdy old mansion, leaving today a forlorn and uninviting picture, a far cry from the proud plantation of colonial days.

orchard was a Scheme I have long intended to putt into Execution."

Equally interesting are his observations of farm animals. He kept a herd of 60 goats at his sawmill, but "to no great advantage." He prepared "Excellent pork & bacon," and preserved the recipe. On experiment he found it took two gallons of milk to produce one pound of butter. Now and then he persuaded a friend to try "an experiment" for him. For example, Mrs. Edward Tonkin found that 23 gallons of milk made a cheese weighing 23 pounds, which shrunk to 18 pounds on drying. Likewise, George Trenchard of Salem, at his request, fed a pair of oxen for six weeks on white clover and found they fattened more rapidly when confined than when pastured. Various bits of information from other friends —on feeding cattle, the weight of dressed animals, and other subjects, he carefully recorded.

From agriculture to meteorology was but a natural step, and his notes recount how he placed a receiver "in a Steady but not fast rain" and observed the rainfall. Going farther afield he related how he made "a barrel of Excellent Metheglin" and how he painted his chaise. A homemade, inexpensive saddle which he discovered in the possession of a fellow traveler, he described in the minutest detail. Homely wisdom suggestive of Poor Richard found expression in the following entries relating to the care of farm implements:

> If yr Plow or other utensils want great repair gett new, for yo will find by adding Each repair, if yo Enquire of yr workman, twill cost more than a new one, as I Experienced by a Plow. The Workman has a great Price for repairs.
> Whenever any thing yo use getts out of order dont delay mend[in]g till its wanted.

Read numbered among his friends a distinguished group of country gentlemen, with Governor Belcher heading the list. Another was Colonel Peter Schuyler, master of Petersborough, on the Passaic River above Newark. In the provincial council were Peter Kemble, John Stevens, and Lord Stirling. Also there were Richard Saltar, Robert Ogden, Tunis Dey, Aaron Leaming, Jacob Spicer, Joseph Warrell, William Hancock, and John and Samuel Smith, all prominent in affairs of the colony. The spirit of cooperation prevailed among these country gentlemen. Agriculture was a common meeting ground. They procured for each other new varieties of seeds and assisted in the exchange of farm animals. They passed along useful information and loaned books on agriculture.

In this community of interest, apparently no one was closer to Read than William Logan, who on succeeding his father as master of Stenton, devoted himself particularly to agriculture and gardening. He corresponded with Jared Eliot, who sent him a new "drill plow," forerunner of the modern grain drill. His collaboration with Read in agricultural experiments is revealed in a letter from Stenton, July 25, 1754, in which he wrote Eliot:

> My Cosn. Cha. Read of Burlington gave me about a Pint of thy foul Meadow Grass Seed which I Sowed for a trial, but it did not Come up—Suppose the Seed might be Gathered before it Was Quite ripe—If it be really a Good Grass for feed or hay, I should be Glad of a peck of Clean Seed, and if possible of this Years Growth for which Shall pay thy order.[21]

In 1756 Logan requested John Smith to get him a bushel or two of the seed of blue grass or blue bent grass, to seed the banks of his plantation.[22] From England he procured the seed of vetches and of St. Foin.[23] Again, finding himself unsuccessful in procuring a pair of large oxen he was very anxious to purchase, he wrote to John Smith, "I should think Charles Read might have known where to have got me a pair."[24] At another time he sent Smith a pair of "Guiney Fowls," advising him to keep them up and feed them about a week "before they

---

[21] *Essays upon Field Husbandry in New England and Other Papers (1748-1762)* by Jared Eliot. Edited by H. J. Carman and R. J. Tugwell (New York, 1934), p. 130. Other letters of William Logan to Jared Eliot are given on pp. 226-234.
[22] William Logan to John Smith, Sept. 1, 1756. Smith Papers, V, 18.
[23] William Logan to Jared Eliot, Oct. 14, 1755. J. Eliot, *op cit.*, 1934 ed., p. 232.
[24] William Logan to John Smith, 1759. Smith Papers, V, 118.

are let at large to prevent their Straying."[25] Thanks to Logan's generosity, English cheese graced the Read table in 1759, when Logan wrote to Smith:

> As the last Bristol ship brot. my Gloucester Cheese I now send thee six of them, of which I desire thine and my Sisters Acceptance. I have also put into the Cask two more which I desire thee when an oppy. offers to send to my Cosn Charles Read in my name.[26]

The Burlington fair each fall, a glorified market day that ran for a week, furnished Logan with an excuse to make a trip to the West New Jersey capital, and pay a visit at the Read home. From Read's letters it appears that he looked forward to these meetings with his cousin when the two country squires could satisfy their curiosity in the trading of livestock, farm produce, and farm supplies. Doubtless, too, they enjoyed the gossiping and visiting with friends who gathered for the festive occasion.

Though ostensibly commercial in purpose, the fairs and market days early established by law at Burlington and other colonial towns, were in effect a social institution. People came to the fairs to learn the news. There was a comparing of notes on affairs of the province and discussion of the governor's administration. There was talk of crops and of trade, and an exchange of recipes. Personal news about new babies and old folks who had died was passed around. There was gossip about whose sons had taken up arms and were training, about whose daughter had married out of meeting and joined the "world's people." The fairs were a medium for spreading information when means of communication were scanty and slow. They broke the monotony of isolation and were a welcome respite from long hours of physical labor. The element of sport also entered the fairs in the middle of the Eighteenth Century with the introduction of horse racing. Neither the discipline of the Society of Friends nor laws enacted by the

---
[25] William Logan to John Smith (undated). Smith Papers, V, 259.
[26] William Logan to John Smith, July 13, 1759. Smith Papers, V, 107.

legislature could wholly suppress gambling at the colonial race tracks.

Like all country gentlemen, Read now and then took a hand at horse-trading. An incident of this kind is recorded in a letter to his brother James, July 16, 1745, when the latter was still in Philadelphia:

On Wednesday last I sent down a horse abt 14½ hands High of a brown colour a Starr in his forehead a Switch Tail I directed the Man to carry the horse to P. Robinsons and order him ½ peck of oats night & morn: & to have him Sold by the Huntsman at Vendue if he would fetch 16 pounds clear of Charge. I Sent wth him a Letter directed to you . . . to desire you would give directions to the Huntsman about it—The same Man carryed down a Darkbrown Mare wch I would have bought if to be had for 20 pounds or less—but I have heard nothing of affair Since & perhaps the man may have moved off wth my horse. . . . please to Enquire after the Mare & what is come of my horse . . . I sent down last night 7 fish . . . to be given to you.

We illuminated above 20 houses & fired above 100 cannon.[27]

Not the least important of Charles Read's agricultural correspondents was his brother James, who was, as we have seen, a disciple of the botanist Peter Collinson and of his American contemporary, John Bartram. James Read's letters reflect the views of a country gentleman of a lively and facetious nature, who was as much at home writing of his experiences in gardening and in criticizing current English and French books, as in discussing politics, education, and religion. In a letter to William Logan in 1765, he described the progress of numerous garden plants he had received from Logan and from his "brother at Burlington." "Some Raspberries my brother sent me did well awhile: But many of Them now droop," he wrote. "My Brother's large Plum Trees were too long out of the Ground, and had received very shabby usage (having

[27] Charles Read to James Read, July 16, 1745. Berks County Historical Society MSS. The celebration alluded to was probably the send-off given the vessel which about that time sailed from Burlington for Boston with a cargo of provisions for the Cape Breton expedition (*N. J. Arch.*, 1st ser., XII, 262; *N. Y. Weekly Post Boy*, June 17, 1745). For a description of fairs in colonial Burlington, see Amelia Mott Gummere, *Friends in Burlington* (Philadelphia, 1884), pp. 43-44.

been much stripped of bark) before they reached Reading. Only two of six thrive: The other 4 seem to be dying. The small Plum Trees all Thrive. I water them often."[28]

In similar vein, William Logan wrote to James the following February. After describing an evening spent with Charles in Philadelphia, he told of his plans and preparations for the approaching garden season:

> Does not ye sap begin yet to Run for Gardening—I made a hot bed last week for sowg Collyflower & Cabbage Seeds & Early Sallad—Intend to plant some few Rows of ye Earliest Pease on the first opening of ye Ground & when they come up If attacked by a sharp frost to Cover with long dry straw their Rows.[29]

James Read's interest in his garden, in which he took great pride, went farther than a mere love of flowers; like his brother he was a student of the art and the science of gardening. Some of his aphorisms on gardening reflect a practical common sense as applicable today as two hundred years ago; for example:

> Dung is a medicine that kills in the overdose.
> Labour is the best manure for the kitchen garden.[30]

On disposing of Breezy Ridge and bringing to a close some of his farming ventures, Read took a fling at the fishing industry. In colonial economy, fishing was closely allied with agriculture as an important source of food, and it is not surprising that it appealed to Read's spirit of enterprise and his flair for exploiting natural resources. In his notebook he drew a diagram of a "fishing waire," and likewise copied directions for making glue from sturgeon which Benjamin Franklin had procured from a correspondent in London.

Just south of Trenton at Lamberton[31] in old Nottingham Township, then a corner of Burlington County, was the plantation of John Douglass, "yeoman." This place attracted

---

[28] James Read to William Logan, June 2, 1765. From private collection of Sydney L. Wright, Philadelphia.
[29] William Logan to James Read, Feb. 4, 1766. From collection of J. Bennett Nolan, Reading.
[30] J. Bennett Nolan, *The Foundation of the Town of Reading, in Pennsylvania* (Reading, 1929), p. 195; Shippen Papers, VII, 55.
[31] Now a part of the City of Trenton.

Read's attention as a desirable site for a fishery. Accordingly, on October 31, 1763, he procured from Douglass a 21-year lease for a river-front tract for the development of a fishery, at an annual rental of £3. From the terms of the lease it appears that he had already begun operations nearby, for reference is made to a well dug "for the use of the Boiling House of sd. Charles Read." The lease also mentioned the dam of a stream, and gave Read the right to use the shores of Douglass' plantation "for fishing and landing with Boats, Netts, or Fish and the liberty of brushing and Cleansing and Improving the same . . . and of laying ashore on the adjacent lands the loggs, dirt, stones or other rubbish necessary for the making of the fishery." Read was also given "the liberty of getting timber and stone on the lands of the sd John Douglass for erecting and repairing buildings necessary." Douglass agreed to build a good wharf fronting the premises "for the convenient reception of Shallops and Merchandize" and Read was to open a public road 3 rods wide at the end of the wharf.[32]

This fishery was one of the "schemes for the improvements of the country" cited by Aaron Leaming as the reason why Read did not succeed in building up a fortune.[33] It is probably a safe inference that it was not a financial success, another example of the futility of business under absentee proprietorship.

Charles Read's enthusiasm for farming, his scientific attitude toward agriculture, his extraordinary knowledge of the agricultural literature of his time, his broad-gauged observation of the practices on the farms of the province—these and other facts about his farming enterprises appear more fully in his notes which follow in Book II.

[32] N. J. Deeds, Liber T, 86.
[33] A. Leaming's Diary, Nov. 14, 1775, *loc. cit.*

Seven: Ironmaster

THE mining and manufacture of iron had begun in New Jersey as early as 1675, when James Grover started operations on lands in Shrewsbury, subsequently known as Tinton Manor, estate of the distinguished Morris family.[1] Between 1700 and 1750 numerous iron mines had been developed in the mountainous regions of Sussex, Morris, and Bergen counties. Owners of land in the southern counties also began to consider exploiting the iron-ore deposits found on their holdings. Charles Read must have foreseen the increasing role iron would play in the development of the new country. As the years passed the lure of the iron industry grew upon him, and with it the ambition to become the greatest ironmaster in the province. In so doing he ran true to his natural inclinations and traditional interests. As previously noted, his father had been one of the founders of the first ironworks in Bucks County, Pennsylvania. His friend Colonel Peter Schuyler had made a fortune from the copper mines owned by his family on the Passaic above Newark. His political colleagues, John Stevens, Richard Saltar, and Robert Hunter Morris, had an interest in a copper mine which was being developed at Rocky Hill, and another copper mine had been opened at New Brunswick. Jared Eliot, whose *Essays upon Field Husbandry in New England* Read had studiously followed, had published in 1762 an essay upon the manufacture of iron from black sea sand.[2] All this marked a development in which Read aspired to play a leading role.

As secretary of the province and the governor's advisor,

[1] Charles S. Boyer, *Early Forges and Furnaces in New Jersey* (Philadelphia, 1931), pp. 196-199. This work presents a detailed account of Charles Read's iron enterprises, pp. 154-190.

[2] *An Essay on the Invention or Art of making . . . Iron, from black Sea Sand* (New York, 1762). See *Essays upon Field Husbandry . . .* by J. Eliot, ed. of 1934, pp. 163-187.

Ironmaster

Read had his attention focussed upon iron by the efforts of the Lords of Trade to suppress the manufacture of iron products in New Jersey. A petition was addressed to the royal authorities in 1741 seeking encouragement for iron manufacture,[3] but nine years later an Act of Parliament prohibited the further erection of slitting mills, plating forges, or steel furnaces. This called for a proclamation from Governor Belcher, signed by Read, which ordered an accounting of the ironworks of the colony.[4] It was the British policy to encourage the exportation of pig and bar iron to the mother country, where it was to be converted into commodities which the colonists would have to buy. In spite of numerous instructions for inspecting the ironworks and enforcing the order which came into the governor's hands, iron continued to be hammered into sheets or plates that were cut into strips for the making of nails.

Then, too, Read found that some of his cedar-swamp lands in Burlington and Gloucester counties were underlain with plentiful quantities of bog ore. These reddish-yellow deposits, a variety of the mineral limonite, were formed by the chemical action of vegetable-laden water sifting through ferruginous strata in places where the water was sluggish; they varied from a muddy consistency to the hardness of stone. The Indians knew of the deposits, and were accustomed to use the material to paint themselves.[5]

Read saw to it that specimens of the ore were taken up and tested. He showed the samples to "judicious iron masters" who had "followed the business many years with reputation and success," and who gave their stamp of approval.[6] Samples of iron made from the ore, after passing trial before competent judges, were certified as to quality and were placed on display at Captain Burnet Richards' place in Philadelphia.

[3] *N. J. Arch.*, 1st ser., VI, 140-142.
[4] *Ibid.*, XII, 674-675.
[5] Bertram Lippincott, *An Historical Sketch of Batsto, New Jersey* (1933), p. 12; C. Boyer, *op. cit.*, p. 2.
[6] *N. J. Arch.*, ser. 1, XXIV, 583.

Satisfied with the quality of the ore, Read proceeded with his plans. Ore alone was not sufficient; it was only one of three essentials for an iron manor. There also must be fuel, and there must be water power. In his land acquisitions prior to 1765 Read may have had an eye toward the future establishment of ironworks. The many tracts of cedar-swamp and timber lands, traversed by sluggish creeks and dotted with the sawmill sites used in his lumbering business, furnished the water for transportation as well as for power, also the lumber for building as well as wood for fuel. With the addition of many new tracts during the next two years, he consolidated his holdings around four centers which he singled out for the erection of ironworks—Taunton, Etna, Atsion, and Batsto.

Sometimes his intention to erect a dam or an ironworks was definitely set forth in the deeds which were passed. In other transactions he acquired the rights to charcoal, wood, and iron on adjacent lands for a nominal price of 5s. "in order to encourage the erection of iron works,"[7] or "in order to promote a work so beneficial to the public."[8] In another deal he acquired the right to take timber to erect "Coalers and Miners Hutts."[9]

Among the principal purchases in 1765 was a large tract of pine land in the forks of the Atsion and Batsto rivers, together with a one-half interest in a sawmill, which Read acquired from Richard Westcott of Egg Harbor for the consideration of £200.[10] This and adjacent lands acquired from time to time became part of the extensive Batsto manor. In order to provide power for an ironworks Read petitioned the legislature for permission to erect a dam across Batsto Creek, after ascertaining from Joseph Burr that the dam would not interfere with the operation of the latter's sawmill at the head of the creek. In reply the legislature promptly passed an act which provided that, "whereas the Honourable Charles Read

[7] N. J. Deeds, Liber X, 265.
[8] Ibid., Liber AB, 72.
[9] Ibid., 68-69.
[10] Ibid., Liber U, 289.

Site of Taunton Furnace, Taunton Lakes

Site of Etna Furnace, Medford Lakes

... sets forth, that he hath proved to Demonstration good Merchantable Bar-Iron may be drawn from such Ore as may be found in plenty in the Bogs and ... in such Parts of this Province which are too poor for Cultivation, which he concieves will be of publick Emolument,"[11] his request should be granted.

Just about a month after the act received Governor Franklin's approval, Read announced his intention to establish ironworks on the four sites and invited the interest of prospective investors. An advertisement in the *Pennsylvania Journal* read in part: "Charles Read of Burlington, gives notice to the public, that he is possessed of several tracts of land, having in them streams of water, as constant and governable as can be wished. ... There is at all these places, plenty of food for the cattle from the middle of May to the middle of October." After describing other resources for iron production, the advertisement continued: "As Mr. Read's situation renders it inconvenient to him to take upon himself the expence or care of works so extensive, he notifies to the public that it will be agreeable to him to let the conveniences to any gentleman of credit reserving a share of the produce, or to enter into a partnership with any persons of good dispositions, fortune and integrity." An alluring prospect was pictured in the following summary:

The goodness of the Iron, the visible quantity of the ore, the extraordinary situation, joined to the very easy land and water carriage, and its vicinity to Philadelphia, and easy carriage from the two last mentioned works to New-York, give works erected here a preference to any on the continent.[12]

Charles Read's enthusiasm for his iron enterprise was noted by William Logan, who early in 1766 wrote to James Read, of Reading:

I spent this evening with thy Bro. Charles who is just come

[11] *New Jersey Session Laws* (May 21-June 20, 1765), Chap. XXXII, p. 80.
[12] *Pennsylvania Journal*, no. 1181, July 25, 1765; *N. J. Arch.*, 1st ser., XXIV, 581-583.

down from Burlington to stay a day or two—He is quite hearty & very full of spirits & his Iron Works scheme—I asked him whether he Inclined to write to thee. He says he . . . intends it soon pretty largely—I suppose to make some Inquiry about Iron Ore, or Works ——"[13]

Even before Read made his public announcement he appears to have taken the initial steps in erecting the works. The first plant to be completed was at Taunton, in old Evesham Township. It was situated on what is now called Haines Creek, formerly known as Read's Mill Creek, about three and one-half miles from Medford. On Faden's *Map of the Jerseys* it is shown as "Read's Mill." Here in 1766 was erected "a small furnace with bellows" and a three-fire forge.[14] These two structures were the heart of the works. Into the egg-shaped inner chamber of the four-sided furnace, lined with firebrick and tapering upward approximately twenty feet, were dumped in alternate layers the ore, the charcoal, and the "flux" of oyster shells. The intense heat engendered by the burning charcoal, forced by air blasts from the bellows, melted the ore, and the iron thus freed was drawn off into sand moulds in the form of castings or "pigs." For further working into rods, plates, or other forms, the "pigs" were taken to the nearby forge, whose fires were blown by bellows connected with the water wheel. From the forge fires the red-hot iron was placed under a huge hammer, also operated by the water wheel, which pounded it into the desired shapes. Around the furnace and the forge at Taunton arose other buildings—a "coal-house" with a capacity of 400 loads of charcoal, a dwelling for the manager of the works, and "convenient outbuildings" for the workmen. Fifteen hundred acres of adjacent woodland furnished charcoal for the furnace, and rights to the wood on nearby tracts assured an ample supply. Meadow land provided pasture "during the coaling season" for the oxen, which were depended upon for hauling. In winter the

[13] William Logan to James Read, Feb. 4, 1766, *loc. cit.*
[14] *N. J. Arch.*, 1st ser., XXVIII, 612.

oxen were "boarded out" at 28s. per pair.[15] As the land was cleared, a portion of it was set in fruit trees. Nearby was a sawmill, handy source of lumber for construction at the works. Timber not fit for the sawmill could always be used at the ironworks, which also provided a good market for the produce of nearby farms. The cost of transporting iron to Philadelphia was but 12s. a ton.

The iron manor was a hive of activity, a highly organized industrial unit almost self-sufficing in its daily routine. Besides the managerial staff, several classes of laborers were required, some of them skilled mechanics. There were the wood cutters, the charcoal burners, and the teamsters; there were the colliers, also the forge men, who included "finers," and the "hammer men." In 1769 Read advertised for "Master-Colliers, Moulders and Stock-takers," likewise for "a good Keeper."[16] There was need also for carpenters and blacksmiths.

To procure the needed supply of laborers, Read purchased the time of numerous indentured servants, some of them newly arrived immigrants who may or may not have worked as mechanics abroad. He also used negro slaves—in 1767 he bought a "negro forge man named Cato" for £100[17]—and doubtless free labor as well, wherever he could procure qualified workmen. Such a motley array brought with them a train of labor troubles, as the frequent advertisements for runaway servants testify. Substantial rewards offered for the runaways did not always assure their return. The more reliable ironmasters, apparently, were bound by a code of honor not to harbor or employ a runaway from another ironworks, but this precaution was not always observed by the less scrupulous operators. A report that in 1771 a deserter from another master had found refuge at Charles Read's works, was indignantly denied by Read's son, who wrote:

We have always made it an invariable rule at our Works

[15] *Pennsylvania Journal*, no. 1467, Jan. 17, 1771; Aug. 25, 1773. *N. J. Arch.*, 1st ser., XXVII, 345; C. S. Boyer, *op. cit.*, p. 161.
[16] *Ibid.*, XXVI, 361, 368.
[17] N. J. Deeds, Liber X, 259.

never to be assistant in robing a Person of his Property by Secreating his Servant the Contrary Conduct is base and unjust as well as ruinus to the Interest of Iron Masters.[18]

Liquor was a major source of worry. As long as the men remained on the premises, purchased their supplies at the company store, and received their allotted portions of rum, matters could be kept under control. But visits to taverns resulted in drunkenness, quarrels, running into debt, loss of time, and desertions. With commendable foresight, at the beginning of his venture, Read had petitioned the governor, the council, and the general assembly to prohibit the licensing of taverns within three miles of any ironworks without the approval of the owners of the works, and to place a limit of 5s. upon the amount of debts contracted at taverns by employees of the works, which would be recoverable from the owners.[19] Although somewhat slow to respond, the assembly in 1769 enacted a law obviously intended for the benefit of Read's enterprises, which authorized owners of ironworks in Evesham and Northampton townships in Burlington County to furnish their employees with strong liquor "in such Quantity as they shall from Experience find necessary," but prohibiting any other person or persons residing within four miles of the works to supply the employees with strong drink. Fines collected for this offense were earmarked for the repair of the public roads.[20]

Almost simultaneously with the erection of the Taunton furnace, Read built the second of his ironworks, which he named Etna Furnace, on the present site of the summer colony of Medford Lakes. A tributary of the southwest branch of Rancocas Creek furnished the water power. The furnace was in full swing by 1768.

Etna was surrounded with 9,000 acres of well wooded land.

[18] Charles Read (IV) to James Pemberton, Jan. 23, 1771. Pemberton Papers, XXII, 94.
[19] Petition of Charles Read and Peter Bard, June 11, 1765; N. J. Public Record Office MSS.
[20] *N. J. Session Laws*, 1769, Chap. XIX, p. 109.

Like Taunton it had a three-fire forge, also a store "for the Supply of the Workmen and Labourers only," and the usual buildings. Nearby was a gristmill for "supplying the Persons and Teams necessary at said Works," and a "swift going sawmill," which yielded a rental of nearly £200 a year. The law of 1769 which permitted the sale of liquor to employees, gave further encouragement to the enterprise by exempting from taxation the Etna store, also the gristmill, if used "for the sole Purpose of Supplying the Persons and Teams necessary at said Works, and not taking Toll, or manufacturing Flour for Exportation."[21]

In addition to producing bar iron the Etna plant was equipped for manufacturing iron products such as flatirons, wagon boxes, and iron handles, which were "in very high esteem at foreign markets." It had a small "stamping mill," for crushing the hardened ore mass, "a variety of nice patterns and flasks for casting ware, and equipment "to grind and polish iron ware by water."[22] Consequently, it required a force of skilled laborers, including moulders, stone cutters, grinders, "keepers," or persons to handle casting. Labor difficulties seem to have been in direct proportion to the requirements for skilled workers. The runaways were numerous and frequent—some took stolen clothes with them; the more intelligent who could write a good hand, made their recovery more difficult by their ability to forge passes. The negro slaves apparently gave less trouble than indentured servants. The loss of time from desertions and the rewards paid for recovery, laid a heavy tax upon the profits.

Read placed his son Charles in charge of the works at Etna. In 1773 he conveyed an interest in the works to the son in consideration of stock and wages due him as manager, with accrued interest, amounting to £595.[23]

The Batsto works were built about 1766 on the Batsto

---

[21] *Ibid.*
[22] *Pennsylvania Journal*, no. 1453, Oct. 11, 1770; *N. J. Arch.*, 1st ser., XXVII, 286-288.
[23] N. J. Deeds, Liber AF, 179.

Creek, a branch of the Mullica River, near the Indian summer village Nescochague, later known as "Sweetwater" and "Pleasant Mills." Water power here was abundant—capable, Read calculated, of driving "four bellows and two hammer wheels."[24] Wood was available from 8,000 acres of surrounding land, which Read controlled. The location had the advantage also of being only two miles from navigable water—a scow of ten tons could come to the works and carry the products to vessels sailing to New York, New England, and Philadelphia. Read estimated the cost of mining the ore at not more than one shilling a ton. Because the country was level and the roads good, a team of six oxen could haul a load of two tons.

Batsto possessed another advantage in having available a supply of free white laborers. A mile below the furnace site was a settlement of lumberjacks and fishermen, squatters who were gaining but a scanty livelihood from the sparcely settled region, and who were glad of an opportunity for employment. They proved to be a valuable aid in getting the venture under way. Also, a church erected here about 1760 attracted worshippers from the back country and helped to bring additional workers for the new industry. Pig iron apparently was the principal product of Batsto during the early years of its operation.

Although Read is credited with having built the Batsto works, he had the financial aid of four partners in the enterprise: Reuben Haines, Philadelphia brewer; Walter Franklin, New York merchant; and John Cooper and John Wilson, both of Burlington County. Between December, 1767, and August, 1768, he disposed of most of his own interest to three of these partners—a half interest to Haines for £387-10-0, and an eighth interest each to Franklin and Wilson for £87-10-0.[25]

[24] *N. J. Arch.*, 1st ser., XXIV, 582. See also B. Lippincott, *op. cit.*, and Charles F. Green, *Pleasant Mills, New Jersey, Lake Nescochague, a place of Olden Days*, pp. 5-7.
[25] N. J. Deeds, Liber Y, 131, 146, 183; Liber Z, 186, 198. Also Elias Wright, "A Short History of the several Tracts of Land once Containing Iron Furnaces . . ." (1898). Batsto is the locale of Charles J. Peterson's novel, *Kate Aylesford* (Philadelphia, 1855). The tradition is current, though seemingly unfounded, that the heroine is a

Old Store, Batsto

Old Hotel, Atsion

Ironmaster

The Atsion works were erected in 1767 and 1768, on the Atsion River, near a favorite bathing ground of the Indians, called by them "Atsayunk." The forge was a large one, equipped with four fires and two hammers. The river when dammed provided abundant power. Four thousand acres of woodland supplied the fuel. The bog ore at Atsion contained 45 to 47 per cent of metallic iron, and some of the samples ran as high as 53 to 56 per cent. Great masses of ore were taken from the bed of the pond during the winter, when the fires of the furnace and forges were out, and the water was drawn off. Pig iron produced at the Batsto furnace eight miles away was hauled upstream or overland to the Atsion forge to be converted into bar iron, which was the principal product of the works during their early operation. The portage cost to Philadelphia was 17s. a ton, to New York 15s. Equipment was installed for weighing the loads of iron as they were sent out, usually of 1½ tons burden.

The labor problem here, it seems, was not so acute as it proved to be at Taunton and at Etna. Atsion and Batsto were not so far removed but that one carpenter, one smith and one clerk would suffice for both plants. Also, the Indians on the reservation at Brotherton only three miles distant were available and many of them were employed at the works.

Read succeeded in enlisting outside capital for the development of Atsion. On January 26, 1768, he conveyed for £50 a 249/1,000 interest each in the Atsion property to David Ogden, Jr., and to Lawrence Saltar, son of the late Justice Richard Saltar.[26] By thus shading each quarter share, Read retained majority control.

The management of four large ironworks, many miles distant from the provincial capitals, coupled with pressing duties of state, was too much for one busy man. Then, too, falling prices laid a heavy impediment on industry. Repeated illness

---

portrayal of Honoré Read, daughter of the ironmaster (*New Jersey, A Guide to its Present and Past* . . . , New York, 1939, pp. 606-607). No evidence has been found that Read had a daughter by this name.

[26] N. J. Deeds, Liber AG, 248, 251.

further prevented Read from giving the ironworks the needed personal attention. It soon became evident that he had plunged too deeply, and he sought to dispose of the works. Most of his interest in Batsto, as we have seen, was sold promptly. It was not until 1773 that he sold his remaining interest in Atsion—to Abel James and Henry Drinker, of Philadelphia, for £1,200.[27] Taunton was not so readily sold, and it became a part of the estate which in 1773 Read turned over to his trustees. It was leased to Isaac Evans, and the following year the trustees put it up for public sale. Etna was taken over by Charles IV.

Although the peace-loving Read probably did not realize it, in erecting his ironworks he was placing a major weapon at the disposal of the American colonies in the coming struggle for independence. The ironworks proved to be an important source of munitions for Washington's forces. Taunton is reputed to have produced quantities of cannon and shot. Batsto, which shortly came into the possession of John Cox, manufactured cannon balls, and also pans for the evaporating of sea water to procure the salt indispensable for provisioning the army. Because of the importance of these works to the army in the field, its employees were exempted from military duty except in case of invasion by the enemy.

After the Revolution, the ironworks continued for many years, manufacturing an ever increasing variety of the gadgets of civilization—iron pots, kettles, Dutch ovens, skillets, stoves, pestles, sash weights, forge hammers, and the like. Batsto in particular prospered. Under the ownership of William Richards, it became the center of great activity, a veritable feudal manor with its "big house," its extensive grounds, buildings, and outbuildings. Toward the middle of the Nineteenth Century, however, the bog-iron industry was compelled to give way to the ironworks in northern New Jersey which used the magnetic ores, and to the Pennsylvania works in the anthracite region.

[27] *Ibid.*, Liber AH, 95.

Eight: Secretary

ALTHOUGH READ'S capacity for private business enterprise was truly extraordinary, nevertheless it was excelled by his penchant for public office. While occupying the posts of court clerk and collector of customs at Burlington, in 1743 or earlier he was appointed clerk of the circuits at an annual salary of £20.[1] Chief Justice Robert Hunter Morris, son of Governor Lewis Morris, was instrumental in securing the appointment for him. Richard Partridge, the New Jersey provincial agent in London, also began to pull wires in Read's behalf. As early as 1740 and for several years thereafter, in cooperation with Peter Collinson, he tried to get Read appointed to the council. In the manner of the time, he acknowledged, there was "money already ordered . . . for that purpose," but it did not accomplish the desired result.[2]

Partridge, however, was persistent and after taking "a pretty deal of pains," procured Read's appointment to the important post of secretary of the province.[3] The secretaryship was a patent office, that is, appointment was directly by the Crown. It was a common practice for the appointee, who customarily remained in England, to farm out the office to a deputy in the colony. Archibald Home had been made deputy secretary in 1741 and on his death in 1743 he had been succeeded by his brother James. When the following year James Home announced his intention to remove to another province, it was important to fill the secretaryship promptly, "to the Intent that the business of the Said office may not be interrupted for want of a Person Sufficiently authorized for the Due Execution of the Same." Thanks to Partridge's efforts, Read received at the hand of Governor Morris a royal commission to serve as

[1] *N. J. Arch.*, 1st ser., XVI, 75, 79.
[2] Richard Partridge to Richard Smith, April 1, 1745. Pemberton Papers, III, 164. See also Ferdinand John Paris to James Alexander, Feb. 10, 1746, Paris Papers.
[3] Richard Partridge to John Kinsey, June 4, 1744. Pemberton Papers, III, 126.

secretary of the "Province of Nova Caesarea or New Jersey . . . for and during our Pleasure." On November 10, 1744, he took the oath "for the Due Execution of the Office of Secretary and Clerk of the Council."[4] The same month Governor Morris, reposing "Especial Trust and Confidence" in his "Integrity, Learning and Ability," named Read one of the surrogates of the prerogative court.[5] Also in 1744 he was commissioned justice of the peace for Monmouth County. In 1746, through Chief Justice Morris, he received a royal commission of *Dedimus Potestatem* which granted him authority to administer oaths to all officers within the province.[6]

Meanwhile Governor Morris had become involved in a quarrel with the assembly. Although he came of a wealthy family and was master of two large estates—Morrisania in New York and Tintern (Tinton) Manor in Monmouth County, up to this time he had been a firm friend of the common people during the long struggle against the executive authority. But with his accession to power he appeared to the colonists arrogant and selfish, for he was under oath to sustain the English constitution and the rulings of the Crown, even though arbitrary and contrary to the interests of the common people. The governor's attempt to oust the squatters and frontiersmen from their land claims in the northwestern part of the colony resulted in a renewal of rioting and other disorder. The sum appropriated by the assembly in 1739 the governor considered inadequate for the support of the government, and after a sharp dispute, the cranky old man ordered the assembly dissolved.[7] With the next assembly the quarrel was resumed; the governor refused to sign any bills passed by the assembly, and that body persistently declined to vote any support for the government. Morris' invective against the rural representatives, who without apology admitted they were "Farmers and Plowmen," served but to stiffen their resistance to the "ill-

[4] N. J. Commissions, Liber AAA, 258.
[5] *Ibid.*, 259.
[6] *Ibid.*, Liber C2, 103, 122.
[7] *N. J. Arch.*, 1st ser., XV, 79-84.

natured and superannuated Governor."[8] The assembly became increasingly distrustful and vindictive. It was more powerful than the council—it controlled the appropriations, and its members were not averse to showing a royal governor what they considered his proper place.

Charles Read's letters during this period reveal a close bond of friendship with the Morrises and sympathetic cooperation in problems of state. Of the election of a new assembly in 1745 he wrote:

> The City and County of Burlington have returned their old members on view—so will all the counties to the westward; there is some stir in Hunterdon, but it will be to no purpose. I heartily wish better times.[9]

A prolonged illness in 1746 prevented Read's attendance at some of the sessions of the circuit court. To Chief Justice Morris he wrote:

> I have scarce had one hours health since we parted & was so bad abt. 8 days ago that Dr. Shaw was not easy to Act on his own judgment but Dr. Cadwalader also attended me by the blessing of Providence & their Care I gott a little recruited but . . . the Fever seems unwilling to submit even to the Costese. . . . Dr. Cadwalader is of Opinion that it will be of Dangerous Consequence for me to attempt any Journey this winter.[10]

In spite of his illness Read had followed with understanding interest the governor's efforts to persuade the assembly to vote the operating expenses for the government, and concluded with the following comment:

> You know my Opinion as to the Arreages & I have told you the method I thot. most likely to Obtain them I Assure you I heartily wish them paid & have in all private Conversations spoke of the not paying them as Dishonourable to the Province.[11]

The deadlock between governor and assembly was finally

---

[8] A. C. Myers, *op. cit.*, p. 73; *N. J. Arch.*, 1st ser., XV, 413.
[9] Charles Read to James Alexander, Jan. 21, 1745, *Papers of Lewis Morris, op. cit.*, p. 201n.
[10] Charles Read to Robert Hunter Morris, Oct. 6, 1746. Boggs Collection.
[11] *Ibid.*

broken by the sudden death of the former at Trenton on May 21, 1746. Read, now an accepted member of the official family, served as pallbearer at the funeral.[12]

The fifteen months intervening between the death of Governor Morris and the coming of his successor, Jonathan Belcher, was a period of uncertainty in affairs of state which must have caused the secretary much anxiety. On Morris' death, John Hamilton, president of the council, assumed the executive chair. His death a year later brought to an early close an administration marked mainly by New Jersey's support of the Canadian expedition in King George's War. Thereupon John Reading, succeeding president of the council, occupied the governorship temporarily.

During the interregnum Read found it difficult, in his capacity as secretary, to assemble a quorum of the council. Delays and discomforts of travel, illness and the fear of illness deterred several of the members from answering his summons. In a letter to James Alexander in 1746, Read wrote: "Mr. Rodman has never had the smallpox, therefore will not venture."[13] It taxed his powers of ingenuity and persuasion to get the council's business transacted.

Available records indicate that Read performed the duties of secretary continuously from 1744 to 1767, either as secretary in his own right or as deputy. Although his commission in 1744 appears to have been to the secretaryship proper, rather than as deputy, in 1748 he wrote, "The Secry. Office I hold as Deputy. . . ."[14] From available records it is not easy to ascertain when, if at all, his status may have changed in this respect. It seems frequently he was addressed as "secretary" when actually he was "deputy secretary." The appointment of a non-resident secretary and the delegation of duties to a deputy had not always been conducive to good administration. Read, however, whether secretary or deputy, gave close per-

[12] *N. J. Arch.*, 1st ser., VI, 368n.
[13] May 27, 1746, N. Y. Historical Society MSS.
[14] Charles Read to R. H. Morris, July 9, 1748. Boggs Collection.

sonal attention to the office, and utilized it as a position of importance and of influence.

Through the changing political scene he did not always feel secure in the post. For example, in 1748 he wrote James Pemberton, then in England, that he was in danger of being supplanted in office by "an Irishman from Amboy," and asked Pemberton to help him with his influence in London.[15] After the death of Governor Belcher, it seems he again feared he might be dropped. In October, 1757, he asked James Pemberton to write Richard Partridge "to renew my Lease of the Secretaries office which is near Expiring (as a neglect of it may lose that office)."[16]

Again in 1759 he wrote Pemberton in London, about "renewing the Contract" which was about to expire—presumably the contract for the deputy secretaryship. He enclosed a letter to a Mr. Hunt with power of attorney, adding, "The little Acquaintance I have at London putts it out of my power to procure security without the Assistance you have been so kind as to afford me & wch I beg you will again render to me." He asked that Mr. Hunt be informed "if by the chance of Warr He should at any time by accident be in advance I will allow him 20 pct more than the Exchange."[17]

Possibly at this time Read was serving as deputy under Christopher Coates, whose commission as secretary was renewed in 1761. Coates may have been appointed several years before this, but the record is not clear.[18] Whatever the length of Coates's term, Read received from Governor Hardy on February 17, 1762, a new commission as "Secretary of Our Province."[19] Presumably this was a renewal of the deputy secretaryship.

The position embraced a multiplicity of duties, each yield-

[15] Charles Read to James Pemberton, Nov. 17, 1748. Pemberton Papers, IV, 156.
[16] Charles Read to James Pemberton, Oct. 17, 1757. Pemberton Papers, XII, 71.
[17] Charles Read to James Pemberton, June 15, 1759. Pemberton Papers, XIII, 85.
[18] N. J. Arch., 1st ser., IX, 257; Edgar J. Fisher, *New Jersey as a Royal Province, 1738 to 1776* (New York, 1911), pp. 45-46.
[19] N. J. Commissions, Liber AAA, 366. Sometimes Read signed documents as "Deputy Secretary," but usually his title appeared as "Secretary."

ing a fee fixed by law in lieu of a salary. The fees law enacted in 1752 (while Read was speaker of the assembly) named a schedule of fees for twenty-four different functions of the office of secretary.[20] They ranged all the way from £3 for issuing the patent for a township to 4½d. per page for copying deeds and other papers. Issuing orders and warrants for the governor, as well as commissions and licenses for various purposes, were services required of the secretary, as well as attending courts, searching records and drawing certificates. The fee for an attorney's license was 20s., and it was 6s. for the commission of a captain of militia, or a justice of the peace, or for a license to operate a boat or ferry.

With travel slow and laborious, the shifting back and forth between Burlington and Perth Amboy as the meeting place of the assembly, as well as the separation of the records between the two capitals, added substantially to the burden of the secretary. In 1748 Read wrote to James Alexander that "the Distance of the Offices renders it difficult & tedious to prepare any thing where both Offices are to be Searched."[21]

As clerk of the council, Read received a salary, usually £30 a year, plus fees for numerous services. For every petition to the governor in council, his fee was 4s., for "reading every private Bill, first time, second time and third time, each time" 5s., for "Each Bill Engrossed in Parchment," 18s., and for "Affixing the Seal and Wax for each Bill," 6s. Also, for "Attending the Committee of Council on a Reference on such Bills," the clerk received 10s.[22]

The offices of court clerk and surrogate also yielded fees for numerous services. The clerk of the sessions and common pleas received for "every Bond to prosecute," 2s., "Sealing every Writ," 1s.; "Entering the Defendants Appearance," 6d.; "Entering every Verdict," 6d.[23] As clerk of the supreme court he received a fee for each of twenty or more activities—

[20] *Nevill's Laws*, I, 338-352.
[21] Charles Read to James Alexander, Sept. 16, 1748. Theodore Stevens Collection.
[22] *Nevill's Laws*, I, 343.
[23] *Ibid.*, p. 345.

Secretary

making out processes and writs, filing indictments, entering verdicts and judgments, reading evidence, swearing the jury and constable, searching the records, filing the roll in each action, and other routine services. The surrogate's duties, among others, included taking depositions to and recording wills, drawing orders for administration, engrossing letters of administration, drawing administration bonds, filing wills, recording inventory, swearing executors, auditing accounts of administrators and executors, and issuing marriage licenses. The fees ranged from 7d. to 9s. for the various items. These offices and others which Read subsequently held "furnished him with a constant flow of Cash."[24] His income from his public offices, added to his private earnings, was a substantial one.

JONATHAN BELCHER arrived to begin his duties as the new chief executive on August 10, 1747. He had received this commission largely through the good offices of Partridge, who happened to be his brother-in-law. On the occasion of Belcher's appointment, Charles Read was involved, perhaps quite innocently, in a peculiar incident. As we have seen, Partridge tried to secure Read's appointment to the council as early as 1740. Now, when the governor's commission was drafted, Charles Read's name appeared in place of that of James Alexander among the councillors listed—reputedly through a clerical error. When Ferdinand J. Paris, the proprietors' agent, discovered this he had Alexander's name restored and Read's dropped.[25]

Less than a year later Belcher recommended to the Duke of Newcastle, secretary of state, and shortly thereafter to the Duke of Bedford, who succeeded him, that Read be appointed to the council. He described Read as a "Gentleman Quallified, in all Respects,"[26] and expressed the desire that no one be recommended for appointment to public trust "but persons

---

[24] A. Leaming's Diary, Nov. 14, 1775, *loc. cit.*
[25] F. J. Paris to Jas. Alexander, Feb. 10, 1746. Paris Papers, X16. Also Donald L. Kemmerer, *Path to Freedom* . . . (Princeton, 1940), p. 209.
[26] *N. J. Arch.*, 1st ser., VII, 137-139.

of Loyalty to His Majesty & his family, of good Estates & of Sober life."[27] But Belcher had begun to lose favor with the home government, and could expect less from the Duke of Bedford than from his predecessor. Although it was an unwritten rule for the Board of Trade to accept a governor's nomination to the upper house, in this case Belcher's wish was not granted.[28] Paris had more influence in London than had Partridge. Through some subtle maneuvering by Paris, Read's appointment was blocked, and Richard Saltar was commissioned instead.

Belcher came to New Jersey a seasoned statesman, of better than average calibre, having served a long apprenticeship in public life and learned by unhappy experience that to antagonize is not to rule. New-England-born, son of a Boston merchant and graduate of Harvard, he had been named a member of the Massachusetts council. Subsequently he had served as governor of Massachusetts and of New Hampshire, through a stormy administration until he was removed in 1741. Now in his middle sixties, and after an extended period abroad, he was awarded this new commission which he filled prudently and mildly, and with a substantial measure of popular success.

Governor Belcher represented an uncommon admixture of personal traits. Thanks to his wealth, he had enjoyed special advantages of education and travel. Still he remained in many ways a man of small views and mediocre abilities. He was overtly religious, yet capable at times of decidedly unchristian conduct. He was an opportunist, vain, boastful and self-seeking, yet few of New Jersey's colonial governors contributed more to the public welfare.

Mellowed with age, yet not lacking in youthful enthusiasm, he led more the life of a genial country gentleman than of an officer of state. Of agriculture he possessed a practical knowledge, as well as a keen interest in its development, born

---

[27] Charles Read to R. H. Morris, April 16, 1748. Boggs Collection.
[28] D. L. Kemmerer, *op. cit.*, p. 222.

no doubt of experience at Milton, his 300-acre estate near Boston which he had left in charge of his son Andrew. In his rural tastes, as well as in his political philosophy, he found a basis for congenial association with Charles Read. A glimpse of Belcher's private life reveals the sort of man who made Read his close advisor and suggests the interests and activities which at times occupied their mutual thought.

After arriving at Burlington the governor lodged for a time with Richard Smith, and he later leased for £100 a year a residence which apparently belonged to Isaac Conarro, of Trenton.[29] He described the place in letters to friends:

> I have taken a house in this Little City standing on the Banks of the fine River Delaware—where from my Bed Chamber I have a prospect for 10 miles up and down the River. . . .[30]
>
> I have a pretty garden of near an Acre inclos'd with a handsome brick wall—an Orchard of 6 acres and about 60 acres more of good pasturing and mowing Land—a Large barn—Coach house and Stabling for 6 Horses. This place will give me many of the Necessaries of Life for my Family as well as for my Horses, Cows, Sheep and Poultrey, and the managing of it may be an innocent Amusement as well as Contribute to my health. . . .[31]

Again, he wrote:

> I have a Small Collection of Books that I shall divert my Self with as the Publick affairs may allow.[32]
>
> I have made a Short Visit to New York and another to Philadelphia which are pretty Little Cities for North America especially the latter and to which I shall every now and then make an Excursion when Tired with my Library to find some Agreeable Conversation, and as I keep 4 good Trotters and the Road pleasant but 20 miles Riding may give the Blood a better Circulation.[33]

There were in Philadelphia, he added, "a Number of Gentlemen of good Sense and Reading for America," among them his "Dear & Worthy Friend Mr. Tennant." This was doubt-

---

[29] Jonathan Belcher to Isaac Conarro, June 20, 1751. Belcher Papers, VII, 161.
[30] Jonathan Belcher to Mr. Cradock, Sept. 27, 1747. *N. J. Arch.*, 1st ser., VII, 58, 59.
[31] Jonathan Belcher to Secretary Willard, Oct. 2, 1747. Belcher Papers, VI, 86-89; *N. J. Arch.*, 1st ser., VII, 81-83.
[32] A. C. Myers, *op. cit.*, pp. 33-34n; Belcher Papers, VI, 42.
[33] Jonathan Belcher to Mr. Cradock, *loc. cit.*

less the Reverend Gilbert Tennent, pastor of the Second Presbyterian Church of Philadelphia from 1743 to 1764, with whom he proposed now and then to spend the Lord's Day.[34]

A little later he mentioned his "Conveniences for Geese—ducks—Dunghill Fowls and Rabbits and a fish pond about 150 Rods from my door" and added that "next Spring I shall have a Paddock for 15 or 20 Deer."[35] Again he described his rural environment:

> The place where I live is a short mile from the Town and no house near me. No body comes to the Governour but on business or perticular friends—Bass—Pike—Perch and Eels pass in Schools by my door and water Fowl in Flocks—Pigeons—Partridges and Quails in their Seasons with Rabbits and good Venisons ... Have Cows and Muttons and Porks grace the stable and Barn and God gives us the fat of Kidneys of Wheat and the pure blood of the Grape in plenty.[36]

During his first fall at Burlington he sent, by way of his friend John Smith, three barrels of apples and a barrel and one-half of red potatoes to his son-in-law, Byfield Lyde, at Boston, who on acknowledging the shipment declared, "The Pippins are the Best I ever saw, and the other Apples exceeding good for Baking."[37] In January following he wrote his daughter, Mrs. Lyde, "The New town Pippins will hardly eat well till next mo."[38] The same month he ordered two or three hundred scions of "Providence Sweetings," which he described as "a fine apple ... for feeding swine and for making Choice good Cider of which I am told 4 barrels will boil into a barrel of good Molasses."[39] In turn he sent to Boston "Trees, Nuts and Cyons," and a supply of seeds for his nephew, Mr. Oliver, and for his son at Milton.[40]

[34] Jonathan Belcher to Secretary Willard, *loc. cit.*; also A. C. Myers, *loc. cit.*
[35] Jonathan Belcher to Colonel Alford, Dec. 15, 1747. *N. J. Arch.*, 1st ser., VII, 92-93.
[36] Jonathan Belcher to Mr. Byles, Jan. 26, 1748. Belcher Papers, VI, 207-208.
[37] Jonathan Belcher to John Smith, Nov. 3, 1747. Smith Papers, III, 14. Also Byfield Lyde to John Smith, Dec. 15, 1747. Smith Papers, III, 17.
[38] Jonathan Belcher to Mrs. Lyde, Jan. 25, 1748. Belcher Papers, VI, 204-205.
[39] Jonathan Belcher to Dr. Gibbs, Jan. 26, 1748. Belcher Papers, VI, 206.
[40] Jonathan Belcher to Mr. Oliver, Mar. 14, 1748. Belcher Papers, VI, 268-269.

Secretary

His letters during the next five years indicate anxiety about the affairs at Milton. To his son and his son-in-law he wrote instructions and advice, and asked for reports of progress. He inquired about the "new orchards," the nursery of walnuts, the chestnuts and "the two Cargo of Trees I sent you from London—they Cost near Twenty Guineas and are a Choice Collection."[41] He directed that a swamp be cleared and drained, and planted with herd grass and clover.[42] He observed that £0-2-8 sterling a cord was a poor price for wood.[43] He complained that the rental he could get for Milton represented only about one per cent of his investment in the place (£3,800), and decided to have it advertised for sale. In one year, he said, he had taken from the place 100 tons of hay—80 of "English" and 20 of "salt" hay—and in another year 600 bushels of corn. Among the 300 feet of outbuildings was a barn 100 feet long.[44]

In April, 1752, at his order "a collection of Choice English Fruit" trees, was shipped from Boston, presumably for planting in New Jersey.[45] To friends in England he sent "a few Shrubs and some Trees of the Produce of this Country," and offered to send other North American "Exotics" that might be desired.[46] He engaged John Bartram to make a collection of native plants and on receiving the bill protested at what he considered an overcharge, offering to settle for £5. When subsequently he wanted to order "two handsome Boxes of Wilderness Plants," to avoid another disagreement he asked Bartram to submit in advance "his lowest Price in Reason and Conscience."[47] Charles Read, too, it seems, was involved in his agricultural correspondence. In 1753 he wrote Read:

---

[41] Jonathan Belcher to Andrew Belcher. Nov. 16, 1747. Belcher Papers, VI, 136-138.
[42] Jonathan Belcher to Andrew Belcher, Jan. 25, 1748. Belcher Papers, VI, 200-203.
[43] Jonathan Belcher to Byfield Lyde, Feb. 22, 1748. Belcher Papers, VI, 252-253.
[44] Jonathan Belcher to Byfield Lyde, June 23, 1748. Belcher Papers, VI, 379-380. Also Jonathan Belcher to Mr. Foye, Dec. 11, 1751. Belcher Papers, VII, 299-300.
[45] Jonathan Belcher to Mr. Smith, Apr. 22, 1752. Belcher Papers, VII, 400.
[46] Jonathan Belcher to Lord Hardwicke, Dec. 8, 1755. Belcher Papers, IX, 506-510.
[47] Jonathan Belcher to John Bartram, Sept. 16, 1751; Belcher Papers, VII, 217. Also Jonathan Belcher to Mr. James, Oct. 6, 1755. Belcher Papers, VII, 405.

"As I hear nothing of the Seeds you mention I doubt they are lost."[48]

Shortly after arriving in Burlington, the governor purchased a coach and a chaise in keeping with the prestige of his office, and while he was equipping his stable with animals of his own choice, he borrowed a pair of horses from Chief Justice Morris, a loan which Read helped to negotiate.[49] His four-wheel chaise is reputed to have been the only one in New Jersey for a time.

The following February he was still looking for horses that suited his taste. In a letter to Colonel Peter Schuyler he wrote:

> I want to come to Newark and Petersburgh to visit you and Coll John—but this I cant do till you find me a pair of best stone horses to match mine in Couler and to be 5 coming Six. If you can get 'Em agreeable to your own Judgement I value not the price.[50]

That his association with Colonel Schuyler was a very warm one, appears in a subsequent letter in which the governor wrote:

> I kindly thank your handsome Present of Venison the Produce of your own Park it is indeed very fine and not to be found in the same way in all the English America . . . When will you and your Lady do Mrs. Belcher & I the Pleasure of coming to our little Cottage to eat a Piece of Mutton with us . . .[51]

The selection of a coachman presented unforeseen difficulties. A man to whom he tendered an offer refused to work for less than £50 a year, whereupon Belcher complained of "such wages, as I never heard of before"; and declared he would write for a coachman from London where one could be engaged for £10 a year.[52] Subsequently, it appears, he ordered coach harness imported from England.

[48] Jonathan Belcher to Charles Read, Dec. 1, 1753. Belcher Papers, VIII, 296.
[49] Charles Read to R. H. Morris, March 31, 1748. Boggs Collection.
[50] Jonathan Belcher to Peter Schuyler, Feb. 18, 1748. Belcher Papers, VI, 248. For a biographical note of Colonel Schuyler, see p. 318.
[51] Jonathan Belcher to Peter Schuyler, Oct. 3, 1755. Belcher Papers, IX, 403, 404.
[52] Jonathan Belcher to Mr. Chubb, Apr. 20, 1748. Belcher Papers, VI, 300.

His household coterie included negroes and indentured servants. Through Samuel Smith in 1750 he purchased the services of a 15-year-old girl for six years for £15, and the next year he procured a Dutch girl whom his wife thought "very dear."[53]

Unhappily the governor's period of residence in Burlington was marred by illness. Up until 1746, he had never been "sensible of any Decay of Nature,"[54] but now, in December, 1751, he wrote Read, ". . . my late Situation subjected me and my Family for more than half the year to continual sickness that my House was become a constant Hospital . . ."[55] Thanks to the then unsuspected malarial mosquitoes that must have infested the area, for four years he was the victim of "Fever and Ague."[56] Accordingly, on the advice of his physicians, he determined to change his residence to Elizabethtown, in the hope of finding the new location near the salt water more healthful. His household goods were loaded on three sloops, and shipped to Elizabethtown by way of the Delaware and the Atlantic coast. For the coming winter at the new gubernatorial mansion, he ordered that a supply of firewood be laid in—"50 Cords of young round White Oak" and 12 cords of "Sapling Walnut or Hicory," together with "300 lb of Choicest Butter and 12 Barrels of the best Cider."[57]

New Year's day, 1752, when the new calendar went into effect, found the governor established in his "humble cottage" at Elizabethtown. To the secretary of the province he sent personal greetings: "This day introduces the new year and the new Stile by Act of Parliament. I give yo joy of the day and wish you may live to see the return of many such."[58]

The ample stocks in the governor's wine cellar were such

---

[53] Jonathan Belcher to Mr. James, Sept. 24, 1751. Belcher Papers, VII, 222.
[54] Jacob Spicer's Diary, Apr. 18, 1755.
[55] Jonathan Belcher to Charles Read, Dec. 2, 1751. Belcher Papers, VII, 280.
[56] Jonathan Belcher to Richard Partridge, July 3, 1751. Belcher Papers, VII, 174.
[57] Jonathan Belcher to Mr. Woodruff, Sept. 23, 1751. Belcher Papers, VII, 221.
[58] Jonathan Belcher to Charles Read, Jan. 1, 1752. Belcher Papers, VII, 316-317.

as to do credit to a gentleman of refined tastes. It is easy to imagine the governor and the secretary of the province finding respite from perplexing problems of state in a lively discussion of the relative merits of claret and old Madeira, of the best season for brewing beer, or of devices for keeping good cider. "I commonly drink (besides Water and small Beer) about half a Bottle of old Madeira a day,"[59] wrote the governor, and again, "If I indulge my taste in any one thing more than another it is in Malt Drink."[60] His choice of liquors was set forth time and again in his correspondence. In 1748 he ordered two hogsheads each of Burton ale and Wiltshire beer, specifying just how he wanted them brewed— "'To be done with the Choicest palest Malt" and put into "Substantial Sweet Iron bound Casks."[61] In 1752 he wrote to Edward Antill, who had opened a brewery in New Brunswick, "I have lately turned into Bottles 2 hhds. of Burton Ale that I reced . . . from England it stood 18 mos in the Cask and I intend it shall stand 12 more in Bottles before I meddle with it."[62] With Antill he discussed the methods used by the English brewers, and subsequently gave him an order for eight barrels of beer. Of a Mr. Woodruff, who had a vessel going to Madeira, he ordered "Three Pipes of the best Madeira Wine—One Quarter Cask of Malmsey & a Box of 20 lb of Citron."[63] Other orders for Madeira, claret, and cider, which the governor placed from time to time, reveal that he was concerned about the tapping of the containers in transit, a practice which apparently was all too common. In one consignment his seven barrels of cider before delivery had shrunk to six; in another a cask of claret had disappeared. This "abuse and Pilfering" he vehemently condemned as "villanous and Imprud[en]t in Boatmen and Waggoners," and as a safeguard

---

[59] Jonathan Belcher to Dr. Cadwalader, Oct. 7, 1751. Belcher Papers, VII, 229-230.
[60] Jonathan Belcher to Edward Antill, Apr. 11, 1752. Belcher Papers, VII, 388.
[61] Jonathan Belcher to Mr. Jackson, July 19, 1748. Belcher Papers, VI, 424.
[62] Jonathan Belcher to Edward Antill, Apr. 11, 1752. Belcher Papers, VII, 388.
[63] Jonathan Belcher to Mr. Woodruff, May 30, 1752. Belcher Papers, VII, 442.

he asked that his shipments be sent by "Carefull honest hands."[64]

Members of the Read family, it seems, were always welcome at the governor's new home. Belcher frequently mentioned Alice Read in his letters—"Good Mrs. Read," he was wont to call her.[65] Realizing that a sleighing snow greatly facilitated travel between Burlington and Elizabethtown, in January, 1754, he wrote: ". . . we shall be glad to see you and Mrs. Read upon the first good Slaying."[66]

In 1753 an epidemic, probably of smallpox, took a heavy toll in Burlington County and caused Belcher genuine concern for Read and his neighbors. In characteristic style the governor wrote:

> I very heartily Compassionate the good People of your County while the Terrible distemper you mention rages to that degree as makes a great Mortality among you may that God that sets bounds to the Billows of the Ocean in pity and Mercy to the poor people say to this distemper hitherto thou shall come & no further.
>
> We shall be heartily glad to see you and Mrs. Read at our humble Cottage and may God in this time of great danger cover you with his Feathers that you may not be afraid of the Pestilence that Walketh in Darkness nor of the Destruction that Wasteth at Noon day.[67]

Unhappily the governor's health was not greatly benefited by the change of residence. A type of palsy which he called "a Paralytick disorder," had been growing upon him for several years, and had now become so severe that he was unable to hold a pen. Greatly concerned, before leaving Burlington he had decided to try electrical treatment with the apparatus devised by "the Ingenious Mr. Franklin," but only after receiving assurance that if used in moderation there was no

---

[64] Jonathan Belcher to Col. Low, Feb. 18, Apr. 12, 1748; Belcher Papers, VI, 247. Also Jonathan Belcher to Mr. Chubb, Apr. 14, 1748. Belcher Papers, VI, 295, 296.
[65] Jonathan Belcher to Charles Read, Dec. 2, 1751. Belcher Papers, VIII, 279-281.
[66] Jonathan Belcher to Charles Read, Jan. 4, 1754. Belcher Papers, VIII, 312.
[67] Jonathan Belcher to Charles Read, Aug. 1, 1753. Belcher Papers, VIII, 181-183.

danger of injury.[68] Franklin, unable to attend personally the treatment of the governor, sent the apparatus with directions for its use. Great must have been the gubernatorial chagrin when the apparatus arrived, hopelessly broken in transit. So the experiment upon the distinguished patient did not materialize. Belcher corresponded with Franklin also on other matters. He was a subscriber to the *Pennsylvania Gazette,* and exchanged ideas with the philosopher on education, particularly with reference to the academy in Philadelphia.

On assuming office Governor Belcher had been delighted with the favorable prospect of New Jersey. He enthused about its natural features in superlative language—"a fat Soil Situated in a fine Climate," "a Land flowing with milk and honey," "abounding in the Necessaries of Life," "a Countrey beautified with fine Lands and Rivers nor have I yet seen anything on this Continent Equal to it."[69] Of the people (then "hardly sixty thousand souls" but fast increasing) although "mostly in low fortune," he remarked that they were "generally sober and have made great Improvements for so young a Province," which "fills apace."[70]

Thus pleased with the new province he was to govern, he looked forward with optimism and confidence, "I think nothing will be wanting in my power," he wrote, "to make Trade & commerce flourish in the province as well as Agriculture and Manufactures."[71] It was his ambition to advance the prosperity of the province through internal improvements and the promotion of education. His zealous efforts to aid the College of New Jersey in its formative years are characteristic. In this enterprise, as in many others, Charles Read was closely associated with the governor and doubtless influenced his action.

[68] Jonathan Belcher to Dr. Cadwalader, Sept. 30, 1751. Belcher Papers, VII, 225.
[69] Jonathan Belcher to Mr. Sergeant, Feb. 23, 1748. Belcher Papers, VI, 254-257. Also Jonathan Belcher to Secretary Willard, Oct. 2, 1747, *loc. cit.*
[70] Jonathan Belcher to Secretary Willard, Oct. 2, 1747, *loc. cit.* Jonathan Belcher to a committee of the West Jersey Society, Oct. 18, 1750. Belcher Papers, VII, 7-8. Also Jonathan Belcher to William Belcher, June 29, 1748. Belcher Papers, VI, 402-404.
[71] Jonathan Belcher to Col. Alford, Dec. 15, 1747, Belcher Papers, *N. J. Arch.,* 1st ser., VII, 92.

Offspring of William Tennent's primitive Log College at Neshaminy in Pennsylvania, Princeton University, as the College of New Jersey, had its origin among the leaders of the Presbyterian faith in the Middle Colonies—the Tennents, Jonathan Dickinson, Aaron Burr, and others—who wanted an institution for the training of young men for the ministry.

When in March, 1745, a subscription drive was launched for the "purpose of Erecting a Collegiate School in the province of New Jersey for the instructing of youth in the learned Languages Liberal arts and Sciences," Read was among the first to pledge a contribution in the amount of £10, "to encourage the same and to promote the said undertaking."[72]

When Governor Belcher arrived in 1747, the college under its original charter of 1746 had made a modest beginning in Elizabethtown, but the trustees had not determined upon a permanent location, and the infant institution lacked financial strength to assure its continuance. The governor, when apprised of the project, took to it with the greatest enthusiasm. Henceforth to him it was a matter of first importance. "The inhabitants are generally rustick and without education (yet Civil and Courteous Sober and Honest)," he wrote in 1747. "I am therefore attemptg the building of a College in the Province for Instructing the youth in the Principles of Religion in good Literature and Manners."[73]

A letter to the West Jersey Society of Proprietors in 1748 revealed the governor's vision of the economic and social advantages which would accrue to the province if educational facilities were provided: ". . . the Building of a College and Putting forward inferior Schools in the Province of New Jersey—Will Promote Trade—Increase the Inhabitants—Make them See the Advantage and Beauty of Government, and all these things must add Considerable Value to your Estate in the Rise of Lands." But, he added, "I am At Present much discouraged about a College Not Seeing where Money will

[72] MSS in Princeton University Library.
[73] A. C. Myers, *op. cit.*, pp. 33-34; Belcher Papers, VI, 43-44.

be found to Build the House and to Support the Necessary Officers for the Assembly (Many of them Quakers) will do Nothing towards it. . . ."[74] Private subscriptions, he concluded, offered the only way to success, and to this end he pledged his best effort.

To the Quaker assemblymen a proposed lottery to raise money for the college was an evil comparable with gambling, horse racing, and cockfighting, judging from a letter written in 1748 by Richard Smith, member from Burlington, to his son John:

> Our house has past a bill to prevent all Lotterys Horse raceing Gaiming & Cokfighting for Money and Makeing all Wagers & Bargains In Gaiming Void the Next day we had the parsons Concerned in the Charter for a Coledge aplying for a lottery which I think will not be granted them for the people here do not aprove their Scheem but think (as I Told the Govr.) that theres to many of the presbaterian Clargey Concerned Indeed all their Trustees Except Andw. Johnston & Jno. Kinsey are Such the one Designd to Decoy the Church. And the other the Quakers to send their Children to be Educated prispetearians So that in time theyl have the rule & governing the province.[75]

In spite of Quaker opposition, Governor Belcher in 1748 secured for the young college endorsement of an amplified charter which was countersigned by Read as clerk of the council. The first commencement was held November 9 of that year. To Charles Gray, M.P., Belcher wrote as follows:

> The Province is small and the People not able to do much for the support of this Society. I am therefore indeavouring to get them help from my friends abroad and have some incouragement of Books for their Library and of materials for the Building. And I shall be studious night and day to bring this infant forward into youth and manhood; being intirely with you, that not Learning or Knowledge ever hurt a Kingdom, State, or People, but riches and their concomitant Luxury.[76]

[74] *N. J. Arch.*, 1st ser., VII, 146.
[75] Richard Smith, Jr., to John Smith, Sept. 23, 1748. Smith Papers, III, 72.
[76] Jonathan Belcher to Charles Gray, Nov. 14, 1748, *New Jersey Historical Society Proceedings*, 3rd ser., VII (1912-13), 100.

Secretary

The campaign for funds went forward, Read cooperating closely with the governor and the other trustees. In 1749 the trustees named Charles Read and Richard Smith the authorized agents in Burlington "to take in the Subscriptions, and receive the Monies of all such publick spirited Persons as shall be willing to promote this worthy a publick Design."[77] Evidently Read or some other person had converted Smith to the project since the previous September.

When funds were finally in sight to erect a building for the college, opinion was divided as to the most suitable site. Some favored New Brunswick, some Princeton. Negotiations were opened with the authorities of both towns. The trustees asked that the local citizens provide £1,000 in cash, 10 acres of land contiguous to the college, and 200 acres of woodland within three miles of the town. In the judgment of the trustees, Princeton made the better offer, particularly with reference to the woodland. This was Governor Belcher's choice, and doubtless his opinion carried weight with the other trustees. From a letter he wrote to President Burr, it appears that the promised supply of firewood from Princeton may have been a deciding factor: "Brunswick can by no means make an offer equal to Prince Town as to the Wood and I am still sedately confirmed in my Self that Prince Town upon all Accounts will be the best place to fix the College in."[78]

In 1752 the trustees voted to locate the college in Princeton, and ground was broken on the new site. But Charles Read was not through with the building. He had been among the first to contribute toward its erection. He had persuaded others to subscribe. He had cooperated with Governor Belcher in promoting the project. He had taken part in the negotiations for the charter. Was he not entitled to some of the legitimate spoils of the enterprise? His sawmill was turning out just the kind of pine needed for window casings in the substantial stone structure, and he was given the contract for this

---
[77] *Pennsylvania Journal*, Jan. 31, 1749; *N. J. Arch.*, 1st ser., XII, 513.
[78] Jonathan Belcher to Aaron Burr, Aug. 3, 1751. Belcher Papers, VII, 192-193.

material. On October 11, 1755, Robert Smith, the architect, signed an order directing the treasurer, Jonathan Sergeant, "to pay to Mr. Chas. Read or order the Sum of thirty-seven pounds three shillings and Nine pence, for Eight thous[an]d five hundred feet of pine scantling for Window Cases to the College."[79]

The trustees proposed to name the building Belcher Hall, in appreciation of the governor's constant and unselfish interest in the college, but he modestly declined, and suggested instead the name "Nassau Hall," in honor of William I, Prince of Orange, Count of Nassau. Belcher's devotion to the college continued undiminished to the close of his career. In 1755, by deed of gift, he presented the college with his library of more than four hundred volumes.[80]

But instead of being free to give his major effort to the socio-economic program he envisioned, during a large part of his administration Governor Belcher was preoccupied with the political and military problems which crowded upon him. Land riots, problems of taxation, a boundary dispute between the Eastern and the Western Divisions of the province, a deadlock between assembly and council, and the issue of paper currency, together with negotiations with the Indians, defense of the frontier and support of military expeditions, overshadowed the more constructive measures of building roads and otherwise improving transportation service, draining marshes, encouraging manufactures, promoting agriculture, constructing public buildings, and advancing the interest of education and religion—all of which, nevertheless, in substantial measure were accomplished. The situation called for a strong hand and a diplomatic mind. Belcher soon discovered that Read could supply some of the essential qualities which he himself lacked, and wisely learned to rely upon his colleague's judgment. According to Aaron Leaming, it was Read who shaped

[79] Princeton University Library MSS.
[80] *Princeton University Library, American Library Association Visit* (Princeton, 1916), pp. 4-11.

the policies of the administration and held the reins of government.

The most troublesome problem of the early years of Belcher's administration arose from the ancient quarrel over the payment of quit-rents to the proprietors. Neither the proprietors nor the royal governors had been able to collect the quit-rents, which had now become almost obsolete. Years of litigation over land titles had failed to clarify the issue. Since the council was allied with the proprietary interests, and the assembly was strongly anti-proprietary, the two houses were at odds over the matter.[81] The proprietors were growing more insistent that all persons who held lands without proprietary title deeds, should either pay rents or purchase title from the proprietors. The people occupying such lands formed associations and resolved to maintain their possessions by force. Some of these persons were arrested, found guilty of disturbing the peace, and committed to jail. This served simply to incite further the ire of the opposition. Bands of jail breakers were formed, who forcibly released the prisoners. Buildings were burned, timber stolen, and fences destroyed. Such "riots" in Middlesex, Morris, Monmouth, Hunterdon, and other counties gave the provincial officers grave concern. Governor and council had appealed to the assembly to authorize the use of the militia to put down the riots, but the anti-proprietary assembly, largely Quaker, would not assent to military force as a means of sustaining the courts.

An important step was taken toward the solution of the problem when in 1747 the assembly passed two acts which promptly received the approval of the governor—one to pardon all persons who had been guilty of riotous conduct, the other for the suppression and prevention of riots and disorders.[82] But, as we shall presently see, this did not spell the end of rioting in New Jersey. Governor Belcher, sensing the economic as well as the legal aspect of the situation, pointed out

[81] See pp. 8-10.
[82] *N. J. Session Laws* (November), 1747, pp. 22-27; 47-51.

to the proprietors that continued disorders would depress the value of the lands and impede their development. To a committee of the West New Jersey Society in 1748 he wrote:

> . . . unless these disorders and Tumults can be Effectually . . . brought to an End Let the Societys Rights and Property be what it will they will Certainly find the Value of the Lands sink every day which is but the Natural Consequence of Controversy and Litigating Ti[t]les in the Law by which People become discouraged from coming to Settle and Subdue Wild Lands. . . .[83]

Shortly thereafter the governor proposed to the committee that they consider appointing Charles Read as one of the agents for transacting their affairs in the province, endorsing him as a person of "good Vertue Capacity and Substance," and of good constitution, active and diligent.[84] Apparently, nothing came of the suggestion.

Taxation was the source of another difficulty. The assembly passed a bill designed to tax "all Profitable Tracts of Land, held by Deed, Patent, or Survey, whereon any Improvement is made."[85] The council refused approval unless exemption of "unprofitable" lands, in accordance with royal instructions, should be declared, and amended the bill accordingly. The amendment, however, was not acceptable to the assembly. In the dispute neither party would yield, and during a three-year deadlock the assembly refused to vote funds for the support of the government.

All this was highly distressing to the amiable governor. On assuming office he found it "a very lean thin Government"; the annual salary was to be £1,000. He was allowed only £60 for house rent although he paid £100. He estimated the perquisites at not more than £150. To Sir Peter Warren he wrote in June, 1748:

> This I am Sure you will readily think a poor Support for the

[83] Jonathan Belcher to a Committee of the West New Jersey Society, June 27, 1748. Belcher Papers, VI, 392-394; *N. J. Arch.*, 1st ser., VII, 145.
[84] Jonathan Belcher to a Committee of the West New Jersey Society, July 25, 1748. Belcher Papers, VI, 431-432; *N. J. Arch.*, VII, 150.
[85] *N. J. Arch.*, 1st ser., XVI, 264.

Governour who is Stiled by the King the Representative of His Royal Person, although this in Such an Obscure Part of the World yet I find it very expensive, and that Little can be Saved out of the above little.[86]

The "Ordinary Expences" of the government were given as £1,375, and the "Extraordinary Expences" about £400 per annum. The annual revenues came from the loan of "Bills of Credit," which through the gradual retirement of the issue as provided by law, had declined in eight years from £3,000 to £1,268 in 1749, an amount insufficient to meet even the ordinary expenses. For seventeen years, there had been no general tax for the support of the government.[87]

Difficult as it was for the governor to make ends meet even with his full salary, his troubles were multiplied by the assembly's refusal to vote any funds. In February, 1751, the bill designed to fix the tax quotas for the several counties was defeated for the seventh time.[88] The tactics of the assembly in holding back the governor's salary "for adhering to his Majesty's Instructions," were severely condemned by Read "as an unwarrantable proceeding & of very evil Tendency."[89] The governor, in a letter to his son, laid bare his consternation and embarrassment:

My Sollicitation—Commission—Clothing—House Furniture—Negroes & horses, I say all things put together I am this day above a thousand pounds Sterlg worse than when I sailed from Boston Nor wou'd I undertake Such another dreadfull fatigue to be made Vice Roy of English America.[90]

In June, 1751, he wrote to Sir Peter Warren:

. . . a disagreement between his Majesty's Council and the late Assembly have prevented me from receiving one farthing Salary

---

[86] Jonathan Belcher to Sir Peter Warren, June 27, 1748. Belcher Papers, VI, 388-389.
[87] Jonathan Belcher to the Secretary of the Lords of Trade, Apr. 21, 1749. *N. J. Arch.*, 1st ser., VII, 243-246.
[88] *N. J. Arch.*, 1st ser., XVI, 269, 276; also *Votes and Proceedings of the New Jersey Assembly*, Feb. 14, 1751, p. 28 (elsewhere referred to as *N. J. Assembly Minutes*).
[89] Charles Read to R. H. Morris, Apr. 20, 1749. Boggs Collection.
[90] Jonathan Belcher to Andrew Belcher, Feb. 1, 1751. Belcher Papers, VII, 82-84.

for near two years past nor shall I for near 12 months to come so I have & must still live near three years upon my private Fortune while doing my duty to the King in the Service of his Province. . . .[91]

His patience tried to the limit, Belcher decided that the only way to break the deadlock was to dissolve the "mob assembly," which he proceeded to do. "The difference between the Council and the late Assembly," he explained, "had risen to that pitch as to produce great indecencies in their Treatment of each &c. &c. which made it necessary for me to dissolve the Assembly and call another."[92] This action by the governor opened the way for a new epoch in Read's public career, that of colonial legislator.

[91] Jonathan Belcher to Sir Peter Warren, June 8, 1751. Belcher Papers, VII, 153-155.
[92] *Ibid.*

Nine: Legislator

SINCE the provincial secretary served also as clerk of the council, Read already had the benefit of several years' experience with the legislative branch of government when Belcher became governor. Familiar with political procedure, acquainted with the political leaders, he was able to give the governor valuable counsel on questions of legislation. After Belcher had dissolved the Seventeenth Assembly in 1751, it is a fair assumption that he prevailed upon Read to become a candidate for the new assembly which was about to be chosen. Read was elected, and when the Eighteenth Assembly convened at Perth Amboy, May 20, 1751, he took his seat as "the gentleman from Burlington." It was probably no accident that, even though this was his first term in the assembly, he was selected for the post of speaker, where he would be best able to control the body.

The principal issue before the new assembly, which was slightly more conservative than its predecessor, was whether or not it should persist in trying to tax the proprietors' unimproved lands. The day after the opening Governor Belcher delivered an earnest message to the assembly, pointing out the "Distressing Circumstances" caused by "an Empty Treasury for near two years past," and urging that measures be taken for the payment of the debts of the province and the support of the government. "Private people are obliged, by the Law of the land, to pay their Debts one to Another," he argued, whereas "the Creditors of the Province have been left without Remedy, and for no Other Cause but from the Difference in Opinion between his Majesty's Council and the Late House of Assembly."[1]

Under Read's leadership the assembly began to function, and within two weeks had passed a bill "for the support of the

[1] N. J. Arch., 1st ser., XVI, 291-292.

Government" covering the period of two years from August 10, 1749 to August 10, 1751. Besides providing the overdue salary of the governor and other provincial officers, it authorized the usual stipend of 6s. *per diem* for members of the assembly while in session, and the same for members of the council. An allowance of 14s. a week was made for "the Use of a Room, Firewood and Candle, for the Council," and Charles Read, as secretary, was allotted £62-13-0 "for Sundry Expresses, and for copying of Acts of General Assembly to send to Great Britain, and other Services."[2]

As a means of raising the necessary public funds, a companion act was passed "to Enable the Legislature to Settle the Quotas of the Several Counties of this Colony in order for Levying Taxes."[3] This represented a compromise between the assembly, which heretofore had fought for a tax levied upon the quantity of the land held by any person, and the council, which had wanted a tax based upon the quality of the land. The act as adopted provided for the taxation of lands according to valuation determined within a fixed range as to both quantity and quality. It was really a surrender by the assembly, but this house proposed the measure itself and thus upheld the principle that the council could not amend a money bill. An important step had been taken in solving the ancient land-tax controversy, and a perennial bone of contention was now removed. It was a signal victory for Read, and a convincing demonstration of his talent as a political leader. It was a sample of the substantial progress that was to follow during the next three years in conciliating the disaffected parties of the previous administration and in curbing the forces that were threatening the integrity of the government. Belcher was delighted. With the two contested issues settled—taxation and government support, he now looked forward to peace and order in the province.

As was customary at the close of the session, Speaker Read

[2] *N. J. Assembly Minutes*, June 1, 1751, p. 16.
[3] *N. J. Arch.*, 1st ser., XVI, 301, 303.

addressed the governor on behalf of the assembly. Of the two major measures adopted, he said:

> We . . . agree with your Excellency it is more peculiarly our privilege to make the Necessary Supplies and in our Opinion to Direct the Method of doing it for payment of the Public Debts, and for the Support of Government and when the true Ends of Government are fully answered, and the officers kept Strictly to their Duty and in the Execution of the Laws, it then becomes a Duty Incumbent on us, in behalf of the People, to make Provision for the Support of the Government, as the gratefull Acknowledgment of an Obliged People.[4]

With these measures disposed of, Read now apprised His Excellency that the assembly would be pleased if he would follow the King's example in the redressing of grievances—to be specific, if he would assume the authority to remove a justice of the peace without the advice and consent of the council. A strange turn, indeed, for the New Jersey assembly to suggest that the governor had placed too restrictive an interpretation on his powers. Was this the successor to the "mob assembly" that was speaking? Had Read charmed them into this state of mind? Was Read using flattery on the governor? Did he have a grudge against some justice whom he wanted removed? More likely, it seems, he was seeking to control all appointments himself through the governor.

Belcher was agreeable to having his independence of the council include appointments as well as dismissals. At any rate, he proceeded to grant commissions to several persons, as sheriff, justice of the peace, and judge of the county court, without the advice and consent of the council. When word of this unprecedented action reached the ears of the council members, they promptly went into session and summoned the secretary to inform them of the facts. Read reported the instances of commissions having been awarded without reference to the council, whereupon the council addressed the governor on the matter. The royal instructions to the gov-

[4] *Ibid.*, 300-302.

ernor issued in 1701 and the subsequent precedent established by a succession of governors were tactfully but forcefully laid before His Excellency.[5] Apparently Belcher, on reconsideration, thought it would be better policy thereafter to have his appointments approved by the council.

At the next session of the assembly the following September, Read, in his speaker's address to the governor, again alluded to the matter:

> . . . we apprehend that the clearest Demonstrations that a Governor's Aims and Views are for the Happiness and Benefit of the People, is not only the passing good Laws, but also to keep the Subordinate Officers strictly to their Duty, in the Observance and Execution of them, and where any Failure or Breach of their Duty appears, to remove such, or order such Proceedings against them, as may effectually prevent them, and deter others from committing the like Fault for the future; which we wish we could say had been effected; and whenever it is done, we are humbly of Opinion will be the best Encouragement to Assemblies to raise a Support the more freely.[6]

This session opened in the spirit of earnestness and harmony. The governor urged "Diligence & Dispatch" to avoid "Tedius Journeyings & Long attendance upon the frequent Sittings of the assembly which must be a ditriment to . . . private affairs: & allso an occasion of making the Taxes more heavey & Burthensome."[7] In reply Read expressed the hope that the meeting of the assemblies in the future would always be attended with "Pleasure and Satisfaction" and that they would be "productive of the like good Consequences."[8] However, when the assembly passed a tax bill, the council objected and returned the support bill to the assembly with amendments. The assembly ignored the amendments, and assayed to present the bill directly to the governor. They had passed a resolution to the effect that all money bills "should be pre-

---
[5] *Ibid.*, 356-360.
[6] *N. J. Assembly Minutes*, Sept. 25, 1751, p. 16.
[7] *N. J. Arch.*, 1st ser., XVI, 311.
[8] *N. J. Assembly Minutes*, Sept. 25, 1751, p. 15.

Legislator

sented to . . . the Governor, by the Hands of the Speaker of the House of Representatives."[9] Precedent for such procedure was cited. The result was up to Belcher, but not finding the precedent valid, he replied that to receive the support bill without the council amendments would be irregular.[10]

In January, 1752, the governor again called the assembly together at Perth Amboy, in order to pass an act for the support of the government. The assembly, through Speaker Read, expressed regret that the governor found it necessary to meet them "at this inclement Season of the year, especiell on so Short Notice, That it was with Difficulty a Sufficient number has been got together," but agreed "to do all in our power (Consistent with our duty to our Constituents) to get the Necessary Supplies into the Treasury."[11] The prospect was not encouraging. The treasury was empty and £8,000 was needed to meet the debts of nearly four years—an enormous sum for the people of a small colony who had not paid taxes for many years. However, to Charles Read, who believed that diplomacy was "the gentle art of letting someone else have your way," it was not impossible to reach an agreement on the bill. Accordingly, a maximum-minimum quota for each county was fixed. The question as to whether tenants should pay taxes on what they leased or on the entire tract on which their lease lay, was debated. A tie on this question was broken by Speaker Read's vote, who favored a levy on the leased portions only, and the bill went to the governor, who gladly signed it.[12]

Governor Belcher expressed his appreciation of the "Alacrity and Dispatch" as well as the "good agreement & Harmony" with which the assembly had discharged its business, and thanked the members of the council for their concurrence.[13] Hence all had reason to be pleased with the compromise: the governor had his salary; the council had protected the pro-

---

[9] *Ibid.*, June 6, 1751, p. 21.
[10] *Ibid.*, Oct. 22, 23, 1751.
[11] *N. J. Arch.*, 1st ser., XVI, 355; *N. J. Assembly Minutes*, Feb. 7, 1752, p. 15.
[12] *N. J. Assembly Minutes*, Feb. 7, 1752, p. 13.
[13] *N. J. Arch.*, 1st ser., XVI, p. 364.

prietary lands from the excessive levies first proposed; and the assembly had succeeded in taxing some unimproved lands.[14]

The following spring the rioting which had marred the earlier years of Belcher's administration broke out anew. This time the central figure was one Simon Wickoff, whom Read himself as judge in the case had committed to the Perth Amboy jail for high treason. On April 13 a band of his friends stormed the jail, and ignoring the protests of the sheriff, broke open the doors and carried Wickoff away in triumph. Upon the revival of lawlessness and the threat of mob rule the governor became violently disturbed. Notwithstanding the fact that Wickoff voluntarily returned to the jail, the governor ordered Secretary Read with all possible dispatch to summon the council in special meeting to deal with this "highest & most Outrageous Insult upon the Kings Authority."[15] When the following December he again called the assembly to meet —because of his delicate health, at Elizabethtown—this time the jail-breaking was one of the matters he placed before the house:

. . . I must Earnestly Recommend to you the passing of a good Law for the better Security of the King's Goals and for the Severe Punishment of such Audacious Offenders for the future, but if after so many flagrant Instances of Mobing Rioting and breaking open the Kings Goal in This Province you will do nothing to Prevent it, no mans life or Property can be safe. . . .[16]

Read, in replying, minimized the dangers and expressed the hope that the rioters would now abide by the law. He voiced the displeasure of the assembly in being summoned at so inclement a season of the year, and also in being called to Elizabethtown, rather than to Perth Amboy or Burlington.

When Belcher again convened the assembly the next May and again appealed for the enactment of a law against the rioters, the assemblymen's reply was a flat refusal. In their opinion no new laws were needed since ample penalties were

[14] D. L. Kemmerer, *op. cit.*, p. 234.
[15] *N. J. Arch.*, 1st ser., VIII, pt. 1, pp. 38-39.
[16] *Ibid.*, XVI, 392.

provided. What was needed was effective enforcement. The wrongdoers should be prosecuted and punished. The responsibility was with the law-enforcing officers, not with the assembly. In a lengthy statement, the policy of the assembly toward the riots was reviewed. ". . . this House have always looked upon those disorders with great abhorrence," said Read. ". . . when Offenders are Speedily brought to Justice it must Certainly deter others from Committing the like Offences. But if suffered to escape with impunity it not only prompts them to persevere but also may Induce others to imitate their Evil practices through the hopes of Indulgence." The unhappy state of affairs, he claimed, was due to the "failure in The Persons appointed to put the Laws in Execution . . . not . . . for the want of Law." Therefore, the assembly thought it advisable for the governor "to Press the Tryals of The Perpetrators of those disorders upon the Officers entrusted with The Execution of the Laws."[17]

A condition which aggravated the problem of suppressing the land riots was the inefficiency of certain sheriffs and local magistrates. As a member of the supreme court Read was doubtless more acutely conscious of the weakness of the provincial law-enforcement machinery, than he would normally have realized as a legislator. His addresses to the governor while speaker of the assembly revealed a constant concern for the character of the services performed by public officials. He vigorously maintained that office holders who were incompetent or negligent of duty should not be tolerated. At times he voiced his dissatisfaction with individual justices or sheriffs. He heartily disapproved of the Sheriff Act passed in 1748 which limited somewhat the governor's power in appointing sheriffs, one of whose duties was to preside over colonial elections.[18]

In a letter to Chief Justice Morris in 1753, Read complained of the operation of the Sheriff Act, and of the unsat-

[17] *Ibid.*, 412-414.
[18] Acts of the General Assembly of the Province of New Jersey, compiled by Samuel Allinson (Burlington, 1776), pp. 156-159 (elsewhere referred to as *Allinson's Laws*).

isfactory state of other laws. "The Sheriff Act," he wrote, "is so productive of Inconvenience & is such an Infringement on the Royal prerogative that I should think the Bare mention of it to the Right Honble the Lords of trade would occasion its Destruction."[19] The sheriff problem was further agitated at a session of the assembly at Perth Amboy in June, 1753. The assembly objected to the appointment of Enoch Anderson as sheriff of Hunterdon County, claiming that he was not qualified under the law which required a prior residence in the county of three years. The governor replied that it was not the business of the assembly to interpret the law. "After an Act is pass'd by the whole Legislature," said he, "no single branch can expound upon it, but that it is always left to the Judges of the Land, and others learned in the Law."[20] However, he was ready to submit the matter to the court, and if it should be determined a grievance, he would immediately redress it. Meanwhile Anderson was removed and the assembly dropped the matter with a face-saving declaration of its right to inquire into and complain of the breach of any law, if not to expound the law.

Belcher relied heavily upon Read in filling the sheriff's and other public offices of the province. In selecting a sheriff for Cumberland County, he informed Read: "I desire you to make inquiry & let me know some person that can come well recommended."[21] Acknowledging Read's suggestion for a similar appointment in Gloucester County, he wrote: ". . . your recomendation has it's weight with me and the reasons you mention to incline me to the matter are not inconsiderable."[22] Again, when Sussex was set aside as a new county, the governor wrote to Read: "I shall be glad of the best Advice and Information I can get for setling the Officers of that County . . . and in this matter I desire you to make Inquiry."[23]

[19] *N. J. Arch.*, 1st ser., VIII, pt. 1, p. 186.
[20] *N. J. Assembly Minutes*, June 8, 1753, p. 50.
[21] Jonathan Belcher to Charles Read, Jan. 4, 1754, *loc. cit.*
[22] *Ibid.*
[23] Jonathan Belcher to Charles Read, Aug. 1, 1753, *loc. cit.*

Legislator

In 1753 the harmonious relations between Governor Belcher and Read were threatened by some misunderstanding, possibly due to the intrigue of political enemies. An exchange of letters, however, gave assurance of loyalty and mutual respect. In August Belcher wrote to Read:

> I take a kind and particular Notice of the great Respect you Express for me in your Letter of the 21: Ult . . .
> Since Sir you are pleased to enter into the matter of your Conduct towards me you must allow me to Speak out with freedom and to tell you I perfectly agree with you that Actions make out Sincere friendship better than words and I do assure you no one has taken any Pains with me to misrepresent you but I have tho't that you have often thwarted and Contradicted me in things I inclind to do while at same time I have been always willing to Multiply Offices upon you and to Serve your Interest in any thing I cou'd and this I have tho't you have done in favour of such who whenever they have Opportunity wou'd gladly snatch the Bread out of the Mouths of you your Wife and Children I . . . desire all things past may be in an intire Oblivion and during my Political or Natural Life I shall be glad to Show you any Respect or Friendship and . . . during my Administration of this Governmt it has been my Constant Care to Support and Vindicate the Kings Authority & Honour and also to seek and Advance the Peace and Happiness of this People. . . .[24]

The following June the governor again called a session of the assembly at Elizabethtown. He urgently pressed the assembly to provide measures for cooperation with the other colonies in repelling the aggressions of the French and the Indians. Unwilling to transact business away from its accustomed place, the assembly sent a committee of two to ask the governor to be dismissed to Perth Amboy. As such messages were usually sent through the council, the governor censured the assembly for pursuing this method of address, and intimated a lack of appreciation for his frail condition. Despite his preference, however, the assembly was prorogued to Perth Amboy, where it met on June 3, 1754.

In conveying the message of the assembly to the governor

[24] *Ibid.*

a few days later, Read departed from his customary diplomacy, and in language more intemperate than a New Jersey assembly had ever used in addressing Belcher, delivered a sharp and uncomplimentary rejoinder. To what extent Read expressed his own feelings or simply voiced those of a majority of the assemblymen, does not appear. The failure of the governor to get action against the rioters was a victory for the assembly. Although Read was close to the governor, in these controversies he would be expected to represent the assembly's point of view. After referring to the reflections the governor had cast upon their conduct, he continued:

> . . . no Harmony can long subsist Unless the same good Dispositions and Interest unite those Concerned in it; . . . we can't look upon several Expressions in Your Last Message . . . any ways tending to Cement the same; and are concerned, that Your Excellency shou'd view our Transactions in a wrong Light: . . . this House is not Guilty of that Disrespect and Ingratitude that You insinuate. . . . Your Kind and Benevolent Intentions no ways appears by Your Charging this House with Want of Humanity and Tenderness to a Governor. . . .
>
> In What Manner Your Excellency has exerted Your Publick and Private Interest for the good of the Inhabitants of this Province, We are at a Loss to know. If it is the removing their Grievances; it is with Concern we are Oblidged to Say, that we don't know, that You ever agreed either with this or any other House of Assembly that any of the Grievances they Complained of were so. . . . If any Representation should be necessary to be made Home, in Favour of his Majesty's most dutiful Subjects of New Jersey; We dont think that Your Constantly finding fault with and blaming our Conduct any Ways tends to place them in that favourable Light with his Majesty that their Loyalty deserves.[25]

Notwithstanding this clash with the governor, the assembly adopted a resolution favoring the union of the colonies to prevent the encroachment of the French, and passed a bill for the support of the government. However, as no provision was

---

[25] *N. J. Arch*, 1st ser., XVI, 463-465.

Legislator

made for aiding the military forces, the governor ordered the assembly dissolved on June 21.

In the ensuing election of the Nineteenth Assembly, Charles Read was again chosen to represent Burlington, but when the body convened at Perth Amboy on October 1, 1754, he was succeeded as speaker by Robert Lawrence, of Monmouth County. Although not now the titular head of the assembly, he remained its real leader. This assembly, which sat through intermittent sessions for six years, had to deal with the weighty problems which attend a state engaged in war. Because the assembly was mainly concerned with the raising and equipping of military forces, and with measures for financing them, the old quarrel over quit-rents became an issue of secondary importance, and the consequent rioting subsided.

As is shown elsewhere, Read was at the forefront in managing the military operations of the province. The financing of New Jersey's participation in the war required several issues of paper currency, or "bills of credit." Read had made a report to the governor on paper currency in 1749,[26] and was now regarded as an authority on public finance. As long as he was a leader in the assembly it was inevitable that he should have a hand in the preparation of nearly every money-raising venture. New issues of currency seemed to offer the only feasible way to finance the government without a heavy tax burden during this critical period, but the English officials had been hostile to paper money, and had already rejected a petition from the assembly for an issue of £60,000. However, the King wanted troops, supplies, and financial assistance and when the Lords of Trade were again petitioned, they gave consent, provided the new bills should not be declared legal tender in payment of debts. The new assembly at its first session promptly resolved by a vote of 20 to 2 (the dissenters were Messrs. Leaming and Spicer) to grant aid to the King against the French, and approved a bill to make current £70,000 in bills of credit. Charles Read was desig-

[26] N. J. Arch., 1st ser., VII, 357-360.

nated as one of the four persons to authorize the printing of the bills.[27] However, believing that if the bills could not be used as legal tender the purpose of the measure would be defeated, the condition named by the royal authorities was not included. The council was disposed to agree with the assembly, and in November Belcher forwarded the draft to the Lords of Trade, expressing the hope that they would report on the bill as early in the spring as possible. Months passed before the reply came from London. Meanwhile the assembly met twice at Elizabethtown, in February and in April, 1755. During the latter session Assemblyman Jacob Spicer recorded in his diary an incident which reflected Read's influence with Belcher:

Mr. Speaker, Mr. Wetherill, Mr. Fisher & Mr. Read were in conference with the Govr who Inquired whether they had recd anything from the agent respecting the paper money Bill—to which Fisher replyed that I had recd a letter from the agent of the 21st of Jany which seemed to read favourable whereupon the Govr smiled I suggested as much in Substance as if the Agent was willing to please us but at the Same time waged that if any body wou'd give him 20s if the bill did not pass he wou'd give £10 if it did, to wch Mr. Read observed that perhaps people wou'd think the Govr was agst it wch he answered with Silence— this in Substance by Information.[28]

The counsel Read offered the governor, however, he did not apply to himself, for that same day he offered Assemblyman John Ladd, of Gloucester County, a wager of £5, "One-half wet and one half dry," that the paper money bill would not be approved. Belcher's premonitions were well founded, for the Lords of Trade finally notified him that they would not approve the issue as outlined. Meanwhile the assembly in April had proceeded with an issue of £15,000, conformable to an earlier act, which made the bills legal tender only in the province. This was followed in August with a second issue in equal amount, both to finance the military expedi-

[27] *Ibid.*, VIII, pt. 2, pp. 36-72.
[28] J. Spicer's Diary, Apr. 22, 1755.

Legislator

tion. Now that the tide of currency was let loose, another issue of £10,000 was enacted in December to provide protection for the frontier. Read served on the committees which drafted all three bills.[29]

The propriety of a person serving simultaneously in both the legislative and the judicial branches of the government came to a test in the Nineteenth Assembly, when it met to organize. Ten years earlier the assembly had declared it was inconsistent with the freedom of the people that the chief justice of the supreme court should also be a member of the council, a measure aimed at Robert Hunter Morris who then held both posts.[30] Subsequently Read, as we have seen, after being appointed to the supreme court, had been elected to the assembly, had presided over it as speaker, and had engineered the enactment of laws which, as supreme court justice, it would be his responsibility to interpret. So far as we have been able to discover, no objection was raised to his serving in this dual capacity. Now, after Read had resigned from the supreme court, Samuel Nevill, who had served in an earlier assembly and was now a supreme court justice, again had been elected to the assembly from Middlesex County. Some of the residents of his home county, probably the followers of a political opponent, protested his seating on the ground that for a justice of the supreme court to serve in the assembly, "is contrary to the Laws of Great Britain," as well as to the colonial law for securing the freedom of assemblies.[31]

With the present assembly under Read's thumb the outcome of the petition could hardly be in doubt. The chairman referred the matter to a committee of five headed by Read himself, "to inspect the Laws of Great Britain, and of this Province, relating to Persons disqualified to sit in the House of Commons and General Assembly, and make report

---
[29] *N. J. Assembly Minutes,* Apr. 26, 1755, p. 21; Aug. 5, 1755, p. 9; Dec. 17, 1755, p. 7.
[30] *N. J. Arch.,* 1st ser., XV, 371-372.
[31] *N. J. Assembly Minutes,* Oct. 4, 1754, p. 7.

thereof."[32] Ten days later Read reported for the committee that they had studied the laws, and finding instances where judges had sat in the House of Commons, they saw no reason why Justice Nevill should not sit in the assembly. Accordingly Nevill was seated with only two dissenting votes.[33] Subsequently Read followed the example of Robert Hunter Morris by accepting appointment as chief justice while retaining his seat on the council.

The question of the place of meeting continued a bone of contention between governor and assembly. Because of his frail condition, the aging governor would call the house to convene at Elizabethtown and the members would respond under protest. At the August session in 1755, by a vote of 14 to 3 the members requested that the governor permit them to reconvene at Perth Amboy. The governor insisted on his right, under the royal prerogative, to call the assembly to Elizabethtown, but in a conciliatory vein added:

However, the old Romans wisely consider'd, that while Hanibal was at their Gates, all Disputes and Contention should subside, and as I look upon the Province in the present Situation of Affairs to be attended with much Difficulty and Hazard, in Answer to your request I have Ordered the Secretary to Adjourn your House, tomorrow . . . to meet at Perth-Amboy.

Accordingly, "by writ under the Great Seal of the Province," Secretary Read adjourned the assembly.[34] When the question was again raised at the opening of a session in May, 1756, Read went on record as acceding to the governor's wishes, and helped to carry the decision for Elizabethtown by a vote of 11 to 10.[35]

The printing of the provincial laws was one of the numerous responsibilities of the busy secretary. It was the practice of the assembly at the close of each session to authorize the printing of a given number of copies of the laws enacted, and

[32] *Ibid.*, Oct. 5, 1754, p. 9.
[33] *Ibid.*, Oct. 15, 1754, pp. 18-19.
[34] *Ibid.*, Aug. 1, 1755, p. 6.
[35] *Ibid.*, May 24, 1756, p. 10.

also of the "Votes," as the minutes were then called. The printer was sometimes selected by the assembly, and sometimes the matter was left to Read, or to a committee of which he was a member. The choice usually lay between William Bradford, of Philadelphia, and James Parker, of Woodbridge. Read seems to have favored Bradford, although his services did not always give satisfaction. In 1755 the governor wrote to Read expressing vigorous disapproval of the "negligent dilatory Way" Mr. Bradford had proceeded in printing the laws and journals of the session. So inconvenienced had he been by the delay, he warned Read that unless the work could be expedited it would become necessary to engage another printer.[36] At the August session that year the choice of the printer was laid before the assembly. Read cast his vote for Bradford, who was chosen by the narrow margin of 10 to 9. At the same session it was ordered that Mr. Read have the "Care and Custody" of "all the Statutes, Laws &c" belonging to the Western Division, and that Mr. Nevill be custodian of the law books belonging to the Eastern Division.[37]

Now and then, as in former sessions, the assembly was asked to grant special privileges to certain manufacturing enterprises. The argument was advanced that the fostering of these industries by exemption from taxes or by granting bounties, would be of public benefit. At the September session in 1751 the assembly received from the owners of ironworks in Morris County a petition for a bounty on iron and tax exemption for their manufacturing plants.[38] The following January the owners of glassworks asked that they be relieved of taxes.[39] In June, 1754, a group of citizens of Cape May requested a similar favor for gristmills in that county.[40] Petitions came, too, for various other purposes. In 1752 "sundry Inhabitants" of Morris County complained of inconveniences

[36] Jonathan Belcher to Charles Read, Oct. 1, 1755. Belcher Papers, IX, 510.
[37] *N. J. Assembly Minutes,* Aug. 18, 1755, pp. 26-27.
[38] *Ibid.,* Sept. 18, 1751, p. 10.
[39] *Ibid.,* Jan. 29, 1753, p. 5.
[40] *Ibid.,* June 7, 1754, p. 15.

and damages suffered by having "great Numbers of Cattle from the neighboring Counties drove up into their County" at certain seasons of the year, praying legislation to provide relief from the nuisances.[41] At the same session a petition of bolters set forth "that it is almost impracticable to comply with the Act concerning Flour, because much Wheat is Threshed on Earthen Floors," and prayed an act to prevent the practice.[42] Most of these petitions received scant attention. Apparently the assembly thought it inexpedient to single out certain industries for special concessions.

On occasion, Read proved willing to sacrifice personal interest for the public welfare, to the extent of deferring his salary payments in order that other treasury obligations could be met. Governor Belcher wrote to Richard Partridge in London that Read had been "much Concernd and very Active in getting your money paid after it was Voted. For as the Treasury was exhausted he Consented to stay longer for Monies due to himself and got others to do the Like that your Money might be paid without delay."[43] Also in the interest of harmony within the official family, he seemed ready at times to overlook questionable acts of the governor. In 1751 Richard Saltar wrote to Chief Justice Morris that Governor Belcher had counterfeited a warrant on the treasury of the Western Division, but added: "I believe I shall meete with Dificulty to prevail on the Secretary [Read] to make affidavit on that affair but to my Knowledge he is well acquainted with itt the Governour making use of him to make up the matter with Mr. Allen [the western treasurer] after he had Discovered the fraud."[44] When pressed on the matter, said Saltar, Justice Read declared that if any other person had been guilty of a like crime and convicted of it in the supreme court, he should think it his duty to order him placed in the pillory.

[41] *Ibid.*, Feb. 7, 1752, p. 14.
[42] *Ibid.*, Feb. 6, 1752, p. 9.
[43] Jonathan Belcher to Richard Partridge, Apr. 26, 1748. Belcher Papers, VI, 309-310.
[44] Richard Saltar to R. H. Morris, Dec. 5, 1751, Boggs Collection.

Legislator

Whatever doubts may be cast upon Belcher's integrity, there is no question of his militant patriotism and loyalty to the responsibilities of his trust during the trying period of the war. In the midst of his efforts to prod a reluctant assembly into more aggressive action, he died on August 31, 1757. He had realized his wish expressed years before: "If God and the King please [I] shou'd be glad to dye Gov[erno]r of New Jersey."[45]

When Governor Belcher died, it was Read's official duty as secretary of the province to notify Thomas Pownall, the lieutenant-governor. Pownall, having recently been appointed governor of Massachusetts, promptly visited New Jersey, but finding it not practicable to retain the administration of both provinces at the same time, he returned to Massachusetts.

Accordingly John Reading, senior member of the council, again occupied the office *pro tempore*. Francis Bernard, the next governor, a cultivated and able graduate of Oxford, arrived in June, 1758. After a brief but satisfactory term of two years, Governor Bernard went to Massachusetts, and was succeeded by Thomas Boone, whose administration was uneventful. Less than a year passed before he, too, was transferred to another post—in South Carolina. The next governor was a London merchant, Josiah Hardy, who arrived in October, 1761. He appears to have been well liked by his constituents, but proved to be too weak a man for the governorship. The following year he was removed by the home authorities because he made an appointment to the supreme court "during good behaviour" instead of "during the King's pleasure." Although Governor Belcher had given Justice Nevill good behavior tenure, it was now considered an unpardonable breach of official etiquette.[46] Hardy's successor, named in 1762, was William Franklin, last of New Jersey's royal governors.

This period of changing administrations must have been

[45] Jonathan Belcher to Sir Peter Warren, June 8, 1751. Belcher Papers, VII, 153-155. See also E. J. Fisher, *op. cit.*, p. 36.
[46] N. J. Commissions, Liber C2, 195.

a trying one for Read, serving as he did under six governors in as many years. Although as secretary he probably knew more about the machinery of government than any other person in the province, under a rapid succession of governors he could not be sure of reappointment. Apparently he was on intimate and friendly terms with the several governors, with the exception of Boone. Aaron Leaming noted a lapse in Read's political power during the brief administration of Mr. Boone, who, he wrote, was "Governor without a prime minister."[47] Governor Bernard in 1759 commissioned Read "Registrar, Actuary & Scribe" of the acts of the prerogative court of the province and "Keeper of the Registry of the Same."[48] Likewise, Governor Hardy, in 1762, renewed his commissions as surrogate and as court clerk for Burlington County, and also as secretary "for and during our pleasure," being well assured of his "Loyalty, Ability and Integrity."[49]

As clerk of the council since 1744 Read was no stranger to the council chambers, and Governor Belcher, as we have seen, for a number of years had wanted him to be a member of this body. Now, after he had served several years in the lower house there were vacancies in the council; James Alexander had died, and John Reading and Thomas Leonard had resigned. On August 24, 1758, Governor Bernard submitted to the Lords of Trade the names of Charles Read and John Smith of Burlington for the two vacancies from West New Jersey. At the same time he mentioned John Stevens and Peter Schuyler for vacancies from East New Jersey. Appointment of Read and Smith followed by royal order, December 12, 1758.[50] However, Read did not at once move up to the council, preferring to finish his term in the assembly which continued for two more years. Although a seat on the council was considered a higher honor, Read could wield greater influence in the assembly. The country was in the midst of a

[47] A. Leaming's Diary, Nov. 14, 1775, *loc. cit.*
[48] N. J. Commissons, Liber AAA, 331.
[49] *Ibid.*, 366-368.
[50] *N. J. Arch.*, 1st ser, IX, 127, 151.

war, and the assembly held the key to the military program. So he remained, and under changing governors kept up the fight for maintaining the military forces and voting issues of paper currency as needed.

At a session of the assembly in March, 1760, Read was named to a committee to draft an address to the governor requesting a longer term of maturity for the bills of credit. In replying, Governor Bernard revealed that he was an economic realist as well as a prophet. In part, he said:

> There has been great Plenty of Money brought into the Country; and in Consequence of it, the Produce of Lands has bore an extraordinary high Price. When the War is over, the Price of Produce will be continually sinking, and Money will gradually grow scarce. Therefore, to postpone the Payment of the Provincial Debt, from a Time of Plenty to a Time of Scarcity (or at least less Plenty), will be removing a light Burthen from yourselves, by laying an heavy One on Posterity.[51]

Although the all-important questions of taxes, appropriations, currency issues, and military operations occupied the center of the legislative stage, the assembly had to consider also measures dealing with public improvements and other domestic problems. For example, in 1755 Read fathered an act, "to preserve the Navigation of Rivers and Creeks," designed to prevent obstructions in navigable streams, since "the Transportation of Timber, Planks, Boards, Hay and other Things to Market by Water" was considered "a great Conveniency to the Inhabitants."[52] He served on numerous committees to which were referred an endless variety of measures as widely divergent as the regulation of lotteries and the relief of persons imprisoned for debt.

Prompted doubtless by his personal interest in the lumber trade, in 1759 he sponsored a law "for the further Preservation of Timber in the Colony of New Jersey."[53] An act "for preventing the Waste of Timber, Pine and Cedar Trees and

[51] *N. J. Assembly Minutes,* March 19, 1760, p. 8.
[52] *Nevill's Laws,* II, 64.
[53] *N. J. Assembly Minutes,* March 13, 1759, p. 10.

Poles," and to lay a duty on the exports of "Pipe and Hogshead Staves," had been on the books since 1714, and had been found "very advantageous," but its "good Intentions" had been defeated by persons cutting and destroying timber on land not belonging to them. The new act provided that, for a period of five years, trespassers who should "cut, box, bore, or destroy any Tree, Saplin or Pole" on lands to which they had not right or title, should be fined 20s. for each "Tree, Saplin or Pole" so treated.[54]

During his last year in the house Read was engaged with a particularly heavy calendar of miscellaneous matters. The assembly, during November and December, 1760, passed a long list of bills which included acts for the support of the government, for the relief of prisoners for debt, for fixing the boundaries of townships and counties, for erecting and maintaining bridges, and for the regulation of roads.[55] Read himself sponsored an act to regulate the size of traps to prevent the trapping of deer.

As secretary of the province, no one had more reason to be concerned about the safety and preservation of the public records than he. Fearful lest important documents be lost by fire, which "would be introductive of the greatest Distress," in order "to prevent as much as possible so great a Calamity," Read in 1760 introduced a bill "for the Preservation of the Publick Records of the Colony of New-Jersey." The bill, which was promptly adopted, provided for the erection of a one-story fire-resistant building in Perth Amboy, and another in Burlington, at a cost not exceeding £300 each.[56] Read saw to it that he was named in the act as one of the six commissioners who were to carry out its provisions, and, incidentally, to receive a 5 per cent commission for their services. The appropriation proved to be insufficient, and in March, 1762, the assembly, after the manner of modern legislatures, voted a deficiency appropriation of £100 for each building. Read,

[54] *Nevill's Laws*, II, 263-264.
[55] *Ibid.*, p. iv.
[56] *N. J. Assembly Minutes*, Nov. 26, 1760, p. 42; *Nevill's Laws*, II, 343-344.

by this time a member of the council, had the pleasure of reporting the bill favorably for passage.[57]

The regulation of lotteries was one more problem which arose from time to time. In 1748 the assembly had enacted a law to prevent lotteries, and other forms of "Gaming for lucre of Gain," as well as "to restrain the abuse of Horse Raceing."[58] Nevertheless, the law was evaded, and petitions were pressed upon the assembly to authorize lotteries for public or charitable projects. In 1758, petitions were presented to authorize lotteries for the benefit of churches at Newark and at Bedminster, and for the erection of public buildings at Perth Amboy. All three were denied. Read voted in favor of the Perth Amboy lottery, but against those for the churches. Apparently he had no objection to this method of raising money for worthy secular purposes, but at that time thought it had no proper place in church financing.[59]

The law enacted in 1758 to authorize the purchase of the land claims of the Indians, described in a later chapter, provided for the sinking of a sum of £1,600 by means of three public lotteries. The first of these was begun, but was not drawn because of the failure of the ticket sale. In November, 1760, a petition to hold a lottery to finance a bridge at Bound Brook was denied.[60] Later in the same session, Read had a hand in drafting a new lottery law, entitled "An Act to prevent the Sale of Tickets, in Lotteries erected out of this Province, and more effectually to prevent Gaming; and to revive Three publick Lotteries appointed by a former Law of this Colony."[61] The former prohibitions were continued, but exception was made for the revival of the lotteries to finance the payment of the Indian claims. The commissioners for Indian affairs, of whom Read was one, were named to supervise the lotteries.

[57] *N. J. Arch.*, 1st ser., XVII, 301-305.
[58] *Nevill's Laws*, I, 405-408.
[59] *N. J. Assembly Minutes*, Aug. 3, 1758, p. 17.
[60] *Ibid.*, Nov. 14, 1760, p. 18.
[61] *Nevill's Laws*, II, 362-365; *N. J. Assembly Minutes*, Dec. 1, 1760, p. 48.

In certain cases after 1760, private acts of the assembly authorized the drawing of lotteries, a rather questionable policy, because the practice that the law sought to curb was kept alive. After Read became a member of the council, bills to legalize lotteries for the benefit of St. Peter's Church at Perth Amboy, St. Mary's Church at Burlington, and the College of New Jersey at Princeton, and also for the rebuilding of Bound Brook bridge were referred to him. By this time he seems to have changed his views concerning the propriety of church lotteries, for he returned the bills with his approval.[62]

With Read in command of the assembly, it was to be expected that he would be in favor of most of the measures which were brought to a vote. The machinery of the colonial law mill was not much different from the legislature of today. Proposals to which Read objected would likely be smothered in committee before reaching the stage of a roll call. Furthermore, if a bill he did not favor should by chance come up for passage, and he saw no way of stopping it, he would prefer not to have his vote recorded.

As speaker of the Eighteenth Assembly, which lasted from 1751 to 1754, he did not vote except in the event of a tie. Through the six years of the Nineteenth Assembly, however, he was usually recorded as present and voting. Of approximately seventy-five roll calls on bills and other questions on which his vote is recorded, only seven times was he found to be in the minority. His losing votes were mostly on measures which commanded his active interest, but lacked majority support. For example, he favored restraining the Indians to the counties where they resided, sending the militia to join the New York expedition, and allowing captains expense money for enlisting their companies, proposals which were rejected in the assembly.[63] He voted for sending relief to Bostonians who suffered from a great fire in 1760, and also, as

[62] N. J. Arch., XVII, 256, 302, 303, 330.
[63] N. J. Assembly Minutes, Apr. 19, 1755, p. 13; Dec. 23, 1755, p. 18; May 27, 1757, p. 7.

Legislator

we have seen, to permit Perth Amboy to hold a lottery, both of which measures were disapproved by the assembly.[64]

TRUE to Governor Bernard's forecast, the artificial prosperity created by the French and Indian War was reflected in the later stages of the conflict by rising prices of farm produce and other commodities. Although the colonial issues of paper money resulted in some degree of inflation, New Jersey currency compared well with that of the other colonies. In 1760 the English traveler Andrew Burnaby wrote: "The paper currency of this colony is about 70 per cent, discount, but in very good repute; and preferred by the Pennsylvanians and New-Yorkers, even to that of their own provinces."[65]

Meanwhile, farm prices were in inverse ratio to the annual production, which fluctuated with seasonal rainfall and temperature. In August, 1755, "the late uncommon dry Seasons, by which the Crops in several Parts of this Colony have in a great Measure failed," was given as one reason why New Jersey could not enlarge its military forces.[66] The influence of changeable weather and the effects of the war upon colonial agriculture were noted in letters written between 1759 and 1762 by Charles Read's friend and sometime partner, Samuel Allinson of Burlington, to the latter's uncle in England. In 1759 he wrote:

> Our Crops last, and this present year have been I Believe generally good, Tho Family necessaries, (Such as Butter, Meat, Cyder &c) were Sold at a Greater Price than I believe ever before here, occasioned, I suppose, by the Great number of Soldiers Quartered throughout our Countrey hereabouts.[67]

A letter written in 1761 was not only weather wise, but prophetic of political and military disturbances to come:

> Last winter was very Cold & Severe, but as the Weather was pretty Setled and an uncommon Quantity of Snow fell & Lay

[64] *Ibid.*, Aug. 3, 1758, p. 17; Nov. 25, 1760, p. 41.
[65] A. Burnaby, *op. cit.*, p. 59.
[66] N. J. *Assembly Minutes*, Aug. 20, 1755, p. 32.
[67] Samuel Allinson to his uncle, 1759. Caroline Allinson Collection.

long, there is like to be very Fruitful Crops of Corn. Tho from the Coldness of the Weather this Spring many things are Injured & much Behind in Growth, It's with us as I have been told It is with you. The prosecutn of the present war has Stript the Country very much of Its Laborers which Lays some Farmers under great Difficulties & subjects them to loss at this annl Season wch is now very near. It has also had another very Considble Effect (to wit) the Enhansing the price of almost every kind of produce ¼ & many things more, but as peace between the contending Powers we here is no[w] in agitation, if that be Concluded & Ratified Its very likely we shall see another Revolution before long after.[68]

Drought, a short crop, and still higher prices told the story of the following year:

Our Crops last Season were in General indifferent of all kinds & in some places very bad, the Winter not being good, the Spring late, and yet the Harvest Very Early Ripe occasion'd by an uncommon dry Summer, such as [?] was scarce ever known by the oldest Men—this makes all necessaries greatly Encreased in Price, but yet We Enjoy Plenty.[69]

In foreseeing "another Revolution before long," Allinson apparently was a better prophet than Read. During the decade of the 1750's, Read was so thoroughly engrossed in the affairs of government and in his business interests he probably did not perceive that more trying times lay ahead. He had been a success as a legislator. He found satisfaction in the settlement of controversial issues and in constructive legislation achieved. The floating of colonial currency, the conciliation of the Indians, the pursuit of the French and Indian War, all testified to his political leadership. When in December, 1760, the Nineteenth Assembly finished its labors, he was ready to assume his place on the council to which he had been appointed two years earlier.

[68] Samuel Allinson to his uncle, June 26, 1761. Caroline Allinson Collection.
[69] Samuel Allinson to his uncle, Dec. 5, 1762. Caroline Allinson Collection.

Ten: Councillor

UNDER his royal commission of 1758, Charles Read was seated in the council when it convened at Perth Amboy, March 31, 1761. Meanwhile, George II had been succeeded by George III, and on June 25 following, new commissions were granted the members of the council under the new sovereign. When the council reconvened December first, Governor Hardy administered the oath of office to which Read again subscribed.[1]

About this time Read was brought into a controversy over a supreme court appointment. Near the close of Governor Bernard's administration Nathaniel Jones, a London attorney, presented to the governor a royal commission to the chief justiceship. Five years earlier Robert Hunter Morris, when appointed governor of Pennsylvania, had written the Board of Trade that he wished to resign the chief justiceship, but his letter was not acknowledged. Now that his Pennsylvania assignment had expired, he insisted upon remaining in the office of chief justice. Thus the governor found himself in a quandary. Morris requested Charles Read and David Ogden, as "two of the ablest and most experienced members of the Governor's Council," to represent him when the matter came up for judicial decision.[2] The outcome was that Morris remained in office and Jones returned to England.

The commissioning of William Franklin as governor of New Jersey followed shortly after Josiah Hardy's recall by the London authorities. Franklin had accompanied his father to London, had studied law at the Middle Temple, had received an honorary Master of Arts at Oxford, and had made himself agreeable in official circles. His appointment, ap-

---

[1] *N. J. Arch.*, 1st ser., XVII, 234, 269; also James Munro, ed., *Acts of the Privy Council of England, Colonial Series*, IV, 1745-1766 (London, 1911), 791.
[2] William Nelson, *New Jersey Biographical and Genealogical Notes. N. J. Hist. Soc. Col.*, IX (1916), 145. See also D. L. Kemmerer, *op. cit.*, pp. 267-270.

parently made without any solicitation by either father or son, may have been designed by the ministry to make his father more tractable to the governor and proprietors of Pennsylvania. In any event, it came as a surprise to the people back home, and was not altogether pleasing to the proprietors. Although personally popular, he was the victim of prejudice because of the humble origin of his father, who never quite succeeded in making himself socially acceptable to the provincial gentry; and there was the added reason of William's alleged illegitimacy. It is reasonable to believe, however, that William Franklin's appointment was not objectionable to Charles Read, with whom there was, reputedly, a family connection. Read may even have shared the secret of the governor's birth. There is good reason to believe that Deborah Read herself was William's mother.[3] At any rate, for another period of years Read continued the same close relationship with the chief executive he had enjoyed through Belcher's long administration.

Franklin's appointment came after several years of war and of artificial prosperity, which was followed by a period of financial depression and growing dissatisfaction with royal domination. New Jersey's population was growing apace. From 61,000 in 1745 the number had risen to 80,000 in 1754, and was to reach approximately 115,000 by 1772.[4] By 1760 Trenton, Perth Amboy, and New Brunswick were described as towns of one hundred houses each, and Elizabethtown as having between two hundred and three hundred houses. Burlington was still the capital of the Western Division, and Newark and Princeton were towns of growing importance. New Jersey farms were producing "vast quantities of grain, besides hemp, flax, hay, Indian corn, and other articles."[5]

William Franklin arrived early in 1763, established his residence at Burlington, and began an administration of thirteen

[3] Charles H. Hart, "Who was the Mother of Franklin's Son?" *Penn. Mag. Hist. and Biog.*, XXXV (1911), 308-314.
[4] *N. J. Arch.*, 1st ser., VIII, pt. 2, p. 84; X, 445-446.
[5] A. Burnaby, *op. cit.*, p. 58.

years' duration which was destined to witness the end of royal dominion in the province. To celebrate the peace of Paris in 1763 the governor issued a proclamation, signed by Read as secretary, appointing August 25 "a Day of Public Prayer and Thanksgiving" for peace.[6] But it was to be only a temporary peace. The close of the French and Indian War marked the beginning of a still more critical period in the life of the province, particularly in its relations with the mother country. The power and the importance of the colonies, and their value to Britain, had for the first time been clearly demonstrated during the war. The British now realized as never before the promise of vast potential riches to be derived from trade in American staples, and with this realization grew the desire for greater control. By the same token the colonies began to appreciate their own strength, and their ability to get along without maternal aid. Regulations and restrictions from abroad served but to foster the spirit of independence. These issues came to a head in New Jersey while William Franklin was governor. His was a trying administration. He chose the unpopular side, remaining loyal to the Crown, but he proved to be one of the ablest of colonial New Jersey's executives. Except during the later years of his regime, Charles Read was at his elbow as councillor and advisor.

Whereas for a decade while a member of the assembly Read had often been delegated "Messenger of the House" to take official word to the council, now as a member of the council the process was reversed, and he was named frequently as "Messenger to the House." Having served one term as speaker of the assembly, and another as a leader without title, he could be a particularly valuable liaison officer between the two legislative houses. The council had to consider all legislation, but it was more than the upper house of the legislature, or senate: it served also as cabinet, court of appeal, vice-president, and buffer between the governor and the people.

[6] *N. J. Arch.*, 1st ser., XXIV, 219-221.

Among Read's associates in the council were a number of extraordinary men. Robert Hunter Morris was president of the council from 1758 to 1764. He was succeeded in the presidency by Peter Kemble, of Morristown, who had served in the council since the beginning of Governor Belcher's administration. Richard Saltar (1749-62),[7] Lord Stirling (1761-75) and John Stevens (1763-75) were among Read's close colleagues in the council. Other councillors from the Eastern Division included Stephen Skinner (1770-75), treasurer of East New Jersey; also Lewis Morris Ashfield (1753-65), David Ogden (1752-75), James Parker (1765-75), Frederick Smyth 1765-75), Richard Stockton (1769-75), and Samuel Woodruff (1757-68). From the Western Division besides Read were John Smith (1762-71) and Samuel Smith (1767-75), the latter for many years (1750-75) treasurer of the Western Division. Also there were John Ladd (1763-70), Daniel Coxe (1771-75), and John Lawrence (1771-75). During these years the Eastern Division had an advantage in numbers over the Western Division, but not in qualities of statesmanship. Governor Bernard after leaving New Jersey referred to "the Friends" of Burlington as "some of the best men I have met with in America."[8]

Illness during the 1760's occasionally prevented Read from sitting with the council; but he was present at more than two-thirds of the meetings. Time and again he served on legislative committess which had to deal with the ever-pressing problem of finances. In 1761 he was named to a committee "to inspect the Accounts in the Eastern Treasury and burn the cancelled Money."[9] Again in 1772 he was assigned to a committee which met at Perth Amboy to inspect bills of credit issued in 1737, and to burn the cancelled bills.[10] At other times he was a member of committees to consider acts

---

[7] The dates which follow the names in this paragraph indicate the term the respective members served in the council.
[8] Francis Bernard to Charles Read, June 1, 1765. Bernard Papers, IV, 51, 52.
[9] *N. J. Arch.*, 1st ser., XVII, 248.
[10] *Ibid.*, XVIII, 341.

"for the support of the government"—the all-important appropriations bills. When the question of additional issues of paper money arose, his counsel was sought repeatedly. That the finances of the colony, on the whole, were well handled is greatly to Read's credit. Benjamin Franklin on appearing before the Board of Trade in 1767 took occasion to mention New Jersey's excellent monetary record.[11]

Continued restrictions upon paper-money issues and duties levied upon trade without the consent of the colonies fed the growing resentment against royal authority. The issue between the prerogatives of the Crown and the rights of the colonists was being more sharply drawn. Representing popular sentiment in New Jersey, Charles Read, together with Samuel Smith and Jacob Spicer, comprised a "Committee of Correspondents" which in a letter dated September 10, 1764, directed Joseph Sherwood, who in 1760 had succeeded Partridge as provincial agent in London, to point out to the authorities how the policy of the mother government was impeding the development of the colony:

If anything comes on the Stage next Session of Parliament either for repealing the Duties laid on the Trade of the Northern Colonies and prohibiting a paper currency at last Session, or for adding any thing new by way of Tax on this Colony, the Committee of Correspondents direct that you will humbly and Dutifully Set forth, in the name and on Behalf of this Colony that we look upon all Taxes laid upon us without our Consent as a fundamental infringement of the Rights and Privileges Secured to us as English Subjects; and by charter. And that our paper Currency hath always kept its value [torn] and being prohibited from having any more but upon Terms of not being Legal tender let the Necessity be ever so pressing we esteem a very great provincial hardship, for these among other Important Reasons that it will not only cramp us in our Business with one another & the English Merchants, but Impede the Growth of the Settlement of this province, and very probably greatly obstruct his Majesty's

[11] William Franklin to Benjamin Franklin, June 10, 1767, *N. J. Arch.*, 1st ser., IX, 625.

Service upon any future emergency You are Desired to enlarge and make the necessary proofs as Occasion may require.

The more active and Expensive part of the Opposition we expect will lie upon the other Colonies who are abundantly more Concerned in Trade, yet it is necessary so far to cooperate with them as to Show the Colonies are unanimously of One Mind.[12]

The breach between the executive policy and legislative will was still further widened the following spring when word arrived of the passage of the Stamp Act. Scarcely a less opportune time could have been chosen by Parliament to levy taxes upon the colonists, suffering as they were from the depression which followed the French and Indian War. New Jersey sent three delegates to the Stamp-Act Congress which met in New York, October, 1765, among them Robert Ogden, speaker of the assembly. Ogden, though opposed to the principle of the Stamp Act, declined to sign the address of the congress to the Crown, believing each colony should take action individually. His refusal, misconstrued as siding with the Crown, met with a violent wave of resentment. He was burned in effigy and was forced to resign the speakership.

The Stamp Act, unwelcome "mother of mischief," dumped a parcel of knotty problems on the doorstep of the governor. He had to devise ways and means of receiving the stamps, of appointing a distributor, of arranging for their use, and of dealing with the rising opposition. It fell to Read's lot to handle some of the official communications about these delicate matters, among them a public statement in 1765 explaining why Governor Franklin had refused to summon the assembly to consider the Stamp Act.[13]

News of the repeal of the Stamp Act, which reached Burlington May 24, 1766, was received with great rejoicing. "The city was handsomely illuminated; Bonfires were lighted, and other Demonstrations of Joy shewn . . . An elegant Enter-

---

[12] Committee of Correspondents for West New Jersey to Joseph Sherwood, Sept. 10, 1764. Spicer Papers.
[13] *Pennsylvania Journal*, no. 1191, Oct. 3, 1765; no. 1242, Sept. 25, 1766; *N. J. Arch.*, 1st ser., XXIV, 640.

tainment was prepared . . . at which his Excellency the Governor and the principal Inhabitants of the Place were present."[14] In all probability Charles Read was among "the principal Inhabitants," who joined in the celebration and drank to the health of *The King, The Queen, and Royal Family*. Among the eighteen toasts which were drunk was one to *Doctor Franklin*; and the concluding one, *Increase to the Manufactures of Britain, and Prosperity to the Agriculture of America,* might properly have been proposed by Read himself.

Although the Stamp-Act repeal heartened the colonial officials, it brought no solution for other problems which perplexed the governor and council. New customs duties laid under the Townshend acts met with non-importation agreements and a virtual boycott of British manufactured goods which led to the repeal of the taxes on all imports except tea. In 1769 an issue of £100,000 in paper currency was proposed as the reasonable way for New Jersey to provide the funds needed for government expenses, but it was disallowed in London. All this did not lighten the troubles of Governor Franklin, who like Governor Belcher, had the worry of a meagre and uncertain salary. He received no fixed stipend from London; he was dependent upon the assembly for a yearly appropriation. In 1763 his initial salary of £1,000 a year had been increased to £1,200. This, however, proved not sufficient to meet his expenses and his private fortune continued to dwindle during his incumbency. In 1773 he appealed to the Crown that he be allowed an increase in salary or transferred to a more lucrative post, all to no avail.[15]

Social and economic discord within the province added to the spirit of unrest. In the depression thousands had suffered financial ruin, and as a result efforts were made to effect legislation which would bring relief to insolvent debtors,

[14] *Pennsylvania Gazette,* no. 1953, May 29, 1766; *N. J. Arch.,* 1st ser., XXV, 122.
[15] William Franklin to the Earl of Dartmouth, Jan. 5, 1773; *N. J. Arch.,* 1st ser., X, 393.

many of whom were imprisoned because they could not meet their financial obligations.

In this situation a smouldering resentment against lawyers for charging excessive legal fees came to a head. An English writer, noting that continued disputes and lawsuits over uncertain land titles had retarded the development of the province, remarked that there were "no men growing rich here so fast as the gentlemen of the law."[16] In suits brought to court in Monmouth County, popular opposition to the lawyers who represented the creditors developed into riots and mob attacks upon the court. The governor in 1770 summoned a special session of the assembly to deal with the matter. The assembly, naturally favoring the masses in preference to the lawyers, passed an act to provide a more effectual remedy against excessive legal fees, whereupon the court riots ceased.[17] Because of his large property holdings as well as his standing as attorney, jurist, and councillor, Read had both a private and an official interest in the whole problem. He was also concerned with special legislation to exempt certain individuals from imprisonment for debt.

Anxiety over the troubled relations with the mother country was temporarily diverted in July, 1768, when the treasury of the Eastern Division was robbed of more than £7,000. The incident caused great consternation in official circles. The assembly was disposed to hold the treasurer, Stephen Skinner, responsible for the loss. The governor, on the other hand, declared that Skinner was the victim of circumstances and refused to remove him, a stand which received the backing of the council. The controversy over the issue between the governor and the assembly was kept alive for several years. In 1772 Read served on a special committee with Lord Stirling, John Stevens, and Daniel Coxe to confer on the governor's message concerning the robbery, but the mystery was never solved.[18]

[16] R. Rogers, *op. cit.*, p. 76.
[17] *N. J. Session Laws* (March, 1770), Chap. IV, pp. 7-14.
[18] *N. J. Arch.*, 1st ser., XVIII, 341.

Though not a violent partisan, Read's sympathies inclined definitely toward the interests of the colonists. Years before, he had seen the necessity for ultimate union of the colonies. Jacob Spicer in 1755 recorded a conversation with Read which revealed this conviction. After commenting on the gloomy prospect of getting approval of an issue of paper money, he added: "Mr. Read appeared also to be of opinion that the plan of union or something of a Similar Nature would Take Effect for how cou'd it be otherwise said he."[19] In his judicial opinion on the office of chancellor, discussed in a later chapter, he indicated the principle of self-determination in advocating the adaptation of government to local conditions. When some of the colonial newspapers appeared in mourning as a protest against the Stamp Act, Read mingled humor with sentiment by presenting James Parker with a pair of wooden shoes "as a proper Badge of the slavery the Stamp Act must soon reduce all Printers to." "I shall wear them sometimes," wrote Parker in describing the gift to Benjamin Franklin, "for the sake of contemplating on the changes of Fortune's Wheel."[20]

The spring of 1767 marked a sharp reversal of Read's political fortune. His monopoly of the several appointive offices he had held for more than twenty years, was broken. The previous year Christopher Coates, the secretary under whom Read held the post of deputy, had died and the Honorable Maurice Morgan had been appointed to the secretaryship. Joseph Sherwood wrote from London in November, 1766, that Morgan was receiving offers for the deputyship, "but he has Declined Entering into any Engagement, and will give Charles Read the Preference of Continuation, Provided his offers are any thing near what is offered by others."[21]

In the meantime young Joseph Reed, Trenton-born and Philadelphia-reared, graduate of the College of New Jersey

[19] *J. Spicer's Diary*, Apr. 22, 1755.
[20] James Parker to Benjamin Franklin, Apr. 25, 1765. A. H. Smyth, *op. cit.*, X, 223.
[21] Joseph Sherwood to Samuel Smith, Nov. 18, 1766, *N. J. Hist. Soc. Proc.*, 1st ser., V (1850), 153.

in 1757, and subsequently to become Washington's adjutant-general, upon being admitted to the bar in 1763 had gone to London for further study in the Middle Temple.[22] During two years abroad he had made acquaintance in official circles and apparently had won the favor of Morgan. Returning to New Jersey he settled in Trenton and began the practice of law. Just what political maneuvering took place is not clear, but it seems he outbid Read, who early in June, 1767, had word that he would lose the secretary's office and perhaps that of registrar as well. This was no mean blow to prestige and pride, and on June 6 he wrote about it to former Governor Bernard, then in Massachusetts. In reply to his inquiry as to whether or not the office of registrar to which Bernard had appointed him in 1759, was embraced in that of secretary, the former governor gave the following opinion:

> I am glad to hear of yr Success in yr Iron Mine. I hope it will compensate yr dissapointmt in the Secry's Office . . . When I granted to you the Office of Registrar, &c, I did not know it (the Office of Regr) was included in the Patent of Secry . . . If the Commn of Sec'ry does not mention the Office of Registrar, &c wch upon a review of yr letter seems to be yr meaning, I think it will not include yt of Registrar. But I wd advise you by all means, if possible, to get a new grant from yr Govr.[23]

Before this message was received, however, on June 18 Mr. Morgan had been granted "the Offices and Places of Secretary, Clerk of the Council, Clerk of the Supreme Court, Clerk of the pleas, Surrogate and Keeper and Registrar of the Records in the colony of New Jersey . . . during Pleasure." And on June 27, Mr. Morgan farmed out these offices by having young Joseph Reed appointed deputy to the several posts.[24]

Justice Read doubtless sensed the irony in the situation when, on October 10, the new deputy appeared before him to take the oath "for the due Execution of the Offices." To

---

[22] *N. J. Arch.*, 1st ser., X, 5-6.
[23] Francis Bernard to Charles Read, June 20, 1767. Bernard Papers, V, 232-234.
[24] *N. J. Arch.*, 1st ser., X, 1-7.

administer the oath to his unwelcome successor must have been anything but a pleasant duty for the political veteran. He did not shed the mantle altogether gracefully. It was difficult to see the offices slip from his hands, with all the power and prestige, not to mention the income, that went with them. He must satisfy himself that there was no recourse—no door left open to recover at least a part of the fruits of office.

Were the offices of secretary, registrar, and surrogate wholly distinct? Would a commission granted by one governor be valid during the term of his successor? He wrote to London, it seems, for an official ruling. Also, on October 24 he again wrote to Governor Bernard for a further opinion. Bernard was sympathetic, declaring he thought Read had been "hardly used, in having no Allowance for so many Years faithful Service against the hasty Offers of a new Bidder." He was not sure that a commission granted by one governor would be binding upon his successor, but was inclined to believe it would be. He was sure, however, the office of surrogate was independent of that of secretary, and added that had he understood the office of registrar also to be "different & distinct," he would not have made that grant to Read while he was governor of New Jersey.[25]

A strange turn of affairs, this, for the man who over twenty years had done "most of the State business" and had done it "in his own way." The supreme court justice, supposedly an expert in the law and its interpretation, found himself inquiring into the functions and relationships of the several government offices, in order to discover ground for keeping himself in them. The political boss had become the humble suppliant.

When the council met at Burlington November 13, Joseph Reed reported and took the oath of secrecy as deputy clerk of the council. The governor then apprised the council that the young man as deputy secretary and registrar of the province had claimed custody of the seals and records of the preroga-

[25] Francis Bernard to Charles Read, Dec. 8, 1767. Bernard Papers, V, 248-250.

tive court, but that Councillor Charles Read also claimed the office of registrar of the court. In the dilemma the governor sought the council's advice. The claims of the contending parties were heard, and the council took the matter under advisement. The next day they recommended that the governor order Charles Read to deliver the court records to the new deputy secretary, reserving to Read the right of further prosecuting his claim if he thought it expedient.[26]

Apropos of Governor Bernard's letter, the question of Morgan's power to deputize the duties of surrogate also was pressed upon Governor Franklin. To place the matter above dispute, the governor commissioned Joseph Reed "Provincial and Principal Surrogate of the Province of New Jersey," voiding all former surrogates' commissions, so long as he should continue deputy to Morgan, but holding him responsible to the governor for the seals affixed in the office.[27] From this record, it appears that Franklin made little or no effort to keep Charles Read in the coveted offices. It was apparently in connection with closing out the deputy-secretaryship account handled by Sherwood that Governor Franklin wrote Read the following June:

> My Father has, I suppose, left England by this Time.—He writes me that he has lately recd. Nine Pounds 19s & 9d. being the Ballance of Mr. Sherwood's Acct. with you, which he desires me to pay you: you will therefore charge me with that Sum. The Acct. is enclos'd. I should be glad to have your Acct. with me settled as soon as you conveniently can.[28]

It developed that Joseph Reed did not give the offices the same personal attention they had received from his predecessor. This was not altogether pleasing to the governor, but he adhered to the policy of non-interference. In 1769 Governor Franklin wrote from Burlington to his father then in London:

> He [Joseph Reed] never comes here but at the time of the

[26] *N. J. Arch.*, 1st ser., XVII, 457-458.
[27] *Ibid.*, X, 8-10.
[28] William Franklin to Charles Read, June 13, 1768; *N. J. Arch.*, 1st ser., X, 28-29.

Courts, leaving his Business of Secretary entirely to Clerks, both here and at Amboy. Mr. Morgan intimates as if he had a design of changing his Deputy, but it is a matter I don't choose to interfere in; all that I shall desire is, that whoever he appoints may be obliged to reside here, and may be properly qualified to execute the Business.[29]

In October, 1769, Deputy Secretary Reed retired in favor of his brother-in-law, Charles Pettit, who had been associated with him in the duties of his several offices.[30]

Divergence of opinion on the colonial policy of the mother country in a measure may account for the apparent weakening of Franklin's support of Read. Nevertheless, Franklin seems to have retained a large measure of confidence in him. In 1768 the governor, in order that public affairs might not suffer by his absence from the province, commissioned Charles Read, Samuel Smith, and John Smith as custodians of the royal seals.[31] Also, Read and the governor surely were in accord on matters of internal improvements. The governor shared Read's interest in agriculture. Like Governor Belcher he was a Burlington country gentleman. On the north bank of the Rancocas Creek, about five miles southeast of Burlington, he bought a 600-acre estate and in 1769 wrote his father in England that he had "entered far into the spirit of Farming."[32]

Doubtless his personal interest in agriculture, as well as the advice of Read, and also of Lord Stirling[33] who had been

[29] William Franklin to Benjamin Franklin, May 11, 1769; *N. J. Arch.*, 1st ser., X, 114.
[30] E. J. Fisher, *op. cit.*, p. 46; N. J. Commissions, Liber AB, 37.
[31] *N. J. Commissions*, Liber AB, 23; *N. J. Arch.*, 1st ser., X, 54-55.
[32] William Franklin to Benjamin Franklin, May 11, 1769; *N. J. Arch.*, 1st ser., X, 113.
[33] Presiding over his estate at Basking Ridge, Lord Stirling was another of those distinguished country gentlemen who found pleasure and satisfaction in transforming native woodlands into fertile fields. To the Earl of Shelbourne, he wrote in 1763: "The making of pig and bar iron, and the cultivation of hemp, are two articles that want encouragement greatly. We are capable of supplying Great Britain with both, to a great extent; but the first requiring a large stock to begin with, people of moderate fortunes cannot engage in it; and those of large ones are as yet very few, and their attention is generally given to the pursuit of other objects. Some few, indeed, in this Province [New York], and in New Jersey, have

urging the encouragement of hemp culture and iron manufacture, prompted Franklin to sponsor measures for the development of agriculture and industry. By 1765 he could write to the Lords of Trade: "I prevailed on the Assembly to grant some Bounties to encourage the raising Hemp & Flax, & the Culture of Silk for Exportation to Great Britain."[34]

The message from the assembly, signed by Speaker Ogden, which apprised the governor of the passage of the act, revealed the underlying purpose of the measure and the ends it was designed to accomplish:

Agreeable to your Excellency's Recommendation, we have adopted the Measure of giving Bounties on Hemp, Flax, and planting of Mulberry Trees, with a View of stimulating our Inhabitants to future Industry and Wealth, in a Way hitherto but little used in this Government, but which, notwithstanding, appears to us to carry a Probability of succeeding, and being in Time considerable in the Article of Remittance to the Mother Country. On this Occasion our Acknowledgments are justly due to the Society for Arts, Manufactures and Commerce, on Account of their disinterested and benevolent Care respecting Improvements in the Colonies.[35]

Although the results of the bounties, in so far as they stimu-

lately erected excellent works, the success of which, I hope, will encourage others to follow their example. As to hemp, our farmers have got into a beaten track of raising grain and grazing cattle, and there is no persuading them out of it, unless by examples and premiums; and these it would be well for Government to try—a few thousand pounds expended in that way might have a good effect."
After describing his experimental vineyards, he concluded: "It is in these vineyards, my Lord, and the clearing a large body of rich swamp lands in New Jersey, and fitting it for the cultivation of hemp—settling a good farm in the wilderness, and bringing to it some of the productions and improvements of Europe, that are my present employments. They have taken place of the pleasures of London, and I sometimes persuade myself that this is the happier life of the two." (William A. Duer, *The Life of William Alexander, Earl of Stirling*, N. J. Hist. Soc. Col., II, New York, 1847, pp. 75-77). As an example of his interest in rare varieties of crops, in 1767, Stirling procured "a small parcel" of Morocco wheat through Gerard Bancker who wrote from New York: "It is said to grow different from our wheat—that the Stalk has a pith in it. It is in England preferred to the Sicily Wheat." (Gerard Bancker to Lord Stirling, Dec. 11, 1767. Correspondence of Lord Stirling, 1765-67.) (See also p. 264.)

[34] William Franklin to the Lords of Trade, Aug. 8, 1765; *N. J. Arch.*, 1st ser., IX, 491.

[35] *N. J. Arch.*, 1st ser., XXIV, 554-556.

Councillor 159

lated production, were disappointing, the enactment of the law represented an achievement for the governor in dealing with an assembly opposed on general principles to public expenditures which meant higher taxes. Franklin found himself handicapped in carrying out his plans for public improvements, as the assembly refused to appropriate funds, while the Crown would not approve new issues of currency. When some years later the governor pointed out that New Jersey's taxes were very low compared with those in other colonies and in England, Assemblymen Theunis Dye, of Bergen County, and Robert F. Price, of Gloucester, delivered a reply which vigorously took issue with His Excellency and presented the following gloomy picture of the economic status of agriculture in the province:

> Lands have continued and still do continue to sink in Price, and are sometimes sold for less than one third Part of the Value they were sold for a few Years ago.
> The high Price of Wheat is owing in Part to there not being enough to Supply the Demand, occasioned by the Failure of the Crops, the Consequence of the Land being much worn, and the Badness of the Seasons; so that a Farmer notwithstanding this high Price does not get as much now for his Year's Labour as he formerly did when he had a full Crop. But there are not one fourth Part of the Housholders in New-Jersey that raise Wheat to sell, most of the Rest buy that necessary Article; and the high price that Bread Corn, and other Provisions now sell at has reduced large Numbers of the Inhabitants of this Colony to great Distress, and is very sensibly felt by many more.[36]

In discussing the status of the provincial treasury later in the message, it was estimated that above £2,000 was "liable to be drawn by the Eastern Proprietors Bounties on Hemp and Flax."[37]

Franklin argued that New Jersey was backward, that long neglected public improvements must be promoted. "This Province," he said, "which has equal Advantages with any

[36] *Ibid.*, X, 251-256.
[37] *Ibid.*

of the Neighbouring Colonies, is, tho' one of the most Antient, shamefully behind all the others in its Trade, Roads, Bridges, publick Buildings, and such other Improvements as denote a Sensible and Spirited people."[38] Though falling short of his goal, Franklin did effect a measure of progress in internal improvements, in which Charles Read gave substantial aid. After all, the people of New Jersey were mainly concerned with domestic affairs, such as the taking up of lands, the building of houses and barns, the damming of streams, the erection of mills, and the building of roads and bridges. A lottery was invoked in 1765 as a means of raising funds to improve the highway from New York to Philadelphia,[39] and the following year, a two-day trip between the cities was advertised. With further improvements the time was reduced by 1771 to a day and one-half, a speed so great that the stage coaches were advertised as "flying machines."[40]

Although from the Twentieth-Century point of view colonial travel in both mode and speed seems to have been highly primitive, the progress during the period of Read's public career was substantial. The King's highways and branch roads as well were being developed, well supplied at crossroads and ferries with picturesque taverns for the accommodation of hungry and thirsty travelers. Because of her geographical position New Jersey enjoyed a greater amount of intercolonial travel than any other part of America. She was among the first to witness the passing of the frontier. By the middle of the Eighteenth Century she probably had more ferries in operation than any other colony. She took the lead in the development of the wheeled vehicle from crude "Jersey waggon" to stage coach; in the speed of travel on through trips; and in the general diffusion of stage lines. Such improvement in transportation did much to knit together into a unified commonwealth the diverse racial and social units which comprised New Jersey. It effected wider tolerance and

[38] *Ibid.*, XVIII, 299.
[39] *Allinson's Laws*, pp. 273-275.
[40] W. J. Lane, *op. cit.*, pp. 86-87.

a greater interest in affairs beyond the immediate community. It did much to consolidate East New Jersey and West New Jersey into a compact political unit by breaking down the sense of provincialism and leveling off the differences in the cultures of the two divisions.[41]

Read's committee assignments in the council indicate the diverse activities in which he was engaged. He was often appointed to committees having to do with the bridging of creeks, the draining of swamps, and the banking of meadows. At other times he was involved in legislation relating to the navigation of rivers and creeks, the ownership of salt marsh, proprietors' estates, and land claims. Other committee assignments were concerned with road work in Gloucester County and the clearing of obstructions from the Passaic River. Again, he was on a committee to review a bill which would provide for a new courthouse in Middlesex County.

Other services resulted from his efforts. On his initiative a bill was passed for collecting and securing books of mortgages formerly kept by loan offices in the several counties. Several times he served on committees dealing with the naturalizing of citizens. Because of his familiarity with the law, he was a logical person to consider the status of acts of the assembly passed at previous sessions, and to make recommendations as to their expiration, renewal, or amendment. Thus in the council, even as he had done in the assembly, he rendered an important service in the clarification and revision of the laws.

Franklin's administration witnessed another step in the advancement of education. On November 10, 1766, while Read was a member of the council, the governor signed a royal charter establishing Queen's College, forerunner of Rutgers University. Just as Governor Belcher had negotiated a royal charter for a group of Presbyterians for the founding of the College of New Jersey, subsequently Princeton University, so now Governor Franklin was the sponsor of a royal charter

[41] *Ibid.*, pp. 44, 55, 93.

granted to a group of the Dutch Reformed faith for the founding of a college at New Brunswick.

When minor defects were discovered in the charter of 1766, a petition was presented to the governor by Hendrick Fisher, president of the Board of Trustees of Queen's College, that the charter be amended. Read was present at the meeting on November 24, 1769, when the governor laid the petition before the council, which "advised His Excellency to grant the Prayer of the said Petition, so far as Relates to the Distinction of Residents and Nonresidents."[42] Out of this action by the council came the second charter granted March 20, 1770, for the founding of a college "to promote learning for the benefit of the community" by "the education of youth in the learned languages, liberal and useful arts and sciences."[43]

Thus to New Jersey came the honor of being the only colony to have more than one college, and to Charles Read possibly belongs the distinction of being the only person to have had an official role in the founding of two of America's colonial colleges.

It may not be more than a coincidence, but in the traditional struggle between the assembly and the council, the alternating ascendancy between the two bodies corresponds approximately with Read's membership in them. During the decade of the 1750's, most of which Read spent in the assembly, the lower house, as we have seen, gained substantially in prestige and influence. After 1760, however, particularly under Franklin's administration, the council of which Read was now a member, became increasingly powerful. In the important local issues which arose between council and assembly under Franklin—the assembly's sole control of the provincial agent, the Crown's payment of the patent officer, the enactment of a satisfactory loan-office law, and the method

[42] *N. J. Arch.*, 1st ser., XVIII, 24.
[43] "Rutgers University—Federal and State Relations," *Rutgers University Bulletin*, ser. V, no. 1 (1928), pp. 9-10. It is under this charter that Rutgers University still operates. The first charter was probably surrendered when the new one was granted; all trace of it has been lost, and no copy is known to exist. It may have been destroyed when the deposed governor made his hasty exit from New Jersey in 1776.

of granting barracks supplies—the council prevailed over the assembly. In the question of the removal of the treasurer the assembly barely held its ground.[44] It would be claiming too much to attribute the increased prestige of the council to Read's presence. It is doubtful that he added greatly to its strength, especially after 1766. Rather it seems the assembly was weakened by the loss of his leadership when he went over to the council.

At no point have we found Read involved in a partisan way in the controversy over the growing encroachment of the Crown. He was too much of the diplomat and practical politician to align himself irrevocably with a warring faction. Had he lived until 1776, he would have had to declare himself one way or the other, but he left the scene before matters reached a crisis. By showing an appreciation of all points of view, he was best able to serve the true interests of the province.

[44] D. L. Kemmerer, *op. cit.*, p. 315.

Eleven: Colonel

UNLIKE the Revolution, the colonial wars brought little bloodshed to New Jersey soil; nevertheless they did interrupt the course of economic and political life in the province, and occasioned considerable military activity. Three times during Read's career New Jerseymen were called upon to join in protecting the American colonies against England's enemies and their Indian allies. Each time he was involved in an official way.

The time of his coming to New Jersey coincided with the outbreak of hostilities with Spain in the third inter-colonial war in 1739. New Jersey was asked to contribute troops and funds toward a proposed naval attack on the Spanish West Indies. Read was recommended by the council for a commission as major, and in the spring of 1740, by proclamation of Governor Morris, he was appointed to enlist the militia in Burlington County.[1] It is doubtful that he actually joined the expedition which, when it finally started in the fall, was doomed to failure.

When four years later war was declared against France, Governor Morris summoned the assembly to make provision for attack against the French and the Indians; but little was accomplished. Notwithstanding considerable activity across the river at Philadelphia, where privateers were being built and fitted out to press the conflict at sea, it was difficult to engender in a Quaker-dominated assembly any enthusiasm for a war on distant territory which New Jersey had no part in starting. Upon the death of Governor Morris in 1746, Acting Governor Hamilton summoned the assembly to consider aid for an overland invasion of Canada then being planned. This time the assembly acted; it voted £10,000 in bills of credit to finance the raising of five companies of 100

[1] *N. J. Arch.*, 1st ser., XV, 104, 117.

men each. Six commissioners—three from the Eastern Division and three from the Western—were appointed to purchase the necessary provisions, and six others were named "for the Arming and Cloathing the Forces."[2] Charles Read was one of the latter group. Other commissioners from the Western Division were his associates Jacob Spicer and Thomas Shaw. For their services the commissioners were to receive 5 per cent of the value of the supplies handled. Colonel Peter Schuyler was placed in command of the New Jersey forces.

The story of the expedition is one of gross mismanagement, criminal negligence, and corruption, only partially offset by the high-minded integrity and liberal spirit of Colonel Schuyler. When after months of delay the troops finally set off for Albany, they were under-fed, under-clothed, and ill-equipped. The guns, which cost 30s. each, were "so rusted and rotten as not to be of the value of old iron."[3] The cutlasses, costing 9s. each, could be bent like lead; the corned beef was so bad that barrels of it were condemned. Some wits suggested that the arms were purposely incapable of killing the enemy, in order that the Quaker commissioners from West New Jersey might have easy consciences. While the army complained of the poor supplies, the public protested at open graft. Why, asked the critics, should the provincial government pay £30 a ton for transporting supplies from New Jersey to Albany, when they were being taken from New York for £10?[4]

While waiting through the long winter at Albany for the British forces to arrive, the New Jersey troops became restive; many deserted, and the remainder threatened mutiny unless arrears of pay were forthcoming. Trouble was averted only when Colonel Schuyler paid the troops out of his own pocket —"upward to £4,000" in all.[5] For this patriotic act he received the reprimand of Governor Clinton of New York, on the

[2] *Nevill's Laws*, I, 313-331.
[3] Joseph F. Folsom, "Colonel Peter Schuyler at Albany," *N. J. Hist. Soc. Proc.*, n. s., I (1916), 160-163.
[4] *New York Weekly Post Boy*, Sept. 8, 1746; *New York Gazette Revived in the Weekly Post Boy*, Feb. 16, 1747; *N. J. Arch.*, 1st ser., XII, 322-323, 331-336.
[5] Peter Schuyler to R. H. Morris, May 11, 1747. Boggs Collection.

ground it was a bad precedent for the other troops. The New Jersey assembly, moved to unprecedented generosity, finally reimbursed Colonel Schuyler in the amount of £623.[6] At last the expedition was abandoned, and the British government allowed New Jersey £2,231 as the proper compensation in place of the £5,302 spent by the commissioners. The assembly named a special commission to draw and receive these funds from the Crown. On this commission were three Burlington County assemblymen, all Quakers—Richard and Daniel Smith, and William Cooke, who several years earlier had been opposed to the expansion of the military forces.[7]

To what extent Read himself was responsible for the mismanagement of this expedition cannot be easily determined. The affair, it seems, may be regarded more properly as an example of the inefficiency of commission management, rather than as a reflection upon any individual member.

A hitherto unpublished letter by Read, written in 1748 to Robert Hunter Morris, reported naval activity on the Delaware Bay that caused no little excitement along the New Jersey shore, and even moved to action the latent militaristic spirit of peace-loving Quakers. For example, four years earlier William Hancock, of Salem, had opposed an act for expanding the military force of the province. Now he ordered his sons to take up arms. Of the public reaction Read wrote:

I have been exceedingly busy all Day on a Letter sent by Mr. Gibbon from Salem the People there are under Arms I hear that our Frd. Hancock appeared with Isaac Sharp at ye Council of Warr at Salem town & that Hancock ordered two of His Sons to Arms His Exy. has sent down Comms to Officers recommended by Gibbon for Salem & Cumberland & Cape May is made a Seperate regiment whereof Mr. Young is Collo: no other feild officer is appointed Orders are given for a military Watch in both Counties & also at Cape May. The Salem Express brings ansr. that four French privateers came up as far as Cohanzy last Sunday & lay [at] that Place when He came away [torn] the Spannish bright. who fired into New Castle last week is gone

[6] N. J. Arch., 1st ser., XVI, 77.
[7] E. J. Fisher, op. cit., p. 329.

down the bay. My kinsman Wm. Logan has appeared hearty in fitting out ye privateers at Philada. to attack them I Expect He will be dealt wth The Man of war is heaving down & goes on slowly her men all run away or not to be found.

I heartily rejoice that you are returned to Tinton & hope you are in health. I have your Warrants for Several Sums of Money wch I shall Deliver to your order. There is no money in this Treasury.[8]

The peace of Aix-la-Chapelle in 1748 but laid the seed for future conflict by leaving the western boundaries undetermined. A few years, and hostilities again broke out. The New Jersey assembly declined the invitation to send delegates to the Albany Convention in 1754, at which Benjamin Franklin's scheme of union was proposed. But the new assembly, elected in July of that year, was more favorable to active participation and voted to aid His Majesty in the operations against the French. On January 12, 1755, the assembly ordered Read and Stevens to wait upon His Excellency and advise him that New Jersey could not further augment the troops from the other colonies that were being raised for General Shirley's command. However, at the session of the assembly in February, it was resolved to provide £500 to provision the royal troops should they march through the province, and Read was named on a committee for the purpose.[9] He was also appointed to a committee to bring in a bill which would prohibit exports to the French possessions,[10] a measure subsequently enacted.

As a matter of military expediency, Governor Belcher asked the members of the assembly, when they convened in April, 1755, to take an oath, as the members of the council had done, to keep confidential any plans relating to military operations which might be disclosed to them. The assembly did not oppose the object sought but frankly resented the method chosen. Hence the proposal was rejected by a vote

[8] Charles Read to R. H. Morris, May 31, 1748. Boggs Collection.
[9] *N. J. Assembly Minutes*, Feb. 25, 1755, p. 6.
[10] *Ibid.*, Feb. 26, 1755, p. 7.

of 12 to 8, Read voting with the majority. Perhaps they thought such a step might be construed as an encroachment upon the basic right of freedom of assembly and of speech. However, each assemblyman could certainly be considered on his honor, and it was resolved that if any member should be so devoid of a proper sense of responsibility as to divulge a military secret, he would be called to account, and if found guilty, would be "subject to the severest Censure of the House." Charles Read and Samuel Smith were delegated to wait on the governor and deliver the answer, which he received as "very acceptable."[11]

At this session the assembly caught some of the contagious enthusiasm of Colonel Schuyler, and authorized the enlistment of 500 volunteers under his leadership to join the forces from New England and New York. At the same time they voted an issue of currency to support the venture.[12] But Braddock's defeat and death early in July completely changed the picture. In letters dated July 20 and 23 Read submitted to the governor an official account of the "melancholy and shocking Affair."[13]

The pall of gloom which fell with the news upon the executive chambers quickly gave way to a plan for determined action. Belcher wrote Read at length and instructed him to call the assembly into session to deal with the emergency. He informed Sir John St. Clair, quartermaster-general of the British troops, who had been wounded at Braddock's defeat, that "our Spirits shoud rise with our difficulties," and that the number of troops should be doubled.[14] The smug sense of security which New Jerseymen had been enjoying now disappeared. As long as the forest expanses of Pennsylvania and New York separated them from the French and the hostile Indians, why should they worry about someone else's war?

---

[11] *Ibid.*, Apr. 8, 1755, pp. 1-2.
[12] *Nevill's Laws*, II, 49-63.
[13] Jonathan Belcher to Charles Read, July 25, 1755. Belcher Papers, IX, 331.
[14] Jonathan Belcher to Sir John St. Clair, Sept. 3, 1755; *N. J. Arch.*, 1st ser., VIII, pt. 2, p. 133.

Now they realized the unpleasant truth of Governor Belcher's warning that the province was "in a miserable Defenceless State."[15] Not a fort or blockhouse stood guard at the northern or western frontier.

So preparations went forward. Schuyler's forces had sailed for Albany. As we have seen, Read was at the forefront in the enactment of campaign measures in the assembly. On August 9 he published a proclamation of the governor which placed restrictions on shipping likely to fall into the enemy's hands.[16] The assembly authorized another issue of paper money[17] and mobilization of militia for the protection of the New Jersey frontier was ordered. Read, now commissioned a colonel, headed the regiment of Burlington and Gloucester Counties. On November 6, 1755, a proclamation of the governor directed Colonel Charles Read together with the commanders of eight other regiments to keep their troops in readiness to repel an anticipated attack on the northwest boundary.[18] Within a month he was ready for the field, and so reported to the governor who replied: "I am well pleased at the Accounts you gave me of your having two hundred men of your Regiment in Readiness to march upon my First Orders, & that if necessary, you will go and Command them yourself."[19]

Profiting by the lesson of the Canadian expedition, better attention was given to provisioning the troops. The feeble but indefatigable Belcher wrote to Joseph Yard, one of the commissioners for procuring supplies for the troops, urging all dispatch possible in procuring "for each man a day, a Pound of meat and a pound of Biscuit (or Flower)," in addition to wagons, horses, cannon, and ammunition.[20] Another commissioner of supplies was honest Jacob Spicer, who

[15] Jonathan Belcher to R. H. Morris, Nov. 10, 1755; *N. J. Arch.*, VIII, pt. 2, pp. 158-159.
[16] *N. J. Arch.*, 1st ser., XIX, 529-531.
[17] *N. J. Assembly Minutes*, Aug. 16, 1755, p. 25.
[18] *N. J. Arch.*, 1st ser., VIII, pt. 2, pp. 157-158.
[19] Jonathan Belcher to Charles Read, Dec. 12, 1755; *N. J. Arch.*, 1st ser., VIII, pt. 2, pp. 191-192.
[20] Jonathan Belcher to Joseph Yard, Sept. 27, 1755, *N. J. Arch.*, 1st ser., VIII, pt. 2, p. 140.

traveled from town to town, contracting with butchers, tailors, tanners, and cobblers for the beef, shoes, shirts, and jackets needed to equip an army. "Mr. Ogden," of Elizabeth (presumably Robert Ogden), he found, could put up 98 barrels of beef, at 55s. per barrel. On November 11, Spicer reported "Inspecting the Beef Mr. Ogden is packing for the publick's use," adding that Mr. Ogden had given 3s. "light money" for his barrels.[21]

In December the governor instructed Read as secretary again to summon the assembly to consider the accounts received daily "of the Distress & Danger the Inhabitants of the Frontiers are in from the near approach of the Enemy & of Numbers of them withdrawing from their Habitations."[22] One-half of Colonel Schuyler's forces were recalled from New York and stationed along the Delaware River during the winter to protect the New Jersey frontier.

The fears of the governor and his advisors were well founded, for hostile Indians did actually come down upon the settlements on the west bank of the Delaware and terrorized the scattered settlers, who at their approach fled to New Jersey. Probably only the presence of the troops prevented an invasion of New Jersey soil with its inevitable atrocities. Late in December Belcher wrote that there were "between two & three thousand men traversing & patrolling the whole length of our Frontiers."[23] That Colonel Read saw actual service in the field at this time is doubtful. Although Belcher announced his intention to send Read to the frontier at the head of his detachment and Read indicated he was ready and willing to go, the governor subsequently reported that "the present Face of Affairs does not require putting the Scheme in Practice."[24] Then, too, he doubtless wanted Read to stay close to the executive and legislative chambers.

[21] J. Spicer's Diary, Nov. 11, 1755.
[22] Jonathan Belcher to Charles Read, Dec. 3, 1755; *N. J. Arch.*, 1st ser., VIII, pt. 2, p. 181.
[23] Jonathan Belcher to Richard Partridge, Dec. 20, 1755; *N. J. Arch.*, 1st ser., VIII, pt. 2, p. 197.
[24] Jonathan Belcher to Charles Read, Dec. 12, 1755, *loc. cit.*

Colonel 171

Meeting at Elizabethtown in May, 1756, the assembly decided to push military operations with greater vigor. A new currency issue was proposed, and Read served on a committee which waited on the governor to plead for an extension of the time of maturity. The governor replied that his hands were tied by "the King's Royal orders," but he would not object to the assembly applying directly to His Majesty.[25] A few days later, Read introduced a bill "for the better Regulation of the Forces upon the Frontiers," which was passed, as was another issue of paper money, this time in the amount of £17,500, to mature in the required five-year period.[26]

When at Belcher's direction the assembly convened again in July "on business of Great Importance,"[27] Read was named on the committee to draft an address to the governor.[28] About this time Colonel Schuyler rejoined the expedition in New York State. A month later came the distressing news of Montcalm's capture of Schuyler and his forces at Oswego on August 14, 1756.

After more than a year in captivity at Quebec, Schuyler was permitted a visit home on parole. At New York, at Newark, and at Princeton he was received with great enthusiasm. On his return to Canada he was set free through an exchange of prisoners. While in captivity he had been allowed considerable freedom, and devoted his attention to his fellow prisoners, feeding and housing the women and children captives. When finally released, he purchased a number of these prisoners from the French and Indians "at a very high price," and brought them back to their people. For thus relieving the distress of his fellow countrymen, he spent from his own purse more than £1,333, of which only £211 was refunded by the Crown.[29]

In the spring of 1757 the necessity of another paper money

[25] *N. J. Assembly Minutes*, May 25, 1756, p. 15; May 26, 1756, p. 18.
[26] *Ibid.*, May 28, 1756, pp. 30-31, 33.
[27] Charles Read to Jacob Spicer, July 2, 1756. Spicer Papers.
[28] *N. J. Assembly Minutes*, July 22, 1756, p. 6.
[29] Charles H. Winfield, *History of the County of Hudson, New Jersey* . . . (New York, 1874), pp. 540-541.

issue became evident. True to form, Read was named to the committee to prepare a bill which would provide £60,000 and to draft a petition to His Majesty to accompany it. In praying that the King would give his assent to the proposed money issue, the committee pointed out how paper currency had contributed to the value of lands, and had been otherwise beneficial to the colony. Its proposed use for furthering the military campaign was urged as additional reason for the bill.[30] No issue in so large an amount was approved; but at a session the following October Read helped to prepare a bill for an issue of £30,000, which was enacted into law.[31] He likewise introduced a bill which was enacted at the March session for raising the New Jersey regiment to its full strength of 500 men.[32]

In spite of the assembly's best efforts to maintain adequate forces in the field, Governor Belcher was not satisfied and repeatedly goaded the house to greater activity. Lord Loudon asked that a portion of the militia stand ready to reinforce the regiment in the Albany campaign. Read was in favor of sending the militia, but the assembly refused by a vote of 10 to 9, one of the rare instances when Read voted with the minority.[33] Read had been asked in 1754, as a member of a special committee, to examine the militia act, and had then recommended no change.[34] It was now agreed to amend the act, and, of course, Read was named on the committee to draft a new bill.[35] Messrs. Read and Johnston, duly authorized by the assembly, waited on the governor to explain why more liberal provisions had not been made. In a brief address that was more a rebuke than an apology, they expressed regret that they had been unable to satisfy the governor:

That although this House esteem it their Duty, to pay due

[30] *N. J. Assembly Minutes,* March 17, 1757, p. 7.
[31] *Ibid.,* Oct. 11, 1757, p. 5.
[32] *Ibid.,* March 21, 1757, pp. 12-13.
[33] *Ibid.,* May 27, 1757, p. 7.
[34] *Ibid.,* Oct. 17, 1754, p. 20.
[35] *Ibid.,* August 24, 1757, p. 7.

Deference to his Majesty's Commands; yet they expect to enjoy the Freedom of judging from the Words, how far those Commands extend, without any Comment upon them, and of their own Abilities to execute them.

That it is with the greatest uneasiness the House perceives the utmost Exertion of the Colony falls short of giving Satisfaction, and that they have of late scarce ever been parted with, till they have received his Excellency's Disapprobation of what has been done.[36]

In the raising of the military forces, Read stood firm with the Quakers against conscription. When at the session in March, 1757, Governor Belcher announced that he favored the drafting of men in order to augment New Jersey's troops, the assembly countered with a 12 to 7 declaration that it was "determined not to oblige or compel any of the Inhabitants by Force to serve as Soldiers."[37] At the same time the assembly went on record as opposed to enlisting apprentices or servants.

The war dragged on. The aged fighting Belcher did not live to see its finish, but Read remained in the midst of the war activities. At a session of the assembly in March, 1758, he was appointed to a committee to inspect the accounts of the several expeditions and to bring in a bill for the reimbursement of Colonel Schuyler. The next day, without even waiting for Schuyler to render a full accounting, Read introduced a bill "to supply Colonel Peter Schuyler with the Sum of Six Thousand Pounds," which was promptly passed, and received the approval of the council and the governor. At the same session Read's vote helped to carry another issue of £50,000 in bills of credit.[38] Again, in August, 1758, Read brought in a bill providing a currency issue of £10,000 "for the further defence of the Frontier."[39] In 1759 the assembly provided for 1,000 men under the command of Colonel Schuyler, now ready to return to the war zone in New York.[40]

[36] *Ibid.*, June 3, 1757, p. 18.
[37] *Ibid.*, March 26, 1757, p. 16.
[38] *Nevill's Laws*, II, 187-188; *N. J. Assembly Minutes*, March 28, 1758, p. 9.
[39] *N. J. Assembly Minutes*, Aug. 11, 1758, pp. 34-35.
[40] *Nevill's Laws*, II, 239; E. J. Fisher, *op. cit.*, p. 354.

An act for augmenting the New Jersey forces passed in 1758 named Read as one of a Committee of Correspondence to draw upon the treasury for expenses incurred in equipping troops, in anticipation of promised reimbursement from the mother government.[41] For the committee on March 14, 1760, he laid before the assembly a letter from London announcing that the Lords of the Treasury had allotted £9,166 of the sum appropriated by Parliament to reimburse the colonies in part for their expenditures during the war.[42]

Meanwhile, the presence of royal troops in the principal New Jersey towns had become wholly distasteful to the inhabitants. Because no military quarters were available to house the soldiers they were billeted in private homes. For each foot-soldier quartered, householders received 4d. a day.[43] Such a system was bound to be repugnant to the townspeople and otherwise unsatisfactory. It was beyond the power of the officers to discipline and control troops scattered throughout a town. Not only did they infringe upon the privacy of the homes, but members of the families were frequently subjected to indignities by soldiers whose manners and morals left much to be desired. The annoyance was further increased by the prevalence of scurvy among the royal troops.

Formal complaint was made to the assembly in 1757 by Joseph Yard, of Trenton, whose family had suffered from the presence of troops in his home. The assembly declared by resolution that quartering in private houses was contrary to the acts of Parliament.[44] This, however, did not lighten the citizens' burden of housing the troops, nor alleviate its pernicious consequences to family life. When the assembly met in March, 1758, it was confronted with petitions from several communities that measures be taken to abate the nuisance. It was pointed out that poor people with small houses and

[41] *Nevill's Laws*, II, 181.
[42] *N. J. Assembly Minutes*, March 14, 1760, p. 6.
[43] *Nevill's Laws*, II, Appendix, 19.
[44] *Ibid.*, Mar. 25, 1757, p. 15.

"numerous families were sometimes obliged to entertain ten, twelve, or fifteen soldiers, to their unhappiness and inconvenience."[45] The erection of barracks to house the troops was urged. The assembly acted promptly, and as usually happened in matters of first importance, Charles Read played a leading role. To take care of those who had billeted soldiers during the previous winter, an act was passed for augmenting the New Jersey regiment to 1,000 men, which named commissioners, one of whom was Charles Read, to defray the expenses to which the inhabitants were subjected in the quartering of regulars in the colony.[46]

For the purpose of meeting future needs a committee, of which Read was a member, was authorized on March 31 to prepare a plan and estimate the expense of building barracks for 1,500 men. Apparently the matter had been thought through in advance, for the committee, with Read's name heading the list, presented its report the same day. The committee proposed that barracks to accommodate 300 soldiers each be erected at Burlington, Trenton, New Brunswick, Perth Amboy, and Elizabethtown. It was recommended that three responsible freeholders be appointed in each of these places, and empowered to draw upon the treasury for any sum up to £1,400 for the building of the barracks "in the best and most substantial, most commodious and frugal Manner."[47]

The assembly agreed to the report unanimously, and authorized a committee of five, which included Charles Read and Joseph Yard, to bring in a bill to authorize the building of the barracks. The bill, which contained a clause "for Preventing Spirituous Liquors being Sold to Common Soldiers without leave from authority," was introduced on April 7 and passed the next day. Read, with Samuel Nevill, was named

[45] E. J. Fisher, *op. cit.*, p. 347.
[46] *Nevill's Laws*, II, 163-182.
[47] *N. J. Assembly Minutes*, March 31, 1758; *Nevill's Laws*, II, 183-187.

to deliver the bill to the council, which promptly approved it, and on April 15 it was signed by Governor Reading.[48]

Among the fifteen freeholders named in the five communities were Read's acquaintances Hugh Hartshorne and William Skeeles, of Burlington; Joseph Yard, and Robert Ogden. The act also provided that nine men, among them Charles Read and Richard Saltar, be appointed trustees for the colony, in whose names the respective deeds of grounds should be taken. Work on the Trenton barracks began on May 31. By December the building was sufficiently advanced to house some soldiers, and it was completed in March, 1759—within ten months of the beginning.

As Read's committee had intimated, the cost, including furnishing, varied widely in the different communities, from £2,932 at New Brunswick to £4,055 at Perth Amboy, while the Trenton barracks cost £3,487, those at Burlington £3,384, and those at Elizabeth £3,589.[49] As the final cost was two to three times the original estimate, additional funds were subsequently granted by the assembly.[50]

Quebec fell into British hands in September, 1759, and Colonel Schuyler was present at the surrender of Montreal in 1760. Read was instrumental in effecting the final settlement of Schuyler's accounts.[51] It was 1763, however, before the war

---

[48] E. J. Fisher, *op. cit.*, p. 347; *N. J. Assembly Minutes*, Apr. 7, 8, 10, 15, 1758, pp. 19-20, 24.

[49] William S. Stryker, *The Old Barracks at Trenton, N. J.* N. J. Hist. Soc. (1885), p. 3; E. R. Walker, "The Old Barracks, Trenton, N. J." *Penn. Mag. Hist. and Biog.*, XXXVI (1912), 187-208; *N. J. Assembly Minutes*, Dec. 3, 1760, pp. 52-55.

[50] *Nevill's Laws*, II, 210, 262. Of these five colonial barracks, only the one in Trenton now remains. Few older military structures in the United States have been preserved or restored. During the Revolution it was occupied in turn by both the American troops and by their enemies, the British and the Hessians. It figured prominently in the Battle of Trenton. This substantial two-story structure of gray stone stands hard by the present State Capitol, site of the "modern battles of Trenton." It is more than a relic of early days; more than a beautiful example of colonial architecture. At three world expositions it has been the model for the "New Jersey Building," symbolizing perhaps better than any other building the life and the traditions of the state. Splendidly preserved, and widely cherished, it reflects the spirit and the quality of the colonists, and stands a monument to Charles Read and his colleagues who planned and supervised its construction. (Trenton Historical Society, *A History of Trenton*, Princeton, 1929, I, 301.)

[51] *N. J. Assembly Minutes*, Nov. 29, 1760, p. 46.

Old Barracks, Trenton. From a pamphlet by William S. Stryker, Trenton, 1885

Old Barracks, Burlington. A sketch by Henry S. Haines

officially ended with the peace of Paris. During the later years of the war, New Jersey was happily free from Indian troubles, thanks in large measure to the wise policy of conciliation and the fair treatment accorded the Indians.

Troubles with the Indians, however, were not all cured with the peace of Paris. A renewed outbreak on the Pennsylvania frontier again sent refugees across the Delaware for protection, and in 1764 New Jersey was asked to raise 600 militia for service in the field. Read was delegated by the council to wait on the governor and advise him that the council did not concur in an act for raising this number, but it was finally agreed to raise the full quota, provided the province of New York did likewise, and £25,000 was appropriated for the purpose.[52] On March 6, Governor Franklin notified the Lords of Trade that inasmuch as New York had raised only half of its quota, New Jersey would proceed immediately to raise a force of 300 men.[53] Read served on the legislative committee for recruiting the volunteers. In 1765 Governor Franklin addressed a message of congratulation to the council and the assembly upon "the happy Termination of Hostilities with the Indians."[54]

The continued presence of British troops in New Jersey remained a source of friction. At the close of the French and Indian War in 1763, the equipment of the barracks had been sold. When instructed by the governor in 1766 to make provision again for quartering the British troops in the barracks, the assembly passed an act appropriating £100 for each of the five barracks and authorizing commissioners to refit the buildings and purchase furniture, bedding, blankets, firewood, and other supplies for the purpose, but specified that the commissioners should be responsible to the assembly and not to the governor.[55] Read served on the committee of the council

---

[52] *N. J. Arch.*, 1st ser., XVII, 369, 373; *Allinson's Laws*, pp. 267-268.
[53] William Franklin to the Lords of Trade, March 6, 1764; *N. J. Arch.*, 1st ser., IX, 428-430.
[54] *N. J. Arch.*, 1st ser., XVII, 384.
[55] *N. J. Assembly Minutes*, June 25, 28, 1766, pp. 39, 53.

to which the assembly bill was referred.[56] As might be expected, the act was disallowed by the Crown.[57]

The assembly, which regarded the quartering of British troops at the expense of the province as a form of unjust taxation, refused to appropriate funds for the barracks supplies if they could not control the expenditures. The issue waxed hot between governor and assembly for several years, but Franklin's political strategy finally won out. The barracks act of 1770 provided that the money should be spent on warrant by the governor with the approval of the council.[58] This precedent prevailed for a time, and thereafter the matter ceased to be a major political issue.

[56] *N. J. Arch.*, 1st ser., XVII, 417.
[57] *Allinson's Laws*, p. 296.
[58] *N. J. Session Laws* (Sept. 26-Oct. 27, 1770), Chap. II, p. 8.

Twelve: Indian Commissioner

LIKE the forests which were their abode, the native Indians gradually gave way before the rising tide of western culture. New Jersey was the home of three subtribes of the Lenni Lenape nation, commonly called the Delawares: the Minsis, or wolf tribe (men of the stony country), a warlike group who inhabited the glacier-formed region in the northern counties; the Unami, or turtle tribe (fishermen), who lived mainly in the central region south of the Sourland Mountains to Camden; and the Unalachtigo, or turkey tribe (people living near the ocean), in the region farther southward to the Delaware Bay. Their numbers are estimated to have been about 5,000 to 7,000 at the middle of the Seventeenth Century, but as the white settlers increased they were gradually crowded out. Charles Read was only a small lad when, about 1725, the Unalachtigos forsook the New Jersey shores for the Juniata Valley, on the way to their ultimate destination in western New York. He had been in New Jersey but three years when the Minsis left northern New Jersey for the vicinity of Wyoming and Wyalusing, Pennsylvania. Sometime in between these two migrations Tedyuscung, the great chief of the Unamis, led his people northward to abide a while at the forks of the Delaware, the present site of Easton, and in 1754 to move on to the Wyoming Valley.[1] Disease and drunkenness, too, had taken a heavy toll. Many of those who survived smallpox, tuberculosis, and the social diseases that came in the train of civilization were demoralized by the white man's whiskey. By the middle of the century only a few hundred Indians remained in New Jersey, a pitiful remnant of a once proud and vigorous people.

[1] William Nelson, *The Indians of New Jersey* (Paterson, 1894), p. 97; Charles A. Philhower, "The Aboriginal Inhabitants of New Jersey"; *New Jersey—A History*, ed. I. S. Kull (New York, 1930), I, 20-21, 24, 52.

Despite the official policy of fair dealing, the New Jersey Indians of Read's time had just cause for complaint. Later settlers were not always so careful as their predecessors had been about compensating the Indians for the lands they occupied. Although the sale of intoxicating liquor to the natives for many years had been prohibited by law, bootlegging was common, and unscrupulous traders frequently took advantage of the Indians while under the influence of liquor. The Indians, too, complained that the erecting of dams on the creeks prevented the free passage of their canoes, and protested against the catching of deer in large steel traps. At the outbreak of the French and Indian War, matters came rapidly to a head. Under the threat of invasion on the northwestern frontier, the public became panicky, and the fires of prejudice spread.

Public officials feared that some overt act might incite bloodshed. The jailing of certain Pennsylvania Indians at Trenton in 1755 caused Charles Read grave concern, and he vigorously protested the act. The Indians belonged under the jurisdiction of Pennsylvania, and such ill-advised treatment, he argued, was a step which might involve New Jersey in an Indian war, "which of all Others is the most alarming and Ought to be Studiously avoided."[2] Governor Belcher upheld Read in the matter. He ordered the Indians delivered without delay to the governor of Pennsylvania, and requested the chief magistrate in Philadelphia "to treat them kindly in all Respects."[3] Read promptly sent word of the case to William Logan in Philadelphia, whose Indian sympathies were well known.

Other prominent Quakers, public officials, and religious leaders counseled with Read on the vexing question of what to do with the Indians. Foremost among these was the Reverend John Brainerd, Connecticut-born Presbyterian and Yale graduate, who had succeeded his brother David as mis-

[2] Charles Read to William Logan, Dec. 14, 1755. Logan Papers, XI, 38.
[3] Jonathan Belcher to Charles Read, Dec. 12, 1755; *N. J. Arch.*, 1st ser., VIII, pt. 2, pp. 191-192.

Indian Commissioner

sionary to the Indians near Cranbury. Brainerd's missionary efforts were sponsored by The Honorable Society for Promoting Christian Knowledge, which had been established in Scotland in 1709. Out of the combined efforts of these friends of the Indians was evolved a workable moral and economic substitute for war, a program in every phase of which Read played a leading role.

A series of conferences gave the Indians opportunity to present their grievances and provided common ground for a frank understanding. In December, 1755, the assembly named Read, Richard Saltar, and Samuel Smith as commissioners to treat with the representatives of the several tribes.[4] At the first conference, which was held at Crosswicks, January 8, 1756, definite progress was made toward an amicable settlement. On recommendation of the commissioners, the assembly in March, 1757, adopted an act introduced by Read which embodied in large measure the agreements entered into at the conference. The act prohibited the use of steel traps weighing more than 3½ pounds, declared that an Indian could not be imprisoned for debt, and placed rigid restrictions upon the sale of liquor to the Indians. It also provided for the regulation of future land purchases.[5]

The question of land claims was not so easily settled. The long list of claims presented by the Indians had to be studied. Accordingly, Read and Saltar, together with John Stevens, Andrew Johnston, and William Foster, were named commissioners to investigate the claims. For this purpose a second conference was called for February 20, 1758. By public notice Read requested "All persons who have any Indian deeds respecting the lands in said colony . . . to get the same proved and recorded, or to transmit them to the clerks of the council of proprietors at Amboy or Burlington that they may be

[4] *N. J. Arch.*, 1st ser., XVI, 585.
[5] *Nevill's Laws*, II, 125-128; *N. J. Assembly Minutes*, March 18, 1757, p. 9. Read's bill of expenses to the conference was only £49-01-11 (*N. J. Assembly Minutes*, March 15, 1756, p. 12).

made use of by the commissioners at the conference."[6] The conference convened in "The Great Meeting House" at Crosswicks; Chief Tedyuscung himself was present; and further progress was made toward reaching an agreement. The assembly voted to reimburse Read for conference expenses in the amount of £27-12-05.[7]

Meanwhile, other efforts were going forward. On December 8, 1755, Read proposed to the governor the establishment of an Indian reservation. How long he had entertained this idea does not appear; perhaps he was prompted by the Reverend Mr. Brainerd, who had advocated the plan as early as 1754.[8] In this proposal he had in mind a joint project with Pennsylvania; to provide a fertile tract where the Indian wives and children would be kept at small expense to the colony, and the braves might be enlisted, with white hunters, to form "a Company or two of Excellent Rangers who . . . would . . . Keep the Inhabitants of the frontier in peace to proescute the Improvement of their Lands."[9]

The following year Brainerd launched a drive to raise £880[10] for the Society for Promoting Christian Knowledge to purchase the site of a reservation. Charles Read headed the list of subscribers with a pledge of £2-14-0. Other subscribers were Jacob Spicer (£2), who recorded the matter in his diary,[11] Joseph Yard (£1-17-0), Robert Ogden (£3), and Stephen Crane (£1-10-0). The project received further impetus in April, 1757, when the New Jersey Association for Helping the Indians was founded, largely on the initiative of a few West New Jersey Quakers. Samuel Smith, who is reputed to have drafted the constitution of the association, and his brother Daniel each subscribed £20, and John Smith generously

---

[6] *Pennsylvania Journal*, no. 750, Jan. 26, 1758; *N. J. Arch.*, 1st ser., XX, 172-173.
[7] *Nevill's Laws*, II, 186.
[8] *N. J. Arch.*, 1st ser., IX, 355n.
[9] Charles Read to James Logan, Dec. 14, 1755. Logan Papers, XI, 38.
[10] Proclamation money. It was estimated the land would cost £450 sterling. (Thomas Brainerd, *The Life of John Brainerd*, Philadelphia, 1865, pp. 71, 294.)
[11] *N. J. Hist. Soc. Proc.*, 1st ser., III (1848-49), 196.

Indian Commissioner

pledged £50. Altogether, more than £180 was raised,[12] but it later proved not to be needed.

For further consideration of the Indians' claims, a third conference was held at Burlington on August 7, 1758. The native conferees now requested that a tract of land in Evesham Township, which Brainerd had suggested, be acquired for the Indians living south of the Raritan River, in exchange for which they agreed to release all claims to land in southern New Jersey. Within a week Read brought in a bill which provided an appropriation of £1,600, half of which was for the purchase of the reservation, the other half to satisfy the claims of the northern Indians. The assembly promptly enacted the measure, and empowered the commissioners, to whom Jacob Spicer was now added, to proceed with the purchase and hold title to the lands to be used for the reservation. On August 29 the commissioners bought the reservation tract for £745. Also, according to the accounting subsequently rendered, they paid £492 for the settlement of the claims in the northern region and £49 for those in the southern region. The expenditures under the act reported by Charles Read and Jacob Spicer in all amounted to £1,310-06-0.[13]

The three tracts acquired for the reservation, 1,983 acres in all, were located in the vicinity known as Edge Pillock, on Bread-and-Cheese Run. The commissioners were authorized to erect dwellings, mills, and schoolhouses, and to exercise general supervision over their wards.

Read envisioned agriculture playing an important role in the new community. In a letter addressed to Israel Pemberton, he remarked:

> . . . We have purchased an tract of Land for them [the Indians] extreamly Convenient for them abt. 2000 or 2500 acres & have this day sent a Surveyor to Survey a parcell of Wild natural meadow near the place where they can cutt their Hay & directed him to take up 500 Acres of it. They can in a day come from the

[12] *Ibid.*, 2nd ser., IV (1875), 33.
[13] *Allinson's Laws*, pp. 220-221; *N. J. Assembly Minutes*, Aug. 11, 1758, p. 35; Dec. 1, 1760, p. 62.

Sea within 5 Miles of this place wth Clams & oysters There are 300 bearing apple trees on it & 24 Acres of good Indian Corn wch We propose to lay down wth Rye & this will be their first Years provisions on their removal.[14]

The following spring Governor Bernard and the commissioners visited the site and began operations for its development, which he reported forthwith to the Lords of Trade:

It is a tract of Land Very suitable for this purpose, having soil good enough, a large hunting country and a passage by water to the Sea for fishing. It is out of the way of communication with the Wild Indians; & has a saw mill upon it which serves to provide them with timber for their own use & to raise a little money for other purposes. To this place I went with 3 of the Commissioners for Indian affairs, where we laid out the plan of a town, . . . & saw an house erected being one of ten that were ready prepared; & afterwards ordered lots of land to be laid for the Indians to clear & till, the land allready cleared being to remain in common till they have acquired themselves separate property, by their own industry. We also made an appointment of an house & lands for a Minister, I having engaged Mr. Brainerd a Scotch presbyterian for that purpose, for which he is most peculiarly suited. The next day I had a conference with the chiefs, at which they expressed great satisfaction at what had been done for them, & I assured them that the same care of them should be continued & exhorted them to order, sobriety & industry. The whole Number of them at present does not amount to 200, & when We have gathered together all in the province they will not be 300. If I can but keep them from being supplyed with rum, for which there are laws strict enough, I shall hope to make them orderly & useful Subjects.[15]

Upon being appointed superintendent of the reservation, Brainerd erected a log meeting house and in due time a village emerged which Governor Bernard appropriately called Brotherton, and was later known as Indian Mills. In 1760 he sought government aid to build a schoolhouse, a gristmill, a blacksmith shop, and a trading store. The governor, the coun-

[14] Charles Read to Israel Pemberton, Aug. 15, 1758. Pemberton Papers, XII, 141.
[15] Francis Bernard to the Lords of Trade, June 15, 1759; *N. J. Arch.*, 1st ser., IX, 174-176.

cil, and the speaker thought well of it, but the Quaker-dominated assembly tabled the petition by a vote of 18 to 1.[16] At the same session a memorial was presented from the Society for Propagating Christian Knowledge on behalf of work with the Indians, which met with an equally cool reception. The assembly, by a vote of 13 to 4, resolved not to make "any Provision for any Person to live among the Indians."[17]

Read and other leaders in the project, however, seem to have overlooked the fact that Indians were not by nature adapted to the white man's agricultural and industrial ways. Gradually the natives declined in thrift and in numbers. The concluding chapter of this, believed to have been the first government-established reservation for the Indian in America, was not a brilliant one, but the state's liberal policy was maintained to the last. In 1801 the Brotherton Indians, then less than one hundred in number, petitioned the legislature for permission to accept the invitation of their relatives, the Mohegans, who lived near Oneida Lake, New York, "to pack up their mat" and "come and eat out of their dish." The lands were sold and the proceeds used partly for the Indians' removal and partly for their future needs. When the legislature in 1832 appropriated $2,000 to settle a final claim of the survivors, even though its legality could not be established, old Stephen Calvin, the Indian who had acted as schoolmaster and interpreter at treaties, wrote in acknowledgment: "Not a drop of our blood have you spilled in battle—not an acre of our land have you taken but by our consent. These facts . . . place the character of New Jersey in bold relief, a bright example to those states within whose territorial limits our brethren still remain. Nothing save benisons can fall upon her from the lips of a Lenni Lenappi."[18] A beautiful benediction, this, upon the zeal for brotherhood,

[16] *N. J. Assembly Minutes*, Oct. 31, Nov. 12, 1760, pp. 5, 16.
[17] *Ibid.*, Nov. 12, 1760, p. 16; T. Brainerd, *op. cit.*, p. 318.
[18] Samuel Allinson, "Fragmentary History of New Jersey Indians," *N. J. Hist. Soc. Proc.*, 2nd ser., IV (1875), 49-50.

justice, and conciliation that marked all of Charles Read's dealings with the Indians.

After the negotiations with the southern New Jersey Indians had been successfully closed with the purchase of the Brotherton site, there still remained the claims of the Minisink and Pompton tribes in the northern part of the province, which because of the absence of recorded deeds, were especially perplexing. It was therefore decided to hold a general conference at Easton, at which the commissioners of Pennsylvania and New Jersey would meet jointly with the several Indian groups concerned, to conclude a bargain for the release of all claims. Since this involved tribes that had been guilty of hostilities in this area earlier in the French and Indian War, it presented a more critical problem in diplomacy than the previous conferences. For Pennsylvania, too, it promised to be one of the most important and one of the most difficult of their conferences with the Indians.

Read appears to have kept his hand closely upon developments, uneasy, not free from suspicion of the Indians, and constantly alert to forestall discord. To insure the protection of the natives concerned in the negotiations, and to prevent any outbreak which might jeopardize the success of the conference, he addressed to the Sussex County officials an order from the governor, which carried both a rational plea and a veiled threat, entreating them to exercise the utmost care for the kind treatment of all. The order, dated at Burlington September 6, 1758, read in part:

. . . it is his [the Governor's] pleasure that if any of the Indians who Shall come to Easton at the Ensuing Treaty Should cross the River Delaware in a friendly Manner that they be treated kindly and by no means Insulted as such a Behaviour from Us may greatly obstruct the Good Ends proposed by the Said Treaty and give them Reason to Complain of our Want of Faith and good Intentions towards a General Peace with them.

. . . the Government has power to punish Such as Shall disobey his Excellencys Orders but Would give me great Satisfaction if the People on the Frontiers Would be Convinced of how much

The Council of the Commissioners from Pennsylvania and New Jersey with the Indians at Easton, 1758. Mural by Richard Blossom Farley in the Library of the New Jersey State Teachers College, Trenton, Courtesy of Dr. Roscoe L. West

Indian Commissioner

a Peace is Preferable to the Excessive Charge of a Defense which We find ineffectual, And that Indians (the Savage in their Natures) can take the Side where they are convinced their Interest lays and they Can be best Served. And in this they act only like other Nations—If any P[e]rson Should abuse them it is the Duty of the Officers of your County to put Stop to it and bind over the Offenders.[19]

The conference was held in October, 1758. William Logan was one of the commissioners from Pennsylvania; Read was on the New Jersey delegation.

The situation was fraught with hazards, and called for the utmost of patience and tact. More than five hundred Indians were present from several nations and tribes, including the Delawares, Wapings, Minisinks, Senecas, and Pomptons, who among themselves were not wholly in accord. Israel Pemberton came, representing the Friendly Association for Regaining and Preserving Peace with the Indians by Pacific Measures, an organization which had provided a liberal supply of presents for negotiating a bargain with the Indians. Pemberton was at odds with the Pennsylvania commissioners, who were not so strongly disposed to conciliatory measures as were the New Jersey commissioners. Moreover, Governor Denny of Pennsylvania offended Chief Tedyuscung. Some days conferences could not be held because most of the Indians were intoxicated. Governor Bernard was inclined to go along with the Friends' association, influenced doubtless by Charles Read in the role of intermediary. On behalf of the Indians Thomas King, the Oneida chief, addressed Governor Bernard as follows:

> Our Cousins the Minisinks tell us, they were wronged out of a great deal of Land, and the English settling so fast, they were pushed back, and could not tell what Lands belonged to them. . . . We say that we have here and there Tracts of Land, that have never been sold. You deal hardly with us; you claim all the wild Creatures, and will not let us come on your Land to hunt

---

[19] Charles Read to John Anderson, Sept. 6, 1758. New Jersey Historical Manuscripts, 1664-1853, p. 149.

after them. You will not so much as let us peel a single Tree; this is hard, and has given us great Offence. The Cattle you raise are your own; but those which are wild, are still ours, and should be common to both; for our Nephews, when they sold the Land, did not propose to deprive themselves of hunting the wild Deer, or using a Stick of Wood, when they should have Occasion. We desire the Governor take this Matter into his Care, and see Justice done in it.[20]

The negotiations dragged on for days. Governor Bernard delivered an address which aroused the ire of some of the Pennsylvania commissioners, among them Benjamin Chew, who kept a journal of the conference. Of this incident, Chew wrote in terms anything but complimentary to Read. "We found that Israel [Pemberton] had seen and approved this speech of Bernard's, and that Charles Reed, the Secretary of the Jerseys was his tool and creature, and was governed by the House of Israel. We therefore resolved to be on our guard against him, and found he had acted a false part by us."[21] A fortnight passed, and the spirit of conciliation finally prevailed. Nichas, chief of the Mohawks, "thanked Governor Bernard for making up all the Differences between the Government and the Minisink Indians, so much to their Satisfaction." Belts and strings of wampum passed hands to seal the contract. "Some Wine and Punch was then ordered in, and the Conferences were concluded with great Joy and mutual Satisfaction."[22]

For the consideration of 1,000 Spanish dollars the Minisinks released all their claims to New Jersey territory, and deeds were signed by the chiefs of the Minisinks and Pomptons. The settlement of the land claims cleared the way for the treaty of peace thereupon concluded between the governors of New Jersey and Pennsylvania on the one hand and the Indian chiefs on the other. Subsequently the frontier guard was disbanded and the countryside again enjoyed tranquillity.

[20] Julian P. Boyd, ed., *Indian Treaties Printed by Benjamin Franklin, 1736-1762* (Philadelphia, 1938), pp. 229-230.
[21] *Ibid.*, p. 316.
[22] *Ibid.*, pp. 241, 243.

Indian Commissioner

New Jerseymen had cause for feeling gratified with the result. Samuel Nevill, publisher of the *New American Magazine*,[23] wrote that "never any treaty held upon the continent of America, was so extensive, as to the number of nations met together, so full, as to the particular affairs there in discuss'd and agreed upon, nor so effectual in its consequence and conclusion. . . ." Bernard was an able governor, and the Easton treaty has been called his greatest accomplishment. But a large share of the credit belongs to Charles Read, without whose diplomatic services the governor would have been seriously handicapped.

Read had labored tirelessly for a peaceful agreement—assembling the tribal chiefs for conference, seeing that proposals were properly interpreted and understood, at all times insisting upon fair play and sympathetic treatment. The conference was a great victory for Read's policy of conciliation, and the achievement was the greater for the reason that it was gained "against very popular Prejudices."[24]

The Easton treaty, however, was not the end of Read's activity on behalf of the Indians. In subsequent years as a member of the council and as supreme court justice, he frequently had occasion to deal with Indian problems, which again took a serious turn in 1763 with the outbreak of Pontiac's war, in the northwest. The previous year he had received from James Pemberton an account of how the latter, to satisfy an Indian who had been wronged in New Jersey, had made the Indian a gift of goods and wampum valued at £15-06-0; Mr. Pemberton expressed the hope that the governor would be willing to reimburse him from the New Jersey treasury.[25]

In 1764 Read issued an outstanding pronouncement of Indian policy in an open letter to "Hon. John Ladd, Esq.;

[23] I, No. XI (Nov., 1758), pp. 290, 292.
[24] *N. J. Assembly Minutes*, Mar. 14, 1759, p. 14. Read's expense bill was £108-03-07 (*ibid.*, Dec. 1, 1760, p. 62).
[25] James Pemberton to Charles Read, Sept. 15, 1762, N. J. Public Record Office MSS.

And his Associates, Justices of the Peace for the County of Gloucester." It was written at Burlington and published in pamphlet form in Philadelphia.[26] The occasion was the settlement near Woodbury of a band of peaceful Pennsylvania Indians, who had fled to Philadelphia to escape destruction by hostile groups, and on crossing New Jersey under military escort provided by Governor Franklin, had been refused permission to enter the province of New York. It came as a sequel to an episode at Lancaster, Pa., in which some friendly Indian guides who had undertaken to pilot the "Paxton Volunteers," had been treacherously killed. Against this outrage, Read registered his high resentment:

I know of no Law to oblige them [the Indians] to remove from one Place to another; while they comply with the Laws in Force, their Treatment ought to be the same with other Subjects in like Circumstances; their Persons and Effects equally claim the Protection of the Laws; and to murder or assault one of them, is a Crime equal to the doing of the same to another of His Majesty's Subjects.[27]

Furthermore, he argued, beyond the consideration of justice, it is good public policy to win and keep the friendship of the Indians:

To consider them in a true Light, we should recollect, that

[26] This pamphlet is comparable with Benjamin Franklin's *Narrative of the Late Massacres in Lancaster County*, published the same year (A. H. Smyth, *op. cit.*, IV, 289-314). It appears there were at least four editions of the pamphlet, one from Franklin's press, the others printed by Andrew Steuart. In the work-book of the Printing House of Franklin & Hall, is entered the following charge on January 20, 1764: "Mr. Israel Pemberton—for printing sixty copies of a Letter from Charles Read, Esq. to John Ladd, Esq., and others—£1-10-0." No copy of this edition is known to the present author. In the New York Public Library is a copy of one of the other editions bearing the imprint; "Printed and sold by Andrew Steuart, at the Bible-in-Heart, in Second-Street, (Price 3 old Pennies) 1764." Since the letter was dated January 7, 1764, and came from Franklin's press at some date prior to January 20 of that year, it seems probable that the sixty copies printed for Pemberton by Franklin and Hall were the first edition, and that subsequently Steuart copied from them. Pemberton's order is further evidence of his association with Read in the pacific attack upon the Indian problem. It is likely that he distributed the original copies of the pamphlet where they would do the most good. (George S. Eddy, *A Work-Book of the Printing House of Benjamin Franklin & David Hall, 1759-1776*, New York, 1930, pp. 6-8.)

[27] pp. 4-5.

COPY OF
LETTER

From *CHARLES READ*, Esq;
now Chief Justice of New Jersey

TO

The Hon: JOHN LADD, Esq;

And his Associates, Justices of the Peace for the County of GLOUCESTER.

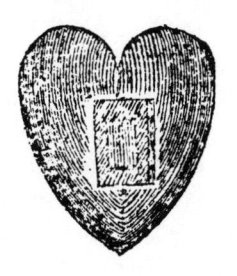

PHILADELPHIA: Printed and sold by ANDREW STEUART, at the Bible-in-Heart, in Second-street, (Price 5 old Pennies) 1764.

Title page of Read's Letter on Indian Policy

on this Continent their Numbers are very great; that every Individual will purchase some Article of British Manufacture; that it is Trade only that enables Britain to maintain Armies, and send forth Fleets, which terrify the World, and raises the British Name in the most distant Parts of the Universe; and the Indians will (as soon as they are undeceived, and Peace is established with them) make us Masters of the most valuable Trade for Furs and Peltry in the World, and thereby contribute to the Riches and Glory of the Nation.[28]

If for no other reason, the Indians of their day deserved kindly understanding treatment in return for the assistance which their forebears gave to the earliest settlers:

There must be many now in your Country, who have heard their Ancestors recount the Kindness with which the Indians treated them when it was in their Power to have destroyed the whole Number of English Settlers in a Day. They then fed them, and gave them all Kind of friendly Assistance; I hope their Descendants will now let Humanity, let Christianity prevail over them to return the Kindness, and not lay to the Charge of these poor distressed People the Actions of the remote Nations of Indians, to whom they are as much Strangers as we are.[29]

A challenge to Read's declaration promptly appeared in the form of a pamphlet entitled *A Letter from a Gentleman at Elizabeth-Town to his Friend in New York*, over the initials "W. P.," February, 1764.[30] The opponent pointed out the impropriety of Read in his capacity of secretary addressing such a letter to justices, or in the capacity of a judge giving an opinion and branding people as murderers on mere hearsay. He was dabbling with a matter in which the province was not concerned. It was not a time for such a statement "when we are justly arming to scourge them [the Indians] for their Insolence and Cruelty: He would have found that such Stories might discourage the inlisting Men for the Campaign." In conclusion he invoked "the Regard" he had "for Jersey" and

[28] p. 7.
[29] p. 8.
[30] Copy in the Library of the Historical Society of Pennsylvania. This also came from Andrew Steuart's Printshop.

expressed his "real Sorrow to see the Authority of its Government prostituted."[31] No record of a reply by Read has been found. He probably regarded it as all in the day's work of a public official in an era when the pamphlet was a commonly accepted weapon for political duelling.

Realistic evidence of Read's determination to see justice done the Indian in the courts was revealed in one of his letters concerning the murder of an Oneida Indian in Sussex County. Read presided at the trial at which the murderer was convicted. He took great pains to see that a representative of the Indian nation was present at the trial, and that word of the murderer's execution got home to the victim's people. Even the squaw was not overlooked. In the letter, apparently addressed to one of the leaders of the tribe, he wrote:

As I have the Honor to be one of the Justices of the Supream Court of this Colony I came three days ago into this County wth a Commission of Oyer & Terminer to try Seamor the person who murdered the Oneida Indian He is Condemned & this day executed. I found with difficulty one Abraham an Indian of Housetonnack & procured him to be present at the tryall of Seamon or Seamor & to Attend his Execution By him I send you this Intelligence that you may make it known to the Nation to which the Murdered Indian did belong. Your Letter to the Magistrates at Minisink had this good Effect That they gave the Squaw forty four dollars & her Husbands Riffle Gun.[32]

Like Benjamin Franklin, Read regarded the Indians with the tolerance and natural respect which he felt for a people whose manner of life was different from his own. Complete justice for the Indians was his constant aim, not only for their own sake, but because he believed in the end such a policy would yield the greatest public good. The benefits the colony reaped through Read's services to the Indians can never be accurately measured in economic terms; nor can the accruing social values be easily appraised. New Jersey's Indian

[31] pp. 7, 8.
[32] Charles Read to a leader of the Oneida Indians, Dec. 20, 1766. Gratz Collection (American Judges), case 3, box 32.

problem, of course, was much simpler than that of other colonies which had a more exposed frontier and larger Indian population. Nevertheless, the peaceful and secure status of New Jersey in the later years of the colonial wars, as compared with that of the neighboring colonies which continued to suffer from Indian atrocities, must in large measure be accredited to the statesmanship of Read and his colleagues. All that was thus realized in the saving of human life and of property, as well as in public confidence, and in freedom to follow the normal pursuits of peacetime, placed New Jersey in an enviable position.

Thirteen: Jurist

THE provincial courts of Read's time, based upon the English judicial system, were essentially the same as those prescribed by an ordinance of governor and council in 1704.[1] Although now and then alterations were instituted by royal instructions, governor's ordinance, or provincial legislation, the general structure was maintained until the close of the colonial period.

Altogether there were seven groups of courts. The lowest were the *justice's courts*, held by justices of the peace, with jurisdiction in minor civil cases. Above these were the *county courts* of general sessions and of common pleas which met quarterly at the county seats. The courts of general sessions were conducted by the county judge and dealt with criminal cases; the courts of common pleas, which dealt with civil cases, had three presiding judges, a president and two others. Above the county courts was the *supreme court of judicature*, consisting of a chief justice and two associate justices, having jurisdiction in all pleas, both civil and criminal, which sat alternately at Perth Amboy and Burlington—two sessions of eight days each in a year. In addition to attending the regular sessions, the supreme court justices traveled the circuits once a year, holding general assizes and sessions of oyer and terminer and of common pleas at the county seats, assisted by the local justices of the peace.

In addition to these, there were four courts with which the governor was directly concerned. The *vice-admiralty court*, having jurisdiction in maritime affairs and authority to enforce statutes for regulating commerce, was headed by the governor as vice-admiral. In trying cases he had the assistance of commissioners of his selection, usually designated "com-

[1] Richard S. Field, *The Provincial Courts of New Jersey*, N. J. Hist. Soc. Col., III (New York, 1849), 256-262.

missioners for trying pirates." The *prerogative court,* which handled matters relating to the administering of estates, was presided over by the governor. Associated with him in this function were the surrogates. Cases in equity, in turn, were handled by the *court of chancery,* over which also the governor presided. The final court in the province, as a *court of appeal,* was the council with the governor presiding.

Under an act adopted by the legislature in 1747,[2] judges of the common pleas received 3s. for each action "tryed to the Bench," and an equal fee for acknowledging and endorsing a deed. In actions involving £5 and under, the fee for a summons was 6d., for a judgment 9d., and for swearing or attesting the jury, 18d. For every "Warrant of Appearance," the justice received 1s. and for "Actions tryed in the Sessions," 3s. for each action. Certainly no justice would become wealthy on such fees. Small wonder that, generally speaking, the New Jersey provincial courts were not attractive to the ablest men.

In almost every department of this judicial system Read served his time. From his appointment as court clerk in 1739 until his retirement in 1773, he filled one post or another as a court official. His commission as surrogate was renewed from time to time and his commission as justice of the peace was extended to Bergen, Middlesex, Monmouth, Morris, Burlington, Salem, and Cumberland Counties. His early years as court clerk as well as his duties as surrogate and as justice of the peace, under Governor Morris, gave him a chance to learn court procedure and to observe at first hand the interpretation of the laws of the province.

The inefficiency and suspected corruption which Read encountered in the provincial courts aroused his indignation. In a letter to Chief Justice Morris in June, 1748, he reported the bungling of a trial for counterfeiting and incidentally revealed his concern for the public welfare, particularly the interest of the farmers:

[2] *Nevill's Laws,* I, 338-352.

I am now at Trenton at a Court of Oyer & Terminer &c where one Henry Yager alias Hunter was Indicted for counterfeiting the Six Shillg. Bills of this Province & Convicted on very full & clear Evidence but by some great Neglect or something worse there appears no Name in the . . . Indictment so that We Expect the whole work to go over again. The Affairs have been so strangely managed this Court that for my part I never intend to Enter another where it shall be under no better direction. Your Friends most heartily wished for your presence wch was scarcely ever more necessary & I really believe that Had it not been for Mr. Coxe who interested himself on the side of the Publick Matters of the Highest Importance to the Community would have been managed in the same hidden manner as usual of late. . . .

. . . it Seems to be the Inclination of ye Gent of the Law to lay a New Ind[ictment] before the Gd. Jury I most heartily pity the Farmers who spend so much of their time at this Busy Season of ye year on an Affair wch might have been dispatched wth Decency in two Days.[3]

When, about this time, it appeared that the anticipated retirement of Associate Justice Allen would create a vacancy in the supreme court, Governor Belcher offered Read the post. At first he declined, but on second thought he regarded it with greater favor. To Chief Justice Morris he wrote in July, 1748:

It is become necessary to Appoint a Second Judge, . . . & I should be sorry that any person disagreeable to you should have a Seat on that Bench.

About Six Weeks ago His Exy. offered me this post, wch for many reasons I refused, & then . . . Assured him that I looked upon the Appointment of that officer, to be an affair of great consequence to the Province, & therefore advised him to give no Expectation of it to Any person, but to take the Advice of as full a Council as He could gett upon it.

Your Brother Allen was by at this time, & pressed me much to take it, by telling me that for my Encouragement He would only tax some few Bills of Cost & Inspect rolls & that I should have the Advantage (as He called it) of going the Circuits. Mr. Allens behaviour of late has given great Disatisfaction to the thinking part of Mankind, & His Exy. grows uneasy wth him, so that proba-

[3] Charles Read to R. H. Morris, June 25, 1748. Boggs Collection.

bly He may not continue long; when His Dismission shall happen, If the gentlemen of the Council think proper to recommend me in his stead, & it be agreeable to you, I shall not object, if that place be not incompatable wth what I have at Present. The Secry. Office I hold as Deputy, & my Continuance in it is precarious, & while I hold it a C[hief] J[ustice] of the Supream Court may be Appointed: It cannot be supposed that I would give it up for a Place wch will sett my failings in a more conspicuous light, & lay me open to the Censure of all the World, & where I must be under a necessity of taking great Pains to Qualify myself in a tolerable manner for the Execution of it. . . .

I Assure you that I have more regard to my own reputation than to Attempt to Lodge myself in a Station by Stratagem or to Abuse the Confidence wth wch His Excellency is pleased to Honour me.[4]

Accordingly, when a few months later the supreme court vacancy occurred, Governor Belcher prevailed upon Read to accept, and commissioned him associate justice March 28, 1749.[5] Whereupon Read promptly dispatched a request to James Pemberton in London to send him some law books. It seems that Read found the duties of supreme court justice somewhat distasteful, and after a time resigned his commission. However, his resignation may not have become effective, for his commission was renewed August 17, 1753, and he again accepted, but apparently without enthusiasm.[6] The problems of the bench weighed heavily upon him, particularly when illness caused the absence of Chief Justice Morris. Three months after being recommissioned, he wrote Morris:

I hope this will find you return'd from the Spa & at Leisure to think of your Friends here who I really think have reason to Complain of you on this Account. You are quite Sensible of the Distress you left us in & the Matter is little mended . . . I cannot think the duty I owe the public should forever keep me in an Office so Detrimental to my private Interest I have pressed this thing upon your friends here who gave me no further relief than by insisting upon my Continuance in it till better times

---

[4] Charles Read to R. H. Morris, July 9, 1748, *loc. cit.*
[5] N. J. Commissions, Liber C2, 179.
[6] *Ibid.*, 212; also Liber AAA, 316.

However, the Calls of my Duty to my own family have brought me to a Determination not to Continue longer than next May term whether any provision be made for a Successor to you or to me or not.[7]

Read's resignation was accepted the following spring, and in accordance with his recommendation, Governor Belcher named Richard Saltar in his place. About this time Read also suggested Peter Kemble, of New Brunswick, for appointment to the supreme court.

From this it appears that Read was anxious to develop a private law practice, which promised to be far more lucrative than the supreme court post. When first appointed associate justice he was not a licensed attorney. Whether or not he ever studied law under formal instruction, he had acquired by experience sufficient first-hand knowledge of it to receive in 1747 a royal commission as master of chancery.[8] When in 1753 he applied for a license to practice, it was granted by Governor Belcher on the recommendation of Samuel Nevill, the other associate justice of the supreme court. Read took the oath of attorney and counsellor on the same day that he received his reappointment as associate justice.[9] Two months later he was admitted to practice before the supreme court of Pennsylvania.[10]

Although now relieved of his supreme court duties, Read's numerous other public offices limited the time he had available for law work, and may have been the reason for the remark of Aaron Leaming that he was "not very faithful to his client's cause."[11] Public responsibilities increased with his appointment to the council in 1758. To look after his practice in his absence an assistant was needed. A man by the name of Blackwood, it appears, served him for a time at a salary of £50 a year. In 1761 he proposed a law partnership to Samuel Al-

---

[7] Charles Read to R. H. Morris, Nov. 10, 1753; *N. J. Arch.*, 1st ser., VII, pt. 1, p. 186.
[8] N. J. Commissions, Liber AAA, 291.
[9] *Ibid.*, 316.
[10] C. P. Keith, *op. cit.* p. 187.
[11] A. Leaming's Diary, Nov. 14, 1775, *loc. cit.*

linson, of Burlington, a young Quaker attorney whose ability and integrity had gained Read's "Esteem and Confidence."[12] For taking care of the law practice in Burlington and Gloucester counties, while helping with Read's other business, he was to receive the same salary as Blackwood plus other revenues which would make the annual income about £100. Seeing that such a connection would introduce him to future business, Allinson accepted on Read's promise to ease him respecting "Conscientious Scruples,"[13]—his Quaker code would not permit him to administer oaths nor doff his hat in court—and they signed an agreement for a five-year period. But apparently the restrictions of joint practice did not satisfy Allinson. More than likely, also, he found some of Read's practices, both personal and official, a little too worldly for his Quaker conscience. At any rate, in less than two years he was ready to quit and to pay for the privilege of doing so. An instrument dissolving the partnership was signed January 15, 1763, in consideration of £100 paid by Allinson to Read.[14] This seemingly was done without any personal break, for the two men continued warm friends. At Allinson's wedding on April 25, 1765, the first guest to sign the marriage certificate was Governor Franklin, the second Charles Read.[15] A year later Read wrote to Colonel Henry Young, of Cape May, recommending young Allinson to his notice. Allinson's compilation of *Acts of the General Assembly of the Province of New Jersey*,[16] in which he had the assistance of Samuel Smith and James Kinsey, represented a public service of high order. His association with the secretary of the province furnished a good preparation for this work.

Meanwhile, Read continued as clerk of the court of common pleas of Burlington County until his resignation of this post December 6, 1762, and at the same time he kept the

[12] Charles Read to Henry Young, Oct. 25, 1766. Caroline Allinson Collection.
[13] Samuel Allinson to J. Kinsey, Apr. 27, 1761. Caroline Allinson Collection.
[14] Allinson Papers. Burlington County Historical Society.
[15] Oct. 25, 1766. Caroline Allinson Collection.
[16] p. vii.

docket of the supreme court.[17] In 1761 along with Peter Kemble, Richard Saltar, Lord Stirling, John Smith, and others, he was nominated a commissioner for trying pirates.[18]

Although Read had voluntarily left the supreme court, sentiment in favor of his return persisted. In 1758, Governor Bernard wrote Lord Halifax recommending him for the chief justiceship to succeed Robert Hunter Morris, whose tenure several years earlier had been thrown into confusion by his appointment as deputy governor of Pennsylvania. He had just learned, the governor explained, that Mr. Read would be glad to accept the post; he had not recommended him earlier, thinking he would not be interested. Read had been practicing law with great success, and when the governor had hinted that he return to the bench, he had not indicated a desire to do so. Perhaps "the ballance of acquiring Dignity & lessening his income was too even." Bernard considered Read "on many accounts . . . a much abler Lawyer" than either Nevill or Saltar, the associate justices, as well as a man "of greater consequence both as to affluence & influence." Also, Bernard added,

> He does not propose entirely to quit the Secretarys office, but would take a colleague who might act in such things as may interfere with the office of Chief Justice. This would be extremely agreeable to me; for I had rather not be entirely divested of Mr. Reads assistance, as he is well acquainted with the business of the office & is of considerable Weight in the province. But it would be Very convenient to me to have a joint Secretary residing at Amboy: for tho' Mr. Read does ev'ry thing He can to make business convenient to me, yet there is no avoiding all the inconveniences which must arise from the Secretary's residing 50 miles from the Governor.[19]

The governor's plan did not materialize, for Chief Justice Morris continued in office. Charles Read received a second call to the supreme court, however, on the death of Associate Justice Saltar in 1762. With the endorsement of his colleagues

---

[17] N. J. Commissions, Liber AAA, 373.
[18] *N. J. Arch.*, 1st ser., IX, 282-284.
[19] Francis Bernard to the Earl of Halifax, Dec. 11, 1758. Bernard Papers, I, 159-161.

on the council, he was appointed to the vacancy on November 24.[20] Read had been Saltar's prime advocate nine years before, when he had urged the latter's appointment as chief justice upon the expected resignation of Robert Hunter Morris. Now Read was to succeed the friend he so warmly admired.

A little more than a year, and another seat was vacated by the death of Chief Justice Morris, January 27, 1764. The filling of this post, second in importance only to the governorship itself, presented a major problem. Lord Stirling, in order to forestall another unacceptable appointment from London, urged Governor Franklin to fill it promptly with a temporary appointee, and then to ask for royal confirmation. "The office of Chief Justice," he wrote only three days after Morris' death, "is a dangerous one to leave open; for its being so will be an inducement to the Ministry to fill it up, . . . and if they do . . . it is a thousand to one if it be tolerably filled. . . . And yet the office is of utmost importance to the Crown, as well as to every individual in the Province." The selection of a local man was not easy. "Fit persons are difficult to be found in New Jersey," continued Lord Stirling. "Few, if any, of the gentlemen of the country, have read Law enough to qualify themselves for the bench, and as few of the lawyers fit for it will give up their business." It was Stirling's suggestion that Charles Read, the person who most naturally occurred to him, be elevated to the position of chief justice, and that someone else be appointed associate justice.[21]

This seemed to Governor Franklin like good advice, and on February 20 he assigned Read to the coveted post, during the "Will and Pleasure" of the Crown, "reposing especial Trust & Confidence" in his "Integrity impartiality, prudence and Ability."[22] The next day Read was sworn in as chief justice. John Berrien was appointed associate justice in Read's place.

[20] *N. J. Arch.*, 1st ser., XVII, 323-324.
[21] William A. Whitehead, *Contributions to the Early History of Perth Amboy and Adjoining Country* (New York, 1856), p. 123n.
[22] N. J. Commissions, Liber AAA, 390; *N. J. Arch.*, 1st ser., IX, 424-426.

Thereupon Governor Franklin addressed the Lords of Trade, asking their confirmation of Read's appointment. Read's virtues were lauded in impressive court language. "As there was an absolute Necessity, in order to prevent any Interruption to the usual Proceedings of the Courts of Law, that some Person should be immediately appointed Chief Justice, and as Mr. Read was so well entitled thereto by his Services, I hope his Appointment will meet with your Lordship's Approbation," wrote Franklin, adding that for a considerable time Read had acted as one of the judges of the supreme court "with great Credit to himself." The governor accordingly recommended him "as a Person well qualified by his Character, Abilities and Experience."[23]

But in certain quarters Read's appointment did not meet with favor. William Smythe of New York, violent supporter of the King, poured out his invective upon the Franklin-Read regime in a letter to Major Horatio Gates. His remarks, admittedly colored by prejudice, do violence to the character of both Read and his wife. After praising the late Chief Justice Morris, he criticized the state of the current government and continued:

Franklin has put Charles Reade in his [Morris'] Place upon the Bench, and filled up Reade's with one John Berrian, a babbling country Surveyor, not fit to be a Deputy to any Sheriff in England. Oh! how far is Astraea fled! . . . For want of a salary no man asks for the second Post in the Province, and it falls into the Hands of a Cuckhold. . . . Where is the Spirit of Dignity that seeks to support the weight and Honour of Government Caesar's wife was to be not only innocent but free from Suspicion— The Proconsul of New Jersey differs from the Roman Emperor— Forgive me this Severity—I confess myself to be much influenced by vulgar Prejudice . . . I am a little angry at the Jersey successions. F[ranklin] after Boone—After Morris Reade—Patience, kind Heavens![24]

Whether or not Smythe's protest had any influence in Lon-

[23] William Franklin to the Lords of Trade, Feb. 28, 1764; *N. J. Arch.*, 1st ser., IX, 426-428.
[24] *Wm. Smythe to Horatio Gates*, Mar. 9, 1764. N. Y. Historical Society MSS.

don, Read was not confirmed for the chief justiceship.[25] Opportunity to adjust the situation came later in the year with the death of Samuel Nevill, the other associate justice. Read stepped down to the place of associate justice, taking the oath November 12, 1764,[26] and Frederick Smyth was commissioned chief justice, the last to be named before the Revolution. Read continued as associate justice until his withdrawal from the province in 1773.

The relatively low salary of the justices of the supreme court, as we have seen, made it difficult to keep the bench filled with the ablest men. Efforts were made from time to time to have the salaries increased. Read's salary as associate justice was now £50 a year, double the amount he received during his first term of service. But this, of course, was still only a nominal figure. Shortly after his appointment, a pamphleteer "J.W.,"[27] in making a plea for more adequate salaries for the governor and other provincial officers, praised Read as an exemplary, but underpaid, public servant. The stipends of the justices of the supreme court were "so absolutely low and pitiful," he pointed out, that the people were secured from bribery and corruption only by the "well known Integrity and public Spirit" of the present judges:

... no Lawyer of Practice and Eminence will accept of a Seat on the Bench, by which he must throw up the more substantial Means of Subsistence, unless his Stock of Patriotism rises superior to that of the generality of his Brethren, superior to all private Views of Gain. ... the Hon. Mr. Read, who has lately accepted a Commission, ... relinquishes a profitable Business, in all Probability ten or twelve Times the Amount of the late Mr. Salter's Salary, which was but £25 per Annum. ... this ...

[25] *N. J. Hist. Soc. Proc.*, new ser., L (1932), 267.
[26] N. J. Commissions, Liber AAA, 396.
[27] J. W., *An Address to the Freeholders of New Jersey, on the Subject of Public Salaries.* D. L. Kemmerer (*op. cit.*, p. 280) suggests that the author was John Wetherill, a member of the assembly from Middlesex County, 1749-75, but this seems improbable, since in 1772 Wetherill voted in the minority against increasing the salaries of provincial officials. It may have been written by Joseph Warrell, clerk of the circuits 1764-68, presumably the son of Joseph Warrell of Trenton, Attorney-General 1733-1754.

must reflect great credit on that Gentleman's Character,—and indeed his Promotion . . . whether we remember his acknowledged Abilities, or his long and various Service, is much to the Honour of those who advised it.

The pamphleteer recommended that the salary of the chief justice be £200, and that of the associate justices be £150, but it was not until 1771 that Read's salary as second justice of the supreme court was increased, and then only to £75. The next year the assembly, by a vote of 17 to 10, enacted a law which again raised Read's salary to £100 per annum.[28] For at least a part of his incumbency, in addition to his salary he collected the modest fees to which he was entitled for holding circuit courts in the several counties. As a supreme court justice he received numerous commissions to the courts of oyer and terminer, and spent much of his time holding court in the counties. Now it was an assignment of one month in Morris County, now two months in Monmouth. Again he journeyed south to Gloucester and north to Essex and to Sussex. In 1767 he got assignments of three months each to Somerset and Bergen. The burden was particularly heavy in 1763 and 1764, for Justice Nevill, suffering from a stroke of palsy, was unable to attend the circuits and Read was obliged to perform his duties for him.[29]

Pomp and ceremony attended the justice when he travelled the circuits. Following the old English tradition, the sheriff and other county officials met the justice's equipage at the county line, and escorted him to the county seat. Farmers in the fields paused to watch the justice and his party ride by. Curious eyes followed him from kitchen window and porch. People from the countryside turned out on court day to enliven the streets of the county seat, and to catch a glimpse of wig and gown.[30] The homage thus tendered the exponents of justice in large measure offset the hardships of travel,

[28] *N. J. Assembly Minutes*, Sept. 12, 1772, p. 57.
[29] William Franklin to the Lords of Trade, Feb. 28, 1764; *N. J. Arch.*, 1st ser., IX, 426-427.
[30] E. M. Woodward and J. F. Hageman, *op. cit.*, p. 16.

and made life more tolerable for these self-sacrificing public servants.

For the lot of the justice was not an easy one. Due honor and respect were not always shown the court. Court dockets were crowded with suits growing out of conflicting land claims, a perennial source of trouble that resulted in the land riots. With the post-war depression in the 1760's, suits to recover damages from insolvent debtors tested the patience and the legal sagacity of the men on the bench. Trials for robbery, for murder, for fornication and other criminal acts were not infrequent. Disputes over boundaries and suits for trespass were regular items on the court calendar. In suits over lands, Read is on record as upholding the validity of patents granted by the proprietors over claims based on Indian purchase, confirming, as would be expected, the supreme authority of the Crown and its representatives in the granting of lands to individuals or private corporations.[31]

Justice was speedy and the penalties severe. The ancient supreme court docket is a grim record of New Jersey justice at its harshest. The whipping post, the branding iron, and the gallows were no idle threats; they were commonly utilized to punish minor offenses. For example, in 1754 a Francis Canfield was convicted of larceny. In pronouncing sentence, the court ordered "that he Receive Thirty Lashes at the Publick Whiping Post in the City of Burlington on his Bare Back Immediately after the Rising of the Court and that he Stand Committed till fees are paid."[32] Less severe was the sentence pronounced by Read some years later when several defendants pleaded guilty of setting deer traps. The fine was £2 each, as provided by the acts of the assembly. The stealing of a cow, however, was a much more serious matter. When in 1753 Justices Nevill and Read heard the case of a John Shores who pleaded guilty of this offense, the record reads as follows:

> The Prisoner being brought to the barr and ask'd what he had

[31] Opinion on Brockhurst & Co., Bucks County Historical Society MSS.
[32] New Jersey Supreme Court Docket, VII, 94, May 21, 1754.

to say why Sentence of Death should not pass against him, pray'd the Benefit of his Clergy which was allowed And it is therefore Order'd that he be burned in The Brawn of the left thumb with the Letter T which was accordingly Performed in open Court, and it is further Ordered that he stand Committed to the Custody of the Sherriff of the County of Burlington untill the fees of his prosecution is paid and it is also further Ordered that Restitution be forthwith made to Isaac Horner being the Party injured."[33]

At the May term court at Burlington in 1772, Justice Read was called upon to pronounce the death sentence upon one William Reed whom the jury found guilty of burglary. In the customary sombre language the court ordered "that the Prisoner be taken from hence to the place from whence he came & from thence to the place of Execution on Saturday the 23d of May Instant and there and then between the Hours of Ten and Twelve of the said day be hanged by the Neck till he be dead."[34]

The jurisdiction of the several provincial courts was not always clearly defined, and their functioning was at times thrown into confusion. This is not surprising, in view of the youthfulness of the province, the distance from the mother government, and the difficulties of travel and communication. One of the knotty problems with which Read was concerned during the 1760's was the effect of the death of King George II upon official acts performed after his demise. The justices and practitioners in the supreme court expressed doubts as to the validity in the reign of George III of an ordinance issued under the authority of the late King, pending receipt of new instructions. Accordingly, in 1761 the assembly passed an act confirming the legislative and judicial proceedings in the colony subsequent to the death of George II and previous to the proclamation of George III, and providing for the future continuance of assemblies until six months after the death of the Crown, also applying the six months' provision

[33] *Ibid.*, VII, 15, 22; May 10, 15, 1753.
[34] *Ibid.*, VIII, 181, 183; May 12, 15, 1772.

to the courts.[35] Read was a member of the committee of the council which reported the bill favorably for passage, and it was he who took it to the assembly to ask concurrence.[36] The act, however, was disallowed in London as affecting the royal prerogative.

This problem, it seems, was the subject of a letter in which Read referred to an opinion of Mr. Nevill, and sought a conference with Samuel Smith on the matter:

I cannot help being concerned at ye state of our Laws   perhaps Mr. Nevill may change his Opinion   if so great disturbance will ensue   I find some others incline to Sett them at Nought. . . . If Mr. Nevill is of ye Same Opinion he was they are for the present safe   I shall look into the Books to find Cases something similar.

After citing several relevant cases in English law, he concluded: "Nor can I think any person who wishes the public tranquillity would encourage the Attack on Laws of such Importance."[37]

The royal disallowance of the assembly's action did not help to restore public confidence in the integrity of the courts. Besides, there were other sources of friction and of confusion. The term of appointment of a judge was a choice bone of contention—whether it should be "during good behavior," or as London insisted, "during pleasure only." That the Crown made colonial judges dependent on his will became one of the chief grievances against the mother country which led to the Revolution.

Still another instance of uncertainty had to do with the governor's powers as chancellor. New Jersey had a court of chancery as early as 1684, but during the first thirty years there were periods when it was dormant. Opinon differed as to whether it should be held by the governor alone or by the governor and council. In 1715, however, Governor Hunter assumed to act as chancellor without assistance of the council.

[35] *Nevill's Laws*, II, 390-391.
[36] *N. J. Arch.*, 1st ser., XVII, 228, 229.
[37] Charles Read to Samuel Smith (undated). Smith Papers, VII, 105.

During subsequent administrations the court of chancery functioned irregularly, sometimes requiring the opinion of the council, sometimes not at all.[38] Now, during Governor Franklin's administration, doubts were expressed as to the authority of the governor to execute the office of chancellor. Desiring an impartial opinion, on November 24, 1769, Governor Franklin laid the matter before the council. Considering it to be "a matter of great Importance to the Interests of the Province," the council referred it for study to a committee of five: David Ogden, Frederick Smyth, Samuel Smith, Richard Stockton, and Charles Read.[39]

The question may well have been a delicate one, since it involved the authority of the governor at a time when royal prerogatives were being challenged. The committee took four months for deliberation and reported to the council March 21, 1770, each member delivering his own opinion. The decision was divided—Ogden and Frederick Smyth concurred with Read; Stockton and Samuel Smith dissented. The majority held that the governor did not have duly authorized power to function as chancellor. Read's opinion in the case is of interest not so much in throwing light upon the problem in question as in giving a hint of his legal philosophy. Between the lines may be read a liberal interpretation of British law as applied to the province—a middle ground between local autonomy and dependence upon the British Crown.

Read pointed out that no part of the British constitution, nor of the common, or statute law, of the kingdom from which the colonists migrated could with propriety be said to be "their Rule or their Inheritance any further than as they by their new Regulations may Copy after it." The several colonies, exercising the right of self-government, differed from each other in the mode of government they had developed. Furthermore, and here was a fundamental point of law, "the Governor cannot issue a Commission to himself tested by him-

[38] George J. Miller, *The Courts of Chancery in New Jersey, 1684-1696* (Perth Amboy, 1934), pp. 30, 41.
[39] *N. J. Arch.*, 1st ser., XVIII, 25.

self. . . . As there is not the most distant hint given the Governor either in his Commission or Instructions that he is to take upon him to try Causes in Equity, I cannot think it a Duty incumbent on him nor even Warrantable under the Royal Commission or Instructions to proceed in a Course of Equity, ..." With his usual tact, he concluded with the suggestion: "Altho' I am clearly of Opinion no Court of Equity exists in New Jersey, Yet, as it is a matter of Consequence, I would advise His Excellency the Governor, to have this Matter settled in Great Britain."[40]

Taking all the opinions into consideration, the council advised that a new ordinance be drawn for better establishing the court of chancery, and appointing the governor as chancellor. A satisfactory solution of the problem was provided when, later in the year, the ordinance was adopted and the governor took the oath of chancellor.[41]

On the bench, as well as in the legislature, Read seems to have successfully avoided the bickering and controversy that marred the careers of smaller men. In 1768 dissension which broke out in the supreme court caused him considerable concern. Chief Justice Smyth complained to the governor and council that Associate Justice Berrien's treatment of him was such that he could sit no longer on the bench with him. A hearing on the matter was set for the middle of April. Read wrote from Perth Amboy to Samuel and John Smith, of Burlington, asking them to attend the hearing. "I wish you would think the public so farr concerned in the matter as to be up," he urged. "It may be attended with Consequences very disagreeable."[42] Apparently the quarrel was patched up, for both Smyth and Berrien kept their places on the bench.

When in 1773 Read's public career came to a close, he had served two terms on the supreme court, aggregating more than sixteen years. The first seven of these he had been third justice; then after presiding nine months as chief justice, he

[40] *Ibid.*, 130-135.
[41] *Ibid.*, XVIII, 169; X, 184-186.
[42] Charles Read to Samuel and John Smith, April 15, 1768. Smith Papers, VII, 156.

continued eight and one-half years more as second justice. Although the legislative and executive branches of the government were more in keeping with Read's dynamic character than the less spectacular judiciary, he probably gave more painstaking and conscientious service to the supreme court than to any of his other public offices. He was conscious of the prime importance of the courts of law, and sought to safeguard their dignity and integrity. Despite weaknesses of character which may have appeared elsewhere, his uprightness and ability as a judge were attested both by contemporary opinion and by his record of performance.

Fourteen: Exile

WHILE from year to year Governor Franklin was seeking earnestly to stem the rising tide of rebellion in the province, misfortune was making gradual inroads upon the Read family. Financial loss, failing health, and death followed quickly upon each other, and exacted their inevitable toll of body, mind, and spirit. The "remarkable Fall in the price of Lands"[1] in 1765 that came with the post-French-and-Indian-War depression, was a severe blow to one who had invested heavily in land. At this time, too, Read plunged deeply into his iron enterprises, which absorbed a large capital outlay. As the years of depression dragged on, he became more and more involved in debt. The revenues from his law practice and his public offices were not sufficient to carry his commercial enterprises. A loan of £500 from William Logan in 1768 secured by a mortgage on the Etna tract helped for a time.[2]

Another source of worry to Charles and Alice Read was their second son, Jacob. The older son, Charles IV, in large measure was fulfilling his parents' hopes. Though not measuring up to the stature of his father, he nevertheless had shown capacity for accepting responsibility, had taken over the management of the ironworks at Etna, in 1767 had married Anne Branin, daughter of Michael and Elizabeth Branin, of Burlington County, and was rearing a family. He was in a fair way to succeed to his father's estate. But not so Jacob. He was of a shiftless, intemperate disposition, irresponsible alike as to both money and morals. He did learn the silversmith's trade, and his father set him up with tools and a small shop in Burlington, which he operated in indifferent fashion.

Recurrent illness, too, was becoming a serious handicap to Charles Read. In 1767 he complained that he had "suffered

[1] Richard Stockton to Charles Read, Dec. 2, 1765. Gratz Collection, case 1, box 20.
[2] Charles Read to William Logan, mortgage, May 10, 1768. Allinson Papers.

much."[3] A wound he had received in the left leg about 1758 continued to give him trouble.[4] Illness frequently kept him from sitting with the council. At one time, too, he was reported as very ill at New Brunswick, perhaps seized there while passing through to Elizabethtown or Perth Amboy. Jacob wrote to William Logan for some pearl barley, presumably for a sickroom diet. Logan, who admitted he was "very uneasy" about his cousin, at once dispatched his servant with the desired article and "a few other additionals" for the relief of the sick man.[5]

Impaired health also had at times clouded the life of Alice Thibou Read. Death finally came on November 13, 1769, after she had reached her fifty-first year. In her final illness she was attended by Dr. Cadwallader Evans, prominent physician of Philadelphia who wrote Benjamin Franklin, "I was up at Burlington 2 weeks agoe, to see Chas Read's Wife, Just before she died."[6] As befitted her husband's position, her burial in St. Mary's Cemetery was attended by a large and distinguished group of citizens. "The Corpse was carried to the Grave by respectable Housekeepers of the Place: The Pall was supported by the Gentlemen of His Majesty's Council, the Chief Justice, and Attorney-General." Besides the representatives of officialdom, it was noted in the press that: "The great Number of the most respectable People assembled on this Occasion from the adjacent Towns, manifested the affectionate Regard paid to her Memory."[7]

[3] Charles Read to Samuel Smith (undated but apparently written in 1767). Smith Papers, VII, 105.
[4] Charles Read (IV) to James Pemberton, Jan. 23, 1771. Pemberton Papers, XXII, 94. The wound may have been received during service in the French and Indian War, but no evidence has been found that Read took part in a military engagement.
[5] William Logan to John Smith, July 23, 1767(?). Smith Papers, VII, 106.
[6] Cadwallader Evans to Benjamin Franklin, Nov. 27, 1769. Amer. Phil. Soc., Franklin Papers, II, 201. On this same trip, Dr. Evans appears to have called upon William Franklin, for his letter continues: "The Govr & Mrs Franklin, were in high health, & as much beloved & esteemed, as your Paternal affections can desire." Benjamin Franklin and Dr. Evans corresponded at length about the possibilities of silk culture in America.
[7] *Pennsylvania Gazette*, no. 2134, Nov. 16, 1769; *N. J. Arch.*, 1st ser., XXVI, 565. Alice Read's gravestone bears the following inscription: "This Stone Covers the Remains of Mrs. Alice Read The Wife of Charles Read, Esq. of Burlington who died Nov. 13, 1769 Aged 51 years also of James their Son an Infant of 2 years old."

Death soon struck again in the Read household. Charles' 13-months-old grandson and namesake, son of Charles IV, died on December 6. He was buried in the cemetery of Evesham Friends' Meeting at Mt. Laurel, of which his mother was a member.

Several months before she died, Alice Read had made a will providing for the distribution of her estate, a large portion of which consisted of property in the West Indies. Fearing that any sum left directly to the wayward Jacob would be quickly dissipated, she bequeathed to James and John Pemberton and Samuel Allinson a trust fund of £500, the interest to be paid to Jacob; also to the same trustees £200, the income to be used for the schooling of her infant grandson Charles, who unhappily survived her only by one month. The remainder of her estate was to be divided equally between her husband and her son Charles.[8]

As much as Read needed the assets of his wife's bequest to satisfy his creditors, delay in settling the estate was inevitable. Meanwhile, with advancing years his health continued to decline. After Alice's death, it appears that he went to live with Charles IV at Etna.

During the winter of 1770-71 he was so desperately ill that the son had to carry on his correspondence for him. When he was again able to write the following May, in a letter dated at Etna, he described the "Excessive tedious & dangerous sickness wth wch I have been afflicted the last fall & Winter the most of which I spent in bed and my recovery not expected by my friends. . . ."[9]

Death claimed another grandchild in 1772. On November 25 Read wrote John Pemberton: "This day we are engaged with the burial of my Dear little grand daughter who died of the measells her little brother went thro' them wth great difficulty."[10]

[8] N. J. Wills, Liber 14, 82; *N. J. Arch.*, 1st ser., XXXIII, 341-342.
[9] Charles Read to John Swift, May 8, 1771. Ferdinand J. Dreer Collection of Letters of American Lawyers.
[10] Charles Read to John Pemberton, Nov. 25, 1772. Pemberton Papers, XXIV, 62.

In 1773, pressed by his creditors and physically and mentally unfit, he determined upon a trip to the West Indies, in hope of regaining both fortune and health. Apparently his departure occurred in June. He seems to have slipped away under cover of some degree of secrecy. Perhaps he did so in order not to alarm his creditors, and to save himself embarrassment. The precarious state of his finances was revealed by the following newspaper notice:

> Whereas Charles Read, Esq; for the recovery of his health, as well as for securing and recovering some large sums of money due him in the West-Indies, has lately embarked thither, and, being desirous of preventing any uneasiness among such as he may owe money to, has appointed us the subscribers, trustees, to make sale of such parts of his estate, as may be necessary for the discharge of his debts, which we purpose proceeding to do as soon as possible. We therefore desire all persons who have any demands against him to bring in their accounts, properly proved, that they may be settled; and all who are indebted to him, by mortgage, bond, note or book-debt, are desired immediately to discharge their respective debts to the subscribers, who are authorized to receive the same.
>
>   Daniel Ellis, at Burlington.
>   Charles Read, Junior; Aetna Furnace.
>   Thomas Fisher, Philadelphia.[11]

Apparently he failed to file formal resignation from his public offices. Perhaps when he left he thought he might soon return. Several months passed before official notice was taken of his withdrawal. The following February Governor Franklin wrote to the Earl of Dartmouth: "Charles Read, Esqr., one of His Majesty's Council for this Province, having removed to St. Croix, where he intends to settle, I beg leave to recommend Francis Hopkinson Esqr. . . . to supply Mr. Read's place in the Council."[12] Hopkinson, whose father had held office in Philadelphia under Read's father, soon was to dis-

---

[11] *Pennsylvania Gazette*, no. 2323, June 30, 1773; *N. J. Arch.*, 1st ser., XXVIII, 545.

[12] William Franklin to The Earl of Dartmouth, Feb. 28, 1774; *N. J. Arch.*, 1st ser., X, 426.

tinguish himself as one of the signers of the Declaration of Independence. The Lords of Trade concurred in the governor's recommendation.

Letters from St. Croix during the latter half of 1773 indicated that Read had stopped at Antigua to attend to the settlement of his wife's estate, but remittances were further delayed. Charles IV, as one of the trustees of his father's property, was hard pushed by the latter's creditors, and found his own credit stretched to the limit. Besides the expenditures needed for equipping a household and laying out a farm, young Read had put capital in a shipping venture. In a letter to Israel Pemberton in November, 1773, he referred to "ye long stay abroad of ye Schooner Gull . . . which I had loaded last winter."[13] The senior Read, endeavoring to settle his debts at long range, wrote to his trustee Daniel Ellis about the accounts of J. Lawrence and John Smith, and referred to an order for horses. In a postscript he wrote, "My love to all Frds. I hope no man shall lose by me tho' they may be longer than I choose out of ye money."[14]

A year passed, and Read had not returned. In November, 1774, the trustees published the following notice:

> The creditors of Charles Read, are desired to meet at the house of Joseph Haight, in Burlington, the first day of next Month, in order to consider of some proposals then to be laid before them; such who have not brought in their accounts, are desired to deliver them to the subscribers, previous to that time, in order to enable them then to make a dividend of money in their hands.[15]

The record of Read's movements during the months subsequent to November, 1773, is not clear. It appears that shortly he went to Martinburg, North Carolina, and there opened a store. But not for long. In December, 1774, he died, far removed from family and friends and from the scene of his

[13] Charles Read (IV) to Israel Pemberton, Nov. 25, 1773. Pemberton Papers, XXV, 144.
[14] Charles Read to Daniel Ellis, Nov. 4, 1773. American Antiquarian Society MSS.
[15] *Pennsylvania Gazette*, no. 2394, Nov. 9, 1774; *N. J. Arch.*, 1st ser., XXIX, 520-521.

Exile

public career. A brief account of his death was entered in Aaron Leaming's diary, November 14, 1775:

When I was in Burlington Jacob Read informed me that his father The Honourable Charles Read, Esq. died the 27th of December 1774 at Martinburg on Tar River 20 miles back of Bath town in North Carolina where he had kept a small shop of goods for some time.[16]

A strange and tragic end was this, for a man of such high attainments and commanding influence. Nothing short of a failing mind, it seems, could have led him to open a shop in a backwoods town in what was then little more than a wilderness. Leaming wrote that he was "of unsettled mind especially for some years before his death."[17]

Apparently considerable time passed before word of Read's death reached the members of his family. In May, 1775, Charles IV procured from Joseph Hewes, who represented North Carolina in the Continental Congress then sitting in Philadelphia, a letter of introduction to Samuel Johnston, president of the provincial congress of North Carolina. Hewes was a Quaker and a native of New Jersey, who as a young man had sought his fortune in the south. Young Read may have become acquainted with him since the opening of the sessions of the Continental Congress earlier that spring. The letter read in part:

The Bearer Mr. Reed is in quest of some Effects of his Father's who died lately in Pitt County, he is related to some of the first Characters here, if you can render him any services you will oblige me.[18]

We may safely infer that Charles IV made the trip, presented the letter to Johnston, looked after his father's affairs and returned to New Jersey. In all probability among the effects he brought home was the book that contained the notes on agriculture. From an entry made in 1774, referring

[16] A. Leaming's Diary, Nov. 14, 1775, *loc. cit.*
[17] *Ibid.*
[18] *Colonial Records of North Carolina*, IX, 1247.

to an associate in St. Croix, it may be inferred that Read took the book with him when he left in 1773. By contriving to die in North Carolina, Read had thrown a further shroud of mystery around the close of his career and had introduced another complication into the settlement of his estate. At a meeting on April 23, 1776, the safety committee of Pitt County, at Martinburg, authorized two members of the committee "to sue all people Indebted to the estate of Charles Read, Esq., dec'd."[19] Since no record of a will has been found, it is inferred that he died intestate.

The misfortunes which visited Read in his declining years remained to plague his household. The liquidation of his estate proceeded slowly, and was not completed when the affairs of the province were thrown into confusion by the outbreak of the Revolution. Jacob had made "a great noise" about the handling of his mother's bequest, and proved to be a thorn in the side of his trustee James Pemberton. In response to repeated pleas, Pemberton cautiously advanced him funds, which were more likely to find their way into the tavern keepers' coffers than into Jacob's silversmith shop.[20]

[19] *Ibid.*, X, 499. The spelling appears as "Martinborough" at this place in the record.
[20] Charles Read (IV) to James Pemberton, Sept. 6, 1773. Pemberton Papers, XXV, 107. Under the terms of the will, Jacob's share was received by Pemberton in the form of a shipment of 25 hogsheads of Muscovado sugar from Antigua, which arrived at Philadelphia in August, 1774, on the brig *Three Brothers*. This had probably been arranged for by the elder Read when he visited the island the previous summer. The sugar was not of the best quality, and the market being dull, Pemberton stored it until October, when he sold it for £734-19-3½. On being importuned by Jacob, Pemberton advanced over £100 before the sugar was sold, "from a hope of encouraging Jacob to industry & a reputable way of living." However, "his well meant intentions" failed of their desired result. (Memorandum of James Pemberton, March 3, 1779. Pemberton Papers, XXXII, 173-174.) Jacob was soon in distress again, and was pleading for further funds to provide needed clothes and a stock of silver for his business. Pemberton was now firmer; he reprimanded Jacob for his "weakness & improvidence," and begged him "to lead a life more honorable to thyself and the memory of thy parents" (James Pemberton to Jacob Read, Mar. 3, 1779. Pemberton Papers, XXXVI, 9). Jacob appealed to William Logan to intercede with Pemberton on his behalf. It seems, also, that at one time Jacob sold some of his tools in order to have money to spend, and that he quit Burlington for Philadelphia. This brought a sharp reprimand from his brother Charles, who charged him with being untrue to the memory of his father, accused him of "too great attachment to Liquor & Love of loose Disorderly Company," and urged him to return to Burlington (Charles Read IV to Jacob Read, undated. Elting Papers, Book of Jurists, p. 60). Evidently he did return to Burlington, but otherwise the pleas were of little avail.

Exile   219

Jacob's wayward course came to an unhappy end on September 14, 1782. Bitterness and distress can be read between the lines of the letter which Charles IV, on learning of his death, dispatched to James Pemberton.

The Situation he [Jacob] was in till some few Days before his Death are truly Lamentable [wrote Charles] If his Conduct had not Merited ye Notice that one Could have Wished; the Christian's duty, (If any is left in Burlington) and the Esteem they held my poor Father in, one would have thought would have been Inducement not to have Suffered him to have Perished in his Shop If not taken Notice of by one who could Illy afford it.[21]

Unlike his brother, Charles IV for a time seemed in a fair way to emulate his father as a political, military, and industrial leader. In 1764 he had been commissioned lieutenant of the Burlington County militia[22] and the outbreak of the Revolution found him with the rank of colonel. In 1773 he was added to the commission of the peace for Burlington County,[23] and in 1775 he was appointed to the committee on observation for the City and County of Burlington.[24] The next year he was a deputy to the New Jersey convention which met to frame a new constitution and to draft ordinances for the government of the state.[25] During the campaign of 1776, Colonel Read was in command at different times of militia from Burlington, Monmouth, Gloucester, and Cumberland counties. In November he was placed in charge of a battalion of state troops known as the "Flying Camp" of New Jersey, a five-months levy designated to serve until April 1, 1777.[26] In December, however, he submitted to the enemy, and availed himself of the British general's offer of protection to persons who would lay down their arms. It appears that he retired to his home, but shortly thereafter while traveling between Mt.

[21] Charles Read (IV) to James Pemberton, Sept. 14, 1782. Pemberton Papers, XXXIX, 131.
[22] New Jersey Adjutant-General, Official Records.
[23] N. J. Commissions, Liber AAA, 437.
[24] *Minutes of the Provincial Congress and the Council of Safety of the State of New Jersey, 1774-76* (Trenton, 1879), pp. 51-52.
[25] *Ibid.*, pp. 446, 471, 472, 488.
[26] N. J. Adjutant-General, Official Records, MSS. No. 3809, p. 5.

Holly and Moorestown, he was captured by Col. Samuel Griffin of the Continental army.[27] With other tory prisoners, he was taken to Philadelphia to be held in custody. On January 21 he was released "on giving his word not to quit Phil[adelphia] without leave."[28] Eventually—the time is not clear—he returned to Etna.[29]

Charles IV did not long survive the Revolution. Perhaps his end was hastened by the humiliation he suffered from "the Disagreeable Necessity of taking Protection from the British,"[30] as he described it, only to become a member of the defeated party. It is reported that in his later years he moved "to a small place between Medford and the Cross-Roads,"[31] and that he lived here until his death November 20, 1783. Tradition has it that he was buried in the Friends' cemetery at Medford, but no trace of his grave has been found.[32]

Charles Read IV was survived by his wife and five children: Charles V, William Logan, James, Alice, and Elizabeth.[33] In his will which bequeathed his estate to his wife and children, he directed that his negro wench Rachel be set at liberty, and

[27] Read is reported to have told Griffin that "he was not disposed to serve any longer," and Griffin branded him as a "damned rascal." (William S. Stryker, *The Reed Controversy*, Trenton, 1876, pp. 5-11).

[28] *Penn. Arch.*, 2nd ser., I, 496.

[29] In this incident of defection from the patriot ranks, young Charles has been repeatedly confused with his father, and also with Adjutant-General Joseph Reed, of General Washington's staff. Biographical sketches of the elder Charles have failed to distinguish between father and son on this and other occasions. As eminent a historian as Bancroft has attributed the act to the Adjutant-General, who engaged in a controversy with General James Cadwallader, but documentary evidence subsequently produced, places the affair definitely upon Charles Read IV. (W. S. Stryker, *loc. cit.*; *Dictionary of American Biography*, XV, 451-453).

[30] Charles Read (IV) to General Cadwallader, Apr. 15, 1783. Cadwallader Papers, II, 63.

[31] Franklin W. Earl to John Clement, March 1, 1873. Clement Papers, Liber J, 31.

[32] In the record of his later years, Charles IV again seems to have been confused with his father. There is one report that the elder Charles returned from the West Indies to Evesham Township, where he died in 1786. The same source records that he was "rocked in a cradle like a child for several years before his death." (*Mt. Holly Mirror*, Sept. 7, 1892; Clement Papers, Liber K, 79.) This definitely could not have been Charles III, but the legend, whether based on fact or falsehood, might have originated in the circumstances of Charles IV's death.

[33] In addition to the firstborn Charles and Alice already mentioned, two other children, Samuel and Ann, had died at an early age. The graves of Charles and Alice are still marked legibly in the old Mt. Laurel Friends' Cemetery, and that of Samuel in the Friends' Cemetery at Medford.

that the negro boys, Richard and Frank, be bound out until twenty-one years of age. It was prescribed, also, that several parcels of real estate be sold: a house, mills, a 250-acre farm, tracts of cedar swamp, and other lands.[34] The inventory of personal holdings suggests an interesting picture of a well equipped plantation of the time.[35] Among the farm animals were four teams of horses, a mule, a pair of oxen, seven cows, two young heifers, three sheep, sixteen pigs, and ten geese. Other items included wagons, plows, harrows, and miscellaneous equipment, besides pine boards and cedar rails. That he left his estate considerably involved appears from the number of bills paid by Samuel Allinson as executor. Among the Allinson papers are dozens of receipts, ranging in amount from 11s. to £200 or more, in date from 1783 to 1790, for such varied items as bonds, interest on mortgages and notes, petty debtor's claims, old debts, taxes on land and court costs, the scattered ruins of a once substantial family fortune.

Charles Read of the fifth generation brought to a tragic close the immediate family succession of that name. He appears to have become a hatter in the city of Philadelphia. There, on September 12, 1788, he became involved in a drunken brawl with a boatman named Andrew Homan, and drawing a knife, "of the Value of six pence," stabbed his adversary fatally in the heart.[36] He was confined to the Walnut Street prison and when tried the following January, on the testimony of thirteen witnesses, he was found guilty, and was hanged for the crime. According to unverified tradition, his heart-broken mother obtained the body and buried it in her garden. After her death, we are told, the remains were removed to the Friends' Burial Ground at Medford.[37]

Family tradition was better perpetuated by another son, William Logan Read, born in 1777, through whom subse-

---

[34] N. J. Wills, no. 10902C, Liber 25, 83; *N. J. Arch.*, 1st ser., XXXV, 321.
[35] See Appendix B, pp. 407-411.
[36] Extract from Minutes of the Court of Oyer and Terminer, Philadelphia County, Penn., January, 1789. Colonel William Bradford Papers.
[37] Franklin W. Earl to John Clement, March 1, 1873, *loc. cit.*

quent generations of the Reads are descended. William, stirred by the pioneer spirit of his forbears, in 1817 loaded his motherless children with his personal possessions into a Conestoga wagon, crossed the Alleghenies, and settled in Ohio. Among his children and his children's children, the enterprising qualities which marked their colonial ancestor have reappeared now and then with characteristic versatility. For example, in one of the lines which leads to the present day, James Logan Read (1808-1889), son of William, served several years in the ministry of the Methodist-Episcopal Church, operated successfully a book business in Pittsburgh and in later years practiced medicine. Charles Hamline Read (1847-1935), son of James Logan, after being in partnership with his father in the book business, became prominently identified with iron and steel manufacture in western Pennsylvania. He was especially skilled in the chemistry of steel; also, endowed with strong cultural tastes, he was a well-known music critic and a self-educated authority on Shakespeare. James Charles Read (1869—), son of Charles Hamline, spent several years in active army service with the rank of captain, in the West Indies and in the Far East. Subsequently he engaged extensively in the steel business and is now interested in investment banking. Although none of these descendants has paralleled the extraordinary career of Charles III, among them they have revealed many of his characteristic traits and interests, such as a talent for business enterprise, particularly in the iron industry; aptitude for military affairs; and a devotion to literary and other cultural pursuits.

NOTWITHSTANDING the substantial documentary records of Charles Read's varied activities, his biographer comes to the end of this study with the conviction of a task unfinished. Even the full account of the diarist Leaming is denied us, for in the midst of his biographical sketch of Read, the pages of his journal have been worn away by the ravages of time. The vexing curtailment of this record is symbolic of the close of

Read's career and the riddle of his final years. We are left with our curiosity only partly satisfied.

However, sufficient is known of Read in relation to his times to leave certain clear impressions. At many points his career is not unlike that of his contemporary, Benjamin Franklin. Both Read and Franklin visited London in their youth; both spent a brief time in the mercantile business of Philadelphia. And whereas Read left the city of his birth to become in time the dominant figure in New Jersey, the Boston-born Franklin became increasingly the moving spirit of Pennsylvania.

In many respects Read was to New Jersey what Franklin was to Pennsylvania. In public office and in community service, their lives often followed parallel courses. Franklin learned at first hand the technique of colonial legislation while serving as clerk of the Pennsylvania assembly; Read served a similarly valuable apprenticeship as clerk of the New Jersey council. Subsequently each became the speaker of his respective assembly; also in time Read became a member of the New Jersey council while Franklin was made president of the Pennsylvania council. Both became a part of the judiciary—Franklin as commissioner of the peace, Read as justice of the supreme court. In the French and Indian War they were charged with similar responsibilities. Both were colonels of militia; both were concerned with fortifying the frontier; both were active in equipping and provisioning the British and the colonial troops. Franklin and Read were in close accord on the vital question of policy toward the Indians. Both were commissioned by their respective assemblies to treat with the Indians and both displayed a high order of diplomacy in effecting conciliation at a critical time. Of other public issues which were rife in the two provinces, such as the floating of colonial currency and the taxation of proprietary lands, they held similar views. Each played a master role as political strategist with his own legislature. Each was the chosen scribe for official communications from the assembly to his provincial governor.

In scientific and educational interests, too, they had much in common. Both envisioned improvements in agriculture and in manufacturing as contributing to the development of the American commonwealth. Read became a member of the American Philosophical Society which Franklin had founded. In promoting educational projects, Read in New Jersey followed Franklin's example. As the latter had established the Library Company of Philadelphia, Read some years later became one of the founders of a public library at Burlington. And even before Franklin's academy in Philadelphia had acquired the status of a university, Read was subscribing funds for the College of New Jersey and otherwise working for its interests. Although Franklin did not invest so heavily in lands and industrial schemes as did Read, he was involved in certain land projects in Nova Scotia and in Ohio, somewhat akin to Read's property interests in New Jersey and in New York. But on the eve of the Revolution the analogy ends. For Franklin the train of trying events which climaxed the colonial era were but stepping-stones to greater achievement and enduring fame. Quite to the contrary, they hastened to an unhappy close the career of Charles Read. For this reason, to call Read "The Benjamin Franklin of New Jersey" would be an exaggeration, but so much was there in common in the two careers, they may readily be likened without doing violence to the name of the great philosopher.

Read lived in a period when one of the most critical revolutions of all times was brewing. A new social order was in the making; a new political philosophy was taking form. Soon one of the greatest of political and economic storms was to break. In the midst of milling forces he played many active roles. He knew the key figures of his time, he gave himself constantly to constructive endeavor. Though sensitive to the events which stirred about him, he was so engrossed with his work as not to be wholly aware of the changes they portended. Like people in high office before and since, he was too deeply absorbed with the problems of the day fully to perceive the future. Before he realized it, he was caught by the times in which he lived, times so overpowering that, ultimately, they broke him. With all his learning and his varied experience, he rarely qualified as a prophet.

In spite of his high political attainment and his leadership in military and commercial affairs, it is as an agriculturist that he becomes most significant. In his farming operations he was a realist, and he left an unparalleled record of everyday practices on the colonial New Jersey farm. After nearly two hundred years of obscurity, his farm notebook has come to light and assumes a greater importance than the voluminous state papers which bear his signature. It is one of our richest known sources of information about farming in the American colonies. Paradoxically, Read emerges as an authority in the field with which he probably least expected, if at all, to be identified by posterity. When Benjamin Franklin flew his electric kite on a pasture lot in Philadelphia, he little thought that future generations would associate his name with this simple experiment. In like spirit Charles Read, when draining his marl-bedded meadows or counting the roots of his clover plants, was completely unaware that he was making agricultural history. Interesting as he was in his many roles, and important in the affairs of the colony, it is in agriculture that he will remain a vital figure.

BOOK II

Read's Notes on Agriculture

Introduction

THE agriculture of colonial New Jersey reflected the practices of the several European peoples who contributed to the settlement of the province, grafted upon the native agriculture and adapted to the natural conditions which prevailed in this part of the New World. The proprietary land system in New Jersey was more conducive to the development of the independent farm unit than of the village and community life characteristic of New England. The independent freehold was the most common form of tenure. In the settled regions were found farmers of widely varied means. There were, of course, so-called "poor" farmers, including former indentured servants, whose land holdings were meagre, and whose homes were crude and poorly equipped. On the other hand, there was a very considerable class of well-to-do farmers, with relatively large estates, who were literally "country gentlemen." They kept slaves, both negroes and indentured servants, and they knew how to enjoy a good living. They were described as "good-natured, hospitable, and of a more liberal turn than their neighbours the Pennsylvanians."[1] Their farms were commonly called "plantations" after the style of the southern colonies. In fact, farm life in colonial New Jersey in many respects was closely akin to plantation life in the pre-Civil-War South.

The agriculture of the province was in the main self-sustaining, and produced a substantial body of exports. Wheat was raised in abundance, and the other small grains—rye, oats, barley, buckwheat—in lesser quantities. As the wheat acreage increased, New Jersey, New York and Pennsylvania became known as America's "bread colonies," New Jersey for a time leading the other two. Hunterdon County, which included most of the present Mercer, Warren, and Sussex counties, was called "the most plentiful wheat country for its bigness in

[1] A. Burnaby, *op. cit.*, p. 59.

America."[2] Indian corn, perhaps, was second in importance only to wheat, and hay was harvested in large quantities. Hemp and flax were commonly raised, and no farm was complete without an orchard and a vegetable garden. The growing of improved varieties of apples, pears, plums, peaches, and cherries seems to have been well advanced by the middle of the Eighteenth Century. Apples were the favorite fruit, and vast quantities went into the manufacture of cider and spirits, used liberally both for home consumption and for export. In this early day, the soil was found favorable for a great variety of vegetables, and an English visitor called New Jersey "the Garden of North America."[3] Perhaps this was the origin of the popular term Garden State.

Animal husbandry was an essential part of colonial economy. Horses, cattle, sheep, swine, and poultry all occupied an important place on the typical farm. Horses were bred in large numbers for home use and for the West Indian export trade. That there was some breeding of mules may be inferred from the advertisement of William Kelly's farm, Red Barracks, which appeared in *The New York Gazette and Weekly Mercury*[4] on March 21, 1768. This 2000-acre estate embraced an orchard of 1400 trees (reputed to be the largest in the province), nursery, dairy, greenhouses, fowl-house, pigeon house, and numerous other buildings. Stocked with cattle imported from Holland, and manned with twenty slaves, the owner represented it as "one of the finest Places in America to breed Mules for the West-Indies."[5]

Cattle served a fourfold purpose—as a source of farm power, of meat, of leather, and of dairy products. The ox, after years of hauling cart and plow, met an inglorious but useful end at the butcher's block. Local tanneries converted cowhide into

---

[2] George S. Mott, "The First Century of Hunterdon County, State of New Jersey," *N. J. Hist. Soc. Proc.*, 2nd ser., V (1878), 89.

[3] A. Burnaby, *loc. cit.*

[4] No. 855.

[5] *N. J. Arch.*, 1st ser., XXVI, 88-90. The place was sold in 1772 to Lucas Van Beverhoudt, and was subsequently known as Beverwyck (Mrs. Benjamin S. Condit, "The Story of Beverwyck," *N. J. Hist. Soc. Proc.*, n. s., IV, 1919, pp. 128-141).

Introduction

leather. Milk from the dairy yielded butter and cheese at the skillful hands of the housewife; after the needs for home consumption were satisfied, the surplus was disposed of in the local city markets. Here, too, was an early chapter in American cattle ranging. The marshes along the seacoast were a natural grazing ground that supported large herds of cattle. Range cattle were often taken from the mainland to the island beaches where they were branded and turned loose to roam over the dunes. An English writer in 1765 noted that among the chief exports of New Jersey were "black cattle,[6] which they drive in great numbers to Philadelphia, on whose rich pastures they are generally grazed for some time, before they are killed for market."[7]

Upon sheep the colonial farmer depended for both meat and raiment. A few years prior to the Revolutionary War, according to a report laid before Parliament, New Jersey had approximately 144,000 sheep, a larger number than any other colony.[8] Wool manufacture had made a beginning as fulling mills were erected. Likewise, hog-raising was a farm enterprise of prime importance, particularly in Monmouth, Burlington, and Gloucester Counties. Vast quantities of salt pork were exported to the West Indies, and ham and bacon from New Jersey commanded top prices.

Jonathan Belcher in 1748, a year after he become governor of New Jersey, gave an account of the agriculture of the province in a letter to the Honorable Charles Gray, M.P.:

New Jersey . . . is a fine climate and a good soil, when cultivated makes good orcharding, fill'd with many sorts of choice apples, cherries, plums, and peaches in great plenty. I believe equal to those of their mother country from whence you know, the Latins called the Peach *Persicus*.

The arable lands give wheat, rye, barley, oats, and a grain called Indian corn, all in great abundance.

The gardens, roots of all kinds, cabbages, colliflowers, sweet

[6] Bovine cattle of any color.
[7] R. Rogers, *op. cit.*, p. 78.
[8] Francis S. Wiggins, *The American Farmer's Instructor, or Practical Agriculturist* (Philadelphia, 1840), p. 364.

herbs, pease, and beans, of all sorts, and these things, I think, far better than I ever eat in England.

The face of the uncultivated or wild lands is covered with oaks of many sorts, black walnut trees, elms, maples, birch, white cedars, pines, hickorys, sassifras, &c.; with shrubs, grape vines and wild flowers of numberless kinds.

We have poultry in plenty, better than with you, as turkeys, dunghill fowls, pidgeons, geese and ducks.

Beef, veal, and mutton enough, and very good, venison, rabbits, and wild fowl, partridge, grouse, and quails.

. . . take this Province in the lump, it is the best country I have seen for men of middling fortunes, and for poor people who have to live by the sweat of their brows.[9]

Charles Read's notes presented in the pages which follow, deal with nearly every phase of colonial agriculture, and throw new light upon rural life in New Jersey during the generation immediately preceding the Revolution.

IN THE INTRODUCTORY DISCUSSION of each chapter, where pertinent, comparison is made with Eliot's *Essays upon Field Husbandry*, which were contemporary with Read's notes. There also is occasional allusion to *American Husbandry*, the two-volume anonymous work published in England in 1775, which has long been regarded as the standard authority on agriculture in the American Colonies.

Charles Read's notes are not, like *American Husbandry*, a finished discourse, yet in some respects they are more significant than that work. First, they antedate *American Husbandry* by a considerable time, since they were written over a period of approximately thirty years prior to that publication's appearance. Furthermore, they were the fruits of the observation and the experience of an American-born experimental farmer, who knew at first-hand whereof he wrote. *American Husbandry*, on the contrary, was conceived and produced in England, by a man who for some reason chose to remain unknown. In spite of the pseudonym "An American," the author

[9] *N. J. Hist. Soc. Proc.*, 3rd ser., VII (1912-13), 101-102.

Introduction

is believed to have been an Englishman who never visited America, and who dealt with his subject wholly at secondhand. Then, too, *American Husbandry* was written for publication, and the author might naturally have been influenced in his point of view by the anticipated tastes of his readers, whereas Read's notes were made with no thought of publication, wholly objectively and untrammeled by popular prejudice. In character and in purpose they resemble the diaries which George Washington kept of his operations at Mount Vernon. Both Read and Washington consulted English books on husbandry, both conducted field experiments and recorded the results, both made agriculture the subject of correspondence and wrote of their observations on practical farm problems. Like *American Husbandry*, however, most of Washington's writings on agriculture are of a later date than Read's notes, all of which were written in the colonial period. The greater volume of Washington's contributions to agricultural literature, and the most significant, were written after the Revolution.[10]

It does not follow that for these reasons either *American Husbandry* or Washington's farm diaries lose any of their value as a historical source. Rather, both should gain in interest because they are now supplemented by an earlier, indigenous record from one of the important agricultural regions of the American Colonies. Read's notes reveal a wealth of facts known or about to become known two hundred years ago, which were not then commonly accepted or were subsequently forgotten, only to be rediscovered by the agricultural scientists of the Twentieth Century.

IN ARRANGING READ'S NOTES for publication, the following procedure has been observed:

[10] For a discussion of the authorship of *American Husbandry*, see the Columbia University Edition (New York, 1939), edited by Harry J. Carman, pp. xxxix-lxi. Dr. Carman presents strong evidence that the author was Arthur Young (1741-1820), secretary of the British Board of Agriculture. See also *The Diaries of George Washington, 1748-1799*, edited by John C. Fitzpatrick (4 vols., Boston, 1925).

1. The notes are reproduced in full, except certain odd items not germane to the general subject and a few statements made in repetition.

2. The order has been rearranged to permit grouping according to subject matter, as indicated by the chapter titles and sub-titles.

3. The spelling, capitalization, punctuation, paragraphing, and sentence structure are reproduced as in the manuscript, except for minor changes which seemed desirable for the sake of clarification and readability.

4. Where neither punctuation nor capitalization appears in the manuscript to indicate a break between two sentences or separate topical statements, a single *em* space is introduced for the convenience of the reader.

5. Superior letters, while commonly used in abbreviations throughout the manuscript, have not been reproduced as such.

6. The page numbers of the manuscript are omitted. A copy of this volume, with the manuscript page numbers noted in corresponding position on the margins of the printed pages, is filed for reference, together with the manuscript, in the Rutgers University library.

7. In the footnotes describing persons referred to in the manuscript, general biographical information is kept to a minimum, but agricultural data from original sources, where available, are given in detail.

8. In most instances, the meanings of unusual words are given in the Glossary.

9. Topical headings and the titles of books, whether or not underlined in the manuscript, are italicized in the text.

10. Brackets are used to indicate the following:
    (a) Marginal notes transferred to the body of the text.
    (b) Parenthetical references or insertions.
    (c) Missing words or letters supplied.
    (d) Misspelled or ambiguous words explained.
    (e) Items transposed as to position.
    (f) Comment by the author or the editor.

One: The Husbandry of the Soil

THE careless, wasteful farm practices which prevailed in the American Colonies were the subject of critical comment by European travelers. Peter Kalm, the Swedish botanist, traveling through the Middle Colonies in 1748 and 1749, remarked upon the slovenly appearance of the farms, the luxuriant growth of weeds, and the exhaustion of the soil. Finding a rich and fine soil, the Europeans on coming to America, said Kalm, soon learned they could grow abundant crops without manuring and with little effort. "This easy method of getting a rich crop has spoiled the English and other European inhabitants," he wrote, "In a word, the cornfields, the meadows, the forests, the cattle, etc., are treated with great carelessness."[1] A quarter century later the author of *American Husbandry* noted that, in spite of a substantial degree of wellbeing, the same faults of husbandry were present. "The American planters and farmers," he wrote, "are in general the greatest slovens in christendom."[2] After commenting on the weedy acres in New York, and the carelessly kept cornfields in New Jersey, he wrote, "There is nothing can give a man, that only travels through a country, so bad an opinion of the husbandry of it, as to see two circumstances; first, the fences in bad order; and, secondly, the corn full of weeds."[3] He pointed out the need for the larger use of manure, for crop rotation, and for growing of root crops and legumes as means of restoring fertility to the "butchered soil."

A wasteful husbandry was but natural, if not inevitable, in the American Colonies, where land was plentiful and cheap, and labor scarce. The observations of Kalm and other Europeans present a picture that probably was substantially correct.

[1] Peter Kalm, *Travels into North America* . . . (Translation by John R. Forster, 2nd ed., London, 1772), II, 47, 48.
[2] *Op. cit.*, I, 145.
[3] *Op. cit.*, I, 168.

But, it seems, they failed to give adequate recognition to the forces even then at work to correct these faults. Not only the planting of the soil, but its conservation as well, was in the minds of at least some of the leaders responsible for the founding of the colony. The patent issued by Governor Nicolls in 1665 for the Monmouth tract named as its purpose "to the end the said Land may be planted, manured, and inhabited," and it was added that "said Patentees . . . shall within the space of three Years . . . *manure* and plant the aforesaid Land."[4] Among the American farmers of Read's time were men of intelligence and vision, educated and well read, who were mindful of the importance of maintaining soil fertility, who were aware of the defects in the prevailing farm practices, who sought to correct them and were constantly alert to ways and means of persuading the land to yield more generously. A group of such persons in New Jersey and neighboring colonies is revealed in Charles Read's notes in the chapters which follow. Among the prominent citizens of the time, who according to Read's comments gave thought to soil improvement, were Peter Kemble, of Morristown; Judge William Hancock, of Salem; Edmund Woolley, architect-builder of Independence Hall in Philadelphia; William Cook, of Chesterfield, member of the provincial assembly; and William Hugg, innkeeper and merchant of Gloucester.

The description of crop growth in Read's notes indicates that New Jersey soils were richer in his time than they are now, but even at that early day they had lost much of their primeval fertility. Read and his associates seem to have understood this, and realized that soil fertility plays a vital role in successful farming. An appreciation of soil conservation and reclamation that would do credit to Twentieth Century agriculture is indicated in the experiments conducted and the measures adopted by these progressive citizens to increase the productivity of

[4] A. Leaming and J. Spicer, *op. cit.*, pp. 661-662. Italics not in the original. In the Seventeenth Century *to manure* was construed as embracing the process of cultivation by manual labor.

The Husbandry of the Soil

their acres—by spreading mud and rotting hassocks on the poor sandy soils, by sanding marsh lands, by burning old grass, and by the use of dung. Reference to "blue mud" suggests the tapping, at that early day, of New Jersey's vast store of greensand marl, rich deposits of which were on Read's farm, Sharon. Read found it advantageous to mix farmyard dung with mud, preferring this to the use of dung alone, and apparently preferring composted manure to fresh dung. Attention is given to ashes, soot, burnt clay, lime, salt, and the various types of animal manures.

The scientific turn of Read's mind is brought out in his mathematical computations of the business of soil improvement. For example, the labor costs of mudding an acre, with teams and farm hands, are minutely figured—£14-6-0 if spread four inches deep, £10-14-6 at three inches, and £7-3-0 at two inches. Ditches and hedges receive due consideration, as does the control of weeds, specifically of dock, rushes, and elder.

Apparently Read had thoughtfully studied the writings of the English authorities on soil management, noting when his own observations coincided with their claims, and taking exception if his experience differed from theirs. On the improvement of marsh land by the application of sand and gravel, reported by Mr. King to the Royal Society of London, Read comments: "To see this verifyed you may go to the Cove of Meadow next above Mountholly saw mill the marsh . . . is now so improved by Sand that it bears excellent White Clover as I have seen."

Read also quotes Jared Eliot who, in his *Essays upon Field Husbandry*[5] . . . , stressed the need for soil conservation and recommended the wider practice of rotation. He notes Eliot's discussion of drainage and swamp reclamation; of the use of salt as a fertilizer, and of a way to get rid of elder bushes. He

---

[5] Eliot's six essays were published at intervals between 1748 and 1759, and an edition which included all the essays was published in 1760. They were again reprinted in *Essays Upon Field Husbandry in New England, and Other Papers, 1748-1762,* by *Jared Eliot,* edited by Harry J. Carman and Rexford G. Tugwell (New York, 1934). The page references given in the footnotes which follow are of the 1934 edition.

goes farther than Eliot, however, by presenting a comparison of different kinds of fertilizers, dealing much more fully with practical measures of soil management. By pooling the experience of his associates with the observations of Eliot and of several English authorities, he approached thoughtfully and effectively the problem of soil management as it was unfolding before the progressive farmers of the time.

MANURE

[Mudd] Anno 1756 Mr. DeNormandie[6] putt 1000 Load of Mudd on Twenty Acres of Land inclinable to a Sand it was poor & 2 years before it had been tryed wth a Crop wch was not worth reaping & by this 50 cartload of Mudd to an Acre he had this Year a fine Crop upward of 25 Bushel of wheat to ye Acre. He plowed it in.

10 Acres of this was done wth blew mudd out of an old Ditch bank & 10 more wth loose mudd of an Inland swamp a little sandy & ye last proved best ye first year how afterwards remains to be tryed.

Joshua Fenimore[7] putt a load of Hassocks on four square Rod. [40 Load pr acre] they lay 2 yrs to rott & in 1756 cutt them in peices & spread them then plowed them In & had 30 bushl. of Indian corn pr Acre from very poor Dry Sandy Land on Ancocas.[8]

[6] In 1747 Charles Read sold John A. DeNormandie, merchant, of Bristol, Pa., a tract of 21 acres in Northampton Township, Burlington County (N. J. Deeds, Liber H, 127). About the same time Mr. DeNormandie also purchased farm land in Springfield Township and invested in an iron works at Mt. Holly. In 1749 his son Dr. John A. DeNormandie (1713-1803), physician of Bristol, Pa., acquired an interest in the New Jersey enterprises. Dr. DeNormandie was a member of the American Philosophical Society and was the first president of the Burlington Society for the Promotion of Agriculture and Domestic Manufactures formed in 1790. In a presidential address he discussed the value of plaster of Paris as a manure, and urged the members of the society to experiment with it on their farms. Again, in a paper "on promoting the raising of Hogs and improving the Dairies of Burlington County" he recommended the growing of more root crops and of clover to feed hogs and cattle. He advocated soil conservation by the more systematic practice of crop rotation, advised more frequent ploughing, more thorough cultivation and the destruction of weeds, and stressed the importance of domestic manufactures. (*Burlington Advertiser or Agricultural and Political Intelligencer*, I, nos. 1, 3, 4, 40; II, nos. 84, 88; Apr. 27, May 4, 1790; Feb. 15, Nov. 15, Dec. 13, 1791.)

[7] Joshua Fenimore lived in Willingborough Township, Burlington County. Upon his death in 1771, the inventory of his personal estate, valued at £286-2-5½, was typical of a modest general farm. Among the items listed were a pair of oxen and 11 other head of cattle, 11 sheep, 29 head of swine, a cider mill and press, a harrow, ploughs, "Rye in the Barn" (£20), "Indian Corn in the field" (£9) and "turnips in the ground." (*N. J. Arch.*, 1st ser., XXXIV, 172; N. J. Wills, 8899C-8904C.)

[8] The Rancocas Creek.

The Husbandry of the Soil

Hassocks if laid in a heap will rott tolerably in 2 yrs if Spread at once they roll about a field several years.

Wm. Cooke[9] (Crosswicks) putt a Large Shovel full of Mudd after it had been mellowed wth the Frost into his Corn hills it produced equall to Dung & when plowed abroad enriched for many Years next year he carried out & used the same Quantity of Mudd from an adjoining bank but it did no perceptible service tho' on ye same kind of land, The reason was ye 1st Mudd was loose & fibrous & resembled rotton leaves & wood, the Last was blue Mudd.

Wm. Hugg[10] at Gloster on tryall likes horse dung mudd ye best.

Michl Newbold[11] carried out ye black mellow mudd of the Springfeild Meadows on his rich black sand & Saw no alteration. Saml. Davis[12] at ye Bridge of Pensaukin & Mr. Morgan at ye

[9] William Cooke, Quaker, of Chesterfield Township, Burlington County, was a member of the New Jersey assembly 1738 to 1754. Upon his death in 1760, his estate was inventoried at £3772-17-0. It included "books and deerskins" valued at £4-17-6; 8 hives of bees valued at £3; and a negro man, who, according to the terms of the will, was to be set free. His "Husbandry Utenstils" were appraised at £19-6-6, and his "Horses and Cattel" at £85-10-0. (N. J. Arch., 1st ser., XXXII, 68; Burlington Co. Wills, 6613C-6616C.)

[10] William Hugg, innkeeper and merchant, was a man of considerable substance, with a "plantation" and shipping interests in the town of Gloucester, and lands elsewhere in Gloucester County. His inn was a center of commercial activities, where public vendues were frequently held. In 1772 he advertised for sale a house in Gloucester on the water front "very convenient for the reception of both wet and dry goods, being well provided with cellars; a salt house, and a large store on a wharff; it is very convenient for taking in pork, being within 4 miles of Philadelphia, and also in a part of the country, where a great quantity of the best pork may be had." He was a member of the Board of Chosen Freeholders of Gloucester County. In his will probated in 1775, he bequeathed his ferry house, wharves, and stores to his son William. At a public vendue of his property, the articles offered for sale included "A great number of feather beds and bedding, . . . one repeating and one alarm clock; a considerable quantity of walnut furniture, of several kinds; a great quantity of pewter; horses, cattle, hogs, hay, Indian corn, a large, strong, ironbound waggon, husbandry utensils, and sundry other articles too tedious to enumerate." (N. J. Arch., 1st ser., XXVIII, 230-231; XXXI, 223; XXXIV, 260; Penn. Gazette, no. 2279, Aug. 16, 1772; no. 2449, Nov. 29, 1775; Frank H. Stewart, Notes on Old Gloucester County, New Jersey, Philadelphia, 1917, p. 119.)

[11] Probably Michael Newbold III (1706-1763), who inherited from his father a 300-acre farm subsequently called Earlham, in Chesterfield Township, Burlington County, together with a negro, "one mare and Colt." Here in 1736 he built a house which is still standing. On his death, his estate included several farms and other lands in Springfield Township, in Hunterdon County, lots in Philadelphia and a cedar swamp on the Rancocas Creek. His personal estate appraised at £6,350-18-8, included "Hay and Grain in the Barn, &c, £290; Horses, Cattle, Hoggs and Sheep, £403-10-0; Husbandry Utentials £80-10-0; Cheese, Meat & Bees £60-8-0." He was a great uncle of Charles Newbold (1764-1835), born on a nearby farm, who in 1797 patented the first cast-iron plow. (Philadelphia North American, January 24, 1909; N. J. Arch., 1st ser., XXXIII, 306; N. J. Wills, 971C; 7505C-7510C.)

[12] Probably Samuel Davis, "yeoman," of Chester Township, Burlington County,

Mouth of the Creek[13] have tryed Mudd on their dry lands wth Extraordinary Success—They carried it on many years ago.

Mudding

An Acre being as before 660 ft. by 66 ft. Suppose 2 small horse Carts 3 horses & 5 men employed they will carry up 120 load ye two Carts & Suppose their Loads to be 3 ft. long 3 feet wide & 1 ft. deep it will take 70 load in length of an Acre to spread it 4 Inches deep & 3 ft. wide & it takes 22 such loads to spread ye width & 70 multiplyed by 22 produces 1540 load to an Acre & if they carry up 120 load pr day they will be 13 days about it the Cost of wch is as follows

4 inches deep	£14. 6.0	2 Teams pr day . . .	£0. 7.0		
3 do . . .	10.14.6	5 Hands & Diet @3s .	0.15.0		0.22.0
2 do . . .	7. 3.0	Multiplyed by 13 days			13.-
					286
					14. 6.-

besides Spreading & Levelling

[*Mudd 1757*] *Extracts from a letter from Mr. James Matthers at Chester:*[14]

That He had several acres of a very *poor Clay* wch bore Cinkfoil on wch he putt 300 to 350 Load of Mudd of the blue kind & solid as much to a Load as 5 horses could draw as it was fresh thrown out of ye Marsh

---

who died in 1773. The inventory of his personal estate revealed modest holdings, including 19 sheep, 14 lambs, "Green Corn in the ground" valued at £12, and "Geese and Turkies" valued at £1. (*N. J. Arch.*, 1st ser., XXXIV, 131; Burl. Co. Wills, 9393C-9398C.)

[13] Probably Benjamin Morgan, of Gloucester County, who in 1751 inherited the home farm from his father, Alexander Morgan, also a 23-acre tract and 34 acres of swamp in Burlington County. In 1753 he was living on a plantation, in Waterford Township "situate on the mouth of Pensonken creek," which he advertised for sale. It was described as "containing 200 acres, about 100 whereof is cleared, of which there is 40 acres of good drained meadow, . . . ; there is on it a good orchard, good large stone dwelling-house, stone kitchen, barn, stables, coach-house, and several other convenient out-houses." In 1752 he offered £3 reward for the return of a runaway servant—a Dutch lad of 22. He was one of the managers of the Silk Filature of Philadelphia, which in 1771 and in 1774 offered premiums for silk cocoons, and sent a specimen of American silk to Benjamin Franklin in London. (N. J. Wills, 485H; *N. J. Arch.*, 1st ser., XIX, 272-273; XXIX, 128-129; *Penn. Gazette*, no. 1280, July 5, 1753; A. H. Smyth, *op. cit.*, VI, 10n.)

[14] James Matthers was an innkeeper at Chester, Pennsylvania. His name, spelt occasionally as Mather, Mathews, and Matthews, appears on the Chester County tax lists between 1765 and 1780. In 1769, for example, he was assessed for 107 acres of land, 3 horses, 28 cattle, 20 sheep and 3 servants. (*Penn. Arch.*, 3rd ser., XI, 115, 202, 288, 396, 536, 680; XII, 154, 253.)

The Husbandry of the Soil

into ye cart. It was laid on the ground after it had been plowed about October & Novr. & being Spread laid there all winter, in the Spring He Harrowed it & cross harrowed then plowed it, then harrowed it both ways as before. Then plowed it a second time—all this was perform'd as quick as possible, Then He sowed it wth Barley & harrowed it in, He sowed 8 quarts of this Country red clover seed pr Acre. To do this He struck out his Lands 12 ft. wide wth a plow only so as to make a mark & crossed it at ye same distance. Sowed one half of ye Seed one way & ye other ye other & bushed in [a Good Way of Sowing Cloverseed] after ye Barley was cutt he had very good fall feed. That the Grass has Continued good ever since being helped at times a *Little* wth dung. That He never suffers horses or Sheep to go on it at all, Nor any Creatures from the time the fogg is eat off in the fall till after next mowing time.

That his mud would bake in the Sun like a brick but dissolve by Frost—He thinks the Black mudd will *not answer* so well. His furthest Distance for halling was ¼ of a Mile that as ye Distance hapned to be, he halled from 40 to 60 Loads pr day wth 2 Teams. He employed two to fill & 2 to Spread when he halled 40 load—3 to fill & 2 to Spread when 60 & one Driver   That He performed part of this work about 15 years ago & on the other part of the Land 10 yrs. It continues still good as ever. The Mudd covered ye Land 4 inches Deep. He think[s] you should not putt more than 80 load for Corn.

☞ Suppose a Cart 6 ft. Long 3½ ft. wide & 18 in Deep 117 Load covers 1 Inch & reckoning 3 men 2 Teams 3 days to an Acre.

A Ditch 10 ft. wide 3½ deep yeilds 18 Such Loads to a rod or 7 Rod to 126 Load   133 rod of Such Ditch cutts 80 square rod or ½ an Acre & will Cover 20 Acres of Ground one Inch Deep.

Mudd taken out of a pond wch had not been clean'd for 30 yrs. & putt on Sward allowed a good Dressing for 7 yrs it being of a greazy tenacious nature. [Ellis—77][15] but gravelly or Clay mudd rather hurts ye Ground. Mortimer [69] Advises mixing it with Dung.[16] *Compleat Body of Husb.*[17] The same.

[15] William Ellis, *The Practical Farmer; or, the Hertfordshire Husbandman* (London, 1732), p. 77.

[16] John Mortimer, in *The Whole Art of Husbandry* (5th ed., London, 1721; I, 106), wrote, "All Manner of Sea-Owse or Owsie Mud, and the Mud of Rivers, Lakes, or the bottom of Ditches, the wash of Pastures, Fields, Commons, Roads, Streets . . . are of very great Advantage to all sorts of Land, especially dry Gravels or Sand, that are mixed with Dung." Note that the paging in the edition of 1721 which was used for checking, in this and other references to Mortimer's work, differs somewhat from that in Read's notes. Evidently he quoted from another edition.

[17] Thomas Hale, *A Compleat Body of Husbandry* (London, 1756), p. 53. Hale advised mixing; (a) pond mud with well rotted dung for loose soils; (b) river mud with poultry or sheep dung for exhausted mellow soils; and (c) ditch mud with chalk and rotten dung for clayey soils—in each instance for dry pasture.

[*Clay*] At Tollerton in Yorkshire they putt 200 load of Clay pr Acre on their Sandy Land abt Midsummer it Continues in Clods for three or 4 yrs hurts ye 1st Crop but mends all Succeeding ones & is a Manure for 40 or 50 yrs & makes it fitt for Peas or any grain whereas Sand with any other Manure bears only Rye. [Mortimer *Art of Husbandry* 64][18]

[*Vitrioletick Earth*] Mr. Edmund Wooley[19] says that Mr. Cole[20]

[18] Mortimer (*op. cit.*, I, 99-100) remarked: "The Clay is of a bluish brown Colour, and is close, fat and ponderous, and burns well for Bricks. They lay an hundred load upon an Acre." In his notes Read wrote "200 load." The discrepancy may have been due to an error in transcribing, or to a difference in the edition used.

[19] Edmund Woolley, architect-carpenter, of Philadelphia, had part in the erection of the old Pennsylvania State House (Independence Hall). In 1736 he submitted to John Penn a bill for £5 for "drawing the Elivation of the Frount, one End the Roof Balconey, Chimneys and Torret of the State House, With the fronts and Plans of the Two offiscis and Piazzas allso the Plans of the first and second floors . . ." (*Penn. Mag. Hist. and Biog.*, XXXIV, 1910, p. 498.) In 1746, with other residents of the city, he signed a petition to the House of Representatives of Pennsylvania, asking that a powder house be not erected in the vicinity, but that the establishment of a market place be encouraged (Papers of Hist. Soc. Penn., Petitions and Memorials, Box 3). Peter Collinson, of London, in correspondence with John Bartram (1737-1742), occasionally mentioned "our friend Woolley." It appears from these letters that Woolley was interested in iron manufacture, since there is reference to "the richness and quality of his ores." (Darlington, *op. cit.*, pp. 97, 103, 154.)

[20] Possibly Samuel Coles, "yeoman," of Waterford Township, Gloucester County, who died in 1764. The executors of his estate advertised the plantation "late the property of Samuel Coles, Junr.," for sale. It was described as containing 750 acres, "situate on the southerly branch of Racoon creek, . . . having a good stone dwelling house and kitchen . . . also a good barn and other out houses, a young bearing orchard, about one hundred acres of cleared land, and forty acres of meadow, and near as much more may be made. The woodland well timbered, consisting of a good deal of hickory, chestnut, poplar and white oak; situate pretty convenient to a landing: Also a very good stone quarry thereon, and appears to be great bodies of lime stone, out of which great quantities of lime hath been made." (*N. J. Arch.*, 1st ser., XXIV, 444; *Penn. Journal*, no. 1143, Nov. 1, 1764.) More likely, however, the reference is to Samuel Coles of the same place who died in 1772. His will provided for the disposal of a large estate. Besides the home plantation, among the several holdings mentioned were a plantation in Evesham Township, Burlington County, a sawmill, a grist mill, and lands on "Coll. Read's Mill Creek." That Read was obligated to him at the time of his death appears from the bequest to his son-in-law, David Davis, of "£50, out of the money that is owing to me from Charles Read, for the use of the Preparative Meeting at Haddonfield," and also "the rest of the money due from Charles Read to my daughter Martha Davis." The inventory of his personal estate amounted to £3489-18-8. Among the items of interest listed were: a "riding Chair," a sleigh, a "Chair Horse," "an old riding Churl," an iron-toothed harrow, grindstone and crank, "Shelvens," oxcart, 22 deer skins, barrel churn, dough trough, cheese press, "a Mill to Clean Wheat." The livestock included 13 milch cows (£65), 2 bulls, 3 "Steers for Beef" (£15), 14 young cattle (£35), 4 "Cows for Beef" (£16), 36 sheep (£18), 5 horses (£10 each), "Brown Mare and 2 Colts" (£30) and "a Quantity of Swine" (£30). Two unusual items were "Some Deer in the park" (£12), and a "Nursery of Young Apple trees" (£2). "All the Negroes belonging to the Estate" were appraised at £60. (*N. J. Arch.*, 1st ser., XXXIV, 101; N. J. Wills, 1048H.)

at Racoon Cr. first tryed this in abt one-half ye Quantity that he usd of Dung or less greatly to Advantage. That he himself gave a Sprinkling to a cold gravelly pasture near Philada & it produced greatly.

A Receipt from Ellis's Practical Farmr or Hertfordshire Husbandman [fol. 79][21]

To Burn Clay

With three or four pecks of a Mattock the whole ant hill will come out like a core a few wheel barrows of this Earth may be easily fired with the help of a little Brush, wood or Sticks. But care must be taken that vent is not given to the fire which is done By throwing Earth on where the Smoke comes through so may you increase it into a round Body in the Summer time till as big as you please

when you have fired about three Loads put on ye Clay a little at a time with Shovel and the fire By degrees will bring a red or other Clay or flint Stones into an ash or powder which certainly is a most excellent dressing for any Grain or Grass about forty Bushels sows an acre by the hand out of a seed cot and harrowed in with Barley and Grass seeds does vast service I had it Burnt for a half penny a heapd single Bushel, [½d, pr bushl burng.] others get six or eight loads of Clay cut into Spits about as thick as a brick let it be pretty well dried in the Sun and make a heap of Coal or wood as big a Small Burn-fire, like a pyramid Bring the spits of Clay and lay them round the Same two or three spits thick Leaving room to light the fire the Clay with the heat soon will take fire and as the fire advances outwards Lay on more Clay so that the fire may be pent up in the heap and not go out for if you do that, your Labour will be lost and you must begin again, after you have burnt up your six or eight Loads of Clay wch was dry you may then lay on green Clay from ye Pitt laying it on so as not to smother the fire. You may spread out ye heap at bottom as it swells & stages of boards may be made for the Labourers to go on ye heap to place ye clay. Watch your fire night & day the larger the heap grows the Easier your clay is to burn. This is a Cheap Dressg & agrees wth all sorts of Land Fruit trees or Gardens.

[40 bushells to an acre]

[*Sand* Mortimer 65 This was Jno Edmunds of Bowden Cheshire][22] Six hundred load of Sand laid on two Acres of Marshy

---

[21] The following paragraph "To Burn Clay" was taken from Ellis' *The Practical Farmer* . . . , pp. 79-81. The first half of the paragraph was copied almost verbatim, the remainder is materially condensed, but contains all the essential facts of the original. Eliot also gives the substance of this passage from Ellis (*op. cit.*, pp. 50, 51).

[22] Mortimer (*op. cit.*, I, 101) gave "An Account from Mr. John Edmonds of Bowden

ground whch was overRun wth rushes & putt on wthout plowing & the next Winter dunged & he had 14 load of Hay thereon   that he had not Dunged it since wch is *24 years* & that He can Lett that Land for £6 per ann. wch before would not Lett at 10*s*.

Mr. Kings Experiments comunicated to the *royal Society* & in their *transactions* [No 170 p. 948] proves that Sand & Gravel vastly improves the Boggs in Ireland—Gravel the best.[23]

☞ To see this verifyed you may go to the Cove of Meadow next above Mountholly saw mill   the marsh next the bank was very cold & Springy & is now so improved by Sand that it bears excellent White Clover as I have seen   tis only that part of ye cove nearest ye Mill that is so improved.

[*Burning Grass*] Michl Newbold tells me May 1758 that by Several Experiments He finds that the Crop is better for Burning the Grass off than if it was plowed in, but then yo must soon plow in the Ashes.

On a tryall I made it will take 60 bushells of ashes to cover an Acre as thick as Sand on a floor wch covers it from yr Seeing the boards plain & so thick it should be & it brings a peice of Ground into Clover the sonest of any Manure whatever but lasts not above 2 yrs. if more than One.

Ellis says coal ashes are best.[24]

[1756] Mr. Kemble (of Morris County)[25] tells me that he putt ½

---

in Cheshire, of his improving of Land by Sand." The first season after applying the sand, without plowing, Mr. Edmonds sowed "Oats and Fitches, which yielded an extraordinary Crop." After the land had been improved, he "could have two Crops of Grass upon it every Year, if he could be sure of fair Weather to make it in."

[23] William King, Fellow of the Dublin Society, in a paper "Of the Bogs, and Loughs of Ireland," reported that in Ireland the boggy lands when drained were much improved by burning, but not so much as by treating with sand or gravel. (Royal Society of London, *Philosophical Transactions* . . . , XV, 1685, no. 170, pp. 947-960; abridged, 4th ed., II, Chap. LXXVIII, pp. 732-737.) This experiment was cited by Mortimer (*op. cit.*, I, 102), and apparently was copied by Read from this source instead of the original.

[24] This marginal note refers to the portion of Worlidge's text (p. 72) which reads as follows: "The Wood-ashes are the best, and very useful; yet after they have been used in the Bucking of Clothes, they are worth little, unless it be in cold and moist Land, where I have known them also to avail much." Ellis' statement to which Read refers is: "Ashes of Coal are esteem'd much beyond Wood-Ashes, and are sold for 4d. per Bushel thirty miles from London (*The Practical Farmer* . . . , p. 83).

[25] Probably Peter Kemble (1704-1789), merchant, member of the provincial council (1747-1775). After residing several years in New Brunswick, about 1750 he acquired several hundred acres of land near Morristown where he erected a mansion, calling his estate Mount Kemble. During the Revolution he sided with the British, and his estate was occupied as the headquarters of General Anthony Wayne in 1777, and again in 1779-80 while Washington's army was in winter quarters at Morristown. (*Collections of the New York Historical Society, Kemble Papers*, II, New York, 1884,

The Husbandry of the Soil

pint of ashes into ye Hills of 1 row of Corn & that it Exceeded ye rest Greatly [Stiff Land tis 11 bushl pr acre.]

Ashes used in bucking of Cloaths not good.

[*Lime, Animal Manures, etc.*]

[*Lime*] [Pr T. Bowlsby.[26] July 1756 who kept Limekilns in England for sale—they sold @ abt 2½d. pr bushl.]—Lime is used both on Sandy & Clay land they putt 32 bushl on Sand 50 bushl on Clay Land for a Dressing reckoning ye bushls in Stone for they increase by Slacking it last 3 yrs. in Sand & 4 in Clay. They carted it 20 & 25 miles. They putt it on Sand for Turnips in June or July & for Grain after tis plowed up, then Harrow the Ground & Sow & plow all in together He says it saddens sandy Land, & this agreed wth a Story Mr. Hains[27] told me.

Lime may be made of any Stone not Sandy or too cold as freestone, &c. That tis best for Sandy or Gravelly Land & best mixed wth Dung &c. [Mortimer *Husbandry* 55, 56][28]

Mr. Charles West[29] tells me that most of the Jamaica ballast is very good Lime.

---

pp. xiv-xv.) That Mr. Kemble led the life of an aristocratic gentleman is indicated by the inventory of his personal estate at the time of his death, appraised at £2002-15-5, which listed 19 negro slaves. The inventory listed also a bookcase and 120 volumes of books. Only a few agricultural items were recorded, among them 16 sheep, 9 head of cattle, an "old Covered Waggon" and a "Corn Mill" (N. J. Wills, 792N). He was buried in the family plot on the lawn in front of his house, where rest also the bodies of his favorite slaves.

[26] Mr. Bowlsby may have come from England to Burlington County. A Thomas Bowlby is mentioned in connection with an estate in Mansfield Township between 1765 and 1771 (*N. J. Arch.*, 1st ser., XXXIII, 440). Read refers to him elsewhere as a "good farmer."

[27] It is not clear whether the author here refers to Reuben, George, or Caleb Haines, all of whom are mentioned in the text, or to some other member of the family.

[28] *Op. cit.*, I, 84-86.

[29] Probably Charles West, of Deptford Township, Gloucester County, son-in-law of Isaac Cooper. He was proprietor of a large plantation, and also owned a tract of over 1000 acres of pine land in Greenwich Township, Gloucester County. On his death in 1767, his personal property was inventoried at £6003-18-2. Among his possessions listed were 14 negroes. The items of livestock and farm produce given suggest the size of his agricultural enterprise. There were 25 "Horse Kine" and 3 pair of "Working Oxen"; also 130 "Horned Cattle" valued at £613-9-0; 335 sheep at £207-18-0; and 65 hogs at £39-0-0. The produce included hay valued at £223-0-0; 300 bushels of Indian corn at £40-16-8; 100 bushels of "Rie" at £17-10-0; 14 bushels of wheat at £4-4-0; 50 bushels of potatoes at £2-0-0; and 2 bushels of flaxseed. Among the livestock products were beef and pork listed at £21-15-0; 384 lbs. of tallow and lard £9-10-8; and 123 lbs. of cheese at £1-15-10½. "Cider Spirits and Mollasses" were listed at £79-1-8, and "Husbandry Utensils and Sundrys" at £101-1-0. (N. J. Wills, 957H.)

Sandy Land best for Lime. 96 bushells or 112 bushl [14 qur.] [of lime to an acre].[30]

[*Salt*] too much destroys grass Elliott[31] 5 bushl to an Acre *Compt. body of Husbandry* says—2 bushl.[32]

Hitts *Treatise of Fruit Trees* [p. 16][33] says He mixed 1 oz of Salt wth a gallon of Water & watered a Spott of ground in a dry summer (mentions not ye quant. of ground) it did well & better than when mixed wth 2 & 3 oz or than Com[mon] water that mixed wth 3 oz did best ye following summer (he watered ye several Spotts 9 days).

[1754] I weighed at Mr. Hulings[34]—1 bushl of Lisbon Salt 80 lb.—dry ashes 60 lb.

[*Dung*] About this time [1753] I stood by & counted how many Fork fulls of old Dung went into a Waggon body wth only a tail

[30] Worlidge's text (p. 65) opposite this marginal note reads: "One Author saith twelve or fourteen quarters [a quarter equals 8 bushels] will Lime an Acre; another saith 160 Bushels: the difference of the Land may require a different proportion. The most natural Land for Lime is the light and sandy, the next mixt and gravelly; wet and cold gravel not good, cold clay the worst of all." Also in the text on this page, Read had underlined the portions given in italics below:
". . . upon Land not worth above one or two Shillings an Acre well husbanded with Lime hath been raised *as good Wheat, Barley, white and gray Pease as England yields*. . . ."
"Also that by the same means from a Ling, Heath, or Common naturally barren and little worth, *hath been raised most gallant Corn worth five or six Pounds an Acre*.
". . . some men have had and received so much profit upon their Lands by once liming, *as hath paid the purchase of their Lands*. . . ." In subsequent quotations from Worlidge's text, certain portions are italicized in like manner to indicate that they were underlined by Read.

[31] *Op. cit.*, p. 89: "For want of Dung common Sea-Salt may be made use of; a Gentleman at Middletown, who came from the West-Indies, bought a Piece of poor Land, put on it five Bushels of Salt to the Acre, sowed it with Flax, a small Strip through the whole Piece he put no Salt upon it, the Consequence was, the salted Part produced fine tall Flax, and the small Strip was poor and short."

[32] Hale, *op. cit.*, p. 62: "They [the old writers on husbandry] prescribe the spreading of this [salt] in the Manner of Corn, thinly and evenly over the Ground: . . . the Quantity . . . to use is two Bushels to an Acre."

[33] The experiment described was conducted "in an extreme dry summer, upon a bare piece of pasture land, out of which the cattle were all taken for want of grass." The results noted were the effects of the different treatments upon the growth of the grass. (Thomas Hitt, A *Treatise of Fruit Trees*, 2nd ed., London, 1757, pp. 16-17.)

[34] Probably Abraham Hewlings, merchant, who had "a brick store on the wharf" at Burlington (John W. Barber and Henry Howe, *Historical Collections of New Jersey: Past and Present* . . . , New Haven, 1868, p. 95). He served at times as chosen freeholder and as collector of Burlington, and as a member of the colonial assembly (1769-1771) (E. M. Woodward and J. F. Hageman, *op. cit.*, p. 126). In 1774 he was taxed for 350 acres of land in Chester Township, together with 26 head of cattle and horses, and one slave (Burlington County Tax Lists, 1774). In addition, he had other land holdings in Burlington County. The will of Abraham Hewlings, Sr., was probated in 1795. His estate was inventoried at £4827-7-2. (N. J. Wills, 11608C.)

The Husbandry of the Soil

board, Common Locked, & there went into her moderately heaped 475 forks full & supposing 3 hills of Corn on a rod there would be 120 in Length & 12 in breath in an Acre or 1440 hills of Corn & a Spadefull in a hill would be very little more than 3 Load of Dung pr acre to Dung your Corn but as ye 1st Row of Corn is twice reconed & the Corn is not generally so thick planted I think 3 Load will do it well.

Abt 3 Load [of cow or ox-dung] will give 1 Spadefull to a hill of Corn on an Acre.[35]

Dung rotts sooner on Levell ground than in Pitts. [Ellis's *practl farmr* 81][36]

10 or 12 Swine [yield] from 40 to 80 Load of Manure.[37]

This [pigeons' or hens' dung] T. Bowlsby says is sold in England @ 10d. pr bushl.[38]

Lay not more than 20 load of Dung on clay nor 15 on Sand tis better to renew their Surfaces the oftner. [Mortimer 74][39]

A Folio Vol. Entitled *Compt. Body of Husbandry* printed 1756 Treats largely on this Subject [page 68 to 77.][40]

[Horse Dung] fermented wth Straw is its hottest State, even dry has fertility.

[Hoggs Dung] 1 load is good as 2 of any other beast but too hott to use alone.

[Cow Dung] has less Effect than any other used alone—None better mixed.

[Sheeps Dung] Loses its Virtue the soonest of any if Exposed—best for Cold Land.

---

[35] This marginal note is opposite the portion of Worlidge's text (p. 69) which reads "Next unto the Horse-dung is Cow-dung. . . ."

[36] Ellis' text (*op. cit.*, p. 81) reads: "Pits are often made use of, to throw in Horse-litter and other Stuff, in order to rot; but this is a mistake, for these Pits rather prevent it. Laying on level Ground, and often turning, furthers it best."

[37] In Worlidge's text (p. 70) Read underlined the following passage referring to swine's dung: ". . . *being esteemed the coldest of dungs*. . . ."

[38] Worlidge's text (p. 71) opposite this marginal note reads: "Pigeons or Hens-dung is incomparable, one Load is worth ten Load of other Dung. . . ."

[39] The passage by Mortimer (*op. cit.*, I, 114) to which Read refers reads: ". . . whereas the greatest quantity that should be laid upon Clay-lands, if sown with Corn, (which are the best Lands to bear Dung) ought not to be above twenty load upon an Acre, and the red hazely Brick-earth or light Sands, that are subject to Weeds, and that spend themselves in producing of Straw rather than Grain; ten or fifteen load upon an Acre is enough at a time, because by that means you may be often dungning of them, and renewing of their Surface with fresh Dung."

[40] Read here gives abstracts from Hale's discourse on animal manures in *A Compleat Body of Husbandry* (pp. 68-77). He also includes some comment on the use of mud as a fertilizer, and of mixing it with manure. He seems to have taken some material from Hale's chapter of "Of the Use of Mud as a Manure" (pp. 52-53), and to have introduced his own observations.

[Pidgeon & Poultery] 40 bushl to the Acre sown after grain.
If yo use fresh Dung of any sort putt less than when rotten.
[Mudd] River mudd gives fertility, that of Ditches by ye roadside as being nearer a Clay Solidity.
River Mudd has great Fertility   Mix 1 Load of Dung wth three Load of River Mudd by wch yo have 4 load each as good as the Dung wthout mixture.
Sheep in a Covered fold wth a fresh quantity of Earth carried into it every week produces the richest manure art can procure.
Dung fresh has the greatest fertility on pasture Land—but on Corn land brings weeds and ye corn runs by to great a Quantity to Straw.
Old Markham[41] first proposed mixing dung & says one load is Equal to 5   latter writers say to 20. The Author think[s] 1 will make 4.

A Course of Manuring will alter ye most barren Soil so that it can never be so again. The best way is to Spread this compost just before Plowing.
Spread new Dung but plow directly.[42]
[Ellis's *practl. farmer*][43]

[*Salt*] [81-83] Salt [is] a good manure for moist ground but best mixed wth other Dung lasts several yrs [*Compl. Body of Husb.*] 2 Bushl to an Acre.
[*Soot*] Soot 15 bushl to an Acre,—Coal Soott better than wood. [Mortimer 79.][44]
[*Lime*] [74] Lime 40 bushl. to an Acre, best mixed wth the bottoms

---

[41] Hale here quoted from Gervase Markham's *Farewell to Husbandry* . . . (10th ed., London, 1676, p. 47) in which he advises the mixing of manure with other refuse. ". . . this will so strengthen and enrich your manure, that every load shall be worth five of that which wanteth this help." Hale suggests mixing manure with earth in the ratio of 1 to 3.

[42] Read is not in agreement with Worlidge on this point. The above comment is written opposite the following statement in the text (p. 84): "I know it is a common way in most places, to lay Dung in heaps till it rots, and then spread it on the Land, which is much better than to spread it whilst it is new. . . ."

[43] The notes which follow this reference were taken mainly from *The Practical Farmer* . . . (the numbers in brackets refer to the pages of this book), but information and comments are introduced from other sources. Ellis' statement on salt reads in part (*op. cit.*, pp. 81-82): "All sorts of Salt Dressings from the Sea, . . . as their Sea-weed or Oreweed, . . . also their Snailcod from deep Rivers, and Oyster-shells; . . . will, after two or three Years . . . mix with the moist Land, and do a great deal of good for several Years after."

[44] Mortimer (*op. cit.*, I, 121) wrote: "Soot also is very good both for Corn and Grass, especially what grows on cold Clays or Lands much run over with Moss; but Sea-coal Soot is the best by much. They commonly allow forty Bushels to an Acre, but some Lands will require more: It produces a mighty fine sweet Grass, and destroys Weeds and Trumpery." Although Mortimer reports 40 bushels commonly allowed per acre, Read has written "15 bushels."

The Husbandry of the Soil

of ponds & lye a yr or two—Mortimer [56] say[s] 160 bushl to an acre. Lays in a heap till slacked, & then sowed hott.[45]

100 bushl to ye acre last wth small addition 12 yr. [*Compl. Body of Husb.* 82.][46]

[*Burnt Clay*] Burnt Clay 40 bushells to an Acre.

[30, 60] Buckwheat[47] plowed in a Dressing for 3 years, on Clay it will do. J. Morris[48] says not on his Sandy Land wch He tryed.

[75] Lime Ashes, Brine, Soot, kill flies on Turnips &c.[49]

[76] Urine good for Vines.

[*Ashes*] [83] Ashes of Coal best, then Wood Ashes brings Clover & Kills Moss. Ashes of Coal or Wood 4*d*. pr bushl. 30 ms from London *Compl. Body of Husb.* says they do best alone 80 bushl sown pr acre.[50] [NB Ashes brings in grass for 2 or 3 yrs but the ground after turns

---

[45] Taken from the following passage by Mortimer (*op. cit.*, I, 86): "But the Stone-lime is much the best for Land, and indeed for all other Uses; which in many Places they carry out upon the Land, and lay in heaps, allowing a Bushel to a Pole-square, or a hundred and sixty Bushels to an Acre, which they cover with Earth, letting of the heaps lie till the Rain slacks it, and then they spread it: But they reckon that if 'tis carried out upon the Land hot from the Kiln, that 'tis best."

[46] "If the Land be very sedgy . . . set Fire to the dry Stuff on the Surface. . . . The Lime is then to be sprinkled over these Ashes, about an Hundred Bushels to the Acre, and all plowed in. . . . The Dressings of Lime and Mould, are the most durable of all: and with a little Refreshment of the Land, at Times, will last ten or twelve Years." (Hale, *op. cit.*, p. 82.)

[47] Called "French-Wheat" in *The Practical Farmer* . . . (pp. 28-30, 60). The reference to clay is apparently Read's, by way of comparison with Mr. Morris' observation.

[48] Because only the initial "J" is given, it is not possible definitely to identify this reference. It might reasonably apply to "Justice" [Robert Hunter] Morris, with whom Read was intimately acquainted; or it might apply to another member of the prominent family connected with Tinton Manor in Monmouth County. For example, John Morris of Shrewsbury, in 1768 advertised for sale a farm of 110 acres at Long Branch. A foretaste of New Jersey's seashore resort business is suggested in the advertisement, a portion of which reads: "It is well wooded and watered, and in very good fence, has good fresh and salt meadow, sufficient for any stock a judicious farmer would keep on it. Its situation is healthy, and commands a fine prospect of the Western Ocean, and as fine a country view as any the county of Monmouth affords; it is within a mile and a half of a fine river, that abounds with fish and wild fowl, and where there is a public landing. The sea bass banks lie before the door, which afford fish of the best kind, winter and summer. The advantages of such a situation must recommend it to those who want a farm for profit, and the pleasantness of it, to those who want a cool air, and fine prospect. It is very conveniently situated for any person that would take lodgers, or keep a tavern; or any gentleman that has an inclination for a summer seat on the seaside, where he will be troubled with no mosquitoes or other vermin, that render most places, near the salts, disagreeable." (*N. J. Arch.*, 1st ser., XXVI, 12-13; *Penn. Gazette*, No. 2038, Jan. 14, 1768.) Although the evidence is not conclusive, the reference in the text to J. Morris' sandy land fits the character of this vicinity.

[49] ". . . Slug, Caterpillar, Fly, and Worm . . ." (*The Practical Farmer* . . . , p. 75).

[50] Hale (*op. cit.*, p. 89) recommended 80 bushels to an acre for pasture land, but 50 bushels for an acre of wheat.

mossy if ye ground is wett.[51] Soap ashes should be putt on in Novr.[52]—*Compt. Body of Husb.*[53] says The Ashes used in bucking of cloaths of little Value.]

[84] Blood or any Sort of Carrion good[54]
[86] Cleaning of Ponds or Ditches good for Hops
[81, 96] Marl or Clay on Sand—Sand on Clay[55]
[87] Marl must Mellow.[56] [Mortimer 61 & 63.[57] 500 load on Sandy Land 200 Load pr Acre & some 400 load.]
[78] Sheep dressing by ye fold best for all sorts of Land in Surrey they fold only in Winter.

That in flanders they putt earth & Sand to their Sheephouses & thereby 3[00] or 400 Sheep will make 1000 Load of Manure in a year. [Mortimer 67][58]

Ellis [*Pr. Farmer* 85] says that the Sandy Loom is the best Land in England for Corn, Carrotts, Turnips &c.[59]

[51] Evidently Read's own comment.

[52] Probably taken from Worlidge's text (p. 73), where Read underlined the following italicized passage: "It's best to lay them [soap ashes] either on Corn, or Pasture, *or Meadow in the beginning of Winter.* . . ."

[53] The complete passage (*op. cit.*, p. 89) reads: "Some Farmers think themselves very frugal in buying up the Wood-ashes, after they have been used in the bucking of Linnen; but they deceive themselves in this, for those Ashes have altogether lost thei. Salt, and are little more than so much barren Dust."

[54] ". . . dug in at the Roots of Trees" (*The Practical Farmer* . . . , p. 84).

[55] Ellis (*ibid.*, p. 95) wrote, "red Clay . . . will not answer in Lent-Grain, without the help of Chalk or Sand." As bearing on this point, Read underlined the following passages from Worlidge's text (p. 67,) in which the author referred to "light sandy ground which was good for little or nothing, *cured by laying thereon a great quantity of stiff Clay-ground which converted it to good temperament*," and "stiff Clay-grounds that would seldom be fruitful unless the season of the year proved very prosperous, *to have been cured by laying thereupon a great quantity of light sandy-ground.* . . ."

[56] "Marl is accounted one of the best Improvers of Land, but not of itself to be good for any thing; . . . For if we take marl fresh out of the Pit, and sow the Seeds most familiar to it, they will not grow upon it; but let it remain some time in the Air to mellow, it will bring any seed to perfection." (*Ibid.*, pp. 86-87).

[57] Mortimer (*op. cit.*, I, 94, 97) reported that in Staffordshire the usual application of marl was 200 loads per acre, except upon a black, loose, or a sandy mould, or a "wormy" ground, in which case 300 to 400 loads were used. On sandy or gravelly soils an application of 500 to 600 loads was not considered excessive. In Worldige's discussion of this question (p. 66), Read underlined the passage in the text which reads: *"in medio virtus"* [In moderation is virtue.] *It's better to erre by laying on too little than too much.*

[58] Mortimer (*op. cit.*, I, 103) here quoted Samuel Hartlib (*A Discours of Husbandry used in Brabant and Flanders*, London, 1650, p. 13), which gives an account of Sir Richard Weston's observations of agriculture in the Flanders region. Also see *Hartlib's Legacie* (London, 1651, p. 45), and his *Legacy of Husbandry* (London, 1655, pp. 34-35).

[59] Ellis' text (*op. cit.*, pp. 85-86) reads: "The Sandy Loam is one of the best Soils for Corn in England, and indeed for any thing else, where it happens to be deep enough; as six or seven Inches, 'tis admir'd for Carrots, Parsnip, or Turnips."

The Husbandry of the Soil

The Seldomer yo plow sand the better. [Ellis 68 & Home's *prin Agr* Contra Elliott.]

Plow swardy ground shallow [Ellis 69-71.][60]

Experiments

An Extract from Home's principles of Agriculture & Vegetation[61]

He is of Opinion That ye faults of Sandy Soil are having too few Nutritious particles & letting the Water pass thro' too Easily.

That Woolen raggs help in both cases.

That he thinks Moss (wch is a spongy Marsh mudd) would be an excellent manure & that it had been in part tryed & did well.

He made ye following Experiment on Sandy Land in p[lot]s 3 ft. square

		July ye 2d it proved	Aug: 13 how proved
No. 1	Covered & mixed 2 inchs of Clay	bad	bad
No. 2	Do wth 3 inches clay	bad	Bad
No. 3	Do wth 4 inches clay	bad	Bad
No. 4,	2 Inches clay wth com Quant. of Lime	bad	Bad
No. 5	3 Inches Clay wth com Quant. of Lime	bad	Bad
No. 6	4 Do - - - - - - Do	bad	Bad
No. 7	2 Inchs Clay & usual Quant of Dung	very good	Exceeding good & heavy grain
No. 8	3 Do — & Do	exceeding good	
No. 9	4 Do — & Do	exceeding good	
No. 10	6 Inches of Clay	very bad	Quite witherd
No. 11	Do—wth usual Quant Lime	very bad	
No. 12	Do wth usual Quant of Dung	exceeding good	Exceeding good
No. 13	The poor Sandy Soil wthout any addition	worst of all	good for nothing
No. 14	The usual Quantity of Lime	bad	
No. 15	The Usual Quantity of Dung	pretty good	pretty good

[60] Ellis advised against the frequent plowing of "Gravels, Chalks, Sands and such light Grounds" and recommended shallow plowing of "grassy Crusts" (*ibid.*, pp. 68, 69). Eliot was an advocate of deep and frequent plowing (*op. cit.*, pp. 64, 103). Sir Francis Home cautioned against plowing sandy soils too often and thin soils too deep: "The farmer . . . must take care not to go below the soil in plowing, else he will bury what has been benefited by the air, and expose to the air what perhaps cannot. . . . The stiffer the soil, the oftener should it be ploughed. Clay cannot be ploughed too often; lighter soils perhaps may. . . . gravelly soil may be hurt by too frequently plowing." (*Principles of Agriculture and Vegetation*, Edinburgh, 3rd ed., 1762, pp. 166, 167.)

[61] Read here has made a complete outline of Home's experiment, and has noted the essentials of the descriptive text. (*Ibid.*, pp. 24-27, 37-38.)

252   Ploughs and Politicks

The Conclusions. That Dung alone fructifies Sandy Land, but Clay alone or Lime alone does very little good. That a mixture of Clay & Lime does little good, but clay [&] dung enrich to a great degree. Oats did best where only clay & Lime was [used] but where clay & Dung Barley & pease did best.

Mossy Land ie (Spongy rooty wett Soil) from Homes principles of Agriculture &c [37]

Experiment. Peat burnt in ye open Air gave 1/32 Alkaline Salt whereby it appears to afford ye same principles as other Vegetables.

The Only Method to make this Vegetable undergo a degree of putrefaction is by plowing & thereby killing ye plant all vegetables turn to a rich soil when corrupted, even oak bark will do so by Long lying As Moss is an Enemy to Corruption ye parts will be seperated sooner by adding some other Earth. Shell Marl or Lime he thinks would be best Lime is a great dissolvent of Vegetable bodies.

Hedges & Ditches

It is said that if a hedge is to be planted on a Ditch near a Wood you should have the Ditch between ye Bank & ye Wood & that thereby you Escape having ye Growth of the hedge by the Nutriment the trees draw from ye Ground.

The *Transactions of ye Scotch Society* advise hedging wth Sweetbriar & that it grows sooner than any other Hedge.[62]

A kind of Plumb hedge at ye E End of Long Island is by some recommended.[63]

Weeds noxious to Destroy

[Elder] Mr. Elliot[64] says that he has Experienced that cutting them 5 times a year will Destroy them.

[62] Robert Maxwell, *Select Transactions of the Honourable The Society of Improvers in the Knowledge of Agriculture in Scotland*, Edinburgh, 1743, pp. 140-141. In a letter to Mr. Hope, of Rankeilor, Mr. Fullarton, of Gallarie, gives a detailed description of his method of growing a hedge of Sweet-briar or Eglantine plants. According to Mr. Fullarton, in four or five years they "make a Fence, that no Sheep, Cattle or Horses, can pass."

[63] Benjamin Franklin, in a letter written to Jared Eliot in 1750, asked for full information about "a new kind of Fence we saw at Southold, on Long Island, which consists of a Bank and Hedge." "We were told," he added, "that this kind of Fencing had been long practic'd with Success at Southampton and other Places, on the South Side of the Island. . . ." (A. H. Smyth, *op. cit.*, III, 31.)

[64] Eliot (*op. cit.*, p. 23) wrote: "Elder Bushes are stubborn and hard to subdue, yet I know by Experience that Mowing them five Times in a Year will kill them." John Bartram, the Quaker botanist of Philadelphia, in 1758 wrote of the elder to Philip Miller, author of *The Gardener's Dictionary* as follows: "Our Elder is exceedingly troublesome in our meadows. The roots run under ground and spread much; and I do not know that mowing will ever kill it; and grubbing will kill little

The Husbandry of the Soil

☞ In cutting up Weeds cutt but little dirt wth the roots & when yo have gott ye top off thus pass yr hoe again among ye roots if occasion be for by cutting up but little dirt wth the weeds they die sooner.

[*Docks*] Solomon Shinn[65] says that in May soon after the Dock grows if yo dig a hoe or spade round them & make a Circle abt as big again as a Dollar the[n] cutt up ye Dock & the small fibres will bleed to Death wch he thinks they will do at no other time. 1760 I tryed ys with success.

[*Rushes*] William Hancock Esqr[66] tells me that he thinks cutting Rushes 3 times a yr destroys them better than cutting up Some other Experienced Farmers say that if yo set ye edge of yr Scythe so as to cutt them up just in ye Ground in hott dry weather it will Effectually snubb them & being twice so served destroy them better than cutting them up.

Mow Rushes 2 or 3 times a yr better than cutting them up. Wm. Hancock of N[ew] S[alem].

---

more than the mattock takes up, for if there is but a little bit of the root left in the ground, it will grow. I have had a root growing in my kitchen garden about thirty years. It was ploughed once every year, and generally grubbed and hoed once or mostly twice, every summer; yet, last summer, two stalks put up, and if there is an inch of root left in the ground, if it be two feet deep, it will put up again. In short, I believe there is not a shrub in the world, harder to eradicate than our Elder." (Darlington, *op. cit.*, p. 387.)

[65] Between 1759 and 1782 four transfers of land in Burlington County to Solomon Shinn are on record—in Burlington, and in Springfield, Evesham and Northampton Townships. In 1779 he was assessed for 480 acres of land in Springfield Township, together with 9 horses, 22 cattle and 10 hogs. The will of Solomon Shinn of Northampton Township, was probated in 1785. His personal estate was inventoried at £3236-18-6. It included a barrel of dried beef, 4 tons of hay, 4 cows, 9 sheep, 28½ bushels of rye (£19-12-6), and 24½ bushels of Indian Corn (£4-18-0). (N. J. Arch., 1st ser., XXXV, 350; N. J. Wills, 11062C; Burl. Co. Tax Lists, 1779.)

[66] William Hancock about 1725 inherited from his father one of the largest landed estates in Salem County. In 1734 he erected a substantial brick dwelling on Alloway's Creek, near the bridge built by his father, known as Hancock's Bridge. He was appointed justice of the peace and served in the colonial assembly from 1733 until his death in 1762. His estate passed to his son William, popularly known as Judge Hancock, who was justice for Salem County 1767 to 1778, and was a member of the Salem County committee of observation. In 1774 he advertised for sale a tract of marsh as "very fine pasture for cattle, the whole summer through." He died in the massacre of Hancock's Bridge, March 20, 1778, when he and other Quaker noncombatants who occupied the Hancock House, were slaughtered by a raiding party of British and Tories. His personal estate was appraised at £1079-12-10. The inventory included books valued at £4-12-0, wool at £12-8-0, wheat at £24, flaxseed at £4, a Dutch fan at £2; hay, corn, and oats at £35-0-2; 13 head of cattle, £119; sheep and hogs, £70-10-0; 4 mares and colt, £140. (N. J. Arch., 1st ser., XIX, 392; XXIX, 203-204; XXXIV, 224; *Penn. Gazette*, no. 2351, Jan. 12, 1774; N. J. Wills, 1769Q; Salem County Historical Society, "Colonial Roof-Trees and Candle Ends," 1934, p. 22; *Manual of the Legislature of New Jersey*, John P. Dullard, compiler, Trenton, 1931, p. 174.)

Two: The Husbandry of Plants

IN THE NOTES here presented Charles Read reveals a broad studied knowledge of a large variety of crops. As with other branches of husbandry, his own practices and observations are added to the experiences of divers acquaintances and are compared with the recommendations of writers on agriculture, both British and American.

PASTURE AND FORAGE CROPS

Of pasture and forage crops, he treats a long list: red clover, white clover, timothy, herd grass, fowl meadow grass, rye (ray) grass, lucerne (alfalfa), St. Foin (sainfoin), trefoil. He quotes at length from Eliot, Ellis, Bradley, Hartlib, Duhamel, Hale, Ball, and Mortimer. He writes of his own experience with red clover and timothy, and indicates that he also grew fowl meadow grass. He records that in May, 1756, a Captain Reeves brought him from London 50 pounds of red clover seed, 15 pounds of trefoil and 3 bushels of ryegrass, and "it grew well." Direct evidence is lacking that he actually grew any of the other crops, although his notes on lucerne are extensive. It is not always clear from his notes whether the observations recorded are his own or others'. His mathematical approach is again revealed in his computation of the distribution of red clover seeding: 38 clover seeds in one grain (Troy); therefore 218,880 seeds in one pound; therefore 2,188,800 seeds in 10 pounds, or the amount applied to an acre, which contains 6,272,640 square inches; therefore the area per seed would be a little less than 3 square inches.

He is impressed with the value of red clover, introduced into England from Flanders in the Seventeenth Century. He experiments with seeding different amounts of this crop per acre, at different times of the year and on different soils. He finds it will not do well on poor, worn-out land, but on rich meadow

The Husbandry of Plants

land he gets a yield of 3½ loads per acre at the first cutting. His observations of the growth of red clover from its earliest sprouting are recorded in the notes, and in similar phraseology he wrote of them to Jared Eliot. He gives attention to harvesting the crop, both for hay and for seed. After noting the methods of threshing followed in other lands, he tries "dew-rotting" timothy seed before cleaning it, with success.

He is interested, too, in the most effective method of feeding clover. He finds that soiling, or feeding the forage green, gives a larger yield of milk than pasturing, but concludes that it is not always economically practical. The method "will do well for a large Quantity but breaks in too much on a mans work to tend on a small Quantity."

In dealing with red clover, he quotes extensively from Jared Eliot's *Essays,* and in addition cites Eliot as authority for certain procedures in haymaking that are either missing from the *Essays* or do not coincide strictly with the text. The marginal note, "Elliott Advises to Cock yr hay by 4 o'clock in the Afternoon," he probably received from Eliot by letter.

So carefully had Read consulted the English authors, he discovered that Ellis, in *The Practical Farmer or Hertfordshire Husbandman,* had copied almost verbatim the passage in Worlidge's *Systema Agriculturae,* relating to Hartlib's comment on clover.

GRAIN CROPS

Of the grain crops, Read comments briefly on rye, buckwheat, millet, rice, wheat, and speltz, and deals at length with Indian corn. While Eliot's treatment of corn is devoted mainly to the matter of tillage, Read's notes abound in other practical facts about this crop which in the late colonial period was grown in larger quantities in New Jersey than in Pennsylvania. He presents homely advice on the time of planting, on hoeing and harrowing, on topping and harvesting, on devices for protecting the young plants from predatory birds and the standing stalks against damage by the oxen's yoke.

Red Clover Grass

I know this is true

Mr. Elliott allows 10 lb to an acre. that it will not grow on old worn out Land, that 2 quarts will raise to a good burthen but that it will not be thick at Bottom

Ellis's practical farmer & the more purple ye Seed is the riper

That an Acre of Clover will yield 2 bushells of clean seed. That the second crop yields best seed That if you Depend on ye 2d Crop for seed the Land must be very rich & the 1st Crop taken off early

Bradley says the Second is the Seed crop

Elliott says that some of their drained Lands in Conecticut yield 4 load of Clover hay in a season —

Hartlib says that it was brought from Brabant in Flanders soon after the Civil Warr & the use of it soon put our Farmers in a flourishing condition

Clover Seed 2 bush w. 120 from an acre vide also page 2

I sowd ¼ of an Acre of a rich Meadow & it was not too thick & I cutt 3½ Load of hay of ye 1st Crop —

1753 NB it kept them from April to Octor & Increasd their milk cows in their milk ⅓

Mr. Warrell at Trenton fatted Eight Cattle well on something less than 2 Acres I have seen the Ground & it appeared to be a Maple run & not very rich, He cutt for the Cattle 3 times a day & fed them in Racks vide this book folio 27

This Method will do well for a large Quantity but breaks on a man's work & bends in too much on a small Quantity

1756 ½ Acre will keep at this rate 2 Cattle days

Extracts from Mr Trenchards Letter to me of ye 8. Septr 1756 respecting an Experiment I made w. two oxen at my request, That he fed 2 oxen 6 weeks of White Clover that they eat 3 square perch in 2 days that they fattd more by this Method while confined at that diet mowed & given 3 times a day than in double y. time in y. weeks at pasture. & well watered that he thinks each ox will make 1 load of Dung pissd in the Litter. That by pasturing an Ox destroys more in feet dung and Urine than Ho eat &c Know in a hott Day destroy even y. roots of y. Grass where it falls

Typical Page from Read's Manuscript

The Husbandry of Plants

In 1756 he made a trial of millet in his garden. Planted April 29, it tasseled July 12 and was cut August 24, yielding at the rate of 67 bushels per acre. This was more than twice the yield that Eliot got on poor land. Read concludes that millet is best cut green for cattle feeding.

By the same method of computation as he applied to red clover, he finds that wheat seeded at the rate of one bushel to the acre supplies but one grain in 35½ square inches—which is apparently too sparse for the best yields. He underlines in the text the Latin quotation: *Triticum luto hordeum pulvere conferite*—"Sow wheat in damp ground, barley in dry." Considerable space is given to notes on the culture of rice.

To the steeping and brining of seed grain, he gives more space than to any other phase of these crops, in the main quoting from Ellis, Bradley, Mortimer, and Hale. He critically scans their recommendations, noting that Bradley failed to state the quantities of water and of salt needed to steep a given amount of grain. Then to these English recipes, he adds a Dutch recipe from William Logan.

By grinding or boiling, the nutritive value of grain is enhanced, and he notes the proportionate gains for rye and Indian corn thus processed. For reference he gives the standard weights of grain per bushel for New Jersey and Pennsylvania, and he records the grain prices in London in 1753. From an experiment made in 1767 he notes that 70 pounds of rye flour made 89 pounds of bread.

FIBRE CROPS AND DYE PLANTS

The fibre crops, hemp and flax, played an essential role in colonial economy. Read treats the former in detail, the latter but incidentally. Nearly four manuscript pages are occupied with "The Method of raising Hemp in New Jersey"—a thoroughgoing practical treatise covering the choice and preparation of soil, planting, cultivation, harvesting, storage, and manufacture. Jared Eliot likewise, in his *Essays*, gave more space to hemp than to flax. He urged American farmers to grow more hemp. He himself raised it on a small scale, and briefly discussed its culture. But Read's notes provide a wealth of facts on hemp not mentioned by Eliot.

Eliot urged the use of drained meadows for the planting of hemp. In this Read's notes concur, pointing out that "Meadow Land is very Good for it," but the "Tap Root must be above the Water." Eliot refers to "a Man in the Jersies" who had "such a Meadow of half an Acre, which yields him as much Hemp Yearly, as fetches him *Fifty Pounds*," but observes that "this seems incredible."[1] Read adds also that "good uplands well Dunged will bear it well." Of the author of *The Practical Farmer* he notes, "Ellis . . . says wett ground will do well but herein he Certainly mistakes." For best results, according to Read's notes, plowing for hemp should be done in fall or early spring and "The Ground should not be so wett but that it will tumble off the Mould Board." Harvesting requires special care—in order that the stalks be not damaged. An approved technique is described, and the warning given, "if you don't mind the Labourer in this he will deceive you."

Read's treatment of hemp and flax reflects the public interest in these crops engendered in New Jersey during the Eighteenth Century. Lewis Morris, while president of the council in 1719, in order to encourage the raising of hemp and flax, proposed that the settlers be instructed in their culture, but nothing seems to have been done about it.[2] Certain merchants and traders in 1723 petitioned the Lords of Trade for the services of an expert in hemp production, apparently with no better success.[3] Finally, as we have seen, in 1765 the Crown approved an act of the assembly which provided bounties for all hemp and flax of marketable quality produced in the province. Whether because of government subsidy, or in spite of it, hemp culture seems to have prospered. The author of *American Husbandry* in 1775 remarked that there were "no tracts of

[1] *Op. cit.*, p. 17.
[2] *N. J. Arch.*, 1st ser., IV, 443.
[3] *Ibid.*, V, 68-69.

The Husbandry of Plants

good land in this province, without having portions assigned to the culture of hemp; which does extraordinary well here."[4]

Read gives liberal space to the culture of indigo and of madder, and to the manufacture of dyes from these plants—subjects barely mentioned in Eliot's *Essays*.

VEGETABLE CROPS

Read's notes on vegetables suggest the beginnings of the great market gardening and truck growing industry that through the next two centuries was to develop within the Garden State. He gives a valuable seasonal table, with dates for planting, flowering, and ripening of a dozen or more vegetables—radishes, peas, beans, "sallad," spinach, cucumbers, squashes, pumpkins, potatoes, turnips, watermelons, and mallow.

By the middle of the Eighteenth Century potatoes had won a permanent place in European husbandry and their culture had gained favor in the American Colonies. It seems strange that this crop is not treated at greater length in Eliot's *Essays*. Perhaps it was not then so important in New England as in the Middle and Southern Colonies. That potatoes were grown in New Jersey in considerable quantities would appear from Read's notes—several pages of which are devoted to this crop. He treats potatoes in three classes—sweet potatoes, Bermudas or red potatoes, and Irish potatoes.

From the description it is evident that the "Bermudas" were a red sweet potato. Here again Read applies the science of farm management, noting prodigious yields and predicting handsome profits. On Duke Fort's farm he measures off one-seventh of an acre, and computes the yield—326 bushels per acre! A bushel of potatoes weighs 60 pounds. Allowing 1*d*. per pound, "wch is half ye usual price," the income is about £80 (or almost $400) per acre. No wonder this crop captured the author's fancy! But, he reasons, "such a Quantity might not Sell." In that event, he advises, feed them to the hogs. Suppose

[4] *Op. cit.*, I, 137.

3 bushels of potatoes as the equivalent of one bushel of corn. Then the yield of an acre would equal 109 bushels of corn. And Duke Fort had proved they were good hog feed. One year at harvest time he (Fort) had bought two young pigs, weighing perhaps 15 pounds (he had carried them home on his back). They were put on a raw potato diet, and when killed at Christmas time they weighed 150 pounds each.

In treating of turnips, Read quotes Eliot and Hale, and draws upon the experiences of his friends. Thomas Bowlsby, "a good farmer," says one pound of turnip seed per acre is the right amount; Mr. Saltar, on the other hand, says ½ pound is too much. Read doesn't take sides, but is content to set forth the two points of view. To preserve turnips from "the fly," it is advised, treat the seed with flour of brimstone. Or sow turnips between rows of potatoes and "the fly never hurts them."

On carrots Read quotes Eliot, Duhamel, Hale, and Ellis, noting the superiority of these roots to turnips as a feed for sheep. Parsnips, he notes, "may certainly be planted to advantage for Hoggs as a vast Quantity will grow on an Acre." Opposite Worlidge's text that mentions the ease of growing radishes, he writes, "I have proved this."

As to the culture of onions, one of his friends states that on good land a yield of 180 bushels per acre may be expected. Read records that he himself sowed onion seed September 1, 1759, during a rainy period, and the seed came up in 10 days.

The author of *American Husbandry* observed that cabbage was grown by almost every planter in New Jersey; whole fields were common, and quantities were fed to cattle and to hogs.[5] Read, in his notes written several years earlier, discusses cabbages as a cattle feed. They "give butter an ill taste but may be fed fatt[enin]g cattle." He computes that 14,520 cabbages can be grown on an acre. Allowing 12 heads of cabbage a day per cow, an acre will keep a cow 1210 days, or 20 cows 61 days. He computes the cost of growing an acre of cabbages at £17, that is, the cost of feeding 20 cows for two months.

[5] *Op. cit.*, I, 136-137.

The Husbandry of Plants

Hops, Read finds, are little more trouble to grow than Indian corn. A "tolerable" yield in New Jersey would be 400 pounds per acre. This crop in his day sold at 1s. a pound.

Most of the comment on asparagus appears to come from Bradley's *New Improvements in Husbandry and Gardening*, but marginal notes indicate some observations from Read's own experience. From other sources we know that asparagus was a well established crop in Read's time.

Under the heading "Calavance Peas," he discusses what appear to be several varieties of the modern cowpea, quoting extensively from the experiences of friends with this crop. "Tho' they grow well on poor land they will bear better & fuller if they have Manure." That these legumes improve poor soil is noted, but since "nitrogen-fixation" was foreign to the colonial vocabulary, the reason is not given. "Joseph Moore's Land is very high Dry & Poor," reads the entry. Mr. Moore "plowed it & planted Calavance peas several years Successively & he thinks it mended ye Land tho' not manured."

Hotbeds are described for the growing of early plants. Parallel with notes from Bradley's *New Improvements* . . . is the method of making a hotbed recommended by William Masters of Burlington. Both Bradley and Masters employed tanbark and horse manure. Advice is given also on transplanting cucumbers and on watering plants.

We could wish that Read might have written more about a universal favorite, the watermelon. He does not say how commonly melons were grown at the middle of the century, but from the author of *American Husbandry* we learn that a few years later the watermelon patch was a part of every well ordered farm: "Water melons also are in such plenty, that there is not a farmer, or even a cottager without a piece of ground planted with them," he writes. "The country people eat them . . . at all times of the day . . . in the same manner as a labourer in England would drink ale or small beer . . ."[6]

[6] *Ibid.*, I, 140.

Read is likewise silent on muskmelons, although they were grown in New Jersey as early as 1700.

As a source of data on Eighteenth Century vegetable culture, Read's notes are much more informing than Eliot's *Essays.* Eliot treats turnips, carrots, and cabbages at some length, but his *Essays* lack the details of cultural practice given by Read.

FRUIT CULTURE

Read's notes on fruit culture are not as complete or as well balanced as might be desired. With the exception of peaches, he barely touches upon the principal tree fruits, even though they were of major importance in the Middle Colonies. In 1775, for example, the author of *American Husbandry* wrote:

> Every farm in New Jersey has a large orchard belonging to it, some of them of a size far surpassing anything in England. The common fruits are apples and peaches, with some cherries and pears; the peaches are of a fine flavour, and in such amazing plenty that the whole stock of hogs on a farm eat as many as they will, and yet the quantity that rot under the trees, is astonishing.[7]

Nevertheless, Read makes a worthwhile contribution to pomological literature, particularly so because this branch of husbandry, with the exception of mulberry culture, is scarcely touched in Eliot's *Essays.* Added interest is given by Read's references to several prominent figures of the colonies—Robert Ogden, Colonel Peter Schuyler, and William Logan, who in addition to their public duties found time for horticulture.

Notably practical and revealing are Read's comments on peaches, a crop to be closely identified with New Jersey in its development through the Nineteenth and Twentieth Centuries. He alludes familiarly to "inoculation," or budding. He notes the effects of soil, closeness of planting, the proper height for pruning, and relationship of color of stone to flavor. He considers the ancient enemy of the peach—the borer, or the "worm at ye root," and advises that "The Hulls of Walnutts mixed wth the Earth keeps the Worms from them." Here is

[7] *Ibid.,* I, 139.

the germ of the chemical warfare method successfully developed against the peach borer a century and a half later.

He alludes briefly to mulberries. Unlike Eliot, he does not mention silk culture in connection with this crop. Referring to both Worlidge and Mortimer, he records a method of propagating the seed which he himself tried. He advises, too, that horse-radish planted around plum trees "prevents the Bugg from destroying them."

Gooseberries, currants, and apricots, as treated by Worlidge, are the subject of marginal notes. He draws extensively upon Hitt's *A Treatise of Fruit Trees* and Bradley's *General Treatise of Husbandry and Gardening*. Of Worlidge's chapter on "Inoculation," he observes "This Whole Section copied into Mortimer *Art of Husbandry* . . . & nothing more added"—a common form of abuse of which Benjamin Franklin had complained, which now did not escape Read's critical eye. References appear here and there to pears, quinces, cherries, almonds, and apples. Strawberries and cranberries are not mentioned; though important as wild fruits, in Read's time they had not yet attained a status of "culture" to command his consideration.

He shows by diagram how to place stakes about a tree to prevent injury by cattle rubbing against it. He records his own experience in grafting—it is most successfully done in moderately warm weather; thick grafts (not larger than a man's little finger) are better than small weak shoots. Winter fruits should be grafted on winter stocks; otherwise, he observes, "yo will be deceived."

He observes that the operation of inoculating, or budding, fruit trees cannot be adequately expressed except by "Ocular Demonstration." This reminds us of the noteworthy passage in his letter to Jared Eliot, in which he criticized the farmers of his neighborhood for their refusal to adopt improved practices, "where a Gentleman of a more Publick Spirit has given them Ocular Demonstration of the Success." Here was the germ of the idea of farm demonstration, the educational process which is the foundation of our modern extension service in agriculture and home economics.

For grape culture and wine making, Read draws almost exclusively from *The Vineyard*, a pamphlet published in London in 1727, which he abstracts with thoroughness and precision. In 1774 he enters, as a marginal note, the advice of Doctor Vanrow of St. Croix on the planting of slips of the vine.

Although Read does not mention Lord Stirling in his notes, he shared with this fellow member of the provincial council a common interest in viticulture. Stirling for a number of years carried on an experimental study of grapes. It was his conviction that with government backing, wine making would become an important industry in the American Colonies. In 1763 he wrote to the Earl of Shelburne:

> The making of wine, also, is worth the attention of Government. Without its aid, the cultivation of the vine will be very slow; for of all the variety of vines in Europe, we do not yet know which of them will suit this climate; and until that is ascertained by experiment, our people will not plant vineyards;—few of us are able, and a much less number willing, to make the experiment. I have lately imported about twenty different sorts, and have planted two vineyards, one in this Province [New York] and one in New Jersey; but I find the experiments tedious, expensive, and uncertain; for after eight or ten years' cultivation, I shall perhaps be obliged to reject nine tenths of them as unfit for the climate, and then begin new vineyards from the remainder. But, however tedious, I am determined to go through with it. Yet I could wish to be assisted in it. I would then try it to a greater extent, and would the sooner be able to bring the cultivation of the grape into general use.[8]

The culture of semi-tropical fruits under glass was not unknown in New Jersey in Read's time. Colonel Peter Schuyler had a famous garden, which apparently captured Read's interest. The notes include directions for the management of orange trees in a greenhouse, furnished by Colonel Schuyler's gardener.

---

[8] W. A. Duer, *op. cit.*, p. 76.

SHADE AND FOREST TREES

Silviculture for shade and for timber also seems to have received Read's attention, as might be expected of as public minded a person as he, who owned thousands of acres of woodland. From Bradley's *General Treatise of Husbandry and Gardening* is copied almost verbatim a method for transplanting large trees in summer. He notes the relative merits of buttonwood, English elm, locust, maple, and catalpa. He advises the planting of shade trees in meadows, for, he says "Shade in hott weather is absolutely necessary for cattle as food or water." His neighbor Mr. Tonkin advises apple trees for shade in pastures, "Tho' tis Agreed nothing dry up Milk so much as Apples."

The white oak, he writes, is "the best in this Country for lasting," noting in 1761 that a set of oak posts had stood "Strong" around S. Cole's garden for 32 years. The oak also is used for "Waggon work." For fence rails, black ash is better than cedar or chestnut. Shellbark hickory likewise makes good rails.

He notes Worlidge's advice as to the best time for cutting timber, and observes that near Elizabethtown "stave trees" are sawed down in order to save the butt—a measure of economy early recognized midst a plentiful supply of timber. Accompanying these notes is a full-size diagram of a white-oak leaf, with a detailed description—indicating an interest in the botany as well as the economics of forest trees.

ECONOMICS OF CROP PRODUCTION

In his discussions of various crops, it is evident that Read was "farm-management conscious." Eliot's *Essays* are of a philosophical turn; his arguments are developed from the point of view of national economy. Read, on the other hand, deals with the various phases of agriculture from the business point of view—economical management, cost of production, yields, prices, markets, profits. His main concern is with the success of the individual farm enterprise, and the profit of the individual farmer. The mercantile background of his family, his experiences in England and in the West Indies, his contact with shipping, all would naturally tend to develop a concept of farming as a business enterprise. His notes bear the earmarks of the modern textbook on farm management.

As an example of a balanced farming enterprise, he cites Captain Tibout's farm 1½ miles from New York, in 1757. Here is an example of a farm adjacent to a city, which took advantage of the special market demands, supplied it with "garden truck," and took back the stable manure of the city to improve the fertility of the fields. Here is a foretaste of New Jersey's intensive market gardening industry which materialized a century later.

He introduces also references to crop management practices in other states, for example the use of negro labor in Maryland in growing tobacco and corn; and notes on threshing grain.

Though not intended as a treatise on crop husbandry, Read's notes in the aggregate probably present a more complete picture of the fruits of the field in the Middle Colonies than any other known source prior to *American Husbandry*.

Hay and Pasture Crops

Miscellaneous Collections & Observations on Low Lands, Ponds, & Meadows

[March 1748] Mr. Elliott of Connecticutt says in his *1st Essay on Field Husbandry* [page 7] That he drained a pond[9] in Guildford woods wch was swarded over & the ground under it was so intermixt wth Water that it was of the Consistance of Pancake batter there were few bushes in it but abundance of Cranberry vines that when the weather grew warmer he sowed it with red clover, Speargrass, foul meadow grass & Herdgrass which is called Timothy grass wth us. They were all pretty much overgrown wth

[9] Read's notes here refer not to a pond but to a "shaking meadow" which Eliot drained; *i.e.*, "A Man standing upon it might shake the Ground several Rods round him" (*op. cit.*, p. 11).

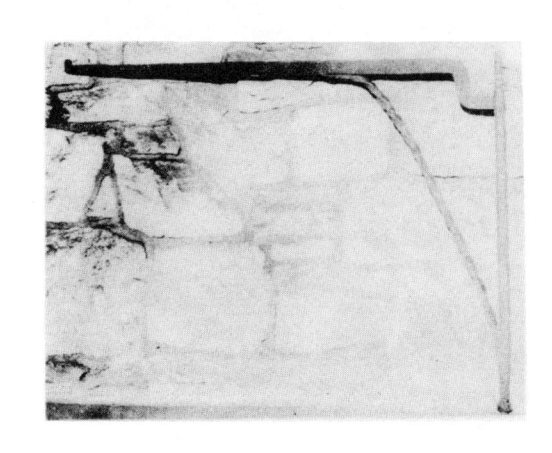

Red Cedar Post from Sharon. Courtesy of Mr. Barclay White
Fireplace Crane from Breezy Ridge. Courtesy of Mr. Joseph B. VanSciver

Colonel Charles Read's Wallet. Courtesy of Mr. James C. Read
Read's Supreme Court Docket. Courtesy of Mr. Nathaniel R. Ewan

The Husbandry of Plants

poor wild grass but that the Clover took good root & where there was no Sward to Choak it grew up midthigh ye 1st year.

J. Hough,[10] A Neighbour of mine tells that in February He hauled three load of Dung & spread it on a small spott of Meadow but did not perceive it did any good the Grass was just the same as formerly but this was a rich meadow 3 or 4 ft Deep in rich Mould.

Red Clover Grass

Hartlib[11] says that it was brought from Brabant in Flanders soon after the Civil Warr & the use of it soon putt our Farmers in a flourishing condition.

I sowed 12 lb pr Acre of a rich Meadow & it was not too thick & I cutt 3½ Load pr Acre ye 1st Crop. [Clover seed 2 bushl or 120 lbs from an acre.]

10 lb. is a good proportion rather exceed than fall short.

in Aug. 1749 I Sowed abt 10 lb to an Acre & where it was thickest it stood the winter best—The Ground was ye deep Meadow & very

---

[10] Two men by the name of Jonathan Hough, probably relatives, lived in Springfield Township, one or both on a farm near Charles Read's Sharon. One of these in 1749 sold Read for £70 a 6½-acre tract of meadow land in this vicinity. Public offices held by one or the other were justice for Burlington County, surveyor of the highways and member of the county committee of observation. The first of these Jonathans to die (in 1778) left a personal estate inventoried at £1110-19-8. Three years later the will of the second Jonathan was probated, providing that his two servants—"Peter" and "Mintes" be set at liberty. In the inventory of his personal estate, totaling £1101-3-7, his "Cattle, Horses and other Live Stock" were valued at £270; hay, green corn, and other grain £155-0-0 and "Waggon, Plows, harrow and other farming utensils" at £46-0-5. (E. M. Woodward and J. F. Hageman, *op. cit.*, p. 22; N. J. Deeds, Liber GG, 429; *N. J. Arch.*, 1st ser., XXXIV, 256; XXXV, 205; N. J. Wills, 10623C-10626C; 10778C.)

[11] In Hartlib's *A Discours of Husbandrie* . . . , (London, 1650) Sir Richard Weston reported his observations of the growth of "clover grass" on the Continent, as a forage crop and a soil improver. In an enlarged edition of this work published in 1655 under the title *Legacy of Husbandry* . . . , appears an advertisement which indicates that the introduction of clover from Flanders by this time was under way. It reads in part: "If any desire to have the great Clover of Flaunders, or the best sorts of Hemp and Flax-seeds of those parts . . . : Let them enquire at Mr. James Long's Shop at the Barge on Billingsgate; and they shall upon timely notice have them procured new, and very good from France or Flaunders at reasonable Rates." Read probably procured this information from Eliot *Essays* (p. 42) rather than directly from Hartlib's work. The statement checks closely with the following paragraph from Eliot: "I had often met with it, that our Nation being much Exhausted and Ruined by the Civil War, retrieved their great Losses by some new Husbandry, and in a little time Recovered themselves and got to a better State than ever; but never could learn what was this advantageous Improvement, till I found by Reading *Mr. Hurtlib's Book of Husbandry*, that it was principally by introducing this Clover Grass, called Flanders Grass; because the Seed was bro't from Brabant and other parts of Flanders."

light it was in a good wett Season & in 4 days it appeared wth 2 leaves, appeared pretty thick & of this shape ♈ the stems of ye leaves did not join directly to the trunk in 10 days some of it took ye appearance of Clover in one leaf ye others not yett opened in ye End of October ye tap root was abt 5 inches long & from it went near 30 horrizontal roots of wch some were 6 inches long, & branched yet it hove out much where thin sown.[12]

In 6 grains of Clover Seed red are 228 seeds wch is 38 seeds in a grain so that in a pound Troy there are 218,880 Seeds & in 10 lb. wch is the Allowance for an Acre there are 2,188,800 seeds.

In an acre are 6,272,640 square inches so that in sowing there falls one seed in something Less than 3 square Inches.

[1753, 8br 25] I sowed red Clover seed on an acre of pretty good orchard ground, wch had been plowed & planted wth Corn I harrowed it twice & then Harrowed in the Seed this ground was not rich & ye seed never came up that Year But Note it did the next.

I also on ye 25 8br, plowed up a good piece of upland & harrowed it, ye Same Ground had born flax that year I sowed at ye rate of 12 pints an Acre on Each it had been dry now for 6 weeks the next day it came rain & in 4 days after a NE Storm

NB this Seed did not Do well ye white Clo: &c eat out this directly.

I sowed ye 23d of August it did well & on ye 14th, 7br it did not gett well leaved before winter.

Mr. Elliott[13] allows 10 lb to an acre that it will not grow on *old worn out land*, that 2 quarts will raise to a good burthen but that it will not be thick at Bottom. [I know this is true]

Sow thin to mow—thick to feed. [Ellis's *Practl. farmr.*][14] 10 lb. or 12 lb. [Ellis's *Practl. Farmr.*][15] [Bradley][16]

[Clover seed pr acre] If yo Depend on yr 2nd Crop of Clover for seed You must mow yr 1st Crop very early & the Land must be very

---

[12] This is the record of planting which Read reported in the letter to Jared Eliot, heretofore attributed to Benjamin Franklin.

[13] *Op. cit.*, pp. 18, 25: "ten Pound of Clover Seed, which is five Quarts to the Acre."

[14] "If it is sown to feed it should be thick; if to mow, thin" (*op. cit.*, p. 47).

[15] Ellis recommended 12 pounds of clover seed per acre when sowed alone, 6 to 15 pounds if sowed in barley or oats (*ibid.*, pp. 43, 45).

[16] Richard Bradley, in *A General Treatise of Husbandry and Gardening* (London, 1726, I, 181), wrote: "The Seed-Time or Season for sowing this Seed [clover], is about March or April; and if we sow it simple, or without other Grain, an Acre will take up about Ten or Twelve Pounds of Seed, for the Seed is small; but if we sow it in Partnership with Barley, Oats, or Rye-Grass, which last they call Ever, or Everlasting Grass in the West of England, then about half the Quantity is enough." This work of Bradley's is hereafter referred to as *General Treatise*. . . .

The Husbandry of Plants 269

rich. It will yield 2 bushl. of clean seed (wch is near 120 lb.) pr acre. They thresh the Hay to beat off ye heads. [Elliotts *1st Essay* 17][17]

Elliott says the same.[18]

Compt. Body of Husb. says 2 bushl.[19]

2 or 3 [bushels of seed per acre] [Ellis's *Pract. Farmr.* 48]

red Clover may be sown wth *Rye Grass* called in the West of England Ever Grass or Everlasting Grass but that it is best sown alone in March or April & that you had better sow it with Corn than any other Grass [Bradley (182)].[20]

Bradley says the Second is the seed crop.[21]

Elliott says that some of their drained Lands in Conecticut yeild 4 load of clover hay in a season.[22]

This copied into Ellis's *Practl Farmr.*[23]

Cows horses & hoggs kept on six acres.

Ys done pr Sr. Tho. Payton in Kent.

[Ellis's *Practl. farmer* (47)[24] The more purple ye seed is the riper]

---

[17] Eliot added: "An Acre of Clover will yield two Bushels of clean Seed" (*op. cit.*, p. 18).

[18] Written in the margin opposite the following passage from Worlidge's text (p. 27): "a rich, light Land, warm and dry, . . . *in which it* [*clover*] *principally delighteth*. . . . It will also prosper and thrive on any Corn-land, well manured or soiled, *and brought into perfect Tillage.*" Eliot's comment is: "If you depend upon the second crop of red Clover for your Seed, the Land must be very rich, and you must Mow your first Crop early." (*Op. cit.*, p. 18.)

[19] Hale (*op. cit.*, p. 435) noted: "Two Bushels of good Seed will generally be had from an Acre of Clover, well managed and well threshed." Read's comment is placed opposite the following statement in Worlidge's text (p. 29) which Read underlined: "*Sir Richard Weston saith you may have five Bushels from an Acre.*"

[20] *General Treatise* . . . , I, 181-182.

[21] ". . . when we have a mind to save Seed from it, we must cut our first Crop in June . . . as soon as the Clover begins to knot or joint, and the Crop following must be left for Seed, because then our Field of Clover will, by means of the Cutting, branch into more Seed-bearing Parts than it had before the Cutting, and consequently will bring a greater Quantity of Seed" (*ibid.*, 182).

[22] "Some of the most forward of those Lands have yielded four Loads of English Hay to the Acre" (*op. cit.*, p. 41).

[23] This note appears in the margin opposite the following passage in Worlidge's text (p. 27): "*In the Annotations upon Mr. Hartlibs Legacie,* we find several Computations of the great Advantage hath been made by sowing Clover-grass, . . . that on four Acres there grew twelve Loads of Hay at twice mowing, and twenty Bushels of Seed; one Load of the Hay mown in May being worth two Load of the best of other Hay, and the after-Pasture three times better than any other; the four Acres yielded in one year fourscore pounds. Another, that six Acres of Clover did maintain for half a year thirteen Cows, ten Oxen, three Horses, and twenty six Hogs; which was valued at forty pounds, besides the Winter-Herbage." These observations, as Read notes, were made by Sir Thomas Peyton, of Kent. (See Hartlib's *Legacy* . . . , 1655, p. 245.)

[24] "To know the true useful Seed of Clover, observe if there be a good quantity of

That an acre of Clover will yeild 2 bushells of clean seed. That the second crop yeilds best seed   That if you Depend on ye 2d Crop for seed the land must be very rich & the 1st Crop taken off Early.[25]

Bradley says when it begins to knott if for Hay.[26]

4 Cows pr Acre[27]

Ellis's *Practical Farmer or Hertfordshire Husbandman* [53] advises Sowing Trefoil wth Clover.

The *Compt. Body of Husb.* says Sow Ray grass wth it.[28]

Hay Making

After Red Clover is mowed & lain in Swarth putt it up that afternoon in parcells or Cocks about the bigness of a Bushel, then turn it upside down several times till it be Carted home, by this means the Leaf that would fall off in Stirring is mostly kept on, & Cart it home when it is moist with the Dew. [Ellis's *practl farmr* 47][29]

The Dutch Wooden fork of Three prongs wch we have here seems the best for this Service.

Elliott[30] Advises to Cock yr hay by 4 o'clock in the Afternoon, & says it makes greatly by its own heat & being left in cock till you intend to Cart needs only be opened 2 or 3 hours   by this much Labour & Expence is saved.

Method of Cleaning red Clover seed

Bradley[31] & Elliott say that in Flanders they clean it By carry-

---

the purple and brown-coloured Seed amongst it; for that shows the Ripeness of the Seed, when the white brighter-colour'd is that which was not come to its full maturity. . . . An Acre is said to afford two or three Bushels, and is a Seed that is hard to get out by threshing. . . ." (*The Practical Farmer* . . . , pp. 46-48).

[25] Apparently Read took this note from Eliot's *Essays* (see footnote 18, this Chapter).

[26] "We may judge of the right Time for cutting it [clover], by examining when it begins to knot and then we may surely go about the Work" (R. Bradley, *General Treatise*, . . . , I, 182).

[27] Written on the margin opposite the following passage in Worlidge's text (p. 26): "Clover Grass . . . is much sowen and used in Flanders and in Holland, *Presidents to the whole world for good Husbandry*. . . . here in England they say an Acre hath kept four Coach-horses and more all Summer long; *but if it kept but two Cows*, it is advantage enough upon such Lands as never kept one."

[28] "Ray Grass may be sown either alone or mixed with Clover, or other of the artificial Grasses" (Hale, *op. cit.*, p. 445).

[29] "Cart it home when it is moist with Dew," seems to have been added by Read.

[30] Read probably got this information from Eliot by correspondence. A search of Eliot's *Essays* fails to reveal a coinciding statement. In substance it is similar, however, to the letter from Eben Silliman to Eliot, printed at the close of the *Third Essay* (*op. cit.*, pp. 73-75).

[31] The description of the machine in Bradley's *General Treatise* . . . (I, 178) cor-

The Husbandry of Plants

ing the Clover Hay to a Tanners bark mill where they use a Stone wheel grind it & clean it from the Chaff with a Corn Fann if the seed is not clean out grind it again till you have gott it all out[32]

Elliott says a man may thus clean a bushl in a day.

Bradley [*Genl. Treatise of Husby. & Gardg.* 178] has another method thus

Make a trough about 10 ft. Long [12 ft. long 2½ over] gutter it wth chizzels & putt in the heads of clover & putt on them a hard oak board guttered as the bottom of the trough the board shorter then the trough & lay a Weight on the board Lay the trough aslant & lett the board be worked backwards & forward till it rubb off the Chaff.

They both agree that these methods will not break the seed.

Bradley [178] says this Engine moves by a water wheel & That He Saw another where the bottom of the trough was a Hurdle & so the top & that most of ye Clean seed came thro' & that in Flanders a Man will clean 6 or 7 bushel pr diem. [*Genl. Treatise* &c (177)].

[October 1753] I took 16 bushells of the heads of red Clover & ground them abt 4 hours under a Tanners bark stone & cleaned 10 quarts of Seed & then putt the Tailings under again & ground them abt 1 hour & cleaned off 3 quarts more & broke no seed but ground the Chaff to a powder.

I weighed 4 Quarts of this Clover seed (wine measure) & it weighed 6¾ lb so that a pint wants considerable of 1 lb.

Pasture

[1753] Mr. Warrell[33] at Trenton fatted Eight Cattle well on

---

responds in general with the description in Read's manuscript, except as to size of the trough which is given in the text as "six Foot long, and about two Foot and a half over."

[32] The first paragraph here seems to have been taken from Eliot's *Essays* (p. 37) and not from Bradley's works.

[33] Probably Joseph Warrell, attorney-general of the province of New Jersey 1733-1754, whose handsome estate, Belleville, stood a short distance west of the present site of the State House in Trenton. After his death in 1758 Belleville, on being offered for sale, was described as "pleasantly situated near the River Delaware, and hath a fine Prospect of the same for some Miles; together with the excellent Gardens, Houses, Stables, Chaisehouses, &c., &c., and Orchards, consisting of the best and finest Sorts of Fruit, as well for the Use of the Table, as for making of Cyder, and is esteemed as good as any made in America. It is the most beautiful and neat Place within many Miles of the same, having every Thing thereunto appertaining to render it commodious and agreeable for a Gentleman delighting in good Gardens, Meadows and Orchards. . . . Also . . . a fine Plantation, of near 300 Acres of Land, within a Quarter of a Mile of the abovementioned Premises, in exceeding good Order, a great Quantity of Meadow, and large Orchard thereon, extremely well watered, and lies upon the River Delaware, and a publick Road, and hath a patent

something less than 2 Acres. I have seen the Ground & it appeared to be a maple run & not very rich. He cutt for the Cattle 3 times a day & fed them in Racks. [NB it kept them from April to Novr. It Increased his Milch Cows in their milk 1/3.]

This method will do well for a large Quantity but breaks in too much on a mans work to tend on a small Quantity.

[1756] Extracts from Mr. Trenchards[34] Letter to me of 4th 8br 1756 respecting an Experimt. he made wth two oxen at my request, That he fed 2 oxen 6 weeks wth white Clover that they eat 3 square perch in 2 days. [An Acre will keep at this rate 2 Cattle 107 days] that they fatted more by this Method while confined at that diet mowed & given 3 times a day than in double ye time in Excellent pasture & well Watered that he thinks each ox will make 1 load of dung pr mo. wth the Litter. That by pasturing an Ox destroys more wth feet, Dung and Urine than He eats, yt Urine in a hott day destroy[s] even ye roots of ye grass where it falls.

Observations on Red Clover & Ray grass from a Folio Entitled. The Compleat Body of Husbandry printed at London 1756[35]

[381] In the Summer 1755 The author saw 12 Acres of Clover wch had a year before been sown wth Barley—On wch in the month of May that Summer turned in 12 Horses, 11 Cows & a Bull 10 Oxen 8 Heifers One hundred Sheep & thirty Hoggs & kept them there till Midsummer 6 weeks at least & then saved it for seed. In September it was finely grown & produced 24 Waggon load of Hay. The whole produce valued at £60—This Field was Mr. Woods at Brockshall near Kelvedon in Essex—He much approves of sowing it wth Barley And also Sowing Ray grass wth it as the roots of Clover will run 16 or 18 Inches down & Ray grass runs near ye Surface [429].

[431] It likes best a Light Rich Land. 8 lb Seed to an acre.

---

belonging thereunto, for the keeping a Ferry across the said River." (*N. J. Arch.,* 1st ser., XX, 248; *Penn. Gazette,* No. 1453, July 20, 1758.) Belleville was acquired by Sir John Sinclair (St. Clair), deputy Quartermaster General, who seems to have made it even more attractive. Subsequent descriptions mention a greenhouse and an icehouse (*Penn. Mag. Hist. and Biog.,* X, 1886, p. 115).

[34] Probably George Trenchard, attorney, of Salem County, who was sheriff, county clerk, justice and surrogate. For a time he seems to have had virtually a monopoly of legal affairs in the county, and was criticized for "holding all the lucrative offices, and not having had any Competitor." He had lands in Elsinboro and Penns Neck Townships. At the outbreak of the Revolution he became a member of the committee of correspondence for Salem. He died in 1785. (Thomas Cushing and Charles S. Sheppard, *History of the Counties of Gloucester, Salem and Cumberland, New Jersey,* Philadelphia, 1883, pp. 322-323, 349-350; *N. J. Arch.,* 1st ser., X, 362-364; XXIX, 422; *Penn. Journal,* no. 1650, July 20, 1774.)

[35] Hale, *op. cit.,* pp. 381, 429-436.

The Husbandry of Plants

[432] The Severest winter dont hurt clover like ye Sumers Heat especially if Sown late in ye Spring.

It will come to nothing in poor Land.

The best way to feed it green in Racks Mown, all Grass rises sooner after the scyth than feeding.

No Creatures should eat it when wett wth dew or rain.

[433] If yo turn in lett it be at high noon & turn out again in an hour.

Feeding Cows wth it gives the butter a taste.

Tis best not to turn yr Hoggs in at all.

[434] Cutt it for Hay when ye hairy buttons or Knotts appear.

No hay shrinks more, if not cutt early it hurts ye 2d Crop.

[435] It will be a month after ye heads appear well before ye Seed is ripe. The Stalks should be quite brown—does not easily shed Seed. The Seed should be quite yellow—2 bushl of seed from an Acre. The Seed is better the 1st than 2d year nor trust it longer.

[436] If you feed off yr 2d Crop wth Sheep twill replenish ye Ground

DuHammel [367][36] Says tis a Biennial & if ye Seed ripens ye 2d year ye roots dye but if Cutt before it may last 3. [368] Says ye 1st Crop of 2d year ye Seed Crop.

[Ray Grass]

Compt. body of Husbandry [Fol 444-5-6.][37] Commends Ray grass as the Earliest feed known   No Frost or wett in winter hurts it, any Soil suits it   3 bushl of seed to the Acre—The roots run near ye Surface.

Ellis's *practl farmer* says ye same. Twill grow among ye most Stubborn clods [54.][38]

Sown in Autumn may feed next June. 2 Bushl on an Acre & roll it in 2 weeks. It will last 7 yr.

Timothy

[1753, 8br, 29th] I plowed up an Acre of Loomy ground wch this year had been sown 1st flax then Buckwheat but it was not rich. I sowed 3 quarts of Timothy for seed & the next day came a NE storm   weeds came up & ye Timothy was poor.

---

[36] H. L. Duhamel duMonceau, *A Practical Treatise of Husbandry* (Translation by John Mills, London, 1759), pp. 367-368.

[37] Hale, *op. cit.*, pp. 444-446.

[38] In Worlidge's text (p. 31), Read underlined the following passage: "Ray-grass, by which they improve any *cold, sour, clay, weeping Grounds*, for which it is best. . . ." In *The Practical Farmer* . . . (pp. 54-55) Ellis says: "Clay, or any other sour and uncultivated Land, is proper for it, nor doth it take up so much Tillage as other Grass-Seeds do, growing well amongst the most stubborn Clods that lie in the way. . . . It will last some Years . . . Being sown in Autumn, it will be fit to graze the next Year for Horses and other black [beef] Cattle, without danger of making it bleed."

I am informed that the Method of Cleaning Timothy is to Dew rott it first. Note I have since try'd this wth Success *probatum Est.*

Lucerne

[from DuHamell 355][39] He Transplanted Lucerne into beds wch including alleys were 3 ft Wide   he planted in single double & triple rows on a bed but those in Single rows yeild most on ye Ground.

The plants should be 6 inches asunder & the Transplantg wch Elsewhere he says may be of roots of any age takes up less time than thinning those you sow with a Drill. Several short Rules he gives in followg pages are That the Beds should be raised in the middle & fine.

The Lucerne sowed in ye Spring should be transptd. in 7br   it may be done later if it does not freeze.

If the weather dry Water the plants after setting.

The Autumn planted yeild tolerably next Sumer. There should be great care in Drawing ye roots out of the Nursery, Cutt them off 6 or 7 inches long & putt them into a Tubb of Water, Make a hole for the plants wth a Stick.

The best Way would be to cutt a Deep furrow & sett ye plants up to the neck—trim off the tops also to 1 Inch.

The plants should be weeded as well as ye alleys & plowed so as to loosen ye ground & destroy weeds, and ye ground should never be suffered to be bound or hard. The Alleys must be hoed.

[But ye *Farmers compleat guide*[40] printed 1760, fol 148, say[s] you may harrow across & so tear up ye grass between ye roots & tho' it hurts ye Lucerne a little twill soon recover it.]

Cutt when the Lucerne as soon as it begins to Bloom. [Lett it be cutt

---

[39] Much of this information about lucerne was contained in a letter to M. Duhamel from M. de Chateau-Vieux, who says: "the green tops should be cut off within about *two* inches of the crown of the root" (Duhamel, *op. cit.*, p. 356), whereas Read's abstract reads, "trim off the tops also to *1* inch" (italics not in the original).

[40] The author of *The Farmer's Compleat Guide* . . . (London, 1760, p. 148), described the process of harrowing lucerne as follows: "When this accident of grass rising in the ground is perceived, the plow must be worked boldly to destroy it. A furrow must be plowed from each side to every row, and then the whole piece must be well harrowed cross-ways: this will not only drag away what grass the plow has tore up, but will tear away also such as grows between plant and plant in the rows. The heads of the lucerne will be a little hurt by this, but there is no other way; and if the grass be fairly cleared off, they will soon recover." The author is believed to have been John Ball (Mary S. Aslin, *Catalog of the Printed Books on Agriculture published between 1471 and 1840* . . . , Rothamsted Experimental Station Library, Harpenden, 1926).

The Husbandry of Plants

high so as not to touch the crown or twill rott & the Scyth must be sharp or yo will lose many twiggs.]

The Hay takes long drying. *The Farmers guide* says 10 days.[41]

Never Lett Cattle feed on it, Sheep in Autumn do least harm if the weather is dry. Tho' ye Hay is good tis best fed green.

The Plants increase in Bulk & size for very many years.

They cutt it 5 & Six times a year, 1st & 2d year smallest crops.

The beds wth a Single row yeild more than where there is 3 rows & Single rows in a bed of 3 ft. wide alley & all [362] yeilded in one Year 15,300 pounds of dry Hay on a French Arpent wch is 1 acre 3/16 English or 7½ English loads of 18 hundred & is 6 & 3/4 Loads of 20€ the 3d year.

[DuHamel 348] It must have a light Soil, a low rich Moist Soil is the Worst next to Clay. The Soil shd be dry Cold will not injure it [351 & 361], Cattle should be fed Sparingly with this Hay, & wth it cutt Short instead of oats His horses did as well as on common Hay & oats [360], Mr. Daincourt had 1 oz of hay in the fall from Spring plants & 7½ load pr Arpent [364.][42]

NB Thus planted is 29,040 plants pr Acre & 43,860 if in 2 ft rows.

Sand, gravel or Hassel mould. Ellis [*Practical*] *Farmr* [65, 66] & has succeeded on stiff moist Land.[43] There have been prodigious Crops on land that would bear no other grass The Dressing is Wood, Peat, clay or fern ashes sown on ye Ground wth or after ye Seed or Soot wch He thinks best. It is Subject to be cut off wth the

---

[41] Ball's text (*op. cit.*, pp. 141, 147-148) reads as follows: ". . . the English farmers . . . are not to judge of the future crops, by those of the first or second year; the third season will be greatly better than these, but even that much inferior to what will be the constant and certain product afterwards . . . the farmer may have constantly four crops of lucerne hay in the year, beginning in May, and cutting the succeeding growths about five weeks distance. . . . In France they have sometimes six crops in a year. When the lucerne is fit for cutting, very careful mowers must be chosen, and they must be directed to cut it a little above the crown. The scythe must be in order, and the plants must be moist with the remainder of the dew, otherwise the work will be irregularly performed, and a great many of the weak stalks left. The very best hay of lucerne is that which is made by cutting it before it flowers. . . . lucerne takes a considerable time in drying . . . in a good season, the hay may be very well made in about ten days." The rotting of the crown noted by Read seems not to have been mentioned either by Ball or Duhamel.

[42] Mr. Diancourt, Captain of the French Grenadiers, estimated 26,400 plants per arpent in double rows, 15,400 in single rows. Data in plantings and yields, given in Duhamel's text (*op. cit.*, p. 364) in terms of French arpents, were apparently computed by Read in terms of their equivalents in English acres.

[43] Ellis says of lucerne (*op. cit.*, pp. 64-66): "Its Crop for the most part is double the quantity of Saint-Foin, and may, if it likes the Ground, be mowed three times in a Summer. . . . The Soil most proper for it is light Ground, such as Sand, Gravel, or Hazle Mould; . . . but it has been tried in stiff Soils and moist Land and has succeeded very well. . . . Eight or ten pounds [of seed] will be sufficient on rich Land, but twelve or thirteen on Land which is poorer. On some poor sandy Land that would not bear common Grass, there has been prodigious Crops." Read appears to have supplied the amount of the dressing per acre.

Frost. Then mow off ye Dead grass & give it a Dressing abt 15 bushl as afd to an Acre.

[Sow] wth Oats in March. [Ellis] Ellis says [lucerne may be mowed] 3 times [a year].

It yeilds double ye Quantity of St. Foin. [Ellis's *Practl Farmr* 65] of seed 8 lb. or 10 lb. [will be sufficient] on rich 12 or 13 on poor Land.

[Saint-Foin]

Sown on dry Sandy or gravelly ground tho' it be overrun wth Heath or Fern. [Ellis's *Pr. Farmr.* 63][44] ye Dressing is 15 bushl Soot every 3 years pr acre.

Ellis' *Practl Farmr* Says ye same.

bring yr Ground to a fine tilth. [Ellis 63]

will stand 20 yrs. [Ellis]

Harrow it in wth yr Corn. [Ellis 64.]

Feed it not ye first Sumer nor early next Spring least it bleed to death. [Ellis 64.]

5 or 6 bushl of St. Foyn seed pr acre. [*Compleat Body of Husbandry* page [4]36.][45]

[DuHamell 339, 340-1]:[46]

It Delights in the same Soil as Lucerne & he repeats from Tull that tho' it thrives best in rich Land yett it will grow where scarce any other Grass can Live, Marshy land chils & destroys it.

[338] Tull says that One acre of St Foin by the New Husbandry will yeild as much as 30 or 40 Acres of common grass.[47]

[339] He reports from Tull that it will yeild 2 Tons pr Acre.

[341] Mr. Tull asserts that He kept a team of horses who worked hard all the Year *in good plight* on St foin. Hay only wthout the addition of oats.

---

[44] These notes from *The Practical Farmer* (pp. 63-64) were made on the margin of Worlidge's text (pp. 29, 30) describing Saint-Foin. "Ye same" refers to the amount of seed per acre recommended by Worlidge, and underlined by Read, viz., *"four Bushels on an Acre is the best proportion."*

[45] ". . . five or six Bushels are commonly allowed in this Way of the broad Cast sowing, to an acre, and less than four is not sufficient . . ." (Hale, *op. cit.,* p. 436).

[46] These notes on Saint-Foin are from Duhamel's text (*op. cit.,* pp. 337-346).

[47] Duhamel here quotes from Jethro Tull's epoch-making work, *The Horse-Hoeing Husbandry: or an Essay on the Principles of Tillage and Vegetation.* The passages here cited apparently are from pages 75, 76, 79, 83 and 84 of the London edition of 1733.

The Husbandry of Plants

[345] Mr. Emma[48] says he had 14,445 at one Cutting of St. Foin on an Arpent [1 Acre 3/16] wch he thinks more than Mr. Chatteau Veiux—15,340 of Lucerne at 5 cuttings, & that he Cutts St Foin 3 times in good years & Horse hoes it after every Cutting.

[344] Very difficult to procure good Seed & next akin to impossible if threshed in the field the Seed will heat & Spoil after threshing.

[344] You should never feed Cattle on it in ye Spring.

[345] It will continue 5 years in the Comon way & much longer in ye New Husbandry That the Crops of the 1st & 2d Year are poor in respect to those that follow.[49]

[338] I see no Directions abt transplanting ye roots.[50]

[345-6] Says yo are to Sow it in beds & that one row in a bed as in Lucerne will produce more than 3 rows in the Same bed Sow in the Spring.

[338] That it will grow 5 ft. high & ye roots run 15 or 20 ft into ye Ground

[*Trefoil*]

Best to mix it wth Clover. Ellis's *Pract. farmr.* [55, 56 &c] praises it much.[51] Cutt it in blossom [Ellis] it is soon made. Trefoil if yo do not turn in cattle till full ripe will seed ye ground.

[*Miscellaneous*]

In Buckinghamshire many people sow Parsley wch is good Sheep pasture. Mortimer [28][52]

NB—It is the Comon Herbage of Bermudas, makes Ext. butter.

Elliott in his *Essay on field Husbandry* is of opinion that laying English Grass under water in winter will in time ruin it for that it needs a winter to refresh it.[53]

---

[48] M. Eyma, quoted by Duhamel (*op. cit.*, p. 345). Elsewhere in the notes, Read alludes to the same passage in the following words: "[345] Mr. Emma at Geneva talks of 14,445 lb on an Arpent in the New Husbandry way at one Cutting."

[49] The last part of this paragraph not found in Duhamel's text; it may have been supplied by Read.

[50] Apparently Read's own comment.

[51] These notes on trefoil, taken from *The Practical Farmer* . . . (pp. 57-60), were written in the margin of Worlidge's text, of which Read had underlined the following: "Trefoyl . . . *will exceedingly mend the Hay, both in burthen and goodness.*"

[52] Mortimer (*op. cit.*, I, 42) says: "I am told that in Buckinghamshire they make good improvement of their Lands by sowing of them with Parsley, and that it prevents the Rot of Sheep."

[53] "Some think that it is good to lay their low Lands under Water in the Winter to inrich them, and practise accordingly: But this will kill your English Grass after a few Years: For English Grass will not subsist without a Winter" (*op. cit.*, pp. 15-16).

Capt. Reeves[54] in May 1756 brot from London for me the following Grass seed, in baggs putt in a Cask betwixt decks & it grew well. The heat of ye hole often spoils it.

  50 lb red clover seed @ 4½d. pr lb.
  12 lb. Trefoil   @ 3
  3 bushels Rye grass @ 4s. pr bushl.

GRAIN CROPS

Indian Corn

[Novr. 3, 1757] An Ear of Indian Corn good & of moderate size having a peice of Linnen sewed around it & drawn out & shelled filled ye bagg of Linnen & being measured was a little above half a pint.

Indian Corn should be planted before the 5th of May unless in cold ground, if you plant earlier & it should be touched wth frost, turn your sheep in upon it and let them bite it down.

A hand will plow 3 acres a day if ye ground is light & Team good & Hoe 2½ acres pr day & Harrow 5 Acres twice in a place.

Tis best in Sandy ground to furrow out, & when the Corn is abt 6 inches high mould it.

It should have two hoeings the first just after ye plowing when the Corn is abt 8 inches high.

[In ye 2d plowing throw up a Shallow furr. next ye Corn raising ye Wing of yr shear so as not to touch ye roots but go as close as possible to ye Corn & then throw ye 2d furrw. deeper close up to ye back of ye first & this in a manner hills it.]

Just before it Tossles it should be plowed & hoed again. Keep yr Harrow going to keep it clean of Weeds. It hurts it to plow after tis in tosle. Hill very Little or none in Sandy Land. Mr. Saltar[55] plows clean & tends wth a harrow.

[54] Probably Captain Peter Reeve (1715-1800), who sailed frequently from the port of Philadelphia, and was for many years port warden. There is evidence that he was on intimate terms with Charles Read and his cousins, the Pembertons. (*N. J. Arch.*, 1st ser., XIX, 102; *Penn. Journal*, no. 465, Oct. 17, 1751; *Penn. Gazette*, no. 2038, Jan. 14, 1768, p. 4.)

[55] Probably Richard Saltar (*ca.* 1699-1762), member of the provincial council and associate justice of the supreme court. He built a large house on Black Point, at the present site of Sea Bright, and subsequently he settled at Bow Hill in Nottingham Township, Burlington County (now Hamilton Township, Mercer County). Besides large land holdings in Monmouth, Somerset, and Hunterdon Counties, he had an interest in a copper mine at Rocky Hill. Mordecai Lincoln, great-great-grandfather of Abraham Lincoln, who lived for a time in Monmouth County, was his brother-in-law. At his death his personal estate, inventoried at £1268-1-10, included 10 slaves, 23 head of cattle, 7 horses, 20 sheep, 4 ploughs, a cheese press, 5 hog troughs, a cutting box, a mill to clean grain, a wood sled, an oxcart, and a "Pleasure Slay & Geers" (*N. J. Arch.*, 1st ser., XXXIII, 370; *N. J. Wills*, 7331C-7342C). When Bow

The Husbandry of Plants

The Work of Plowing & hoeing should be over by 1st July O[ld] Stile

If after that it be necessary give it a Harrowing keep always Stirring in dry Weather & keep ye weeds & Grass from growing.

Don't Top till ye Silk dies & ye Corn grows hard.

Tops yeild 1 Load to 3 acres. The blade ½ as much.

They are Excellent fodder especially for Sheep.

A Hand Can top —— pr day.

[Top Corn] Yo cutt off the tops, lett them lay 24 hours, if dry weather bind them in Sheaves, sett 6 or 8 together so as the Air to pass thro them, in a week if dry putt them in Barn or Barrack they are apt to mould in Stacks. Mr. Tonkin[56] says yo may bind in

---

Hill was offered for sale in 1771, it was styled the "most agreeably situated and valuable farm in New Jersey." "The farm contains," the advertisement reads, "about 360 acres of land, 154 of which are excellent low meadow in full improvement, and divided in eight several fields, well fenced, and the whole dry enough for any kind of grain, or hemp, for which the soil seems particularly adapted; the cleared upland contains about 120 acres, properly divided, and in good fence; the remainder is very good wood and timberland, a very fine out-lot, or range for cattle and horses, . . . the orchard is large, thrifty, and of the best grafted fruit, both for cyder and house use; the garden is large, neat, well inclosed and stored with a variety of the best table fruit. The house, barn, stables and out-houses, are all in good repair; a large quantity of hay, with about 70 acres of wheat, barley, and rye, in the ground, will also be disposed of; together with several valuable farming Negroes, men, women, and children, breeding mares, of the best kind, a number of horses, young and old, about 100 head of cattle, several pair of working oxen, sheep, hogs, and farming utensils of all kinds, &c. boats, and a fishing net, there being some valuable fisheries on the river, within the lines, and the navigation coming up to the very banks, will always render the transporting any produce to Philadelphia market extremely convenient and easy. . . ." (*N. J. Arch.*, 1st ser., XXVII, 600-601; XXVIII, 478; *Penn. Gazette*, no. 2234, Oct. 17, 1771; no. 2311, Apr. 7, 1773; John E. Stillwell, *Historical and Genealogical Miscellany*, New York, 1916, IV, 184-191).

[56] Probably Edward Tonkin, justice of the peace and surveyor of the highway, who lived in Springfield Township, Burlington County, near Charles Read's Sharon. From his will, probated in 1768, it appears that he was a man of considerable substance, with large land holdings, and a house in Burlington. His personal estate, inventoried at £5,749-3-18½, included 7 slaves, 8 horses and colts, 41 head of cattle, 94 sheep and lambs, and 87 head of swine. To a son John he left the home farm, to a son Samuel his plantation in Greenwich, Gloucester County, and to a son Edward he bequeathed the rest of his lands in Springfield and Mansfield Townships. Among the provisions of his will was a bequest of £20 for the repair of St. Mary's Episcopal Church, Burlington. (*N. J. Arch.*, 1st ser., XXXIII; N. J. Wills, 8453C-8459C, 9778C-9784C.)

From Read's notes, together with other evidence, the family seems to have been especially skilled in cattle raising. The *Pennsylvania Gazette* (no. 2153) of March 29, 1770, carried the following news item about the elder Edward's successor on the home farm: "This Morning the large Steer, raised by John Tonkin of Springfield, in Burlington County, New-Jersey, was killed by George and Benjamin Wilport; and on Saturday Morning the same will be weighed and sold at their Stall, No. 45, in the Market, where they will be glad to see their Friends and Customers." The issue of the following week reported that the steer weighed as follows: "the four Quarters 1394 lb. rough Tallow 225 lb. and the Hide 126 lb. in all 1745 lb." (*N. J. Arch.*, 1st ser., XXVII, 120, 130.)

6 hours   putt Six or 8 abt a hill   remove them in 14 days & putt 8 or 10 hills together & tye the[m] atop & putt ym. in Barn in 6 weeks. When yo Plant Pompions wth yr Corn You are to begin ye first row wth Corn 2d Pompions So there will always be a hill of Corn opposite to a hill of pompions & gives them room wthout Interfering.

Some Choose to plant a Corn early ripe & cutt up all together.

[Mr. Waddingham][57] Tis said that in Carolina a Negro will tend ten Acres wth his Hoe.

Some people think it answers well to feed their Hoggs wth the Corn while soft & milky Stalk & all. They chew & suck ye Stalk. They Eat Cobb as well as Corn.

If yo Work oxen among your corn yo should have a hole bored thro' each horn   then have ready a crooked piece of wood wth holes therein to fitt on ye horns like a Batt or Bandy wickett so that it begins to bend off near ye inside horn & runns 18 inches beyond ye outside horn bending away towards his shoulder so that when he presses against the Corn it shall gently move it round clear of his Yoke for tis the Yoke end wch takes ye corn & breaks it.

[Tunis Dye][58] One Method of keeping Birds from pulling yr Corn out of ye Ground is to soak ye Corn then draw a horse hair thro' with a Needle & tye this   When they get such Corn in to their Craws the hair hangs out of ye Bill and pesters them so that they often Starve or tear their throats so as to dye.

[Negro work Maryland] Every Negro in Maryland makes 2000 lb Tobacco & 10 barrells of Corn of 5½ bushells of Ears to ye barrell—2 barrells in Ears keep ye Negro. They use lowlands for tobacco   won't grow above 3 years in a place.

[57] This was doubtless Samuel Waddingham, planter of St. Bartholomew's Parish, in the southeastern part of South Carolina, who in 1766 married Rebecca Shoemaker, daughter of Charles Read's sister Sarah, born about 1723 (C. P. Keith, *op. cit.*, p. 186; *South Carolina Historical and Genealogical Magazine*, XI, 1910, p. 31).

[58] Theunis Dey [Dye] (1726-1787) was a member of the colonial assembly from 1761 to 1776, served in the council from 1779 to 1781, and returned to the assembly in 1783. In 1774 he became a member of the committee of correspondence from Bergen County, and during the Revolution was colonel of the Bergen County militia. In 1764 Dey was one of the managers of the lottery for the College of New Jersey at Princeton. He was active, also, in the founding of Queen's (Rutgers) College, and was named a member of its Board of Trustees in the royal charter granted in 1766. His stone mansion on a handsome farm at Lower Preakness, built by his father Derrick Dey in 1740, was used by General Washington as headquarters in the summer and fall of 1780. The mansion, with outdoor kitchen and sunken colonial gardens, has been restored by the Passaic County Park Commission. (*N. J. Arch.*, 1st ser., XXIX, 411-412; *N. Y. Gazette* and *Weekly Mercury*, no. 1184, July 4, 1774.)

The Husbandry of Plants

Rye

Rye in Jersey should be sown by ye 1st of 7br for Feed or sooner it will do to sow it 1 Novr nay some say in ye Spring but then a double quantity should be Sown. Rye harvest is abt ye 10th July [vid MSS Rye sown for Hoggs to graze].[59]

They Shock 12 Sheaves together each Shock will yeild 1 bushell. It will if ripe thresh well directly out of the field but if it lays a few days in shock it should lay in Stack a month.

[1777 Dec. 17] Carryed twenty Bushells of Rye that I got from Calb Austin to Oliphants Mill and they took No toll and I had 765 lb Meal which is 38¼ lb Meal to ye Bushell[60]

[Wheat][61]

At 32 grains of Wheat to the penny Weight there will be 7680 in a pound Troy & If you reckon a bushell to be 66 lb. Troy wch is abt 60 lb. Avoirdupoize there will be in a Bushell 506,880 grains so that there is but one grain to 3&½ square Inches.[62]

Speltz is of ye wheat kind 2 Kernells in one hull makes good flour but must be hulled in a proper mill for wch they have 1/17th it will grow on worse land than Wheat & bear 30 bushl or more pr A. would do well to feed horses in ye Chaff.

[Threshg] They [farmers in Maryland] thresh out their wheat wth a machine 16 square 10 ft. long tis 37 inches at ye big End & 9 at ye other the small End goes wth a Bolt thro an upright post wch turns. Threshes out 60 bushl pr day.

---

[59] See pp. 351-352.

[60] Oliphant's Mill was erected about 1685 by David Oliphant on a tributary of the south branch of Rancocas Creek, about one mile southwest of Medford. It operated for more than 200 years. (E. M. Woodward and J. F. Hageman, *op. cit.*, p. 370.) This entry appears to have been made by Charles Read's son. The date is subsequent to the elder Charles' death, and the handwriting, clearly different from the preceding entries, corresponds closely with that of letters signed by the younger Charles. Oliphant's Mill was not far distant from the latter's residence. Caleb Austin, of Evesham, Burlington County, owned several tracts of land in this township. His father, Amos Austin, had dealings in real estate with the elder Charles Read, and after the death of Charles IV, Caleb purchased from the estate a part interest in a tract of 3253 acres in Evesham Township. When Caleb died in 1805, the inventory of his personal property included hay and grain valued at $643.50, horned cattle $143, swine $129, sheep $19.50, and farming utensils, &c. $602.50. (Clement Papers, Book 7, p. 64; N. J. Wills, 12200C.)

[61] In Worlidge's text (p. 39) Read underlined the following Latin quotation: *Triticumluto hordeum pulvere conferite* (Sow wheat in damp ground, barley in dry).

[62] See p. 268. Allowing 6,272,640 square inches to an acre, Read here apparently is computing the distribution of seed wheat when sown at the rate of 3½ bushels per acre.

40. *Of Arable and Tillage.*

Wilfulness on the other, unless the Design of *Enclosure* might take effect, for then would the Lands be so much the more enriched, that they would bear other Grain, to a greater advantage to the *Husbandman* than *Barley*; or that a double or treble Tax might be imposed on every Acre of Barley-land, for what it is on other Grain, which would provoke the Husbandman to that which would be most for his advantage; then would there be a greater plenty of all other sorts of Grain and Pulse, and at a lower price, and only good Liquor a little the dearer, which may by House-keepers the easier be born withal.

The Seasons for sowing of *Barley* differ according to the nature of the Soil, and situation of the Place: <u>Some sowe in *March*, some in *April*, others not until *May*</u>, yet with good success; no certain Rule can be herein prescribed: it usually proves as the succeeding weather happens, <u>only a dry time is most kindly for the Seed.</u>

For as before is observed, <u>moist Weather is best for Winter Grain, and dry for any Seeds in the Spring or Summer</u>, because the Grain in the Winter should spring the sooner; and that sown in the Spring more gradually, lest the too suddain drought injure it. Also a moist Seed time in the Spring, too much favours the Weeds, but in the Winter the Cold prevents them.

Difference of Barley. There is little difference observed in *Barley*, only there is one sort called *Rath-ripe Barley*, which is usually ripe two or three weeks before the other, and delights best in some sorts of hot and dry Land.

Rye. *Rye* is a Grain generally known, and delighteth in a dry warm Land, and will grow in most sorts of Land, so that the Earth be well tempered and loose; it needeth not so rich a Ground, nor so much care, nor cost bestowed thereon, as doth the Wheat; only it must be sown in a dry time, for Rain soon drowneth it: <u>they usually say a shower of Rain will drown it in the Hopper;</u> Wet is so great an enemy to it. Therefore dry sandy warm Land, is usually termed *Rye-Land*, being more proper for that, than for any other sort of Grain. It is quick of Growth, soon up after it is sown, and sooner in the Ear, usually in *April*, and also sooner ripe than other Grain; yet in some places is it usual to sowe *Wheat* and *Rye* mixed, which grow together, and are reaped together; but the *Rye* must needs be ripe before the *Wheat*: Neither can I discover where a greater advantage lies in sowing them together, than in sowing them apart. The principal Season of sowing of *Rye* is in the *Autumn* about *September*, and after, according as the Season permits, and the nature of the Ground requires.

Oats. Oats are very profitable and necessary Grain, in most places of *England*: they are the most principal Grain Horses affect, and commended for that use above any other, being of an opening nature, and Sweet; other Grains being apt to stop, which is injurious to Labouring or Travelling Horses; although on the other hand, Oats newly Housed and Thrashed, before they have Sweat in the Mowe, or be otherwise throughly dryed, are too laxative. On such Lands, that by reason of the Cold, no other Grain will thrive, yet Oats grow there plentifully, as many places in *Wales* and *Darbyshire* can witness: <u>there is no Ground too rich nor too poor, too hot, nor too cold for them;</u> they speed better than other Grain in wet Harvests; the Straw and Husks being of so dry a nature, that although they are Housed wet, <u>yet will they not heat in the Mowe</u>, nor become Mouldy, as other Grain usually doe; they are esteemed a pceler of the Ground; the best season for sowing of them,

Typical page showing Read's marginal notes and underlinings in Worlidge's text

The Husbandry of Plants

[Buckwheat]

[Buck-wheat, or French-wheat][63] is Sown abt. 10 July near ½ bushl pr acre.

12 Weeks from Sowg till cutt Mr. (?) V. Vacty (?)[64] tells me he had 480 bushl from sowg 4.

J. Morris says this Won't do on his Sandy Land on Tryall. This they say on Tryall does on Clay Land.[65]

[Millet]

[1756] I sowed Millett 29th April in my Garden it began to tosle 12 July & I cutt it 24 August it yielded at ye rate of 67 bushells pr Acre ye seed weighed 40 lb to ye bushell. The Grain was ripe tho' it does not readily shake out, ye Straw & leaves green.

3d Essay Elliot[66] says it yielded 32 bushels pr acre in poor Land. That Cattle leave good white clover to Eat it green, that 3 quarts will seed an Acre for grain, best sowed thicker for pasture. I think best to mow it green for Cattle it has a sweetness like Indn Corn the stalk is sappy and perhaps wont cure well if it will twou'd be Excellent for ye Cutting box.

Rice & Indigo
from Mr. Jos. Shute—Carolina[67]

[Rice] In the first a Negro Earns £15 Str. ye Last 25 pr ann be-

---

[63] Written on the margin opposite Worlidge's section on Buckwheat (p. 41), with the following italicized passages underlined by Read: "... *it is usually sowen as Barley, but later; it is also late ripe, and yields a very great increase ... after it is mowen it must lie several days till the stalks be withered before it be housed.*"

[64] Not with certainty identified. From this and later mention of "D. V. Vacty" (p. 364), it appears Read may have been referring to Derrick Van Veghten (1699-1781), member of the colonial assembly, who lived on a large plantation near the present site of Finderne. The Van Veghten mansion, now regarded as the oldest house in Somerset County, served as headquarters for General Greene during the winter of 1778-79. Van Veghten's will is unique in the measures prescribed for conserving his soil resources. After providing for the disposal of his plantation, he added: "I also order and direct that the Fields of my said Farm be not sowed with winter Grain more than every fourth Summer And that the Meadows be not ploughed except for two Bushels of Flaxseed which may be sown every year And that no Timber be cut but for the use of said Farm." (*N. J. Arch.*, 1st ser., XXXV, 417; N. J. Wills, 694R; Katharine H. Birdsall, "The Historical Van Veghten House," *Somerset County Historical Quarterly*, I, 1912, pp. 92-97.)

[65] These two notes refer to the following paragraph in Worlidge's text (p. 41): "Buck-wheat makes as good a Lay for Wheat, as any other Grain or Pulse, especially if it be not mowed, but ploughed in: But the best way is, when it is in Grass before it blossom, to feed it with Milch Beasts, who will tread it down, and make an excellent Lay thereby for Wheat."

[66] *Op. cit.*, p. 56.

[67] Joseph Shute in 1748 is reported to have built Laurel Hill, a handsome residence

sides his own provision. Is planted in rich low lands in March, April, May & to 10th June the best planters plant soonest. If yo have an opportunity of throwing water into the peice as it is banked it may be kept up so as to be just below ye beards it prevents hoeing & unless yo lett ye water in yo must hoe it 3 times—a Negro tends & cleans 3 Acres. Good Land yeilds 60 bushl of rough rice to ye Acre wch is 4 bbls of Clean rice of 5-5½ & 600—but 3 barrells is good yeild. Tis cutt in August & 7br threshed out wth Ease the Ears long & heavy. They lett off the Water before they cutt it, When Cutt, it will Sprout from the roots, makes Excellent pasture & fodder being very leafy & stools much—rough Rice good for all sorts of Cattle. The great work is in cleaning wch is performed by the Negroes in winter by pounding in mortars, in wch they often work 18 or 20 hours out of the 24 if in a hurry & break 1/10 or 1/12 the Small is used for ye Negroes, ye Dust for Cattle, the large exported. Some have had mills wch make dispatch but they are disused being so often out of order & workmen scarce.

Steeping & Brining Seed Corn

[Ellis's *Practl Farmr* 17]:

Take yr Seed Corn & boil 1 bushel wth 5 pails of water till it Burst, then strain off your Liquor, while this Liquor is warm putt in 3 lb. of Salt petre to dissolve then add 5 pails full of the Water that drains from yr Dung hill or Urine and in this prepar'd Liquor steep yr grain you propose to Sow 24 hours the Liquor should be 4 Inches above ye Grain because ye grain will swell.

Then dry ye Grain in the shade & sift Lime over it wch will dry it ye Sooner sow 1/3 less than usual & you will assuredly find the benefit *Twenty fold*. I have tryed it (says ye author) with barley & had comonly 30 Ears from one root about 3 lb. of Nitre enough for ye corn that Sows an Acre & what is left of the Water will be the groundwork of another Steeping. Putt yr Liquor warm to ye Corn & stop it close to keep in ye Salts. By this rect. he says he had 1731 as much Barley again as usual from ye Ground.

[Bradleys *New Improvmt in planting & Gardening* an 8 vo 1739 much recommends this receipt fol 231][68]

---

in the present East Fairmount Park, Philadelphia. From his estate it was purchased in 1760 by Frances Rawle of Philadelphia, and subsequently it was known as the Randolph Mansion. This was possibly the Quaker leader who had interests in South Carolina, mentioned in John Smith's diary. (A. C. Myers, *op. cit.*, p. 311; William Brooke Rawle, "Laurel Hill and Some Colonial Dames Who once lived There." *Penn. Mag. Hist. and Biog.*, XXXV, 1911, p. 387.) For the description of Indigo, see pp. 290-292.

[68] Bradley, in his *New Improvements of Planting and Gardening* . . . (7th ed.,

The Husbandry of Plants

Putt into a tubb of Water as much Salt as will make it bear an Egg then add as much more then pound 2 or 3 lb. of Allum & Stirr all well about   Let yr Wheat Steep 30 or 40 hours—the night before yo Sow Sift Slack Lime enough on it to dry it. Then Sow it. This prevents ye Smutt of Wheat. [NB. He does not oblige us by incerting ye quant. of Water or Salt nor say how much a Quant. will Steep.]

[Mortimer's *Husbandry* 41 Extracted from Houghtons papers on *Husbandry*]:[69]

Take a Convenient quantity of rain Water to every Gallon add two pounds of Stone Lime. Lett it Stand 2 Days stiring it 3 times a day, then pour off the Clear Water to Every gallon of Water add 4 oz of salt petre and one pound of Pidgeons Dung Stirr it 3 times a day then Strain off the Liquor for use putt yr Grain into this Liquor lett it lay 18 hours take it out & lay it 24 hours in ye Shade to dry, then steep it again 12 hours lay it again to dry as before then putt it in again Six hours then Drain it to Sow.

He sett it in single grains in his garden 10 inches apart. He had 60, 70 & 80 stalks wth Ears 6 inches long & from 40 to 60 grains. There is a Danger in Steeping too Long.

Mr. Yelverton who had the prize in Ireland 1742 had 668 Stone & 11 lb. wheat off one Acre. [156 bushl @60 lb. to ye bushl on ye irish Acre][70] He changed his seed, sowed not too thick & mowed it. The seed was steeped the Evening before sowing in a Pickle thus made.

Take Rock Lime & Bay salt putt in as much Urine as will Dissolve the Salt & slack the Lime by stirring lett them lay so 24 hours then pour off the Liquor for use. Let the Seed Corn lay 12 or 14 hours in this, then riddle on hott dry Lime just before yo sow it. Urine alone is subject to prevent ye wheat from growing. [*Compleat Body of Husbandry*—Folio printed 1756—page 356][71]

London, 1739, pp. 230-231), says that this "Receipt to prevent Wheat from being smutty" was published by Colonel Plummer of Hertfordshire "for the common Good." The critical and practical turn of Read's mind is indicated by his comment that Bradley failed to state the quantities of water and of salt needed to steep a given amount of grain. This work of Bradley is hereafter referred to as *New Improvements*. . . .

[69] The recipe was given in a letter by Thomas Everard (1692) published in John Houghton's *Husbandry and Trade Improv'd: A Collection for the Improvement of Husbandry and Trade* . . . (edited by Richard Bradley, London, 1727, Paper No. 14, I, pp. 40-41), and copied by Mortimer (*op. cit.*, I, 62).

[70] Computed on the basis of the legal English "stone" of 14 pounds, Mr. Yelverton's wheat yield would have reached the prodigious figure of 96 bushels per English acre (668 stone, 11 lb. equals 9363 pounds, or 156 bushels, at 60 pounds per bushel, of wheat per Irish acre. An Irish acre equals 7840 square yards, or is about 1.62 times the size of the standard English or American acre). But since other values of the stone, varying from 4 to 26 pounds, have been in use, it is probable that a smaller unit was the basis in this instance.

[71] Hale, *op. cit.*, p. 356.

Much approves of Steeping, Says Salt & Lime destroy vermin. [*Ibid* 376][72]

A Dutch Rect from Wm Logan Esqr.[73]

Boil a peck of Sheeps dung in Water, Some say Lye would be better, Strain off the Water, add 4 lb. Salpeter & Stirr it well, pour it when cold on a bushl of good grain. Lett it Soak 8 hours dry the grains on a floor in an airy place. Then Soak it again in the same water 6 or 8 hours dry ye grain as before when dry enough sow it directly. Sow ½ the usual Quantity—The Land thus sowed wants no Dung. (Consult this Book page 56 the whole chapter.)[74]

[Steep for grain][75]
[Sea Salt][76]
By Grinding or Boiling grain much is Saved

	Boiled	Ground
Rye yeilds 3¼ for 1		—fine makes 3 of 2.1 bushl meal weighs 40 lb.
Ind. Corn 2 for 1 not burst		—ground Coarse gains 1/3 or 1 in 3.
Buckwheat 2½ for 1		

[Tryed 1756 Jany. 21]—The Weight of Grain [per bushel] as usual in Jersey & Pensylvania

BuckWheat quite clean . . . .   52 lb to 54
Rye good & Clean . . . .   58 & 59
Wheat good from 63 lb to 60
Indian Corn yellow . . 60 ⎫ Maryland or lowr county Corn
                             ⎬ 50 to 52 Egg harbr dented yel-
   Do. white . . . ⎭ low 57 lb.
Oats pretty good . . . 33 to 35
The Long grained Lower Co Corn clean 53

---

[72] Hale (*ibid.*, p. 376) mentioned brimstone also as an ingredient which destroys vermin.

[73] William Logan (1718-1776), of Philadelphia, Charles Read's cousin, engaged in business with his father, was common councilman 1743-1776, and provincial councillor 1747-1775. He executed the conveyance of his father's library to the Library Company of Philadelphia and acted as librarian until his death. His son, Dr. George Logan (1753-1821), inherited the family estate Stenton and was one of the twenty-three men who in 1785 founded the Philadelphia Society for Promoting Agriculture. (C. P. Keith, *op. cit.*, pp. 14-16, 21-22.)

[74] This refers to the section of Worlidge's text entitled, *Of the preparation of the Seed* (p. 56).

[75] Marginal note opposite a recipe for steeping grain in Worlidge's text (p. 59).

[76] Marginal note opposite the statement in Worlidge's text (p. 59) that sea salt is beneficial for treating grain.

The Husbandry of Plants

[1767 Exp.] 70 lb. Rye flour made 89 lb of bread pr rata 110 to 140.

The prices of Grain at Bear Key at London as I took them out of a Gazzette[77] of May 1753

	s.		s.	
Wheat	30:00	to	34:6	pr Quarter wch is
Barley	16:00	to	19:0	8 bushl.
Brown Malt	19:00	to	21:6	
Pale Malt	20:–	to	24:–	
Oats	12:0	to	15:–	
Hogg peas	18:0	to	20:–	
Do Boilers	21:0	to	24:0	
Tares[78]	27:0	to	19:0	

Fibre Crops

The Method of raising Hemp in New Jersey from Mr. Abner Philips[79] & Mr. John Bainbridge[80]

Choose Land for Hemp not too dry but be sure that it is not Subject to Freshes or Puddles on the Surface or have the water so near the Surface as to touch the Root. The Roots of Hemp run from 6 to about 10 Inches down and this Tap Root must be above

---

[77] A careful search of the *London Gazette*, and of contemporary London publications has failed to reveal the source of this reference. *The Gentleman's Magazine and Historical Chronicle* (London) for May, 1753 (XXIII, 252), gives the following "Price of corn" at Bear-Key:

Wheat	29s.	to	35s. qu.
Barley	17s.	to	18s.
Oats	10s6d.	to	12s.
Beans	14s.	to	16s. 0d.

[78] Apparently a typographical error in this line. Probably intended to be 17:0 to 19:0, or 27:0 to 29:0.

[79] Abner Phillips (born about 1716) was collector for Maidenhead Township (Lawrenceville) 1751-1752, and was appointed justice of the peace in Hunterdon County in 1756. Also an Abner Phillips, probably the same, kept a tavern in Princeton about 1767. (E. M. Woodward and J. F. Hageman, *op. cit.*, pp. 845, 851; *N. J. Arch.*, 1st ser., XXV, 490; *Penn. Gazette*, no. 2030, Nov. 19, 1767.)

[80] There were several "Johns" among the Bainbridges of Hunterdon County, a family prominent in the colonial period. It seems probable that Read here refers to John Bainbridge "of Hopewell," who died in 1765. In 1749 John Bainbridge "of Maidenhead" advertised for sale his farm of 600 to 700 acres on South River, in the neighboring county of Middlesex. It was described as containing a young orchard of 160 grafted apple trees, 300 acres of meadow ditched and hassocked and "made fit for the Scyth." (*N. J. Arch.*, 1st ser., XII, 528-529; *N. Y. Gazette Revived in the Weekly Post Boy*, Apr. 11, 1749.) A John Bainbridge, Jr., was involved in the land riots of 1747.

the Water, Meadow Land is very Good for it, If your Mudd is mixed with a Little Sand, or if it be a Little Turfy it is no worse for that or good upland well Dunged will bear it well. The manner of managing it Best is thus—Plow it in the fall or as early as you can in the Spring by the end of March or 10th of April at furthest, it shou'd be plowed twice The Ground shou'd not be so wett but that it will tumble off the Mould Board & Sow it between the 20th & 25th of April if possible. But as the Least Frost will kill it you should think of that. Seed your Ground from 5 Pecks or one bushel & a half on an acre according to the Strength of your Ground wch should be very finely pulverized. If you cou'd choose your Time to Sow it, it shou'd Be just after a Rain so as to come up together or else its apt part to Come up at one time & part at another; that which is Covered deepest will Come up soonest & if it does not Come up, even that wch starts first keeps the other so much under that your Piece will be very uneven the thicker it Comes up the Better, & the Best thickness for the Stalks is about the Bigness of a Large Goose Quill or Pipe Stem & about 6 feet high. It begins to ripen according to the Season about the 20th July or 1st August. These are the Rules to know when it is ripe. The first is to remark it in the morning before the Dew is quite off, or in a fresh air & you will see a kind off Dust or fogg rise from it, then it is time to Cutt. The 2d observation is that there are 2 Sorts of Hemp in your Piece. One is that wch bears the Seed (this never Blossoms) the other is called the Fimble hemp & Blossoms. The Leaf of the Fimble beginning to fall & the Stalk of it Spotted then the Hemp begins to ripen and will be time to Cutt. The 3d is to take notice that yr Fimble gets its growth Soonest, the Seed Hemp is Shortest till the Fimble has got its Growth, then the Seed Hemp shoots away & about the Time it gets to the Height of the other it is time to Cutt. Dews or Rain in the Spring are not good to rott in; good Seed is Bright & Brown, if whitish & Bright its not good.

The Method of Cutting is with a Hook made in this Shape about 1½ Inch Wide, square at the End, it shou'd have its rounding or Hook'd Part next the handle. The Handle is not Longer than a Sickle Hand[l]e & much Less hooking near the End. You cut Close to the Ground. 1 Inch at Bottom is worth 6 at top. You Gather it on the Back of the Left Arm and on your left Side and thigh, Cut it Close to the Bottom with your Hook, Then to put the Ends plum on the Ground to Lay it Even; they often

The Husbandry of Plants

strike the Hook 10 or 12 Inches from the End whereby they Cutt the Harle of a Dozen or More of the Stalks which Looses the Hemp on those Ends & is a great Loss to the Owner—They shou'd Be obliged therefore to putt the Back of the Hook to it & if you don't mind the Labourer in this he will deceive you. His Swarth shou'd be governed By the Height of the Hemp, for the Heighth of the Hemp should Be the Width of his Swarth. Care shou'd also be taken that the Hemp is not broken in Cutting, which they call Crippling. As you throw the Hemp out of your hands Spread it a Little, wch they call Spreading the Gavvel. Here it lays 3 or 4 days to dry which may be known by the Leaves being Crispy & falling off and when it is so Dry'd as not to Mow Burn you put it in Barns, Barracks, or in Stacks of the Bigness you Like. The Cutter shou'd have his Left arm Side, & thigh, Covered with Leather or Woolen, no Linen will stand it a day. A Cutter shou'd cut near ½ an Acre a Day which is Smart Work. You bind it in Sheaves Less than wheat Sheeves with 2 Stalks; If it has a Shower before it is near dry, it rotts the Kinder But if there Comes much rain you must turn it at least once a day & spread it thin or else it will heat & Hurt the Harle. You Leave the Male Hemp or Seed hemp on the outsides or the Banks, and when its Left Cutt off the Tops, then Like Tobacco it will branch & seed but this is better raised in a Piece by itself, sowed in Ridges at about 2 feet distance & Weed out the Fimble & Top it plow between with a Horse & Hoe it; it will yield near 20 or 25 Bushels pr acre. Put your Hemp in a Barn if Possible or Barrack if you can't do that then Stack it. In Stacking which is very difficult keep it well filled in the Middle & but Little Drooping outwards & when you have carried your Stack to the Bilge go no higher for if you gather in; the Ends will not Shedd the Rain well. Top it well with Straw till you Lay it out which is done by the 20th of October or 1st of November at furthest. Lay it out thin on a Clear field & when it is rotted you will know by the Harle Shrinking which will make the Stalk bow or bend. [Mind to turn it while rotting once for one side will rott sooner than ye other ye upper side rotts soonest] But the only sure way is to try it often by drying it & Carry it to your Break & if the Harle & Stalk part Clear & easy, then take it up directly & put it up Like Corn Stalks, if you have a great Deal Bind it with Rye Straw. This Time will take usually 6 or 8 Weeks according to the Season. The Snow whitens it if it is oblig'd to Lay Long before it is Rotten it is apt to be knitty that is small Pieces of the Stalk stick to the Harle. Dews or Rain in the Spring are not Good to

Rot in. If by a Snow's Coming or other accident your Hemp shou'd be a Little over rotten, House it and Lett it lie a year or two & it will recover itself & Clean well. First you Carry it to the Break then to the Crackle or finer Break, then to Swingle and a good hand for about 2s. 6 will Clean 75 lb pr Day then you have nothing to Do but to find a good Market.

Hemp on good Ground yields from 5[00] to 700 lb pr Acre.

Hemp seed is sold from 5s. to 20s. pr Bushel according to the Demand—& weighs 40 or 42 lb to ye bushell.

Ellis [86][81] says wett ground will do well but herein he Certainly mistakes.

Flax

Wm. Dougharty[82] says he had of a good year & on very good land on an Acre 200 lb of Swingle Flax abt 150 lb hatchelld & 10 bushells of seed—Ordinary flax will hatchell away ½. Heavy Strong land best for Flax.

[Flax] will bear 2 Crops in a Sumer ye last as good flax tho' ye seed does not ripen. J. Yard.[83]

Dye Plants

[Indigo]

Indigo is planted in Rows 18 inches apart & close in ye Rows in high Land the richer ye better in March & April cutt when in blossom in July & August They have a 2d & some a 3d Crop in very good Land but ye last Crop is poor.

There is a Sort grows 3 yrs but in comon but one.

When Cutt they putt it into a Steeper to ferment ye time is

---

[81] In *The Practical Farmer* . . . (p. 86) Ellis wrote: "Where the Loam is too wet, and can't be drained, it will bear Hemp." Read takes exception, again revealing his critical attitude toward the English writers. Writing his comment opposite the following passage in Worlidge's text (p. 44), he underlines the last portion: "Hemp delights in the best Land, warm and sandy, or a little gravelly, so it be rich and of a deep Soil; cold Clay, wet and moorish is not good: *It is good to destroy Weeds on any Land.*"

[82] The name William Dougherty appears on the tax lists of Chester County, Pa., in the 1770's and 1780's. William Dougherty, blacksmith, of Lower Chichester Township, was assessed for cattle in 1771. Also, William Dougherty of Northampton Township, Bucks County, was assessed for horses and cattle in 1779 and 1782. (*Penn. Arch.*, 3rd ser., XI, 672; XIII, 37, 240.)

[83] Probably Joseph Yard (d. 1763), of Trenton, who was clerk of the Hunterdon County Common Pleas, 1733-1734; clerk of the Board of Justices and Freeholders, 1739-1763; and served in the assembly from 1754 to 1761. He gave a part of the site of the First Presbyterian Church of Trenton and in his will bequeathed £100 to the College of New Jersey (Princeton). (*N. J. Arch.*, 1st ser., XIX, 394.)

various according to ye season when Fermented they draw it off & in ye receiver break it wth a kind of breaker being a box hollow 3 inches square at bottom 8 or 9 atop wth a pole in it to move it quick up & down. When broke they grain it wth Lime water after the Water is drawn off and the Sediment is of a Consistance it is putt into coarse Canvas baggs then spread & Cutt out into ye Squares yo see it sold in & dryed in the Shade.

The great art is in Fermenting, & Graining it.

[*Carolina Gazt*[84] says Poke berries will in this way make a good Dye.]

[J. Verree][85] Tis a large Steeper if 2 ft. Deep a receiver abt 4 ft. In breakg it froths very high at abt 15 minutes beatg Then yo Sprinkle on ye Surface a little oyl wth a feather wch directly settles it. This throw into 40 gals of Liquor putt 5 galls strong clear Lime water in 4 hours it will settle & wth one plugg above another draw off ye water.

It commonly steeps 10 or 12 hrs. in a shallow vessell just covered Then draw it into ye receiver, from yt it goes into bags or a

---

[84] Probably the *South Carolina Gazette*, published at Charleston, S. C., 1732-1775. The precise source of the reference has not been found. It may have some connection with a letter written to the Royal Society of London in 1763 by Moses Lindo, who went to South Carolina in 1756 as an indigo classer. The letter in part is as follows: "In August 1757, I observed the mocking bird fond of a berry, which grows on a weed called Pouck . . . , the juice of this berry being a blooming crimson. I was several times inclined to try, if I could extract a die from it; . . . till observing these birds to void their excrement of the same colour as the berry, on the Chinese rails in my garden. . . . I therefore made a tryal in the following manner.

"1st I ordered one of my negroes to gather me a pint of those berries, from which I extracted almost three quarters of a pint of juice, and boiled it with a pint of Bristol water, one quarter of an hour. . . . I then took two pieces of flannel . . . , boiled them . . . with alum a quarter of an hour, and rinced them in cold water. . . . I then dipped the . . . flannel . . . into . . . the juice . . . and left it to simmer five minutes, then took it out, and rinced it in cold water; when to my surprize, I found a superior crimson dye fixed on the flannel than the juice of the berry." (*Philosophical Transactions of the Royal Society of London*, LIII, 1763, pp. 238-239; Barnett A. Elzas, *The Jews of South Carolina* . . . , Philadelphia, 1905, pp. 47-67.)

[85] James Verree lived at Green Bank, the section of Burlington fronting on the Delaware River. In advertising for sale the property of his father who died in 1749, he described one tract on the Delaware at Burlington as containing a "good orchard of choice grafted trees, and gardens, well paled in with cedar posts, and boards in the front." Another advertisement described most of the land as "being improv'd and planted out into an orchard, with a collection of the best fruit trees, which makes the best of cyder, having a nursery of about 1500 of very fine apple trees upon the premises." (*N. J. Arch.*, 1st ser., XIX, 88, 158; *Penn. Gazette*, no. 1181, Aug. 1, 1751, no. 1225, June 4, 1752.) He had land interests also in Gloucester, Cumberland and Salem Counties. During the Revolution he was described as "the wisest head on the bank" (John J. Smith, *Letters of Dr. Richard Hill and His Children* . . . , Philadelphia, 1854, p. 219). On his death in 1796, his personal estate was valued at £4400-7-5½ (N. J. Wills, 11706C).

frame wth holes, in ye bottom covered wth Cloth like a Cheese when pressed turn it out on a thin board when stiff cutt it in squares & dry it in ye Shade.

[Madder]

Extracts from a Letter of Mr. Samuel Mason[86] of Camberwell— to Mr. Isaac Levi at Philada:[87]

To prepare the Ground digg it in Trenches full two spitts deep for altho' that is deeper than they will run ye 1st Crop it will mend it for the Second.

Plant yr setts at about 6 inches apart & when they have shott or strike to be 4 or 5 inches long they must be drawn by putting yr two fingers & thumb to ye bottom of ye plant, in order to gett up as much of ye yellow part wch is the best as being the Striking or most fibrous part— You should have Rain at the time of Planting you must water till they strike wch will be in 2 or 3 days. They are best planted in drills like peas ye ground being left in Ridges & the plants laid along the drill wth the head out of ye Ground. This Method is better than planting deep as in this way it runs more to a Stickey Substance wch is the best part of the Madder—They must be hoed as often as necessary. You are to plant in March or April. Tho' Mr. Mason says he planted 22d May 1756.

They lay in the ground 18 or 20 Months The method of taking it up is

You sett 2 or 4 Men in a trench & to Each Man a Woman to pick out ye Madder. After the first Trench is opened you must pare off the top as that is generally foul And lay it at ye bottom of ye 1st Spitt yo lay in the Trench & throw ye Crumbs on that in order to levell ye Ground for planting again. The Second Spitt you must digg up lay it down in the same place The reason of not laying the 2d Spitt on the first is that the Second Spitt is not so good, in Quality as ye first.

To Manufacture it

After it is taken up [put] it in a Cistern or Trough & wash it in several waters, till you have made it as clean as yo can. Then you must dry it on a Kiln. Then it must be ground to a powder & Sifted thro proper Seives. You must be carefull of it as it grinds That wch comes first is the first You will putt that by itself as being the best part of ye

---

[86] Nothing has been found about this Samuel Mason, other than that, as indicated, he was from Camberwell, a southern metropolitan borough of London.

[87] Probably Isaac Levy, watchmaker, of Philadelphia. He was entered on the Philadelphia tax lists in 1780 for a levy of £20 (*Penn. Arch.*, 3rd ser., XV, 312). A child of Isaac Levy was buried in Christ Church Cemetery November 21, 1759 (*Penn. Mag. Hist. and Biog.*, V, 1881, p. 225).

The Husbandry of Plants

Madder And so for the Second the third Crop is the rine or fibrous part of ye roots.

Truck & Vegetable Crops

	Time of Planting	Flowering	Ripening Prob.
Raddishes	March 25		10 May plenty
Pease & Beans	Early in March		6 June
Poke			10 May
Kidney beans	10 May & same of Carolinas		July 15. from ye 1st
Sallad	10 May plenty plant early		
Water Melons	15 May		July 20th
Cucumbers	Do		July 5
Squashes	Do		July 10
Pompions	25 May to 5 June		
4 o'Clocks & french mallow	abt 10 April		Flower July 20 till fall
Convolvolus	Do		Aug. 20
potatoes	for Early 1st April or sooner late 5 to 10 June for Winter	12 July	15 to 18
Indn Corn			ye last of July
Garden Cress			
Spinage	in Spring runs to seed Sow midle of August rich Ground for fall		
Turnips	in Spring fall		June 25
Peaches			Aug 20 not plenty

Huchleberrys ripe abt 1st July

 Our early pease come abt 6 June
 plant [French or Kidney-beans] 10 May ripe 1st July[88]
 Carolina's plant 10 May in blossom & Sett 10 July
 Plant [Melons & Cucumbers] 15 May Cucumbrs ripe 10 July
 Plant [Pompions] 25 May
 Squashes plant 10 may ripe 10 July
 Poke comes abt 10th May Asparagus a little sooner.

[88] These notes were written on the margin of Worlidge's text (p. 161) opposite the section *Of French Beans*, a portion of which Read had underlined as indicated below: "Of all the soils of Codware, there is none so fruitful, nor multiplies so much, as doth the French or Kidney-bean. . . ."

	Time of Planting	Flowering	Ripening	from 119
Raddishes	March 25		10 May	plenty
Peas & Beans	early in March		6 June	
Poke			10 May	
Kidney beans	10 May & same of Carolines		July 15. from ye 1st	
Sallad	10 May plenty plant early			
Water Melons	15 May		July 20 15	
Cucumbers	Do		July 5	
Squashes	Do		July 10	
Pompions	25 May to 5 June			
4 o'Clocks & french mellow	abt. 10 April		flower July 20 till fall	
Convolvolus	Do		Aug. 20	
potatoes	for early 1st April or sooner late 5 to 10 June for Winter	12 July	15 to 18	
Ind Corn			1 last of July	
Garden Cress				
Spinage	in Spring runs to seed	Sow middle of August rich Ground for fall		
Turnips	in Spring so	fall	June 25	
Peaches			Aug 20 not plenty	
Huckleberrys Ripe abt. 1 July				

Vegetable Gardening Calendar from Read's Notes

The Husbandry of Plants 295

Potatoes

they came originally by Sr. Walter Raleigh from No. America [*History of Cork* 128][89] [*Compt. Body of Husb.* 56[0]].[90] [*Life of Raleigh*[91] & *Transactns. Scotts Society*[92] 157]
Each peice wth an eye will grow but whole potatoes better.
[*Sweet Potatoes*]. The time to take up ye Seed is ye [time] when ye vines are killed by the frost wch usually abt ye 10th of October

[89] Charles Smith, *The Antient and Present State of the County and City of Cork* . . . (Dublin, 1750), I, Book II, p. 128: "It was in this town [Youghall, Cork County] that the first Potatoes were landed in Ireland by Sir Walter Ralegh. The person who planted them, imagining that the apple which grows on the stalk, was the part to be used, gathered them; but not liking their taste, neglected the roots till the ground being dug afterwards to sow some other grain, the Potatoes were discovered therein, and to the great surprize of the planter vastly increased, and from those few this country was furnished with seed.

"It is said Sir Walter brought them together with tobacco into Ireland from Virginia. He also brought the celebrated Affane Cherry at the same time, from the Canary islands."

[90] Hale (*op. cit.*, p. 453) wrote: "We had this Plant [the white potato] originally from North America, and at this Time raise it in vast Plenty, and to a very great Profit. It is in a Manner the Food of the common People of Ireland; and is cultivated in Lancashire, and some other Parts of England, in vast Quantities." (Note difference in page reference.)

[91] Since this reference is indefinite, we cannot conclude that Read is quoting directly from a biography of Sir Walter Raleigh. He may have noted this source in connection with the statement by Smith (*op. cit.*), and the article in Maxwell's *Select Transactions* . . . (p. 157), which attribute the credit for introduction to Raleigh. This comment by Read is of special interest because it touches a subject that for two centuries has been controversial. General confusion has attended the question of the introduction of potatoes into Europe, much of which has arisen from the careless use of the term "potato" as applying to both the sweet potato (*Ipomoea batatas*) and the common potato (*Solanum tuberosum*). Read himself in these notes treats both under the general heading "Potatoes."

T. N. Brushfield in a study based essentially on documentary evidence concluded that the only kind of potato in England prior to 1586 was the Spanish, or sweet potato, brought by the Spaniards to Europe from their earliest expeditions to the New World; that it was through Raleigh's instrumentality that the ordinary potato was brought to the British Isles; and that to Raleigh goes the honor of promoting its cultivation in Ireland, whence it was subsequently transmitted to England. ("The Introduction of the Potato and of Tobacco into England and Ireland," *Report and Transactions of the Devonshire Association for the Advancement of Science, Literature, and Art*, XXX, 158-178, Plymouth, 1898.)

Berthold Laufer has traced the common potato from Peru to the West Indies, to Europe, and from England to Bermuda. He has presented convincing evidence that the plant was not found in Virginia by Sir Walter Raleigh, but that it was introduced into Virginia from Bermuda in 1621. Instead of being a native North American, then, the potato appears as "a naturalized Englishman in the United States" ("The American Plant Migration, Part I: The Potato," Field Museum of Natural History, *Anthropological Series*, XXVIII, no. 1; Publication 418, Chicago, July 28, 1938; J. Henry Lefroy, editor, *The Historye of the Bermudaes or Summer Islands*, London, 1782, p. 285.)

[92] See footnote 101, this chapter.

but a surer rule is to See when ye potatoes in the hills putt out sprouts on their Sides & then they have done growing & tis time to take up ye seed & this is often before Frost when ye Potatoes are ripe ye vines on ye top of the hill turn yellowish.

[1770 May] John Griffith[93] Assures me that Potatoes for Winter of the Spanish Sort are best planted the middle of June are best, that he has had them very good planted in rich ground abt 25 June.

Bermudas or Red Potatoes. [April 28, 1756] They are 3 Weeks or a month coming up & wett & cold rott ym   In 1758 mine came up after 6 weeks.

[Ap. 21, 1767] Duke Fort[94] planted 200 hills on a square peice of Ground wch I measured, it was 24 square perches or near 1/7 of an Acre. 19,600 lb ye weight wch would at ye above rate grow on an heaped bushell of seed potos weigh 45 lb price 6d. pr lb. Hills yeild according to ye Season from 30 lb to 12 lb—suppose 14 lb pr hill they will yeild 2800 lb wch—at 1d. pr lb wch is half ye usual price amounts to £11.13.4 or about £80 pr Acre.

As such a Quantity might not Sell suppose ye whole to be given to Hoggs. Then on an Acre there would grow 19,600 lb.   19,600 being the product of 2800 mult. by 7. as they grow on 1/7 of an Acre. 19,600 lb ye weight wch would at ye above rate grow on an Acre being divided by 60 ye pounds in a bushl of Indian Corn is 326 or equal in weight to 326 bushells of Ind Corn. Suppose 1 bushell of Corn equal to 3 of Potatoes then tis Equal to 109 bushells of Ind Corn pr acre. D. Fort says that He once raised more than He could sell & fed his Hoggs wth them & the pork was very good. That at Harvest He bought 2 Piggs & carried them home on his back   they might weigh 15 lb each and He killed them at Xmas   they weighed 150 lb each. He gave them the potatoes raw. Boiled wou'd have been better.

The Hills are abt 4 ft. Diameter at Bottom & 3 atop are round & raised about 1 foot or 15 inches from ye surface of ye ground

[93] Apparently several persons by this name lived in Philadelphia and the surrounding region in Read's time. The name John Griffith appears on tax lists between 1765 and 1785 in Berks, Bucks, Chester, Lancaster, Philadelphia, and York counties, Pennsylvania (*Penn. Arch.*, 3rd ser., vol. XI-XXVI). However, Read may have been referring to John Griffith of New Jersey, who in 1762 was living "near Gloucester," and in 1781-82 was justice of the peace of Gloucester County (*Penn. Mag. Hist. and Biog.*, XVII, 1918, p. 69). The soil of Gloucester County is well adapted to sweet-potato culture.

[94] This was probably Marmaduke Fort, Sr., farmer, of New Hanover Township, Burlington County. He died in 1795, leaving his plantation of 147 acres in New Hanover Township and a cedar swamp in Northampton Township to his son William. (N. J. Wills, 11600C.)

flatt on ye top. Planted wth 4 or Six peices broke off so that there may be 2 Sprouts in Each, thrust them in 6 inches deep. The Hills should stand so farr apart as that a single Horse may plow among them. [The shape of ye hill ⌐▱⌐] Plowing among them tho' yo tear ye vines hurts not. The Hills produce long small potatoes best for Seed or the Vines covered up or stuck in, in moist Weather will produce them. Mr. Rigden[95] says that in Carolina the Vines stuck in bring the largest potatoes & I have observed ye largest grow in ye Alleys wch I suppose came from Vines.

[Tis said pr Rd Somers[96] that the Seed when packed up shd have a little air or they wont keep] They are kept over in Gums or boxes putt in the Chimney corner packed in Dry Sand or in any place very dry & warm in Sand well dryed, but too much heat gives them a dry rott. You are to Observe as an Unerring Rule that you putt down ye Seed potatoes ye day yo digg them, never lett them wilt at all they keep vastly better as Jos Thorn[97] tells me who raises many   he once kept them under dung in his cow house.

Gentlemans Magazine for 1753 [page 358] has an abstract from

[95] Probably William Rigden, painter, of Philadelphia. In 1769 he was taxed for one horse, one head of cattle and two servants, and his name appears for several years on subsequent tax lists. This was probably the William Rigden who lived "on the east side of Second Street opposite the New Market." He died in 1787. (*Penn. Arch.*, 3rd ser., XIV, 172, 235, 475, 751; XV, 211, 751; XVI, 325, 698, 742; Phila. Co. Wills, Book T, p. 442.)

[96] Richard Somers the elder lived at Great Egg Harbor, old Gloucester County, of which he was freeholder between 1723 and 1740. He died in 1761, leaving a large estate, a substantial part of which, including the home plantation, went to his son Richard. His personal property was appraised at £479-10-10, and included horses, cattle, sheep, and hogs valued at £158-0-0. Among the other items listed were a harrow, a winnowing mill, 4 beehives, and 32 pounds of "Roap." (*N. J. Arch.*, 1st ser., XXXIII, 402; N. J. Wills, 775H.) The younger Richard (1737-1794) was judge of the county court, member of the assembly, and colonel of militia during the Revolution. After his death, the plantation was advertised for rental as "containing four hundred acres of upland and three of meadow and marsh. The dwelling house is commodious, with suitable out buildings, and well calculated for store and tavern keeping. . . . There are four apple and a peach orchards, all in good repair; other advantages from nature, of fish, fowl and oysters unexcelled by any place in the country." (Alfred M. Heston, *Absegami* . . . , Atlantic City, 1904, II, 291; *Claypoole's Daily Advertiser*, Philadelphia, Jan. 10, 1800; T. Cushing and C. F. Sheppard, *op. cit.*, p. 137.) The Somers mansion still stands at Somers Point on the mainland opposite Ocean City.

[97] Probably Joseph Thorn, "yeoman," of Chesterfield Township, Burlington County. He was a weaver and was surveyor of the highways of his township for several terms between 1753 and 1765. He died in 1774, leaving to his son Thomas the farm where he lived. (*N. J. Arch.*, 1st ser., XXXIV, 525; N. J. Wills, 10043C-10046C; E. M. Woodward and J. F. Hageman, *op cit*, p. 281.)

298   Ploughs and Politicks

Mr. LePage du Pratz's Natural History of Louisiana.[98] Says the Vines of Red Potatoe being cutt off the Middle of August about 7 or 8 inches long being bent in the form of a double cross & laid underground in Ridges prepared for them *in a poor Soil.* The potatoes raised in this manner are the most Esteemed for the Excellence of their taste & their keeping better in the Winter. [If this be so then there needs not be such Care about seed wch is difficult to preserve over winter.] The French have made Brandy of ym.

J. Verree. They plant red potatoes in rows at Carolina.

They have 3 Sorts of Potatoes the Spanish the White & the Brimstone the last is best.[99]

Mr. Hilhouse[100] tells me there are 6 weeks Potatoes in Antigua & picked 6 weeks after planting.

Potatoes Irish. Extracted from the *Transactions of the Society of Edenborough* published by Robt. Maxwell 1743.[101]

[157] That Sr. W. Raleigh first brot them to Ireland from Virginia.

[158] That there are 5 Sorts of Potatoes of this Kind Viz 1st the Long red 2d the Round red 3d long white 4th round White and the blew besides an Early kind wch delights in a Light Dry soil.

There are Potatoes for which no Ground is too wet.[102] An Other Sort for which None is too Dry. Some Kinds will thrive in a Deep moss that will not bear Horses & the Long white or Kidney is best for Mossy ground.

[159] In Light dry grounds they are Planted Successfully before Winter & these will be earlier than Others.

The Beds 6 ft. wide he calls Lazy beds & the Process of planting thus Cover ye ground with rotten Litter, rotten Hay, Straw or Other Long

---

[98] In the English translation of M. LePage Du Pratz's work, *The History of Louisiana or of the Western Parts of Virginia and Carolina* (London, 1763, II, 7), occurs the following: "In their substance and taste they [sweet potatoes] very much resemble sweet chestnuts." The article in the *Gentleman's Magazine and Historical Chronicle* (London, 1753, XXIII, 358) reads in part: "The small potatoes are cut into slices, each slice having a bud in it: these slices are planted in a poor soil, in ridges of earth about a foot and a half broad, raised on purpose." The "poor soil" is not mentioned in the translation of 1763.

[99] This paragraph probably refers to the culture of potatoes in Carolina. It appears on a page in the manuscript which is headed by the caption "from Mr. Jos. Shute, Carolina."

[100] Possibly James Hillhouse, who in the late 1760's was living in West Nantmeal Township, Chester County (*Penn. Arch.*, 3rd ser., XI, 339, 484).

[101] The notes on Irish Potatoes here given were taken from Maxwell's *Select Transactions* . . . , pp. 156-172, No. XXIV: "Mr. Maxwell of Arkland's Observations, . . ."

[102] This is expressed somewhat more extremely than is stated by Maxwell in the original (*op. cit.*, p. 158): "I have remarked, that the blue, and long white or kidney Potatoe, thrive better than others in Moss and wet Ground; but especially the blue; and that it does not agree so well with dry."

The Husbandry of Plants

dung most preferred plant your Potatoes a foot distance In [page] 161 he says 6 in.[103]

Near Dumfreis they plant the Potatoes whole only cutt out most of the Eyes which they call Gelding but He advises Planting pieces Cutt with two Vigorous Eyes at least for from these shoot as well the runners which bear the potatoes as the Stalk & where a peice with one Eye has been planted he has known a Stalk without any runner or potatoe, the runner taken up and Exposed to the Air will be a Stalk.

[162] That the Potatoes may be Planted in February or March & if not planted in the Lazy bed way may be planted thus. After the Ground is Plowed make a Trench in Every fifth furrow, lay your Long Dung & Potatoes & cover them up. When they appear they with a hoe plow & Nottched Muzzell plow a furrow off, & the nearer they go the better when they appear to be $\frac{1}{2}$ an inch high they[104] cover them gently with a Shovell, when they appear again they lay a furrow on as close to the Stems as you can without covering them.

You may Cutt the tops when in full flower for Cow feed & tis good & dont hurt the roots but rather helps them as it checks the Stalk & makes the roots grow larger.[105]

[159] In the Lazy bed way when they are $\frac{1}{2}$ an Inch above the Ground you Cover them & So when they appear again for twice but these Last Coverings should only be so as Just to putt them out of Sight at the Planting they should be 6 or 8 Inches Covered.[106]

[160] In Mossy Ground you must raise the beds. If they are free of Weeds & the Stems are Short small and Hard you from thence Judge the Potatoes large & ye contra. Your Dung should be in proportion to the Natural Strength of the Ground.

[168] Advises getting seed from a Soil Different from your own or if you can from a Distant Country.

[*Turnips*]

[Turneps] are sowed for Sheep & Cattle & raised 1 row in a ridge and plowd have produced 640 bushl pr Acre [Elliotts *5th Essay* 36].[107]

---

[103] Evidence again of Read's careful study of practical details. In giving the planting distance as 6 inches, the author was quoting a "Mr. Switzer."
[104] "They" appears to be used as a general term for the purpose of describing practice followed in England.
[105] From p. 171 of the *Select Transactions*. . . .
[106] In the *Select Transactions* . . . (p. 159), the depth of planting is given as "five or six Inches," not "6 or 8" as noted by Read.
[107] Eliot's comment (*op. cit.*, p. 36) reads as follows: "Mr. Tull saith, that his Turnips drilled and well ploughed, weighed from six Pounds to fourteen, did produce Six Hundred and Forty Bushels to the Acre: I should be very glad of half that Quantity: As to the Ease in raising them in this Method, by ploughing the Furrows off and on, I am satisfied by Experience, and that they will grow larger; what I tried were Spring Turnips."

T. Bowlsby a good Farmr say[s] 1 lb seed pr Acre. [Sow from 25 July to 10 Aug. N. S.] Mr. Saltar says ½ lb seed too much pr acre. [*Compleat body of Husbandry* Fol 332 to 335][108] 2 lb. Turnip seed to an Acre ye broad Cast way—3 or 4 oz in ye Drill way.[109] Like a Light sandy Soil a little moist—When ye rough leaf comes yo are past ye fly. If yo could by any method cover some seed 2, some 4 inches it would give 2 Weeks odds & if ye fly took ye 1st ye 2d might Escape   Qu if plowing in one sowing & bushing the 2d wd do   Sow ashes lightly over ye Crop in the Evening when ye dew is on to destroy ye fly. [*Ibid* 452]

To preserve turneps from the fly putt the Seed you intend to sow into a Stone jarr close stopped & putt into it 2 Spoonfulls of ye flour of Brimstone & shake it   the 2d Day 2 Spoonfulls more & Shake it keeping it 3 days Close Stopped & shook   then sow ye Seed & it will be so strongly impregnated wth the Brimstone that it will preserve it [illegible].

Sow yr Turnips between ye rows of potatoes   the fly never hurts them. Colel Young.[110]

[*Compl. body of Husby.* 333-452] The fly never takes the turnip after ye 2d pr. of Leaves putt out. To prevent it mix equal Quantities of Stone Lime & Soot   as much urine warmed to the heat of Milk will make it a thin pap.   when cold putt yr seed in 24 hours   then Sow it. [*Do* 450]

[Turnip Seed to raise 1758] If yr Turnips lye out in ye Ground all winter or are farr Sprouted in ye Spring before planted the Seed will not be good but tis best to plant as soon as yo think they will grow well   a ps. of ground 20 ft. by 10 planted @ 9 inch asu[n]der yeild 3 pints of Seed.

If the Turnip be as bigg as a henns Egg ye 1st of Octobr they are good, they grow best when ye weather is Cool.

[108] To protect turnips from the fly, Hale (*op. cit.*, p. 452) advised: "Wood Ashes are found very destructive of this little Insect; it is therefore a good Method to scatter a Parcel of these thinly over the Crop, just after the Plants are above Ground."

[109] Elsewhere in the notes Read had written: "it takes 2 lb. Seed to Sow an Acre ye Comon way 4 oz. by ye Drill."

[110] Probably Henry Young (1691-1767), of Cape May County, who, as a young man, fled from England to escape impressment. He was justice of the peace from 1722 until the time of his death and member of the assembly 1730-39 and 1744. He also served as surrogate, as judge, as deputy-surveyor of the West New Jersey Society of Proprietors, and as colonel of the Cape May militia. Jacob Spicer recorded in his diary an agreement with Henry Young for producing shingles from the timber in a Cape May cedar swamp to which both laid conflicting claim. After Young's death his personal property was inventoried at £1106-6-9½, including "Cattel and Live Stock" appraised at £36-12-6. (John W. Barber and Henry Howe, *op. cit.*, p. 133; *N. J. Arch.*, 1st ser., IX, 359; XXXIII, 493; N. J. Wills, 274E; *Cape May Co., N. J., Mag. Hist. and Geneal.*, I, no. 3, 1933, p. 115; no. 4, 1934, pp. 169, 170; no. 6, 1936, pp. 240 241.)

The Husbandry of Plants

NB—If Turnips are as big as pidgeon Eggs ye 1st of Octobr They are forward enough they grow most when ye top changes colour.

Compt. Body of Husbandry[111] says that Cutting them prevents ye cattle chewing them.

[Carrots]

DuHamell 320 says carrotts better for sheep &c. than turneps[112]

Mr. Elliotts *2d Essay* [8][113] &c. says Mr. Ellis[114] mentions a farmer who had £60 sterl. pr Acre of Carrots. You plain plow, then Ridge plow, then plant at ye distance below & weed once between ye rows wch are 10 inches one way & 6 ye other then plow a furrow off lett it lay 2 weeks, then plow it on & so 4 times tho' 6 times is better.

raised in Ridges in 2 rows 10 inches distance on a ridge produced 230 bush. pr acre tho' the same ground not tilled in that way produced carrotts no bigger yn a candle Elliotts *5 Essay* 23 his own Experiment.

[Parsnips]

Parsneps may certainly be planted to advantage for Hoggs as a vast Quantity will grow on an Acre. DuHamell 322 say[s] ye seed above a year old will not grow.[115]

[Radishes]

I have proved this[116]—Pull them in ye fall putt them up wth yr Turneps.

---

[111] Hale, *op. cit.*, p. 452.

[112] "One acre of carrots, if well planted, will fatten a greater number of sheep, or bullocks, than three acres of turneps, and the flesh of these animals will be firmer and better tasted" (*op. cit.*, p. 320). (See Philip Miller, *The Gardener's Dictionary*, 7th ed., London, 1759, article under *Daucus*.)

[113] *Op. cit.*, pp. 32, 103-105, 111-112. In recommending the number of times to plow carrots, Eliot writes, "but seven times will do better," while Read notes, "tho' 6 times is better."

[114] "A Person of Note, in Gloucestershire, said he could fatten an Ox with Carrots as much in one Week, as common Meat would do in four. Another made sixty-odd Pounds of all his Carrots that grew on only one Acre of Ground, by selling those that grew in one square Pole, at the Price of eight Shillings, and so for every Pole throughout the Acre." (William Ellis, *The Modern Husbandman: . . . For the Month of November*, London, 1743, IV, 87.)

[115] H. L. Duhamel, *op. cit.*, p. 322. See also Miller, *op. cit.*, article under *Pastinaca*.

[116] Read here refers to Worlidge's text (p. 165): "Radishes are so commonly known, and their Propagation so easie, that here needs no more to be said about them."

[*Beets*]

Qu. if not to be planted to advantage for Hoggs.[117]

[*Oinions*]

Extract of a Letter from Mr. Austin Hicks[118] *Jany 26, 1757:*

That He was not acquainted wth raising them from ye seed. He raised from ye Small oinion bot at 4*s.* ye bushell. Sandy land made rich wth good fine Dung best for them. plant in rows at a ft. Distance from Each other & the oinions four inches apart in the rows when the Seed pipe in ye middle begins to grow large break it down to prevent their going to seed. The time of gathering abt. the middle of July. He dryed them in the Sun. He says if ye Land is good & the onions well tended they will yeild 180 bushells pr acre. Time to plant beginning of March.

[1759, I sowed seed 1st, 7br ye weather many days rainy came up in 10 days] Mr. Adoniah Schuyler[119] who well understands the planting Onions tells me that the only way to Execute that matter well is to Sow yr Oinion seed abt ye 1st of September very thick in rich Ground they come up as fine as hair these are not to be covered in winter & being drawn in the spring & transplanted make large onions by abt ye middle of August—Mr. David Johnston[120] says that yo may Sift fine horse dung among ye Small onions abt ½ inch thick & that ye crop of Onions comes off just in time to trench for Cellery—The fine small onion does not run to seed like those

---

[117] Written on the margin opposite the following paragraph in Worlidge's text (p. 164), *Of Beets:* "This ordinary Plant is by several made use of; it loves a fat and rich Soyl; it's usually sown in the Spring, and will come up several years in the same ground, and may be Planted forth as Cabbages are."

[118] Not with certainty identified. He may have been a resident of Burlington County. An accounting of the estate of Robert Hartshorne, of Burlington, in 1752, shows a debt due to Osten (probably Austin) Hicks (*N. J. Arch.*, 1st ser., XXXII, 148).

[119] Adoniah Schuyler had a farm at Elizabethtown Point and also owned property on the Passaic River. After his death in 1762, his widow and his son John offered these properties for sale. The first was described as an "advantageously situated Farm and Plantation . . . whereon is a good Dwelling House, Barn, Out-Houses, a large Orchard, together with the Ferry, and Road from the Sound, to the Upland of Staten-Island." The second place, it was stated, contained 17 acres "whereon is a new Stone Dwelling-House two Stories high, a good Barn and Orchard, where may be cut at least four Tuns of Clover Hay in a Season." (*N. J. Arch.*, 1st ser., XXIV, 124, 169; *N. Y. Gazette*, Jan. 10, April 25, 1763.) He was a brother of Peter Schuyler.

[120] Not identified. Adoniah Schuyler in his will (1761) named "David Johnston of New York City" as executor. Since both these men are mentioned by Read in the same paragraph, the inference might be drawn that this was the David Johnston in question. There were other David Johnstons in Read's time—in Gloucester, Essex, Middlesex, and Monmouth Counties. (*N. J. Arch.*, 1st ser., XXXIII, 375; N. J. Wills, 531B.)

The Husbandry of Plants

as large as Nutmeggs. Mrs. Saltar[121] says ye onions wch stand in ye Ground all winter seed well.

[Cabbage]

If an Early Cabbage about June yo cutt ye head & leave no more than two sprouts & they will come to Small hard heads as sweet as a Cabbage in June this they do in Bermudas & it will do here [H. Roberts].[122]

Among ye Moravian's they plant 3 ft apart sow in 7br plant out that fall for Early Cabb.

[Savoys] A Mistake for tis a Winter green.[123]

Cabbages to be Planted for Cattle

[NB Cabbages give butter an ill taste but may feed fattg cattle] An Acre of Ground contains 66 ft one way 660 ft the other. Suppose

---

[121] This was probably Hannah Lawrence Saltar, the wife of Richard Saltar.

[122] Probably Hugh Roberts, of Philadelphia, companion of Benjamin Franklin in the Junto Club. Franklin described him as "an ingenious acquaintance of mine . . . one of our most eminent farmers," and "a Man of much Business" (A. H. Smyth, *op. cit.*, III, 53). From Franklin's correspondence with Jared Eliot, it appears that Mr. Roberts presented Eliot with a fork and a rake. Referring to the implements, Franklin wrote: "I shall learn . . . how those Teeth are put in, and send you word . . ." (*ibid.*, 280). In another letter to Eliot, Franklin wrote: "Mr. Roberts promises me some Observations on Husbandry for you" (*ibid*, 59), and again, "Mr. Hugh Roberts . . . tells me, that it appears by your writings, that your people are yet far behind us in the improvement of swamps and meadows. I am persuading him to send you such hints as he thinks may give you farther insight into that matter. But in other respects he greatly esteems your pieces. He says they are preferable to any thing of late years published on that subject in England. The late writers there chiefly copy from one another, and afford very little new or useful; but you have collected experiences and facts, and make propositions, that are reasonable and serviceable. You have taught him, he says, to clear his meadows of elder (a thing very pernicious to banks), which was before beyond the art of all our farmers; and given him several other useful informations" (*ibid.*, 53). While Franklin was in London in 1768, he sent to Roberts, Bartram, and a few other friends, by way of his wife, samples of a new kind of huskless oats, "much admir'd . . . to make Oatmeal of" and of 6-rowed Swiss barley (*ibid.*, V, 182). Roberts was associated with Franklin in the first fire insurance company in Philadelphia, and in the Philadelphia Hospital (*Penn. Mag. Hist. and Biog.*, XLVI, 1922, p. 253). He owned land along the Schuylkill River in Montgomery County (*ibid.*, L, 1926, p. 356). For the growing young city of Philadelphia he was an advocate of wide sidewalks (*ibid.*, IX, 1885, pp. 484-485). It appears that Roberts provided supplies for the maintenance of the Pennsylvania State House (Independence Hall). In an account rendered for 1752 and 1753, for sundry supplies to the State House, he listed the following items: "A large file for the cracked bell, 2s.6d.; Oil for the bell, 10d." (*ibid.*, XXXV, 1911, p. 254).

[123] Another instance of Read's taking exception to a published statement, referring to Worlidge's comment on *Savoys* (p. 164): "There is another sort of Cabbage, commonly called the Savoy, being somewhat sweeter and earlier than the common Cabbage; and therefore to be preferred: It is raised and planted as the other, as also is the small Dutch Cabbage."

your cabbages to Stand 1 ft apart 3 ft betwixt Rows. Then Multiply 660 by 22 wch is 1/3 of yr breadth or 3 ft apart   it produces 14520 Cabbages of wch give a Cow 12 pr day   it will keep a Cow 1210 days or 7 cows 180 days. Now one hand can tend these Cabbages 2 months wth one horse part of ye Time   say Wages & diett . . 7.0.0 wth soiling £5 and plantg £2 7.0.0

£14.0.0

pulling etc. 3.0.0

17.0.0

by This Calculation £17 will keep 20 Cows 61 days
  [@ 3 ft. apart 1 Acre bears 4840 @ 3 ft. one way 2 ye other 7260.]

[*Hops*]

Tolerable Hops in Jersey will yeild pr Acre 400 lb. or better sold at 1s. pr lb. & little more trouble than Indian Corn (T. Shaw).[124]

Asparagus to plant in Beds

[Bradleys *new Improvemnts* fol. 290][125]

About the beginning of March (which is the best time for Transplanting of Asparagrass) Measure out your Ground, allowing four foot for the Breadth of each Bed, and two feet for every Alley between the Beds; Open a Trench at one End of it, and lay into the bottom of the Trench, about the thickness of six or eight Inches of Horse Dung well Rotted: Go on then to Trench the same Quantity of Ground, lying Next to the First Trench, throwing the Earth which comes out of the Second Trench upon the Dung laid in the Bottom of the first: Continue to Work in like Manner till the Whole Piece is Completed and levelled; then in lines, at eight or ten Inches Distance, plant the Asparagus taken fresh out of the Nursery, Spreading their Roots, and covering their Buds with Earth, about 4 Inches Thick.

[124] Probably Thomas Shaw, of Burlington City, "practitioner in Physick," or his son Thomas. Read refers to the former elsewhere as "Doctor Shaw an Acquaintance of mine following the Brewing business." He owned sundry tracts of land in Burlington County, and was a vestryman in St. Mary's Church (E. M. Woodward and J. F. Hageman, *op. cit.*, p. 76). He died in 1750, leaving sons Thomas and Samuel (of minor age), and brothers John and Samuel. (*N. J. Arch.*, 1st ser., XXX, 429; N. J. Wills, Liber VI, 377). Lt. Col. Thomas Shaw (possibly the son) was killed at Ticonderoga in the French and Indian War (R. Wayne Parker, "New Jersey in the Colonial Wars," *N. J. Hist. Soc. Proc.*, new ser., VI, 1921, p. 215).

[125] The first two paragraphs of this passage on asparagus culture were taken almost verbatim from Bradley's *New Improvements* . . . (pp. 290-291). The last was a marginal notation of the main facts from succeeding passages.

When four Rows are Planted, leave a Space for an Alley about two Foot Wide, and proceed to Plant 4 Rows more for another Bed, and so Continue your Work till the Whole is Planted. As soon as that is Done, Sow the Whole Piece with Onions, and rake it over as level as Possible for the Alleys will not be of any Use till After Michaelmas, when the Onions will be off, and the Shoots the Sparagus Plants have Made that summer, must be cut down; then dig up the Alleys, and throw, Part of their Soil upon the Beds, to raise the Earth about five or Six Inches above the Buds of the Plants Supplying the Alleys at the same Time with Dung, or some rich Soil.

[Yr Beds must remain so till March then rake down ye earth cutt of yr haulm or Stalks in 8br & lay on dung wch yo rake off ye beginning of March don't cutt till ye 3 year.]

Calavance peas[126]

Of these there are many Sorts.

1. The small round pea abt as big as a good shott a little of a yellowish cast yeild abt 9 or 10 bushl pr acre—sell best at Markett.

2. one of the same colour of a Small size shaped like a kidney a white Eye.

3. Of the Indian peas—are three or four different kinds some wth white, some red, some black Eyed, are good yeilders larger than the foregoing.

4. A larger & flatter kind of a Deep yellow or a pale red Cast interspersed wth black like the spotts in an apple, shape like a kidney they have a white Eye in a small blackish circle they are not so subject to the worm & preserve themselves ye best if rainy weather comes after gathering & grow in a soil worse than any of the former bear 20 or more bushel on an Acre & from their Colour are called Whip poor Wills from their resemblance to their Eggs in Colour.

Note the 2 first do not Colour ye Meat Ye last turn it purple.

The last ripen most together but it must certainly be best to pick of all sorts Once before you cutt up as they will otherwise open & Shake—To Cutt them wth a hook is better than to pull ym If the weather is dry Thresh them out in the field. if not yo must house them & if wett in ye field be sure yo turn & dry them.

☞ If yo can gett a Sled & keep threshers at work tis best as they will sweat & the Shells grow tough so as not to thresh wthout great

[126] There is evidence that all the persons mentioned by Read in his notes on Calavance peas, apparently the equivalent of the modern cowpea, had farms south and southwest of Mt. Holly, mainly in Evesham Township. This suggests that the crop may have been of special importance in this region. (See also p. 332.)

labr even upon one night's lying, pick'd in a dry day thresh very easy [Wm. White][127]

These peas grow in the poorest of ground & coarsest of Sand—Are good for human food, & all kind of Cattle. The longer you till the ground the better, as ground that is old is not Subject to the Worm as that wch is New. Their Enemies are the worm abt 1 inch long & black like the Indian corn worm & a small vermin wch makes a hole thro ye Stalk of them about as big as a pin. They take them when abt 2 inches high—Hoeing them is a great check to these Enemies.

Once plowing will do for them. Hoe them as ye Season may require. [Yo must plow ym Twice or more as ye grass may grow][128] They should be made clean just before they run. The Drought never hurts them & if yr horse should tread on the Vines unless cutt off they will still live.

Plant about ye 20th May (N[ew] S[tyle]). 4 ft. apart 5 or 6 in a hill. They ripen abt 15th 8br [or sooner if frost comes] & some sooner, tho' there will then be some greenish on ye Stalks. 4 ft. apart allows yo to plow Each way.

Nicholas Stiles[129] plows them clean then furrows double & plants 2½ ft apt in rows thinks this better than in hills—Afterward plows once clean just before they run & hoes them then, ye first hoing ye first time is when they first appear in 2 Leaves & this prevents ye worm.

Abraham Perkins[130] plants 4 ft. one way and 2 ft. ye other.

[127] On August 10, 1774, William White, of Burlington County, advertised his farm for sale, describing it as a "valuable Plantation . . . of about 100 acres. It is situated on naviagable water, about one mile from Burlington Court-house, and about nineteen from Philadelphia: There is on the premises a new house, two stories high, 36 feet by 16, with a cellar under the whole; likewise a good kitchen and barn; the fencing is chiefly cedar, about three acres of orchard, about 20 acres of woodland, and about 15 acres of marsh that lies very convenient for banking, and between 8 and 9 acres of clover meadow; the remainder very good land for grain. The place is commodiously situated, is very healthy, and would be very suitable for a gentleman's seat, or a small family." Likewise, he announced for public sale, "a number of milch cows, a fine bull of the English breed, a number of yearlings and calves of the same breed; likewise a number of swine of the English breed. . . ." (*N. J. Arch.*, 1st ser., XXIX, 448; *Dunlap's Pennsylvania Packet*, no. 147, Aug. 15, 1774.)

[128] Noted in the margin, presumably representing a later, more mature judgment.

[129] Probably Nicholas Stiles, of Chester Township, Burlington County. The Stiles farm was located in Maple Shade on the north branch of Pensauken Creek, a little below the Camden pike (George DeCou, *Moorestown and Her Neighbors* . . . , Philadelphia, 1929, p. 144). On Nicholas Stiles's death in 1804 his personal estate was inventoried at $1345.71, and included oxen, cows, and young cattle valued at $284, besides horses, sheep, hogs, poultry, corn, grain, and hay (N. J. Wills, 12190C).

[130] Abraham Perkins' home plantation stood on the banks of the Delaware in the part of old Willingboro Township, Burlington County, subsequently set apart as

The Husbandry of Plants

Tho' they grow well on poor land they will bear better & fuller if they have Manure, tis ye loosness of the Land yt suits ym.

Joseph Moore's[131] Land is very high Dry & poor opposite my Pltn on Ancocas he plowed it & planted Calavance peas several years Successively & he thinks it mended ye Land tho' not manured. He says he Cleaned 140 bushl from 6 acres.

[1758] Wm. Borton[132] says that ground being clean when first cleared He has sowed them abt ½ bushell to ye acre & it did well but afterwards the crabb grass prevents this way of tilling & sowing them a second time.

[NB Pudding beans ripen ye 5 7br 1758]

Israel Taylor[133] who has been used to these says furrow & cross furrow 6 in a hill they come up in abt 5 or 6 days if Seasonable weather when abt 2 inches high hoe them & plow up clean & you must plow them as often as the Crabb grass comes up Cutt up wth a hook carry them directly to the Thresher & 2 good hands will thresh 100 bushl pr day. He makes his floor of rails abt 2 foot from ye ground & as the peas come out they drop thro' & prevent bruising. NB it must be best to have a floor or cloth spread

---

Beverly, and he also owned other lands in the same township. He died in 1764, leaving the home plantation to his son Joseph. His personal estate, inventoried at £273-8-9, indicated a well equipped farm. The following livestock was listed: 3 horses and a colt, 11 cows, 2 bulls, 3 heifers, 4 calves, 5 sows, 32 pigs, 3 barrows, 1 boar, 6 sheep, 2 lambs, also 2 "Neats." Other items of interest were a negro boy valued at £80, a field of rye (£21-10-0), hay in the stack (£5-10-0), and "a parsel of Books—Catto" (£2-9-6). (N. J. Arch., 1st ser., XXXIII, 325; N. J. Wills, 7691C-7698C.)

[131] Joseph Moore lived on a farm on the south bank of the South Branch of the Rancocas, between the present sites of Hainesport and Lumberton. At the time of his death in 1786, in addition to the home plantation, he owned a house in Mt. Holly, and several hundred acres of pine lands. The inventory of his personal property included "Cows, Horses, Sheep, Hogs, &c." valued at £152-2-6, "Farming Utencils" £51-8-6, and hay £6-0-0. The total value was £299-16-0. (N. J. Wills, 11123C.)

[132] Probably William Borton, of Evesham Township, Burlington County, who died in 1763, leaving his son William 188 acres of his plantation. John Woolman was witness to the will, which was executed in 1760. His personal estate, which was appraised at £472-1-8, in addition to livestock and farming utensils, included the supplies and equipment of a tannery. Following are some of the items: bark, £4-2-6; saddle leather, shoe leather, and skins, £99-8-8; sole leather, sheep and calf skins, £29-15-6; dry hides, £27-0-0; hides in lime, £15. "A servant boy's time" was listed at £10. (N. J. Arch., 1st ser., XXXIII, 52; N. J. Wills, 7415C-7420C.) The ancestral Borton home, Hillsdown, on a 200-acre tract surveyed to John Borton in 1682, was located on the south side of Rancocas Creek below Centreton and extended back to the village of Masonville (G. DeCou, op. cit., p. 48).

[133] Probably Israel Taylor, of Northampton Township, Burlington County, who lived on a farm in Brown's Town (Brown's) near Lumberton. In 1784 he sold to Ruloff Voorhees for £150 two parcels of these holdings amounting to 234 acres (N. J. Deeds, Liber AR, p. 154). Israel Taylor, Sr., of Evesham Township (possibly the same) died in 1816 (N. J. Wills, 12903C).

under them. He thinks tis best pick ye first wch come & send to markett & ye rest will ripen near a time nor has he ever known them mould by being laid together directly after Cleaning

It will take near 2 bushl to plant 10 Acres.

He likes best to plant in rows 3 ft apart or 2½ may do & abt 2 ft. the other way.

[Hotbeds]

The Manner of Making & using a Bark hott Bed; from Bradleys New Improvements &c 1739 fol. 601:[134]

Prepare a place 4 ft. Wide 10 or 12 Long enclose it wth brick 2 or 3 ft. high, if in dry ground ye brick is under ground. Sett yr frame on ye top of ye wall of the height that suits turn up ye Shallow side of ye glass Barrs. Fill to ye top of ye brick work wth tanners bark just thrown out lettg it lay a few days to drain well & having Strewed some rubbish in ye bottom & on that fresh horse Litter about a foot thick when trod down putt yr Coarsest bark in first then ye finer. This Rubbish & dung serves two purposes to give a Quicker heat & to drain the water thro', which falls on ye bark by watering ye plants press yr bark hard down in a week yo may putt yr glasses on putt yr potts or boxes on ye bark but it will be then too hot to plunge them into, in abt 10 days after yo sett yr potts on ye bark yo may putt them in. Such a Bed if made in March will keep heat till ye beginning of 7br. When yr bed begins to Mould atop ye ferment is gone but may be renewed by digging it a foot Deep & a few Springklings of water.

NB. I have seen a small door in ye Ends of such beds to give air instead of raising yr Glasses.

[Bark will not do in smaller beds than 6 by 10 feet less will not heat]

Hott Beds of Horse Dung

[From Wm. Masters of Burlington][135] Prepare yr Bed 10 days

---

[134] *New Improvements* . . . , pp. 598-602.

[135] William Masters (d. 1760), of Philadelphia, was a friend of Benjamin Franklin and a member of the provincial assembly. His 500-acre estate he called Green Spring. He operated the Globe Mill on Cohochsink Creek (C. P. Keith, *op. cit.*, p. 453). Whether or not this is the person to whom Read refers is not clear. Another William Masters appears to have been living in Philadelphia in 1775 (*Penn. Mag. Hist. and Biog.*, XLVI, 1922, p. 168). In 1758 a two-family brick dwelling in Burlington was advertised for sale, described as "in the Tenure of William Elton, and Wm. Masters, . . . fronting the River Delaware," also "a Brick Bake-House, and Storehouses, now in the Tenure of said Masters, conveniently built and situated on said River" (*N. J. Arch.*, 1st ser., XX, 179; *Penn. Gazette*, no. 1523, March 2, 1758). However, from Franklin's letters and Eliot's works we find "Mr. Masters of Philadelphia" an authority on manures and composts. Franklin, in a letter to Jared Eliot, December

The Husbandry of Plants

before yo use it. It is made of Tann or horse dung as fresh as may be & ye litter with it. The dung has ye greatt heat but soonest over but if yo want to refresh it cutt away ye Sides & putt fresh litter. Small heat does for seeds greater for Vines transplantd. putt yr Tann or Dung 2 ft. under & as much above ground in a Case. Lett ye Case have a Division for ye Convenience of taking off tread it hard down after — days sift on ye Mold 6 inches deep. Fresh Sandy Loom & old rotten dung or rotten leaves equal Quantitys [Bradley 262 best earth][136] and yr hand will tell when ye heat has abated then prhaps 3 or 4 days then sow yr Seeds. If yr Plants turn Yellow they are chilled or are too hott if ye 1st. Cover them wth Litter & matts, if ye last give them a little water heated in ye Sun. If the last or they run up spindling water &c and raise yr Glasses, tis air gives a Flavour

[Care of Plants]

The time to transplant Cucumbers is when in 2 leaves as big as a dollar, take up wth a hollow trowell & water directly & Shade ym. Crop off ye main runner if they grow too Luxuriant. [2 Vol. *Nat. Displayd* 181 ye Same][137]

Bradley[138] says ye Same & that the Watering should be plentifull

---

10, 1751, wrote: "It is one Mr. Masters, that makes Dung of Leaves . . . (A. H. Smyth, *op. cit.*, III, 59). Two years later, Jared Eliot (*op. cit.*, 1934 ed., p. 89) called Mr. Masters of Pennsylvania "an ingenous and publick-spirited Farmer," who wrote him about improving barnyard manure by gathering leaves and spreading them in the stable and yard to be tramped by the cattle. This would suggest that perhaps William Masters of Philadelphia and William Masters of Burlington were the same person, who changed his place of residence.

[136] *New Improvements* . . . (pp. 262-263): "It remains now only that I give Directions for composing of Earth proper for our Hot-Bed, which should always be light, fresh, and well sifted, considering how tender the Roots will be of those things which are to grow in it. Take therefore of fresh sandy Loam one half, and add to it an equal Part of Horse-Dung well rotted; let these lie in a Heap together well mixd, to be screen'd or sifted when it is wanted: Rotted Leaves, well consumed, are also good, if Dung is wanting."

[137] There were several English editions of Plüche's *Spectacle de la Nature, or Nature Display'd*. The reference here is probably to the London edition of 1760 (II, 185). A portion of the section on the culture of melons (and cucumbers) reads as follows: "As soon as these tender plants attain to a moderate degree of strength, we draw them out of this nursery and transplant them, whenever we see convenient into a third bed, with the earth that adheres to them, at about the distance of two feet from one another. We retrench the luxuriant shoots, that the remaining part of the plant may grow more vigorous. If, however, we take away the two seminal leaves in its first growth, and after that the male-flowers, improperly called the false ones, we rather obstruct than promote its improvement."

[138] This marginal note appears opposite the following paragraph *Of Watering of Plants* in Worlidge's text (p. 168): "Many curious and necessary Plants would suffer, were they not carefully watered at their first removal, or in extream dry Seasons."

tho' it be not often for if done sparingly ye plants will spread yr roots too near ye Surface.

The produce of 34 Acres of Land belonging to Capt. Tibout[139] about 1½ Mile from New York as given me by Mr. William Parr[140] Jany 30th, 1757.

He sowed 5 acres wth Wheat & had 250 bushell @5s.	£52.10–
Do. wth Oats 250 bushells @1s.6	18.15–
The Winter Cabbage on one Acre sold for	80.0—
He sold Turnips to the Value of	130.0—
Other Garden truck to ye Value of	100.0.0
The Straw of His Wheat & Oats	50.0.0
N. Y. Currency	£431:–:–[141]

His Servants in Winter cutt the straw, & sold it to ye families in N York @ — pr bushel to mix wth Bran. He kept a Cart & 2 horses wch went every day to N. York & carried garden truck & brot from thence the Mudd of the Docks, Street Dirt & Stable Dung of these He made a Compost to manure his Land as it was a Clay. He mixed 1/3 Sand wth the Manure of His turnip ground whereby he had turnips mild & sweet.

. . . Early in the Spring, whilst the weather is cold, be cautious of watering the leaves of the young and tender Plant, only wet the Earth about it." Bradley advised leaving the cucumber and melon plants in the hotbed until the "first Leaf is about the Bigness of a Crown-Piece; and then when the second or third Joint appears, cut off the Prime Leader from each Plant, near the Ear-Leaves or Lobes, and they will each of them quickly put out three other Runners, which will produce Fruit plentifully . . ." (*New Improvements* . . . , p. 263). Also, as to setting melon plants, Bradley remarked that "it is commonly practis'd . . . to pinch or prune off the running Branches two Joints above the Fruit; so that all the Sap . . . may be turn'd into the young Fruit for its Nourishment" (*ibid.*, p. 267).

[139] Not with certainty identified. This person may have been Captain Albartus Tiebout, named as a carpenter, who commanded a company of one hundred volunteers from New York in the expedition to Canada in 1746. The proximity of the farm to New York suggests either Bergen County, New Jersey, or King's County (Long Island), New York. Families of this name (variously spelled Tibaw, Thibou, Tibout, Tibow, Tyebout) are known to have settled in both of these counties. (State Historian of the State of New York, *Second Annual Report, 1897, Colonial Series*, I, 617-619; New York Historical Society, *Collections, 1919, Cadwallader Colden Papers*, III, 1743-1747, New York, 1920, pp. 214, 387.)

[140] William Parr, lawyer, was successively recorder and sheriff of Philadelphia, and was master of the rolls for the State of Pennsylvania between 1767 and 1775. (*N. J. Hist. Soc. Proc.*, 3rd ser., V, 1906-07, p. 6; L, 1932, p. 238). A William Parr (perhaps the same) was early proprietor of the inn, The Sign of the King of Prussia, on the road from Point-no-Point to Philadelphia, about 3 miles north of the city. (*Penn. Mag. Hist. and Biog.*, X, 1886, p. 414; XXV, 1901, p. 443; XLVII, 1923, p. 374.)

[141] These computations are somewhat in error. The value of the wheat should be £62-10-0 and the total £441-5-0.

Mr. Parr tells me that a week ago He saw there a Cellar full of Pompions sound.   He thinks long Neckd.

[Fruit Crops]

Observations on Fruit Trees

[peaches]

Tis agreed that these trees come nearly the same from the Stone if planted on the same kind of Land whether the fruit be taken from a tree innoculated or not but the Soil makes great alteration. The same of all stone fruit.

That Peach stones from the same tree will vary if some are planted in one kind of Land & some in another & that the trees wch stand single produce fairer & Larger fruit than ye same kinds in groves or Orchards.

You should budd them late in 7br & Cherries in June   peaches buded late are not so apt to Gum as those earlier. The Bud is from ye Shoots of that year where a Single leaf grows for where there are 2 or 3 leaves it is ye fruit budd of ye next year & will not do so well nor produce a Shoot.

It has been observed by some that of the peaches wch have red Stones some are sweet some not but that peaches wch have a White Stone are universally sweet.

In Sandy or loose Ground lett yr tree branch low or it will be Endangered from high winds & indeed when the tops are too large for the bodies to Support well they will gett a bend hard to prevent or recover tho' you may help them wth Stakes. A tree trunk never grows much taller if any than where ye branches first putt out from.

A peach orchard was a Scheme I have long intended to putt in Execution and having recommended this to Mr. Robert Ogden he recommended it to ——

Peach Trees are apt [to] breed a worm at ye root wch destroys them. The Hulls of Walnutts mixed wth the Earth keeps the Worms from them.   It is thought that to clean them off ye worms, then dig abt ½ foot Deep   & a foot round & putt in temperd Clay.

dry soot stops ye running of ye Gum in peaches [Hitts *treat* 175][142]

---

[142] Hitt (*op. cit.*, pp. 173-174) wrote, "where a part happens to be wounded, the gum issuing out should be scraped off, together with all dead bark, and dry soot applied till the running of the gum is stopped." The page reference noted by Read does not correspond precisely with that in the copy of Hitt's text examined. Perhaps he used a different edition.

[MULBERRIES]

Rub the Fruit in Sand & Sow it in a bed as soon as the Mulberry Drops Shade it a little from the Sun & the next Winter frosts & they will Spring up very thick as I have experienced.[143]

That they may be raised by ye seeds in ye mulberry or from ye Mulberry itself. [Mortimer 2 Vol. 25][144]

[CURRANTS]

Should be kept on a Single Stem.

[APRICOTS]

May be dwarfd but should then be kept very low. [Hitt *on fruit trees* 64][145]

[PLUMS]

Bear best young.

Plant Horse raddish round plumb trees & tis said it prevents the Bugg from destroying them.

[CHERRIES]

[Hitts *treatise of fruit trees* 53] Lett ye place of graffing or inoculating a Cherry be where the head should push out from.

Hitts *Treatise* [277] advises putting 8 oz Com. Salt to a tree in Wett Season.[146]

This Whole Section copied into Mortimer *Art of Husbandry* [2 Vol. 18] & nothing more added.[147]

Observations on fruit and fruit trees from Mr. Bradley [2d. vol. *Husbandry & gardening*]:[148]

[143] Written on the margin of Worlidge's section *Of Mulberries* (p. 115), a portion of which reads: "They are difficult to propagate. . . ."

[144] Mortimer (*op. cit.*, II, 35) wrote of a Mr. Evelyn: "But the sowing of the ripe Mulberries he prefers."

[145] "The kinds of fruit-trees generally kept dwarfs, are the apple, pear, plum, and cherry, and sometimes the apricot: but this last ought to be kept very low, otherwise it will suffer much, both in blossoms and fruit, from violent winds" (*op. cit.*, p. 64).

[146] Hitt (*op. cit.*, pp. 277-278) wrote, "the more salts the juices contain which form the young branches, the more compact and smooth their leaves will be, and thereby less subject to the penetration of the honey-dews, than when they are composed of juices more watery, whereby the leaves become very porous."

[147] Chapter IV, *Of Inoculation*, by Mortimer (*op. cit.*, II, 260-263) is almost a verbatim copy of Worlidge's section *Of the Time and Manner of Inoculation* (p. 127). The similarity of the two treatises did not escape Read's critical eye.

[148] In these extracts from the second volume of Bradley's *General Treatise* . . . , the pages are as given in the notes, except as indicated below.

The Husbandry of Plants

[386] Prune trees but once a year.

[388] In a dry Summer gather your fruit 3 weeks sooner yn common.

[393] If you find your fruit don't prosper Change your Stocks.

[394] Plant Pear trees shallow in stoney or gravelly ground.

[395] In Graffing grown trees Graff every Limb, or else what you leave of the natural Tree will rob yr Graffs.

[397] In Transplanting mind to set the same Side of the Tree to the north, that stood so before.[149]

Graff melting juicy pears on a quince stock & dry Pears on a wild pear Stock.[150]

[402] An almond Tree likes strong land & does not bear transplanting.

A plumb likes dry Sandy Land.

[410] Graff Cherries in the fall and they will shoot very strong next Spring.[151]

[414] Pulling off the Spring Blossoms the tree will bear in the fall.

[119] Mulberries should be planted in the shade. Trim them up but dont cutt the Top—hard to transplant. ☙ [Plant ye root cutt close to ye Ground] [Mortimer—2 Vol. 25 & Hitts *treatise*, 78][152]

The Cornelian Cherry is often pulled green & put into Salt & water to resemble olives.[153]

[124] A plaister for graffing or a wounded tree, or Roots. Take 2 ounces of bees wax 2 oz. Tallow 1 oz. Rosin & 1 oz. Turpentine, melt them in a pipkin. Apply it warm.

[This plaister runs in our hott weather & hurts]

☞ A plaister of Clean hard Tallow spread on a ragg is much better.

[H. Roberts] After the Planting of Fruit Trees some people think it necessary to Protect them from injuries of Cattle rubbing &c. This may be done in ye followg way. Place two

---

[149] *Op. cit.*, II, 397 (not p. 396, as given in the notes).

[150] The passage in the *General Treatise* . . . (II, 397) reads: "seldom or never use a Quince Stock for the dry Flesh Pears." The specific advice to graft "dry pears on a wild pear stock," was not found in Bradley's text.

[151] *Op. cit.*, II, 411.

[152] Mortimer (*op. cit.*, II, 36) wrote: "It [the mulberry] is a Tree something difficult to transplant, except it be planted in a rich Soil and while young, and be kept well water'd; do not cut off the Head in removing of them, but trim up the side Branches, so as to leave but a small Head on them."

Hitt (*op. cit.*, p. 148) wrote: "They [mulberries] produce their fruit at the lower end of a new-made branch . . . these branches . . . grow . . . from buds near the extremities of branches made before. . . . they will not bear till the studs be more than a year old, without having their ends cut off." Note that the page numbers again do not correspond.

[153] *General Treatise* . . . , II, 121.

pretty strong stakes one on one side of the tree ye other on ye other pointing different ways at abt 6 ınches distant from the tree so that ye Point of intersection be abt 3½ ft. from the ground—This effectually prevents ye rubbing   the posts were all croslocked abt a foot under ground.

[GRAFTING AND BUDDING]

Tis best to Cutt Graffs before yo Use them thus in this country tis best to cutt yr Graffs ye last of Febry & graff the last of March. Cutt yr Graffs from bearing Budds. Stone fruit one month sooner, take the thickest yo can get so they are not larger than a mans little finger   in 1758 I found them by Experience vastly better than Smaller, the tipps of the limbs shoot weak ye largr very strong. The Weather should be moderately warm for many graffs are Lost by cold,   hard tallow spread on a ragg ye best plaister [proved by myself].

This I find to be strickly true by Experimt 1758.[154]

Budd Cherries in June   peaches in 7br [H. Roberts   W. Logan] Peaches budded soonr Gum.

[Budds to Choose] The budds to take are of ye Tender Shoots of one years growth, where grows a single leaf, for where there grows 2 or 3 Leaves, these are ye blossoms of next year & Budd low within one ft. of ye ground on tender trees.

A Natural tree lasts much longer than graftd & the orchards produce ye best Cyder.

Tis found by Experience that Grafted fruit when the trees grow old, have a black rott wch makes ye Cider bitter—Lett yr trees bear before yo graft that yo may know yr fruit & graff Winter fruit on Winter fruit stocks   otherwise yo will be deceived.

A Method of keeping plumbs & other fruit & of Turnips & other things Cabages &c from Bugg or Blight communicated to the Secretary of ye Royal Society[155] & by them Published. Make an infusion of Dwarf Elder putting a parcell of the twigs in a tubb of Water & throw over what is to be preserved wth a hand Engine. He had tryed it on Cabbages whipping them wth the Bushes   it preserved what was whipped & the rest blighted. He whipped the Branches

---

[154] This note refers to the following passage from Worlidge's text (p. 123): "Once for all, the stumpy Graff will be found much Superior to the slender one, and makes a much Nobler and larger shoot."

[155] A Letter from Mr. Christopher Gullet to Matthew Maty, M.D., Secretary of the Royal Society, on the Effects of Elder, in preserving Growing Plants from Insects and Flies, written at Tavistock (Devon), August 11, 1771, in *The Philosophical Transactions of the Royal Society of London*, LXII (1772), no. 23, pp. 348-352.

The Husbandry of Plants

of a plum tree as high as He Could reach   it saved them   all above blighted. That twas done when in full Bloom   twould be better when the Bud & when the Blossom first appear   doing it once a week or fortnight recommended.

[Grapes]

[Vines] [2d Vol. Bradley][156] In the year 1720 I took from a Vine the young Shoots that appear in May which were very tender. I pulled off all the lower leaves & having well smeer'd with Soft Soap all the part I Set in the ground I planted it in fine Earth made into mud by Common water & in less than Six weeks they took root. Note, the water of Dunghills will destroy them.

NB from this it appears that ashes is better than Dung for ym.

Ashes & Soot ye proper manure for Vines [Hitts *treatise of fruit trees* 188].[157]

EXTRACTS FROM A PAMPHLET CALLED THE VINEYARD WROTE BY S. J. DEDICATED TO THE DUKE OF CHANDOIS—CHIEFLY DRAWN FROM YE PRACT IN BURGUNDY:[158]

[12-13] The soil should be sandy light chalky or Gravelly no Clay   you plant them one foot apart & manure wth a Compost made of Hog, sheep or Cow Dung & good Earth sometime mixed & prepared [18]   Horse Dung bad for them [19]

[20] There will at that distance grow 6600 Vines to an Acre. The Black muscadine best & you may buy the Plants @ a pistole pr M plant[s] in Octobr. The white not ye Wine Grape. Tis an Advantage that ye Vineyd be on ye bank of a River   the Dews & Vapours moisten & refresh ye Vines   the Vineyd ground should be dry & not moist land.

[21] Plant ye Vine 1 ft. Deep. They may be Allowed to run 7 ft. or 6 tho' Vines of 4 ft. bear as well.[159]

[23] from ye high Vines of 7 or 8 ft. 1/3 more wine than from ye low ones[160]   the Low Vines ye best Wine   the High yeild 360 gallons

---

[156] This experiment was recorded by Bradley in his *General Treatise* . . . (II, 416), as contributed by one "J. M.," presumably one John Merlet (*op. cit.*, II, 28). In Read's notes, the passage is not copied verbatim; nevertheless the first person is used, which obviously applies to "J. M.," and not to Bradley or Read.

[157] In transplanting grape vines, "lay a covering of soot or ashes upon the hill, . . . this being all that is necessary at the time of planting" (*op. cit.*, p. 188).

[158] *The Vineyard* . . . (London) seems to have been quite thoroughly abstracted by Read. The notes were compared with the original text in both the first (1727) and the second (1732) editions, and found to correspond in all but a few details. Variations from the original, except in a few of the bracketed page references, are indicated in the footnotes which follow.

[159] ". . . about seven or eight Foot high" in the text (*op. cit.*, p. 23).

[160] Underestimated: According to the text (*ibid.*), the yield from the high vines is double that of the low vines.

to an Acre. The little Small black Muscadine grape ye best wine grape.
[24] In November they dig the Vines spitt deep & Manure them.
[25] They prune ye Vines in feby. take off all Shoots abt ye roots & crop ye Tops of ye Vines at 2 ft. High nor leave about 2 Shoots to a root   best to prune Early. [28] They Top again in March & tye ye Vine to its stick wth rushes & in May & June yo nipp off the Shoots & tops. Sticks from 4 to 7 ft. high.[161] [30] Never go among the Vines while the Dew or Rain is on them.
[35] Vines thus planted will many of them if pruned & manured bear a little the 2d Year   more ye 3d & are in perfection @7 & 8 last 50 or 60 years. Then he goes on to making Wine (wch is not unlike Cider pulling off ye rotten grapes & dead leaves) They refine it wth Isinglass 1 oz to 50 gall. Their presses cost from £150 to 300£.
[41] They will turn 150 Vignerons into a ps of 20 Acres & pick all in a day or 6 hours.[162]
[48] Wine of ye 1st Cutting 600 Livres ye Cuve   ye 2nd 450   ye 3d 250 livres ye Cuve.[163]
[57] Best to pick ye Grapes in misty Weather
Never lett yr Wine lay on its lees [70.]
Vines from Layers bear ye 3d year [93]
[98] Vines may be raised from ye Seeds wch should be quite dry, sett in the prepared bed or Ground 4 inches deep. They will often shoot ye same yr 6 or 8 inches. In November cutt them down to 2 Inches above ground just above a Joint. They don't bear till 6 or 7 yrs. The prices of the Vines so raised are 18d. pr hundred   very little Earth keeps them in removing. [101] A spott 10 yards square will hold 150 thousand.
[112] Anoint a Budd wth Nitre mixed wth Water to Consistence of Honey   twill sprout in 8 days.
440 Yards square is 40 Acres & would hold 132,000 plants @ 6d. £330.[164]
[120] To graft ye Vine, choose two of a size & having cleft each down the middle so as not to break out the pith of either only so as to make

---

[161] In the original text, "about four or five Foot long" (*op. cit.*, p. 27).

[162] In the original text: ". . . one Hundred Vignerons, Labourers, or Gatherers of Grapes . . . in about Three Hours Time . . . shall gather all that is fit for the cutting" (*op. cit.*, p. 41).

[163] In terms of English money 13 1/3 *livres* = £1.

[164] The author advocated raising grape vines from seed because of the high cost of nursery-grown plants: "For according to the Method us'd in Champaign, . . . one Vine-yard of four Hundred and forty Yards in length and breadth, being forty Acres of Land, would require at least one Hundred Thirty two Thousand Plants; which at the Rates the Gardeners, and Nursery-Men sell their Vines; if at Six Pence per Root only, would cost three Hundred and thirty Pounds" (*op. cit.*, pp. 92-93). This computation is in error. At 6d. per plant the cost would be £3300, an even more prohibitive figure.

The Husbandry of Plants

it smooth mend so that a Budd of each meet wrap them hard round wth paper bind them fast together Cover them over wth Squills or other Glutenous Substance They will grow well if watered & become one Budd he had divers sorts on one vine.

[129] To keep Grapes. Choose fair Bunches take off all that are decayed & when they are perfectly dry have ready a box well pitched inside lay in a layer of Saw dust or meal then Grapes then meal & so till tis full then fasten down in ye top. Chince it well and past[e] paper over it to Exclude the Air sett them in a dry place & they will keep Long.

Cassianus[165] says Beetroots bruised & putt into Wine will make Vinegar of it in 3 hours to restore it putt in Cabbage roots.

[131] Fronto[166] says ye whites of 3 Eggs beat to a froth add a little fine white salt, 1 oz. to each Egg. 6 Eggs to 12 Galls. twill refine it.

[1774] Doctor Vanrow (?)[167] of St. Croix says that in planting slipps there should be one budd just at ye Surface of ye Earth one abt 2 inches under ground & one near ye lower End & that tis well so to Bend the Vine in a place or two as just to Crack ye Coat of ye Vine they putt out Shoots there rather soonest often & one budd above ground.

---

[165] The translation from the original passage (*op. cit.*, pp. 129-130) reads: "CASSIANUS tells us, that if we put Beet Roots, bruised into Wine, in three hours time it will become Vinegar, on the contrary he adds, that when he had a mind to restore the same, he put in some Cabage Roots, and the same was quickly turn'd into good potable Wine." (Cassianus Bassus *Geoponica*, Book 8, Chap. 33, Leipzig, 1895, p. 226.) Cassianus Bassus, called Scholasticus, one of the *geoponici* or writers on agriculture, lived at the end of the Sixth or beginning of the Seventh Century, A.D. He compiled a collection of agricultural literature (*geoponica*) afterwards revised by an unknown hand, and published about 950 A.D. (*Encyclopaedia Britannica*, 14th ed., 1937, III, 191).

[166] The translation from the original passage (*op. cit.*, p. 131) reads: "FRONTO tell[s] us, that if the Wines be foul, if we put three Whites of Eggs, into a large Earthen Dish, and beat them, till they froth, and then put a little of the finest white Salt to them, they will be exceeding Fine, then pour them into a Vessel full of Wine, the Quantity of Salt to the White of each Egg, is an Ounce, . . . stir it very well together in the Cask with a Stick, that it may incorporate and mix with the body of the Wine, and it will be fine in four or five Days at farthest. . . ." Marcus Cornelius Fronto (*ca.* A.D. 100-170), Roman grammarian, rhetorician, and advocate, was born of an Italian family at Cirta in Numidia. He came to Rome in the reign of Hadrian, and amassed a large fortune. Antoninus Pius appointed him tutor to his adopted sons Marcus Aurelius and Lucius Verus. A number of his contemporaries formed themselves into a school called in his honor Frontoniani, whose avowed object was to restore the ancient vigor of the Latin language. (*Encyclopaedia Britannica*, 14th ed., 1937, IX, 874.) C. R. Haines, in *The Correspondence of Marcus Cornelius Fronto* (I, xviii), stated that "The last author to refer to Fronto was John of Salisbury in the twelfth century." Apparently, he overlooked this reference in *The Vineyard*.

[167] Nothing further has been learned about "Doctor Vanrow of St. Croix." The date of the entry is significant. Read went to St. Croix in 1773, and this notation indicates that he took the book with him and made the entry there.

[Oranges]

DIRECTIONS FROM MR. ——— COLL SCHUYLERS GARDNER FOR ORDERING ORANGE-TREES:[168]

[*Situation*] When planted in Pots or Boxes as soon as the weather will permit, Let them be sett in some proper place in the Garden *not to much Exposed to the Wind*. If in Potts they must be sunk into the Earth up to the rimm. The Pots or Boxes must be covered *with grass* to keep the Ground moist.

[*Watering*] Let them be watered with rain or pond water every two days when the weather is warm & dry, When the weather is cloudy & hazy not so frequent if the water be taken from the Well, it must stand in the sun at least six hours before it is made use of. In the Winter when the trees is placed in the Green House they must be watered more sparingly & in proportion as the weather is colder they require less watering the water then made use of must be kept in a Tub in the House for that purpose.

[*Pruning*] As to pruning it is not often required, except when some Branches overtop the others, the shape of these Trees is fully represented by a mushroom. When any of the leaves dry or wither they are to be cut off & the Young Vigorous shoots are in general to be preserved.

[*Seasons*] At the first approaching Frost which is about the last week in September or the first in October. Sett them in the Green House not shutting the windows until the Frost is more Intense.

[168] Peter Schuyler (*ca.* 1710-1762), who had rich mining interests in the Passaic Valley, played a distinguished role in the French and Indian Wars. His estate Petersboro was on the eastern bank of the Passaic River, opposite the present site of Gouverneur Street and Fourth Avenue, Newark. Andrew Burnaby, describing a visit to Petersboro in 1760, wrote (*op. cit.*, p. 57): "In the gardens is a very large collection of citrons, oranges, limes, lemons, balsams of Peru, aloes, pomegranates, and other tropical plants; and in the park I saw several American and English deer, and three or four elks or moose-deer." His fine conservatory was a structure rare for those days. Colonel Schuyler, it appears, reclaimed some of his meadow lands by diking along the river banks. Benjamin Franklin wrote to Jared Eliot in 1750: "I will write Col. Schuyler, and obtain for you, a particular account of his manner of improving his bank'd Grounds" (A. H. Smyth, *op. cit.*, III, 31). In 1768, four years after Schuyler's death, the place was described as "in a flourishing condition containing 906 acres. Of this 265 acres was covered with timber, 393 was under cultivation and the remainder was salt meadow. On it was a two-story brick dwelling, a greenhouse 70 feet long, coach house, stables, barn, cider house, ice and root house, a fine garden, orchard producing 200 barrels of cider, much cedar timber and a shad fishery. . . ." (Notes by the late Joseph W. Greene, Jr.; see also *Newark Centinel of Freedom*, March 17, 1801; and *Newark Gazette*, March 24, 1801.) Colonel Schuyler also owned the White House tavern in Elizabeth. When advertised for rent after his death, it was described as having "8 good Rooms besides a Kitchen, good arched Cellars, a fine large Garden, well laid out, and stored with Peaches, Pears, Rasberries, Currants, and a fine large Bed of Asparagus, with a Barn and Stable, a good Wharf, that Craft of 20 Ton may come to" (*N. J. Arch.*, 1st ser., XXV, 75; *N. Y. Gazette or Weekly Post Boy*, no. 1214, Apr. 10, 1766).

[*Innoculation*] The season for this is the month of July & the first of August. But the Operation cant be well express'd by words nor scarce otherwise than by Occular demonstration.[169]

When I have said all it will be impossible to give any directions so General but that the management at different seasons & on many Accidents which may occur must as last depend very much upon the Judgment of the person who has the care of them.

Bradley in his *New Improvements* says that ½ bushell of sheep or Deers dung in a hhd. of water is ye best manure & may be filled upon ye Dung 5 or 6 times.[170]

To Transplant Trees in Summer

Mr. Bradley in his *General Treatise*[171] gives the method of Transplanting these trees, was by preparing holes for them before they began to be taken up & the Earth taken out of those Holes was made very fine and mix'd with water in large Tubs, to the Consistence of thin Batter, with which Each Hole was filled, for the tree to be planted in, before the Earthy parts had time to settle, or fall to the bottom. A Tree thus planted in pap, has it's roots immediately Enclosed, & Guarded from the air, and as the season then disposes every part of the Tree for Growth and shooting, we find that it loses very little of it's vigour, if we have been careful of it's roots, to wound few of them at the taking it out of the Ground, or have not let them grow dry in the passage from one place to another. This method must not be follow'd in winter when the pap wou'd chill the roots & kill the Tree.

[*Observs. on transplanting*] Hitts *treatise of fruit trees* [fol 78] says cutting ye branches to 6 or 8 inches often kills them. Leave some of the Extremities.[172]

[SHADE AND FOREST TREES—SILVICULTURE]

Shades

The Button wood or Water Beach makes a tolerable Shade but is subject to worms or Catterpillars   it sheds its leaves in the beginning of October   when You plant it you are to trimm it to a

---

[169] *Cf.* passage in letter to Eliot referring to "Ocular Demonstration," p. 71.
[170] This treatment recommended by Bradley is for orange trees (*op. cit.*, p. 323).
[171] This passage is copied almost verbatim from Bradley's text (II, 162-163).
[172] This comment from Hitt (*op. cit.*, p. 78) is written on the margin opposite the following in Worlidge's text (p. 130): "In the transplanting of your young Trees, you may prune as well the Branches as the Roots, taking away the tops of the Branches of Apples and Pears, but not of Plums, Cherries, nor of Wall-nuts."

pole in very Moist ground it will grow wthout root is a quick growth.

The English Elm is a fine tree grows large a quick growth but fouls ground much.

The Locust grows fast, trim it when planted to a Pole. It fouls ye ground much, & when old the Limbs are brittle, the wood very durable.

The Catoliper[173] introduced from Virginia has a fine large leaf wch Stays long on, is of a quick growth. The flower very beautiful.

Shades should be planted in Lines in your Meadows as Shade in hott weather is absolutely necessarily for cattle as food or water.

As soon as planted they should have 2 stakes placed & Croslocked as in page 50[174] for unless you protect them from ye rubbing of Cattle yr Labour of plantg is Lost. Mr. Saltar Advises Maple in Meadows. Mr. Tonkin Apples, Tho' tis Agreed nothing dry up Milk so much as Apples.

[1761] The Leaf of an Oak the best in this Country for lasting it grows in Swamps a sett of posts stand round S. Coles garden Strong have been there 32 yrs. They choose it also for Waggon work.

It bears very large Acorns. It is by some called Chestnt WO [white oak] others call ye Swamp WO & some Turkey Oak.

Time for cutting Timber[175]

R. Ogden say[s] by Experience that W Oak is most lasting felled in May & barked.

Rails of black ash are better than Cedar or Chestnutt.

Shell barked Hickery at sometime of ye Year makes good rails.

Near Elizabeth Town they Saw down their Stave trees & thereby save ye Butt Cutt or Kerf.

[173] Catalpa.
[174] Page 313 of this volume.
[175] Refers to the following passage in Worlidge's text (p. 110): "Fell not in the increase nor full of the Moon, nor in Windy-weather, at least in great Winds, lest it throw the Tree before you are willing. *I have seen a good Tree much injured by falling too soon.*

". . . fell the Tree as near the Earth as you can, *for that is the best timber.*"

[Oak Leaf]

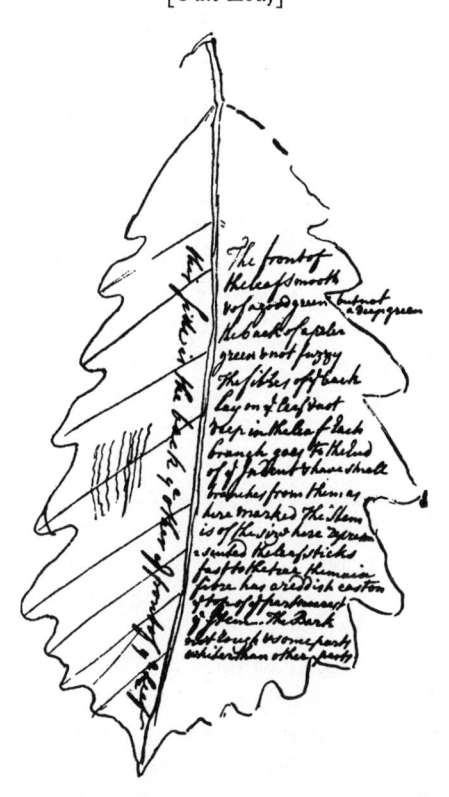

This side [left] is the back ye other ye front of ye leaf.

The front of the leaf smooth & of a good green but not a deep green the back of a paler green & not fuzzy. The fibres of ye back lay on ye leaf & not deep in ye leaf Each branch goes to the End of ye Indent & have small branches from them as here marked. The Stem is of the size here represented   the leaf sticks fast to the tree the main fibre has a reddish cast on ye top of ye part nearest ye Stem. The Bark not rough & some parts whiter than other parts.

Three: The Husbandry of Animals

WERE we to rely solely on the author of *American Husbandry*, we would conclude that animal husbandry in pre-Revolutionary New Jersey was at a low ebb. He wrote that "in no province are all the four-footed animals worse treated."[1] Kalm's account[2] of the care of horses and cattle twenty-five years earlier was scarcely more complimentary. As if corroborating these observations by foreign visitors, John Smith, of Burlington, in 1741 wrote: "Poor country people are almost continually Complaining for want of Hay, Corn, Meat, &ctra and abundance of their horses, Cattle, Hogs and Sheep die for want. One Man hath Lost 5 or 6 Cattle, above 20 sheep & near 40 hogs. Many are forced to give them Wheat to keep them alive. . . ."[3]

Charles Read's notes, however, present the other side of the picture. They disclose a substantial body of progressive farmers who handled their livestock intelligently and profitably, and who were wide-awake to improved methods.

Read's data on livestock are almost encyclopaedic; horse husbandry, swine husbandry, sheep husbandry, dairy husbandry, all receive detailed treatment. Facts noted from his own observation and from the experience of neighbors are matched against traditional practice, and compared with English works on husbandry. Special attention is given to the practical aspects of feeding, of breeding, and of management—more so than to the types and breeds of farm animals. The processing and marketing of animal products, notably the manufacture of butter and of cheese, are discussed at length.

Straightforward advice is given on raising colts. Select mares of good breed with straight limbs, says Read; don't work a

---

[1] *American Husbandry*, I, 139.
[2] P. Kalm, *op. cit.*, II, 45-49.
[3] A. C. Myers, *op. cit.*, p. 341.

mare when suckling a colt; never break a colt until 4 years old; in breaking a colt, "Let him not be used too harshly, it sometimes makes a Horse untractable & Stubborn thro his Life." The time-proven method of telling the age of a horse by the teeth is outlined.

Read's comment on mules is interesting in view of George Washington's role in the introduction of the mule into America. Washington, according to his own account, was the first American to attempt the raising of mules, after receiving a gift of a jack from the King of Spain in 1785.[4] But years before this, the notes reveal, Read was acquainted with these useful farm animals. He describes the method of loading them aboard ship. As to the mules' disposition, he disagrees with Worlidge's text which reads: "The mule . . . is a hardy Beast, . . . and very tractable and capable of much Service." On the margin Read notes: "They are Hardy but very Stubborn & Vicious." We can only speculate as to when and where he became familiar with them. He must have seen them, of course, in the West Indies. But might he not also have had an opportunity to observe them in Pennsylvania or New Jersey? Perhaps some mules were imported into the Middle Colonies. Or, as has been suggested, there may have been some instances of mule breeding in New Jersey in his time, hitherto not revealed.

Of the horse, as of other farm animals, Read has much more to say than Eliot, whose *Essays* deal with livestock only incidentally. Read, however, quotes Eliot on the economy of steeping grain before feeding to horses and swine.

The notes on cattle are remarkable in scope of practical detail. The genesis of America's traditional migratory trade in cattle is suggested (in those days the movement was from south to north): "Cattle from Carolina generally arrive at any time from harvest till ye last of November"; "No Cattle fatt faster than ye Maryland Small Cattle." Common-sense hints and advice are given for selecting an ox or a cow from a herd. Pick a

[4] P. L. Haworth, *George Washington—Farmer* . . . (Indianapolis, 1915), p. 137.

cow with "a good belly," he writes, without a crook in its tail, and one that doesn't "look wild." The "Egg Harbor or Shore Cattle" mentioned were doubtless the cattle which were branded and turned loose to graze on the salt-marsh islands along the coast. Many of the animals were never reclaimed, and became the progenitors of the "wild cattle" which were to be found in Cape May County as late as 1880.[5]

On "fatting cattle," Read cites the experience of a number of acquaintances, among them such prominent citizens as Samuel Smith, of Burlington, and Robert Ogden, of Elizabeth. The weights of the dressed animals of varying ages—of hide, of tallow, and of meat—and the prices received, are recorded. Because of the scarcity of forage during the winter months, the prices for beef steadily rose from fall to the following spring. In April they were practically double those of the previous October. Fat cattle were preferred because "Tallow sells at a higher price than Beef in General." Read also gives the weights of calves killed at the age of 8 to 18 months, by several of his acquaintances, between 1753 and 1756. He estimates the profits that may be expected from the slaughtering of young cattle at the prevailing prices of beef, hide, and tallow. Practices in feeding calves are discussed, touching the merits of boiled milk, and milk mixed with chalk, and barley or wheat flour. The practice of feeding young calves on "hay tea," recommended by the Dublin Society, and copied by Eliot in his *Essays*, is carefully noted. Practical suggestions for foddering cattle with the least possible wastage also are given.

Although in dealing with cattle, Read draws mainly upon his own experience and that of his New Jersey friends, he does not neglect the English authorities. Advice on feeding and management, as given by Hale, Ellis, Bradley, and others, is freely quoted. That the English textbook on agriculture was used with profit by the New Jersey farmer is evident from the story of Daniel Doughty, who saved a cow about to expire

[5] *N. J. Arch.*, 2nd ser., II, 147n.

Charles Read's Sharon as it appears today

The Husbandry of Animals 325

from "wind-puff" by piercing her side with a sharp penknife as directed in Bradley's *Gentleman and Farmer's Guide*.

A scientific approach to agricultural problems that would do credit to the modern agricultural economist is revealed in the notes on milk, butter, and cheese. Not satisfied with the recommendations of the English writers, whom he quotes at length, Read conducts experiments of his own, and has his friends also experiment for him. He collects data on the costs of production and computes the profits from different practices. He gives figures to show the advantage of keeping cattle on "wood range" and feeding rye meal, in preference to pasturing on grass "if you are scant of hay." Records of the income from dairy products received by several acquaintances for the years 1749, 1756, 1757, and 1758 are given. Allowing for butter, cheese, the calf, and milk for pigs, the income per cow was as follows: "My neighbor A. Robins," £6-19-0; Thomas Black (17 cows)—£6-13-7; Wm. Foster (23 cows, "neither feed nor Cows quite so good as the foregoing"), £5-15-2; Rachel Cowgill (10 cows, "she kept an Exact account") £6-15-2; Joseph Biddle (13 cows) £6-14-4. With the exception of Mr. Foster's herd, the results are remarkably uniform.

From the data given, some idea of the annual production of milk per cow can be derived. Read found that one gallon of milk made ½ pound of butter, or approximately one pound of cheese. Mr. Robins' herd of six cows made 1300 pounds of cheese and 200 pounds of butter in a season. This would be equivalent to 1700 gallons (6800 quarts) of milk for the herd, or 1133 quarts per cow. On the same basis, Mr. Black's herd of 17 head produced 1082 quarts per cow. But at best, these figures are rough approximations, and probably do not represent the total annual production. Read recognized the varying quality of milk—"the same quantity of Milk will make more or less cheese or butter according to the Season & pasture." Also the retail prices varied—the usual price at Philadelphia was 3½ or 4*d*. a pound for new cheese, 5*d*. for dry cheese until October first, then 5½ or 6d. In the winter of 1758-59 cheese sold as high as 8d. per pound in New York. Butter was recorded at 7d. a pound. Considerable space is given to the directions for making butter and cheese, and practical suggestions for destroying skippers in cheese.

The notes on swine husbandry are made from the same experimental point of view, and reflect the importance of the trade in pork that prevailed in New Jersey at the time. As to the feeding and fattening of hogs Read notes a wide range of observations. One Nathaniel Thomas fed corn morning and night, and swill containing bran and mill tailings at noon. In 5 months the pigs averaged 140 pounds. Dr. Shaw and "Mr. Hartshorn" fed "grains" (apparently "brewers' grains"), with unsatisfactory results. William Wood of Gloucester County, Mr. Murfin, "an honest neighbor," and Edward Tonkin, another neighbor, pastured pigs successfully on rye. Hugh Hartshorne fed his hogs green corn, "stalks and all, when ye Grains were in milk," and they throve. Read observes that in the West Indies he had seen hogs fed on "potatoe vines," presumably the vines of sweet potatoes. He refers to S. Stokes, Jos. Bullock, Wm. Smith and Edmund Hollingshead as "good Hogg raisers" who were agreed that it was beneficial to pasture hogs two months in the spring. He reports an experiment conducted in 1756 by Joshua Bispham "at my desire." Two pigs were kept on "half allowance" for 38 days. They gained ¾ pound a day. Then their ration was doubled, and they were killed about 3 months later. While on full rations they gained 1¾ pound per day. They lost about one-fifth of their live weight when killed and dressed. At the prevailing price of corn and pork, they "cleared" 12 shillings each. To prevent rooting by hogs Read advises that rings be placed in the pigs' noses as soon as they leave the sow, preferably before.

The dressing, curing, and packing of pork are discussed in great detail, also the procedure in making lard. Recipes are given for treating hams and bacon. The method which "pork merchants" used in cutting up a carcass is given step by step, also instructions for packing pork in barrells for shipment, to conform with legal regulations. As to marketing pork, he observes, "It is better to sell yr Pork for less by 5s. [per barrel] in the Fall than to keep it till Summer," considering the necessary repickling, repacking, and shrinkage. He recommends green hickory for smoking hams. He notes that in 1770 he tried putting pickle scalding hot on his gammons, and it did well. Read presents a detailed accounting of the cost of packing pork, from which is calculated the profit that may be expected, which he says, he had "from a Friend." Allowing for the cost of salt, barrels, labor of packing, freight, and commission, the profit on 1000 pounds, or 3½ barrels, should be £2-1-5.

Apropos of New Jersey's role as the foremost sheep-producing colony, Read gives considerable space to sheep husbandry. He stresses particularly the feeding and the breeding of sheep, and the prevention of disease. He seems more interested in sheep as a source of meat than as a source of wool. A common sheep, he says, will dress 10 to 15 pounds a quarter, or "if large & very fatt," 20 pounds. The butcher expects to clear 5s. on a sheep, and 3s., 6d. on a lamb. He cites the experience of Robert Ogden, of Richard Saltar, and others. His neighbor Mr. Robins lost many sheep from rot when he kept them on a deep meadow, but lost no more after giving them salt. He gives Mr. Skeeles's formula for telling the age of sheep by examining the teeth.

The raising of goats receives a brief paragraph. Of his herd of sixty, many died from vermin, and perhaps also from eating laurel. The larger sorts of goats, he observes, are good for killing brush, but "Scarce any fence will turn them & they are Destructive to fruit trees by peeling the Bark."

The value of white rabbit skins for making hats is mentioned in passing. He gives an instance of 6 skins being worth 25s. for this purpose.

Only sketchy comment is given to poultry. An entry in 1774 states that 19 geese well fed yield 4¼ ounces of down. Guinea fowls, he notes, "should be left to choose their nests and hatch as they please"; adding, "Wm. Logan, Esqr., has proved this." Hens, ducks, and turkeys are briefly mentioned.

The subject of animal husbandry is concluded with a discussion of some of the principles of breeding. The period of gestation of several animals is given, and reasons set forth for occasional change of breed. He cites the example of Joseph Burr who changed his ram every two years, and who restored the productivity of his fowls by introducing some new cockerels. "The Same Rule will hold in all Animals," writes Read, "& the Change of Grain from one Soil or if possible Country to Another is good." Here he is groping for the basic elements of genetics. With scientific evidence at hand, he doesn't hesitate to discard century-old superstitions and traditional farm practices. In the breeding of farm animals, he displays greater faith in the principles of heredity than in the alleged influence of the moon, an attitude which is typical of his approach to all agricultural problems.

[HORSES AND MULES]

The Best Method of Breeding Colts pr Jas. Johnson[6] *& John Hanson*[7]

Get Mares of a good Breed & Strait Limbs   Carry them to Horse at 3 years old abt 1st May. Let her take the Horse till she is well served   this you may know in about 8 days, for if she is served, she will refuse the Horse. If she seem not inclinable to take the Horse Stable her near the horse & she will take him next morning.

Don't Suffer your Mares to Breed every Year. While she breeds she will work at any Light Work, But if you expect good Colts, don't Work her when she suckles. Nor in Less than a Month after taking Horse.

'Tis said, that if you take yr Mare to Horse in the New of the Moon she will bring you a Mare Colt—That Mares of Courage will Cast after the Horse, that is the Colt will be like the Horse. This I Believe has more in it than the former obsern abt the Moon.

[6] There were several persons by this name in New Jersey between 1730 and 1775; and we have no clue as to which one is here quoted by Read.

[7] Possibly Read was acquainted with John Hanson of Maryland, the distinguished Revolutionary leader, but it seems more probable that he here refers to John Hanson of Burlington County, who died in 1759 (*N. J. Arch.*, 1st ser., XXXII, 144; Burlington Administrations, 193).

The Husbandry of Animals

Wean your Colts in the fall if you have good green Corn feed or excellt Hay, if not keep him sucking a Year.

Let him have Tolerable Pasture in Summer, but as Connisseurs differ whether He shou'd run on a good Rowen or fogg or be kept at Hay I wou'd advise to Lett him run out in good days & stable at Night.

Cutt yr Horse Colts in May or June at 2 yrs old.

Give him No Grain till he is Past 2 Years old, it stiffens their Joints. Winter him well and sell him if you can at 3 unbroke. Never Break him till 4 yr old, Heavy weights or hard Usage will destroy him when young.

In Breaking Let him not be used too harshly, it sometimes makes a Horse untractable & Stubborn thro his Life.

[Mortimer 141][8] take care to prevent a mare wth foal from drinkink Snow water twill make her slink foal.

A Mare [goes] 11 months, some 12.

To Know the Age of a Horse from J. Hanson

A Colt sheds his Colt's teeth wch are all Small the middle or inside ones in both Jaws, in the Spring that he comes to 3 Years old. When he is 4 years old he sheds two next & ye 1st two are others in their Place grown out as big as ye Rest all are shed at 5 when he is full mouthed. [T. Douglass[9] says ye middle teeth above & below shed 1st & so 2 every year till ye whole 6 above & below are shed]

A Horse has 6 teeth in the front of his Lower Jaw.

A horse's tushes cutt out of the Lower Jaw at 4 past. A mare has none.

His 2 Middle teeth fill at 6

2 next outside at 7

2 Outside Ones at 8

Then his Tushes are at their length but flat on the Inside & so grow Rounder as he increases in Age and are quite Round at abt 13 or 14 till then they will not be so. The Marks will not be out of the upper Jaw till 15, Unless they go in short Pasture or Gravelly. There are some Horses that have the 2 outside Shell teeth

---

[8] "To prevent a Mare's slinking her Foal—Be sure to take care of her in snowy Weather, and keep her where she may have good Spring-Water to drink, and not drink the melted Snow; which is very prejudicial to her" (Mortimer, *op. cit.*, I, 222).

[9] Probably Thomas Douglass, of Chesterfield Township, Burlington County. In 1734 he married Elizabeth Borden, who was probably the Mrs. Douglass mentioned elsewhere by Read. When he died in 1768 his personal property was inventoried at £334-0-5 (*N. J. Arch.*, 1st ser., XXII, 109; XXXIII, 119; N. J. Wills, Liber XIII, 263).

as they are Called wch will not be out at 14   these may be known by having a hollow larger than Common.

At 5 a horse has 2 tushes also in ye upper Jaw till He is 8 yrs. old   they are Sharp and grows blunter every Year & Shorter. If a Horse is past 15 or 16 whatever Colour he is there will be grey hair under his Mane or Foretop.

Horses

[Elliott][10] Steeping their grain in water will occasion them to fatten Exceedingly, he thinks it saves Every 7th Bushl.

When a horse only Leaps Fence it is said that if you bore a hole thro' the tip of each Ear & tye them close behind his head it will prevent him as he can't jump unless [he] putts his Ears forward or at least does not.

[1746] Mr. Saml Shaw[11] who in the 1st Warr in King George the seconds time was in the Horse Guards says their hay in Flanders was red Clover & that Each Horse was allowed 18 lb. of Hay pr Diem.

[Mules]

Mules never breed.

[Ass] *Compt. Body of Husb.*[12] says the Horse & she Ass & Mortimer [143][13] says they are also between a horse & she Ass. *Qu.*

They are Hardy but very Stubborn & Vicious.[14]

The Method of leading of Mules (as they will not lead by halter) is to putt a person on a white Horse & they will follow anywhere. When they Ship them they tye ye forelegs together & then Sling them with a rope   the Bite about the neck & the ends passed thro' the hind legs & brought forward thro' ye noose around ye Neck & so draw up ye hind legs   then tye them together & fasten ye forelegs & hind legs close & hoist them in as you would a Sheep.

[10] *Op. cit.*, pp. 21, 22.

[11] Dr. Thomas Shaw, of Burlington, named his brother Samuel "of Chester," executor of his will, which was probated in 1750, but whether or not this was the Samuel Shaw who served in the British Horse Guards, does not appear (*N. J. Arch.*, 1st ser., XXX, 429; XX, 368; *Penn. Gazette*, no. 1594, July 12, 1759).

[12] "Beside the Mule . . . bred between the Ass and the Mare, . . . there is another Kind propagated in some Places, raised between the Horse and the she Ass; but this is an inferior Kind" (Hale, *op. cit.*, p. 208).

[13] "Mules are of two sorts, the one between the Horse and the She-Ass, and the other between the He-Ass and the Mare" (Mortimer, *op. cit.*, I, 226).

[14] This comment is written on the margin of Worlidge's text (p. 171) opposite the following passage: "The mule . . . is a hardy Beast, . . . and very tractable and capable of much Service." Apparently Read disagreed with Worlidge as to the mule's disposition.

The Husbandry of Animals

[CATTLE AND DAIRYING]

In the Choice of Cattle from a Flock to feed [pr Mr. Wm. Wood][15]

Observe—If They have come farr & have rested a day or two—drive out an ox or Cow by himself who has a good belly for if they are Girt up & do not fill themselves well They will not answer yr Expectation.

If they start from you & look wild they will be u[n]ruly and apt to run at you & be too restless to feed.

If they have short crooks one or more in their tails or they stand awry, it is a sign that they have been overdone either in their Journey or otherwise the Lameness in their feet will gett well & go off the way to Cure them is to run a hair thro' betwixt ye Claws this breaks the blister and they cure of themselves.

The Cattle from Carolina generally arrive at any time from harvest till ye last of November. They come in tolerable order in ye beginning of ye Season but not so in ye last of ye Season—Young ones best.

If thrifty & Young they winter well.

No Cattle fatt faster than ye Maryland Small Cattle [pr Wright].[16]

Egg Harbour or Shore Cattle keep well till the 1st of Novr. if they have good Sedge to go on something a month longer thriving, then grow worse & worse till 15 or 20 April [D. Knott].[17]

---

[15] Probably William Wood, of Gloucester County, freeholder, surveyor of roads and assessor, who had a plantation on Woodbury Creek (T. Cushing and C. E. Sheppard, *op. cit.*, p. 189). He died in 1762, leaving the home plantation to his son William. The inventory of his personal estate amounted to £1,351-12-9 and included "Carts, Waggon and Other plantation Utencils with Rye Wheat Indian Corn and Oats on the Ground" at £76-16-0, and "Horses Cattle Sheep and Swine" at £281-6-0. (*N. J. Arch.*, 1st ser., XXXIII, 486; N. J. Wills, 820H.) Also, his executors advertised for sale: "A Tract of Land, situated on Mantua Creek, in the County of Gloucester, containing near 400 Acres, about 40 whereof are Tide Meadow, within Bank, and about 20 Acres of Timber Swamp which will make extraordinary Meadow; on which said Land are two Dwelling-houses and two Orchards, about 60 Acres cleared Land, and a good Landing on said Creek, the rough Land mostly well timbered" (*N. J. Arch.*, 1st ser., XXIV, 63; *Penn. Gazette*, no. 1753, July 29, 1762).

[16] Since no initial is given it is difficult to identify this reference. The Wrights were a large family of Burlington County, with some of whom Read was doubtless acquainted. Joseph Wright, of Springfield Township, had a farm at Juliustown, near Charles Read's Sharon (*N. J. Arch.*, 1st ser., XXIV, 82-83; *Penn. Gazette*, no. 1761, Sept. 23, 1762).

[17] Probably David Knott, of Monmouth County. From his father Peter Knott, in 1770 he inherited the home farm at Shark River. At the outbreak of the Revolution he was chosen a member of the committee of observation for Shrewsbury. He died in 1788, leaving a son David. The inventory of his personal estate, valued at £817-17-6,

The wood feed in the pines begins to fail & grows worse & worse from Harvest. Horses find picking ye whole of an open Winter [Jonn. Hampton].[18]

In the back Country abt Paoqualin[19] The Cattle feed is near over & ye Cattle fall off abt 1st Novr & horses live out a whole Winter if open.

Grass Feeding Cattle

[Anno 1749] Mr. Henry Cooper[20] fatted 32 head by Grass feeding & cleared £70 or 45s. head & took them in after Mowing.

[Tend your cattle on Fresh Clover   see the Cure of ye wind puff occassioned by this][21]

When your Cattle first go into a fresh good pasture from short feed, Lett a hand Tend them that they stay but an hour on before

listed among others, the following articles: 2 stacks of rye, £9; 2 cribs of corn in the ear, £15; a yoke of oxen, £6; 3 oxen, £14; 8 cows, £24; 10 young cattle, £20; 27 sheep, £15-16-0; 9 hogs, £11-15-0; 4 calves, 80s.; flax in the sheaf, 8s.; "2 stacks hay and some in cowhouse," 50s.; "salt hay at the shore place," 70s.; 1495 cedar rails, a quantity of oak rails, Calavance beans, and potash kettles (N. J. Wills, 6055M-6062M).

[18] Jonathan Hampton (ca. 1716-1777), of Elizabeth, alderman and judge of Essex County, owned lands in Essex, Morris, and Sussex Counties. He served as agent for Thomas and Richard Penn in large land deals. Prior to 1772 he lived on a farm 2 miles from Elizabeth "on the north side the post road to Philadelphia." He then moved to the town, offering his country seat for rent or sale, described as: "a genteel brick-house, two story high, 53 by 30, compleately finished, . . . there is a good garden, orchard, and all sorts of fruit trees, barn, stable, outhouses, pump at the door, flower garden in the court yard, &c." It was, furthermore, "a Pleasant seat for a gentleman, farmer, or grazier," the house having "nine rooms, all well finished and painted, a cellar under the whole, large kitchen, milk room," and the farm with "fine meadows," and the land "excellent for grass hay and grain." A year later Hampton offered for sale the family carriage: "A Compleat four wheeled carriage made for a large family . . . it is made like a coach, only with curtains round it of good serge, and hangs on iron springs, six grown persons besides one or two on the box may ride in it very comfortably; it is almost new, made strong, and cost 85£. without harness, and is to be sold for 50£." (N. J. Arch., 1st ser., XIX, 280-281n; XXVIII, 109, 425-426, 536; N. Y. Gazette and Weekly Mercury, no. 1068, Apr. 13, 1772, no. 1111, Feb. 8, 1773; Rivington's New-York Gazetteer, no. 9, June 17, 1773.)

[19] Or Pahaquarry, now the northwestern township of Warren County, lying between the Kittatinny Mountains and the Delaware River above Delaware Water Gap.

[20] Probably Henry Cooper, "yeoman," of Northampton Township, Burlington County, who died in 1760. He owned considerable land on the Rancocas in the vicinity of Mt. Holly. The inventory of his personal estate, valued at £590-13-0, indicates that he carried on a considerable farming enterprise. At his death he had 48 head of cattle and 9 horses. His swine were appraised at £30, and his sheep and lambs at £18. Rye, Indian corn, and wheat in the field were listed at £55, and rye in the "Waggon House" at £21. He had 5 plows and 3 harrows, and a Mulatto boy appraised at £35. His "Cheese & Bacon" were priced at £14-10-0. (N. J. Arch., 1st ser., XXXII, 69; N. J. Wills, 6955C-6962C.)

[21] See pp. 341-342.

The Husbandry of Animals

you turn them off again or you may Lose some of your Cattle. I lost one Cow so in Springfeild in 1749.

[1756] I grass feed 20 I gained by them £57 or 57s. Each but then I made ye most of them by selling them among my(?) Labourers & workmen some @ 3d. pr lb some @ 2½ & some @ 2½ all round.

Isaac Cooper[22] never letts them stay long together in their fatting pasture but drives them upon an adjacent peice of upland for wth their feet, lying & Dung they spoil more than they eat.

[Wens in Cattle] If you have any among your Cattle that have wenns on their Chops Grass feed them directly off for they will be in too much pain to feed after ye 1st Year or on Corn at any time well.

[Old Cattle] should be taken in abt January & kept on good Hay for after old cattle have fed on good grass it loosens their teeth so that they never after eat grass but will throw it out of their mouths in great rolls—so yo must not attempt to Stall feed old Oxen.

Foddering Seems to be a great Waste of hay. But tis on all hands agreed that the treading & urine of the Cattle improves Sandy land farr more than double the Quantity of Dung the hay would produce by housing. Nevertheless I have observed that Cattle housed & fed on ye Same hay look vastly better than those abroad. Cribbs made of Poles 16 or 18 foot long keep them from wasting their hay, & in Sloppy wett weather they should never be foddered abroad it wastes the provender greatly I take it to be best on

---

[22] Isaac Cooper (d. 1767) was born in the old Cooper house in Pyne Point Park, built about 1690, the oldest house now standing in the city of Camden. About 1751 he moved to Pomona Hall, built in 1726, now the home of the Camden County Historical Society. (Charles S. Boyer, "Pomona Hall," *Camden History*, I, no. 7, 1935, pp. 2-4; also notes from N. R. Ewan.) On advertising his former plantation for rent he mentioned asparagus and other vegetables and described it as "consisting of about 80 acres of plowing-land; 20 acres of drained meadows; 30 acres of good orchard, and half an acre sparrow-grass; a good dwelling-house; a well of good water, with a pump; a barn; still-house and still, and a good store-house, with a convenient cellar underneath; cyder-house, and utensils for making of cyder; milk house and other out houses; the place is convenient for keeping a dairy, and the soil is good for the raising of vegetables for the market." At the same time Cooper offered for rent a tract at Billingsport on the Delaware River, 14 miles from Philadelphia "containing 100 acres or more, of tide-swamp, and marsh; now in the rough, well bank'd in and drain'd, with a piece of upland adjoining; the land is rich when clear'd, is suitable for the raising of Indian Corn, Hemp, Flax, &c. for mowing or feeding of cattle." (*N. J. Arch.*, 1st ser., XIX, 135-136; *N. Y. Gazette Revived in the Weekly Post Boy*, March 9, 1752.) In 1751 he advertised for two of his horses strayed or stolen, branded with the initials IC on the shoulder (*N. J. Arch.*, 1st ser., XIX, 105-106; *Penn. Gazette*, no. 1194, Oct. 31, 1751).

Sandy Land in dry weather to feed abroad in the Day & at night house ye Cattle.

In foddering abroad Observe to Carry out yr Straw & lay it pretty thick it will prevent ye Hay from getting to ye Ground & unless the Weather be very Stormy the Cattle will lay about ye ground well covered wth Straw when they would take shelter under the fences if they had nothing but the bare ground to lye on & they will also eat part of the Straw [D. Doughty].[23]

The Reason why tis best to fodder on Sandy Land seems to be an Objection to doing it on Clay Viz Saddening of it.

Your Hay is best cutt wth knife or Shovell abt 6 or 8 inches long for the Cattle take great pains to Chew or cutt off ye long Hay wch is foddered in flakes beside when yo foder in flakes yr Cattle gett much that is dryed by the Wind & lost its flavour.

[Feeding Cattle] Cattle will thrive on good Hay foddered 4 times a day but an old ox partly fatened on Grass will not thrive

Directions for Fatting of Calves

[Ellis's *practl farmr* 101][24]

The first Week they allow not the fill to the Calf but afterwards as much as they will suck, because they are at first apt to Scour.

[Michl Loller[25] says boil'd milk fattens fast give it in a horn & if they scour to Each qt. of boiled milk 1 oz of chalk 3 times a day]

They give them powdered Chalk and a Large peice hangs up by them.

Some Cramm them besides their Milk, wth Barley or wheat flour mixed wth milk & crammed.

Keep them in the Dark except when they Suck that they may sleep much.

Lay them on a floor full of Holes, or faggotts the bigness of their penn. This floor should be well Covered wth Straw Changed every morng & Night. Bleed them often, The Butchers when they

---

[23] Daniel Doughty in 1750 purchased Charles Read's Sharon and adjoining lands. Before coming to Burlington County he had served as justice in Hunterdon County. He was twice elected to the New Jersey assembly—1744-45 and 1761-68. His daughter Mary married William Lovett Smith about 1749. Governor Belcher attended the wedding, appearing in a carriage, which according to tradition was the first to be seen in the township. When he died in 1778, his will provided that his negro man "Syrah" be set free. (E. M. Woodward and J. F. Hageman, *op. cit.*, p. 437; *N. J. Arch.*, 1st ser., XXXIV, 149; N. J. Wills, XX, 176.)

[24] *Op. cit.*, pp. 100-102.

[25] Possibly an employee of Robert Ogden or of Jonathan Hampton.

The Husbandry of Animals

are 2 Weeks old cutt off the End of the Tail, at 4 Weeks they bleed in ye Neck, & so Every fortnight Hold yr Calf while He sucks in a Collar & String.

[It is said the Straw should never be Changed but fresh straw laid over it is better Mich. Loller says a plank floor Descending tye ym Short or putt their heads between poles ye rump near ye End of ye floor, bleed @ 14 days old then every 10 days]

Bradleys *Gentleman & Farmers Guide* 137 directs To Keep them cool & dry, never neglect ye Suckling times Lick nothing but Chalk—raise their pens 2 ft. from ye Ground such Calves @9 or 10 weeks old sell for 50s. or £3 at London.[26]

A Calf well fed & white, has the whites of his Eyes clear, if not then red [M. Loller].

The practice of ye fine Veal raisers near Philada. is to take ye Calf off directly & teach it to drink milk for at first it cant suck all ye Cow gives & if milking her is Neglected it hurts her & doing that & Suckling is as much trouble as feeding.

The Method published by ye Dublin Society & in their transactions—for raising of Calves[27]

[Elliotts *4th Essay* 27 new paging 118]

Take ye best english Hay chop & bruise it put it into a Churn or Barrell pour boiling water thereon to be well impregnated with ye Salts & Spirit of ye Hay. Never lett yr Calf suck at all for ye first 10 days mix 3 parts Milk & 1 of Tea next 10 days half one ½ ye other the rest of ye time till yo wean ¼ Milk. The Calves never grow pott bellied.

Observations on fatting of Cattle, communicated April 30, 1756 pr Mr. Skeels[28] *& Mr. Tonkin:*[29]

---

[26] Advice for fattening calves for veal, as practiced in Essex and Hertfordshire. (Richard Bradley, *The Gentleman and Farmer's Guide, for the Increase and Improvement of Cattle,* . . . London, 1729, pp. 136-137; hereafter referred to as *Gentleman and Farmer's Guide* . . .).

[27] Eliot (*op. cit.,* p. 93) wrote, "The Honourable Society for promoting Husbandry and Manufactures in Ireland, published a Way of raising Calves, that appears rational, natural and easy." Apparently he was referring to the Royal Dublin Society. He then briefly cited the method. The source of Eliot's information is not altogether clear. In Henry F. Berry's *A History of the Royal Dublin Society* (London, 1915, p. 51) is the statement: "Another notice appeared on the 9th of December [1749], which advocated a method of feeding calves with a mixture of hay water and a little milk, whereby four or five Calves may be reared in one Season, with the milk of one Cow only." In *The Gentleman's Magazine and Historical Chronicle,* London, XX, 1750, p. 14, is an article on the same subject, communicated to the editor by the Dublin Society. Although in substance the method set forth in this article is the same as that described by Eliot, it differs in certain details of procedure.

[28] Probably William Skeeles, who in 1747 purchased a house and lot on High Street,

1st The Heavier ye Bullock the better he pays Wintering about 6 or 7 yr old best to 9 or 10—if not overworked.
2. Overworked Cattle do not feed well
3. Cattle in good Case in the fall well wintered come in Early
4. Heifers after their 1st Calf fatt soonest & may be putt off in June July or beginning of August next to them Dry Young Cows & Heifers
5. Light Meat if fatt sells best in the Hott months & for ye Highest price, in the fall larger Beasts preferr'd.
6. Buy yr Cattle for wintering as soon as pasture fails for if yo can buy them as cheap at Christmas tis well if they have not been pinched & then are hard to raise.
7. If yo Raise cattle putt off yr Heifers at yr & Advantage [to] abt June or July, for if they are thrifty & the Bull goes wth them tis Much but they will calve ye following Winter when there is much danger & trouble & they will not fetch you so much after yo have wintered ym.
8. If yo raise & take them off the cow in Harvest or fall you should have green corn for them to run on.
9. All Cattle thrive best for being wth Calf.
10. By running wth the Cows in good pasture a yearling will be 1 € pr Qur.
11. Always prefer Cattle bred on feed like your own & never buy wild Cattle, scarce any Creek or fence keeps ym & they are too restless to feed well.
12. Want of water is as bad as want of feed, & Lett them have shade, Drive ym abt as little as possible while feeding.
13. That to Mow burn yr Hay that is to give it the brown Cast not ye white Mould feeds best & makes the Cattle Drink much.

Fatting Cattle

[1755] In March at Elisabethtown I saw two oxen fed by Michael Loller, they appeared very fatt & very full in all the points, I had very convincing proof that they had no Corn, tho' the feeder who was a Sharper, said they had what they could eat, they were killed but did not tallow much, otherwise they were very fatt. The Whole Secrett as I afterwards learnt, was feeding them often in a

---

Burlington. After his death in 1768, an inventory of his estate revealed a long list of personal possessions, among which were 2 negro slaves, 2 horses, 6 head of cattle, 2 barracks of hay, a cider mill and press, and 6 hogsheads of cider. Also listed were 166 gammons valued at £49-16-0, and "in the storehouse loft," 30 barrels of pork "in Bulk," valued at £86-5-0. (N. J. Deeds, Liber HH, 394; *N. J. Arch.*, 1st ser., XXXIII, 388; N. J. Wills, 8443C-8452C.)

[29] Probably Edward Tonkin, of Springfield.

The Husbandry of Animals

day, Carding them well & giving them Water often wch Salt encourages them to drink.

Turnips are good for them to feed on, they are most Expeditiously cut wth an Iron fixed to a Strong sockett & the cutting part of the shape of an S & if putt into a trough & worked wth a Spring pole across a gallows would cutt a vast Many in a Little time. [NB *the Compl. Body of Husbandry* says tis best not to Cutt them[30] it prevents their Chewing well so says ye *transactions of ye Edenborough Society*.[31]]

Turnips Cause Cows to give much milk & feed Cattle but as it gives the beef an ill taste tis best to take them off that feed a fortnight before you kill them. [*compl body of Husby* 334, 452]

[Philadelphia 1755 on Corn only] Reuben Hains[32] bot an Exceeding poor Ox on ye 20th Novr & fed him ye 1st Week on thin Gruel of Indn Corn & some Malt grown on ye floor. He then gave him 1 peck corn pr day wth sometimes a mash of grown Malt. He sold him ye 10th feb. He Eat no Hay nor anything but ye corn & of that 20 bushl. and 4 bushl. Malt. He tallowed 94 lb. & was worth £15 or £6.2 more than his cost wth all ye Grain. [NB He was well curried & Carded & had sometimes Antimony & brimstone]. His Corn by this Experiment cost less than ye hay He wd have Eat. But on good tryall in 1756-7 This will not do tho' well tended & had besides Common hay & Cutt Straw.

[1756 in New Jersey] The prices at Which Cattle usually are Sold by the Grazier to the Butcher & is a good Price for Both

In October & Novr 20 or 22*s*. pr C of 100 lb. or about 2½*d*. pr lb.

In December late abt 25*s*. to 27—3*d*. or 3⅛*d*. pr lb.

In Jany ye Middle abt 27*s*. or 28—3⅛*d*. or 3¼*d*. pr lb.

To Feby ye Middle or End 30*s*. to 35—3¾*d*. or 4¼*d*. pr lb.

End of Feby March & April 38 to 40—4¾*d*. or 5*d*. pr lb.

The Difference is owing at ye same Season 1st to ye Demand 2dly to ye Goodness & tallowing well.

---

[30] Hale, *op. cit.*, p. 452.

[31] "Dont cut the Turnips, but let them [the cattle] nibble them for themselves; for when they are cut, the sharp Corners are apt to choke them" (R. Maxwell, *op. cit.*, p. 178).

[32] Reuben Haines, brewer of Philadelphia, who was in partnership with Read in the Batsto iron works venture. John Woolman was entertained at his home "in High Street near Fourth," and Benjamin Franklin in 1784 referred to him as "our old neighbor." (A. H. Smyth, *op. cit.*, IX, 224; William J. Allinson, *Memorials of Rebecca Jones*, 2nd ed., Philadelphia, 1849, p. 36.)

338   Ploughs and Politicks

They Seldom rise higher unless long fatted & Extraordinary.

May June & July to the Middle of August $\begin{cases} \text{Grass fed 25}s. \text{ to 27} \\ \text{Stall fed over 38 to} \\ \text{40}s. \end{cases}$

In this Calculation the Hide & Tallow Sinks wth ye Butcher. Lent & the Fish Season make the Sale more dull.

An Open Winter when Mutton is plenty hurts ye price of Beef.

As Tallow sells at a higher price than Beef in General so the Butcher can afford to give Most Money for those cattle of Equal weight who will yeild most Tallow & Those Cattle who have been longest fatt tallow best.

[1753] Decembr 20th Danl Smith[33] killed a Calf wch was calved January 1753 & always run wth the Dam of it, it seemed pretty good Meat & fatt its weight 95 lb. pr qr round   the Hide 76 lb. Tallow 16 lb. it was supposed He had rather fallen off the last month for its Dam calved the very day it was killed & ye pasture had been indifferent for a mo.

Beef	@3d.	£4.15 —	Beef	@2d.	£3. 3.4
Hide	@2	0.15.10	Hide	2d.	−.12.8
Tallow	6d.	−. 8 —	Tall.	5d.	−. 6.8
		5.18.10			£4. 2.8
			medium		£5. 0.9

[1754] November 26th Saml Smith Esqr[34] of Burlington killed

[33] Daniel Smith (1696-1769), Burlington merchant, lived in the Smith mansion erected by his father Daniel in 1703. He was a chosen freeholder and collector of Burlington, and served as a member of the provincial assembly from 1745 to 1750. After his death his personal estate was inventoried at £1126-1-8½. Among the items listed were a pair of oxen and yoke (£12), a wood sled, and an "old Waggon & Gears" (£4-10-0), a "Common Strong" wagon (£4), a "Pair of Money Scales & Weights" (£0-7-6), and a "Chambers Dictionary," 2 volumes (£4) (*N. J. Arch.*, 1st ser., XXXIII, 392; N. J. Wills, 8593C). His nephew Daniel, commonly known as "Daniel Smith, Jr.," was a real estate lawyer, and served as surveyor-general of West New Jersey from 1756 to 1774. He, too, for a time occupied the old Smith mansion, and is described as "a man of extensive reading, gentle, affectionate and religious." Charles Read appears to have been acquainted with both Daniels and this reference may apply to either. It was probably the younger who owned farm land in Springfield Township near Read's Sharon. In 1762, for a consideration of £64, Read conveyed to Daniel Smith, "the younger," 800 acres of unappropriated land. (N. J. Deeds, Liber T, 363; R. Morris Smith, *The Burlington Smiths* . . . , Philadelphia, 1877, pp. 99, 117; E. J. Fisher, *op. cit.*, p. 175.)

[34] After engaging for a time in his father's mercantile interests in Philadelphia, Samuel Smith (1720-1776), "benevolent Quaker," settled in Burlington. Besides his town home on High Street, he had a fine country seat, Hickory Grove, on the outskirts of Burlington and a 300-acre plantation near Moorestown. (*N. J. Arch.*, 1st ser., XXXIV, 480-481; N. J. Wills, Liber 17, p. 245.) Samuel Smith was associated with Charles Read in the provincial assembly, where he served from 1746 to 1766, and

Hickory Grove, Burlington, Home of Samuel Smith. From a print by John Collins

First Building of Library Company of Burlington, chartered in 1757
From History of Burlington and Mercer Counties, New Jersey, *by E. M. Woodward and John F. Hageman*

a Bull Calf that had run wth the Cow 11 months. He weighed 97 lb. & 100 lb.—ye fore Quarters & 85 lb. Each ye hinder qurs his hide 65 lb. & Tallow 21 lb. ye four Qurs 367 lb.

[1756] 1 Decr Robert Ogdon Esqr[35] of Elizabethtown killed a Heifer Calf 18 mo old the 4 Quarters weigh'd 610. Gutt fatt 70, Kidneys each 34—price offered £11 proc. [This I had from himself & can depend on it]

[1756] Wm. Forster Esqr[36] of Evesham killed a Bull Calf wch

also in the council which he joined in 1767. He served also as treasurer of the Western Division of the province (1750-1775), and was the author of *The History of the Colony of Nova Caesaria, or New Jersey*. (For a biographical sketch of Samuel Smith, see the second edition of this work, Trenton, 1877, pp. v-vi; also *N. J. Arch.*, IX, 394-395n; and R. M. Smith, *op. cit.*, pp. 115-118.)

[35] Robert Ogdon (1716-1787), of Elizabeth, was surrogate of the prerogative court, clerk in chancery, recorder for Elizabeth, justice of the peace for Essex County, deputy secretary of the provincial council, and speaker of the New Jersey assembly 1763-1765. He seems to have had a special interest in cattle raising. Jacob Spicer, while sitting in the assembly at Elizabethtown in February, 1755, wrote "Robert Ogdon Esqr Informed me that he Killed a 3 year old steer, the 4 quarters of which weighed 1001 lb." (J. Spicer, Diary, Feb. 28, 1755). On March 14, 1763, the following news item appeared in the *New York Gazette*: "We hear that there was killed the Week before last at Elizabeth-Town a Steer, which had been there in keeping, by Robert Ogdon, Esq; about 9 months, weighing, Beef, Hide, and Tallow, about 1700 weight: This steer was bred by Doctor Lewis Johnston, of Perth-Amboy" (*N. J. Arch.*, 1st ser., XXIV, 155). As Elizabeth became unsafe for patriot leaders during occupation by the British, Ogdon removed first to Morristown and then to Sussex County, where he acquired large tracts of land. In 1777 he built a log house at Ogdensburg, which his wife called Sparta, "in the hope that there might be inspiration in the name." At his death in 1787, he left to his wife the use of his "negro man Harry," an annuity of £80, "one good horse and chair," and "two good milch cows." His will also provided 10 acres of land for the use of the Presbyterian Congregation of Hardyston, on condition that the congregation build a parsonage, a school house, and a house for the schoolmaster within 10 years of his decease. (William Ogden Wheeler, *The Ogden Family in America* . . . , Philadelphia, 1907, pp. 78-86; N. J. Wills, 442S.)

[36] Probably William Foster (Forster), Quaker of Evesham Township, justice of the peace, and judge of Burlington County, who was frequently associated with Read in business relations and in public affairs. Newspaper advertisements from time to time indicated his extensive real estate and commercial interests. For example, in 1757: "Wanted immediately a Good Cooper, for packing Pork and Beef, a Person of good Character . . . by . . . the Subscriber, living in Burlington County, who has good Staves and a fine Conveniency to work in." (*N. J. Arch.*, 1st ser., XX, 147; *Penn. Gazette*, no. 1506, Nov. 3, 1757.) Again in 1764 he advertised for tradesmen: "Shoemaker, Taylor, and Wheelwright, who, if they come well recommended for Sobriety, Honesty and Industry, may find good Encouragement by applying to the Subscriber, where convenient Dwelling-houses and Shops may be had at a moderate Rent." This appears to have been the beginning of the village of Fostertown, two miles south of the present town of Lumberton. In 1765 he advertised for "an honest, sober, industrious Man, that writes a tolerable good Hand, that is willing to do the different Sorts of Business about a shop." (*N. J. Arch.*, 1st ser., XXIV, 314, 666; *Penn. Gazette*, no. 1835, Feb. 23, 1764; no. 1923, Oct. 31, 1765.) He died in 1778, dividing his land between his sons William and Josiah. The inventory of his estate, appraised at

run wth its dam abt ye 1st November calved ye 1st of March before. He weighed 85 lb. pr Quarter hide 60 lb. [Received from himself] Jany 26th 1770 had of John Inskip[37] One Beef wt 399 lb. & in Weighg it out in small Drafts it lost 16 lb. or one/25th.

Salted Meat [Beef] 180 lb. beef Salted not very bony lost 10 3/4 lb. or 1/18.

[Dryed Beef] A Bull from 3 to 5 yr old fatt & killed abt Christmas the best for drying.

Ye Method of Curing beef in Summer

Salt it while hott & bulk it   lett it lye 9 hours   take it out & having taken the flesh off the bones hang it in the Sun & Air. If the weather is dry it may hang all night. It is Carved out thinn & keep the peices from closing by Splitts of cypress.

The 1st Wrinkle appears on a Cow's horn at 3 yrs old, But if you stroke down ye Hair you will see it at 2. & one Wrinkle every year after so that she is 1 Year older than the wrinkles show.

A Cow goes 9 months & some say a heifer not so Long an old Cow Longer.

A Heifer springs bag 3 mo before calving   an old cow 3 weeks. [Ellis's *Pract farmer* 101][38]

When yr Cow has taken Bull throw a pail of water on her behind & confine her because she will be riding ye Cows & perhaps miss her bulling thereby.

A Cow brings in, in England abt £4 or £4.10 pr annum in butter & Cheese or Suckling for ye Butcher @ 2s.6 pr week in sumer & 3s. in Winter. [Ellis's *Practl farmer* p. 99] He says a Cow kept up will eat 200 lb. of Hay in a Week wch would sell for 4s. per Ct. 27 mis from London. Grains give much milk but will rott ye Cow wthout plenty of Hay.

£4647-16-4, included cows and horses valued at £574, sheep at £60, salt in barrels £22-10-0, hogshead and molasses £60-0-0, and law books £30. (*N. J. Arch.*, 1st ser., XXXIV, 188-189; N. J. Wills, 10603C-10612C.)

[37] The John Inskeep (Inskip) (d. 1810) here mentioned was probably the third of that name, of a substantial family, who lived near the present site of Marlton, in Burlington County. "He farmed, ran a saw-mill, kept a country store and manufactured plows. He was evidently a man of versatile talents, as he varied the foregoing occupations by occasionally half-soling shoes, mending scythes, and in an emergency could pull a tooth." Among his extensive real estate holdings, were the old Long-a-Coming Tavern near the present site of Berlin, Camden County, and also cedar swamps, pine lands, and a deer park. (Clement Papers, Book K, 61; *West Jersey Press*, Sept. 5, 1888; N. J. Wills, 12456C.)

[38] Ellis mentioned a quaint practice on English dairy farms: "The Country-maids commonly observe to dry a Cow of a Sunday-morning, and then she will always calve in the day-time, as they say" (*op. cit.*, p. 101).

The Husbandry of Animals

Suckling fatigues a Cow more than milking.

Tis better in good weather that yr Cow calve in a field than in a Stable [Ellis 101]

If ye Bag of a Cow yo intend to fatt Gangreens Cutt off her tits & lett her bleed by driving her about when the bleeding is over Anoint wth hogs lard [Ellis]

When yr Cow first calves take a hand full of Salt & rubb on the sides of ye Calf wch She will lick off & it helps her gleaning.

Keep yr Cow well a month before Calving that She may be Strong.

NB A cow's having a small bagg or String hanging to her Bearing is a sign she is wth Calf, tis also said that after taking Bull if She Stand to it her Bearing is smaller yn before.

Rects from Ellis's practicl farmr or Hertfordshire Husbandman

[A Cow who can't Glean] Boil Pennyroyal in a quart of Ale Strain it & putt in a little Saffron or 2 oz flower of Brimstone in a quart of Milk—Twill do in 1 or 2 days. Some give Southernwood, Treacle & 1 oz Brimstone [p. 105]

[The looseness in Sheep] Take Salt, allum or Chalk give it to them in small drink & it will help them presently. When [they] Eat Laurell wch they seldom or never do but in Snow, give Hoggs Lard & Mellasses [p. 110][39]

[Red Water in Cows] Put a hott iron in 2 Quarts of Milk & give it [p. 104]

The Blain is a Distemper that will make a Cow Stare & foam & kills them in a few Hours—They hang out the tongue. [p. 104]

[Remedy] Lett her blood, Then give a handfull of salt in a pint of water. Sometimes they have it behind under ye tail when a blister will appear, thrust a hand into ye fundament Close fingered and draw it out open wch will break ye blister & Blain.

[Swelled Bags] Take rue, Adders tongue & butter before Salted and Anoint it.

The Wind or Puff or Middle Spring in a Cow or Ox

[Bradleys *Gent. & Farmers Guide* 207][40]

This Distemper is fatal unless ye relief is Speedy. Their Bodies appear blown up as a bladder ye Breath Short & shortens perceptibly every quarter of an hour & the Hide swells in proportion.

The Remedy is to *take a Sharp pointed Penknife and strike it into ye Pannell on the Left side of ye Cow about 4 Inches from ye Loyn &*

---

[39] The antidote for laurel poisoning of sheep seems to have been supplied by Read.
[40] For dressing the wound, Bradley recommended a "Plaister of Shoemakers Wax" (*op. cit.*, p. 208).

the same Distance from ye Short ribbs you will then easily perceive the Wind press out from the body of ye Beast & the Creature will immediately be Easy. Shave of[f] ye Hair & Dress ye Wound.

I take this to be what some call ye Hoves on Eating too much green Clover. Mr. Danl Doughty had a Cow who having been kept wth little water a few days She gott to a plenty of Water & fresh clover. She had this disorder and was just Expiring when He Stuck in a penknife as directed above abt 6 inches before ye Hip bone & Cured her. I have heard Gent. of veracity say that they have known ye Intestines of beasts to be cutt without Injury only the Ends would stick to ye sides of ye belly & heal there.

Milch Cattle

If you are Scant of Hay & yo have good wood range Then prhaps it may be best to give yr cows Slop 2 qts. of Rye Morng & Evening. The Calculation is thus—Memo that 2 bush. Rye is 3 bushl Meal.

A Cow @ 4 qts. pr day is 28 qts. pr wk.

12 Cows thus kept wth wood range pr Day 48 qts. of meal wch is 1 bushl of Rye.

1 bushl of Rye pr day for 12 Cows is in 8 wks—56 bushl @ 3s.—£8.8–.

This will save 12 Acres of Grass to Mow @ 1½ load pr Acre & 20s. pr load clear of cost is £18 & will make them give as much milk as good pasture, & if it does so then yo Save by this process, £9.12–.

Tis said That Milch Cattle are likliest to be good if they breed young, but it hurts their Growth. To prevent that as much as You can leave off milking yr Heifers in ye fall & they will Spread & Grow.

Observations & Extracts Respecting a Dairy

[Bradleys *Genl Treatise of Husbandry & Gardening* page 72][41] That from 384 galls. of Milk may be made 250 lb. of Cheese & 100 lb. of whey butter or 200 lb. of butter & 100 lb. Skim milk cheese.

---

[41] One of Bradley's correspondents, "A. B.," quoting from Sir William Petty's calculations of profits from dairying, observed: "By this Account the Profit of a Cow's Milk in a Year may be about Five Pounds" (*General Treatise* . . . , I, 72-73). In estimating the production of a dairy cow, Sir William wrote that the cow will be dry for 90 days (*Political Survey of Ireland* . . . , 1719, pp. 51-52). Read revises this to "95 days," the actual number of remaining days in the year after three 90-day periods have expired, indicating again the precision of his thinking. The computations and the conclusion in this paragraph were apparently made by Read.

The Husbandry of Animals

And Sr. Wm Pettys *Anatomy of Ireland* [page 51, 52] says the Cows there give 3 galls pr day for 90 days, one gallon for 90 days more, One quart for 90 days more & for the other 95 days are dry—So that it takes by the experiment first mentioned six quarts of milk to make 1 lb. of Cheese & 5 oz of whey butter & 7 quarts one pint of Milk to 1 lb. butter & 5 oz of skim milk Cheese. A Cow yeilds in Ireland butter enough by this calculation @ 7d. pr lb. to fetch ye owner £5.16.8 & if turned to Cheese £5.4. giving the whey butter & Skimm milk cheese for Labour. By this Experiment the Milk is not worth a penny pr Quart.

Bradley [84] says that some Cows in England give 31 quarts or 7 gallons 3 quarts in a day.[42]

I tryed an Experiment & it took 2 gallons of Milk to 1 lb of butter.

My Neighbor A Robins[43] began the first Week in April to make cheese wth Six Cows & left off the last week in Septembr. They produced 1300 lb. of Cheese & in ye Spring & 5 weeks in the fall made 200 lb. Butter. He used 70 lb. butter & 180 lb. cheese in the family included in this amount.

Each Cow made in Cheese @5d.		£4.10.4
butter @	7d.	0.19.6
Calf		0.14.0
Pigg say		0.15.2
		£6.19.0

The same year 1749 Mr. Thomas Black[44] kept 17 Cows. They

---

[42] "I have three or four times been Witness, that a large Cow has given in one Day, upwards of Thirty One Quarts; but such Extravagance soon declines, and the Cow is unprofitable during a good part of the Year, unless we let her Calf go along with her" (*General Treatise* . . . , I, 84).

[43] A map of Sharon drafted in Read's hand shows Aaron Robins' house standing near the eastern boundary of Read's property. An Aron Robins, of New Hanover Township, Burlington County, died in 1759, leaving a son Aron. One of these may have been Read's neighbor at the time he was the owner of Sharon. (*N. J. Arch.*, 1st ser., XXXII, 271.)

[44] Thomas Black, "yeoman," lived on the 600-acre estate in Springfield Township, Burlington County, established by his grandfather, William Black, in 1690. The original unit of the residence, Locust Hall, is believed to have been erected in 1693. It has passed from generation to generation of the Black family to the present owner, Harry Black, of Mt. Holly. Thomas Black died in 1751, leaving an estate inventoried at £1129-17-11. Apparently his farming operations were of substantial scope, for his horses and cattle were valued at £173-1-0, his swine and sheep at £91-16-3, his grain and hay at £207-2-6, and his "Husbandry Utentials" at £38-11-0. The appraisers also listed a "Servant Girl" at £14. (*N. J. Arch.*, 1st ser., XXXII, 31; Burlington Wills, 4779C.) Locust Hall though modernized still retains much of its original charm. Near the house stand giant tulip poplars, a catalpa tree more than

made—in Butter 500 lb @ 7d. £14.11.8 Cheese 3600 lb. @ 5d. £75 wch with the Calf @ 14s. & ye benefitt to ye Pigg @ 15s.2 made:

Each cow	
Butter	0.16.2
Cheese	4. 8.3
Calf & Pigg	1. 9.2
Each Cow	6.13.7

Wm. Forster Esqr. of Evesham 1749 kept 23 Cows neither feed nor Cows quite so good as the foregoing He had besides 30 in family as He was building the Cows made Each [£5.15.2].

George Haines[45] 1756 kept 15 Cows—1 farrow—2 Heifers wth 1st Calf she made 170 Cheeses 17 lb. on an Average or 2890 lb. @ 5d. £60.4.2 [Cheese only Each Cow £4.0.0]

[1756] Mrs. Hains's Daughter Jones[46] kept 8 Cows her Cows excellent & her pasture also, her cheese weighed 14 lb. on an Average & suppose she made 170 as her mother did there would be 2410 lb.

2410 lb @5d. is	£50.4.2
Cheese only	£ 6.5.6
Calf & pigg	1.9.2
beside butter	£7.14.8

Jos. Biddle[47] kept 13 Cows his Cheese 2720 lb. @ 5d. £56.13.4.

---

200 years old, also an ancient box tree, and a box bush planted about 1800, now 25 feet in circumference. In a meadow, untouched by a plow for more than 150 years, is a great white oak, aged 250 years or more, whose trunk measures 30 feet in circumference (*Newark Evening News*, Sept. 19, 1925, no. 12,977, p. 2X).

[45] George Haines, "yeoman," in 1753 inherited from his father, Thomas Haines, the home plantation in Northampton Township, Burlington County. At his death in 1760 he provided that the plantation (400 acres) be divided between his sons George and Thomas. His personal estate was appraised at £891-6-7. The inventory included rye, oats, green corn, Indian corn, buckwheat, hay, flax, and lumber. His swine were valued at £24, his sheep at £14-4-0, his 35 head of cattle at £98-16-0, and his 4 horses at £18. Other items were "9 Deer Skins and Stilyerds" £9-1-6, and a "Negro Man," £50. (N. J. Arch., 1st ser., XXXII, 137; N. J. Wills, 6664C-6667C.)

[46] Probably the wife of Hezekiah Jones, of Northampton Township, Burlington County, who was the daughter of George Haines. The inventory of Mr. Jones's estate made after his death in 1806 totaled $2114.52, but did not include any farm items (N. J. Wills, 12318C).

[47] Joseph Biddle, "yeoman," lived in Springfield Township, Burlington County, on lands adjoining Sharon. In 1771 he published the following notice: "The Subscriber, being in Years, purposes to leave off Farming, and has now to dispose of a likely Negro Man 21 Years of Age, has been in the Country seven Years, understands Country Work . . . very handy about a House. Also a Mulattoe Lad . . . 15 Years

So that Each Cow made in cheese only £4.7.2, Calf & pigg £1.9.2 & if they made only 18s. each in butter it would be [Each Cow £6.14.4.]

[1757] Rachel Cowgill[48] tells me that the first year of keeping a Dairy for herself she kept an Exact account & from 10 Cows sold £53 wch is £5.6. pr cow besides what She used.

Butter & Cheese sold Each Cow made	£5.6.0
Calf	14.0
Pigg	15.2
	£6.15.2

Dairy

Mrs. Tonkin[49] tryed an Experiment for me last year and 23 gallons of Milk made a Cheese wch weighed a few days after 23 lb. in drying by ye middle of Novr it weighed 18 lb. & so lost near ¼. Note that the same quantity of Milk will make more or less cheese or butter according to the Season & pasture. Two gallons pr day for 180 days would make 344 lb. of green cheese allow ¼ for drying brings it to 258 lb. @ 5d. £5.7.6.

Mr. Willm Smith[50] of Springfield 1758 kept 27 Cows to ye pail

---

of Age . . . understands Plantation Work well, is a good Hand among Horses, and drives a Team well." (*N. J. Arch.*, 1st ser., XXVII, 663; *Penn. Gazette*, no. 2241, Dec. 5, 1771.) He died in 1776, providing in his will that his negro Zilpha "be well Clothed and Set free." To his son Joseph he left the home plantation, and other lands and cedar swamps. His personal estate was valued at £405-8-11. His farm appears to have been well stocked with animals, and with food and drink as well. His cattle and horses were valued at £117-5-0, and his 21 hogs, 28 sheep, and 22 lambs at £39-2-0. "Rye in ye mill and Indian Corn in ye Cribb" were listed at £13-15-0; cheese in the cheese room at £4, and bacon and hams in the "Smoak-house" at £11. Also there were 2 hogsheads and 2 barrels with "Syder" and "a barrel of Spirits." Flaxseed and hay were listed at £7-16-5. (*N. J. Arch.*, 1st ser., XXXIV, 42; N. J. Wills, 10077C-10083C.)

[48] Rachel Cowgill was the wife of Isaac Cowgill, of Chesterfield Township, Burlington County. She survived her husband who died in 1766. In the inventory of her estate, made after her death in 1782, "Sundry Cattle" are mentioned. (*N. J. Arch.*, 1st ser., XXXV, 98; N. J. Wills, 10810C.)

[49] Probably the wife of Edward Tonkin, of Springfield Township, Burlington County, nee Mary Coles, who married Mr. Tonkin in 1733 (*N. J. Arch.*, 1st ser., XXII, 398).

[50] Probably William Lovett Smith (1726-1794), son-in-law of Daniel Doughty. He was the son of Richard Smith, Burlington merchant and a brother of Samuel Smith, the historian, and of John Smith, who married Hannah Logan. After engaging for a time in his father's business, he lived near Sharon on a 325-acre plantation which he called Bramham, after the family estate in England. He subsequently built a house on the place long known as the Red House which was destroyed by fire in 1850. Read elsewhere refers to William Smith as a good hog raiser. Evidently Bramham was in a center of pork production. In an advertisement of a farm offered for sale in 1771 by William Smith and others as executors of an estate, Springfield

many of them Calved early[51] they made 2 Cheeses of abt 22 lb. each pr day or 44 lb of cheese till the 1st of July then made one of abt 28 lb.

Mr. Smith on ye 30th August sent off 1000 lb. to Philada sold at 5½d. The Usual price is 3½ or 4d. for New Cheese—5d. for dry till 1st Octobr then 5½ or 6d. In the Winter 1758/9 it sold @ 6½ & 7d. at Philada & 8d. at N. York.

TO DESTROY SKIPPERS IN A CHEESE

Mrs. Smith wife of W. Smith tells me that when the skippers are deep in yr Cheese carry it & raise yr Stack of hay & putt it 3 or 4 feet in & it perfectly cures them. It will putt the cheese a little out of shape but mellows it finely—She preferrs a Stack sometime made but has putt them into ye stack when making & has carried them out when so soft there was no removing them but in a Dish.

Tis said that putting fresh Cheese on old Shelves gives ye red Coat & that they should be of pitch pine.

The following Experiment I had from Mrs. Lucy Hartshorne[52] of Shrewsbury anno 1758 as soon as the flies begin to be troublesome She used to open the Doors of her Cheese house make a Smother then brushed the Walls & Shelves very well & shutts up her house very close all day & if the night is pleasant & dry She opened her house to dry her cheeses by the air. She followed this practice till fall wth great Success tho' She had as well her fresh Cheese as that some time made in the same Room.

Some People pick out the Skippers then fill the hole very well & tight & then they coat ye wound wth a hott Iron.

Some persons Colour their Cheese by Expressing the Juice of Orange coloured Carrotts & putting it to ye Milk.

Tis said that as soon as ye Coat of yr Cheese hardens wch will be in abt a fortnight They melt butter & putt some Tarr into it &

Township was described as a place "where great quantities of pork being raised, a trader might be advantageously seated, it being in a healthy pleasant country and in a good neighborhood" (*N. J. Arch.*, 1st ser., XXVII, 653; *Penn. Gazette*, no. 2240, Nov. 28, 1771). Upon his death his personal estate was appraised at £2791-13-10. The horses, cows, hogs, and farming utensils, with household goods, kitchen furniture, &c., were listed at £2580-6-2. (N. J. Wills, 11630C; R. M. Smith, *op. cit.*, pp. 115-118; E. M. Woodward and J. F. Hageman, *op. cit.*, pp. 437-438.)

[51] "Cows to ye pail" presumably means "milking cows."

[52] This was probably Lucy Saltar Hartshorne, daughter of Richard Saltar, who married John Hartshorne of Shrewsbury, Monmouth County, brother of Hugh Hartshorne. Lucy Hartshorne died in 1783; John Hartshorne, who died in 1813, provided in his will that the plantation Black Point, near the present site of Sea Bright, be sold, and the proceeds divided among his heirs. (J. E. Stillwell, *op. cit.*, III, 289, 411; N. J. Wills, 9029M.)

putt some wth yr hand over ye whole Coat lightly this will keep off ye fly & give yr cheese a red Coat.

A Receipt to make Stilton Cheese[53]

[Bradley 1 Vol. 118 says Stilton is in Lincolnshire Lawrence says it is in Huntingtonshire][54]

Take ten Gallons of morning milk & five Gallons of Sweet cream, & beat them together, then put in as much Boyling Spring-water as will make it warmer than milk from the Cow; when this is done, put in Runnet made Strong with large mace, and when it is come (or the milk is set in Curd) break it as small as you would do for Cheese Cakes, and after that Salt it, and put it into the Vatt, and press it for two Hours.

Then Boil the whey, and when you have taken off the Curds, put the Cheese into the whey, and let it stand half an hour; then put it in the Press, and when you take it out, bind it up for the first fortnight in Linnen Rollers, & turn it upon boards for the first month, twice a Day.

[88] The Parmesan Cheese is made at a Town in the Milanese where the country is flatt and floated three or four times a Year with fresh water.

[87] In Somersetshire near Chedder where the famous Chedder Cheese is made the Country is flatt & low often watered & the Grass is free and vigorous.

[Chedder near Oxbridge] The Isle of Ely and other Fenny Countries always Produce good butter, but dry Grounds never produce butter which has any richness in it, nor will it Keep three days without Changing to such a relish, as a nice Taste cannot bear.

[*Skimm Curds*] Skim curds are thus made, Take of Whey any quantity putt it over the fire when it boils throw into it buttermilk or milk changed. If yr Milk be thickned it will not do. To abt 12 gallons of Whey add 2 quarts of Milk or butter Milk if ye Buttermilk not sour add 1 qt more You are not to Stirr yr Milk in. After ye Milk is thrown in the time for doing wch is when the Scum wch rises thick upon boiling blubbers thro' in the middle & the time when ye Curds are come is when it first blubbers up after ye milk is thrown in yo then wth a skimmer full of

---

[53] Copied almost verbatim from Bradley's *General Treatise* . . . (I, 118). In the edition of 1724 (II, 367), Bradley remarked, "Stilton is in Lincolnshire, . . . and as I am inform'd, the Ground lies high." See also *Gentleman and Farmer's Guide* . . . (pp. 141-144) for "Observations relating to the making of the famous Stilton-Cheeses."
[54] Referring to Bradley's article on Stilton cheese, John Laurence, in *A New System of Agriculture* . . . (London, 1726, p. 142), wrote: "I hope the Receipt is more accurate and just than the Description of the Place given us by Mr. Bradley: For Stilton is not in Lincolnshire, but a great Way off in Huntingtonshire; neither doth it stand on high Ground, but on a Flat and Level near the Fens."

holes take ym off. The best Curds are next ye sides. If it boils hard they will break & sink to ye bottom. The Greener ye whey the better for Curd. Green whey is made when yo putt in much rennett & yr Cheese Curd comes tough & hard.

[*Compt body of Husbandry* 568][55] Mentions the making butter from New Milk—The Butter he says is very good for Comon use but will not keep.

[NB *Compt Bod. Hus.*] To Make Salted butter fresh is to beat it up wth new milk but a better is to Cutt it in thin slices & putt it into ye Churn when the butter begins to come this is ye Exact time sooner or latter not so well & being beat & work'd up together a good Quantity may be putt in.

If yr Milk sours or turns to Bonny claber ye Cream can't rise well, it makes much less butter, & ye butter white & will not keep—[Ra. Cowgill].

[Stirr yr Cream] Every time yo putt in fresh cream to yr pott Stirr it up well or else yr butter will be streaked or speckled, And by Standing it will separate so that there will subside to ye bottom a Whey or Water wch obtains a fetid or Stinkg smell & prevents your having good butter—[Barsheba Tonkin].[56] *The Compleat Husbandman* advises pouring daily to a fresh pott.[57]

[Bradley 1st Vol. Page 90][58] Computes, that Butter Cheese and the produce of milk amounts to more than an eighth part of the money gained by Farming in England.

Observations From a Folio Vol. Entitled a Compleat Body of Husbandry printed at London 1756:[59]

[557] Those are the best Cows wch Continue longest in Milk at a moderate quantity—not those that give most after Calving.

[55] As to placing salted butter in the churn with fresh butter, Hale's text reads: "It must be put in when the other Butter begins to come, otherwise it will pervert and disturb the Operation; but in this Manner it goes on very well with the rest, and if not too long kept will, on being washed with the rest, pass with it as very good fresh Butter, not at all debasing the Price" (*op. cit.*, p. 569).

[56] Bathsheba, daughter of Edward Tonkin, of Springfield Township, Burlington County, who married David Clayton, of Gloucester County *(N. J. Arch.*, 1st ser., XXXIII, 435; N. J. Wills, 8453C-8459C, 980H).

[57] This refers to Hale's *A Compleat Body of Husbandry* (p. 564), not *The Compleat Husbandman.* The text reads: ". . . let her [the housewife] every Day change the Vessel in which it is kept, pouring it daily into a fresh one well cleaned and aired."

[58] "One of my Correspondents computes, that Butter, Cheese, and the Product of Milk, amounts to more than an Eighth Part of the Money gain'd by Farming in England" (*General Treatise* . . . , I, 90).

[59] In his instructions for churning, Hale says, "see it is done briskly, with swift, sharp Strokes," "Laziness is the Devil in the Churn, that sets the Spell upon the

The Husbandry of Animals 349

A Cow which does not give 2 Gall. pr day should be Changed.
Milk if properly managed yields 1/6 Cream & each quart of Cream 1 lb. of butter.

[559] Lett ye Milk maid take ye near side begin gently & moisten the Nipples well this will give the Cow pleasure & not pain nor teach ye Cow ill Tricks.
Be sure of all things you strip her Clean nothing makes her go dry so soon as a Neglect herein.

[560] Keep everything clean, wch is to be done by scalding every day & exposing to the Air—He preferrs Earthen vessels next wooden lined wth lead.

[561] Tis an unerring Rule that ye broader & Shallower ye Vessel is the better the Cream rises & ye longer ye Milk Keeps from Souring.

[562] in sumr 10 hours standing is enough in England, standing long getts ye Cream a thick head & gives ye butter a bitter taste the best Vessell for cream is an Earthen Vessell well Leaded wth a Cover.

[563-4] This Short time is meant of fresh markett butter, but butter for potting may be kept & many say is better for Souring—If yo choose to keep it sweet Take what yo gather in 3 days, hang it over ye fire till it once boils, then putt it in a Clean vessell and adding every succeeding Skimming Shift it daily into a Clean fresh Vessel & ye boiled Cream will keep ye rest Sweet for a Week.

[565] In Devonshire they putt the pans of Milk after standing 10 hours into Water kept moderately warm by a Stove till the Cream is perfectly risen Skimm it wth a dish full of holes to lett ye thin milk pass. This heating causes the Cream to keep nor will the butter be bitter. Bitterness in butter he attributes to letting ye milk stand too long unskimmed & not to any kind of food as this happens as well in ye fens where there are no trees, as where there are.
Butter may be kept in Lumps of 40 lb. & a little more salt than usual putt to it & placed in binns of flour, they keep a year round wthout damage.

[566] In Churning The Chief thing is to strain ye Cream well in & move quick choosing a Cool time of the day in the Coldest place yo can gett in summer & one warm in Winter, as Moderate weather promotes Churning And the Temperature of the Air may be Corrected by placing ye Churn in a tubb of Hott or Cold water as occasion may require.

[568-9] If butter to be soon used work it wth water, if not no water.
Hair it before yo Lump it by passing a knife often thro' it.
A pint of Salt to 20 pound of butter for Comon Salting for Markett [NB more must be used for salting] if yo propose firkinning yr Butter Work out ye milky particles well—Lay in a layer of Salt pack down

---

Butter," and "let the Mistress first examine the Manner of working of those who complain" *(op. cit.,* p. 566).

yr well worked butter well wthout water. Then pour a Strong brine over it Till the firkin be full—some perce holes quite thro' ye butter & lett in strong brine wch preserves it then a layr of salt or Strong brine on ye top & head it & keep in a very cool place.

[569] 1½ Firkin pr Week from 10 Cows in summer—1 in Winter. The Worst time to make good butter is between Hay & grass.

[571] Next to Crow Garlick, Cabbage leaves gives milk a bad taste & it will taste of them a week after they have done feeding on them.[60]

To take off a bad taste he advised a box abt 6 inches deep[61] diameter at pleasure perferated wth Small holes. In the middle there should be a tight pipe come up above ye milk into wch yo putt the nozzle of a pair of bellows & by 40 minutes blowing thro' it when just Come from ye Cow twill be perfectly sweet. To Clear it of Garlic the Milk must be kept warm while blowed & twill quite clean up.

[SWINE HUSBANDRY]

Hoggs

Elliott advises the Soaking their corn & says that He thinks it saves every 7th bushl.[62]

Compt. body of Husbandry [74] says they keep Sows before the[y] farow wth raw & boiled turnips & have a hole to lett ye piggs on green clover.[63]

Nathl Thomas[64] who is curious in this way says that He has kept a hogg & charged every grain of Corn and as he gave him corn 1 pt. morng & swill at noon & 1 pt. corn at night he charged for the bran & tailings of the Mill wch he mixed with the Swill one half the price his bread corn cost him wch He thinks (& wth reason) would have made sufficient allowance for the Slop of ye House &

---

[60] John Bartram wrote to Philip Miller, author of *The Gardener's Dictionary*, in 1758: "Crow Garlick is greatly loved by the horses, cows, and sheep, and is very wholesome early pasture for them; yet our people generally hate it, because it makes the milk, butter, cheese, and indeed the flesh of those cattle that feed much upon it, taste so strong, that we can hardly eat of it; but for horses and young cattle, it doth very well. But our millers can't abide it amongst corn. It clogs up their mills so, that it is impossible to make good flour." (W. Darlington, *op. cit.*, p. 384.)

[61] Hale specified "a round Tin Box of six Inches Diameter, and two Inches in Depth" (*op. cit.*, p. 570).

[62] *Op. cit.*, p. 21.

[63] In Hale's text there appears to be no mention of holes in the sties to let the pigs into the clover; this seems to have been Read's idea (*op. cit.*, p. 74).

[64] Probably Nathaniel Thomas, justice of the peace, of Burlington City, who appears to have owned considerable land in the vicinity of Mt. Holly. A will drawn in 1746 mentions "Nathaniel Thomas, sadler," of Burlington County. His death occurred in 1754. Among the beneficiaries of his will was his intimate friend William Skeeles, of Burlington, whom he named executor of his estate. (*N. J. Arch.*, 1st ser., XXX, 74; XXXII, 321; N. J. Wills, Liber VII, 502.)

The Husbandry of Animals 351

that the pork did not cost him more than 7s.4 pr lb. that He once made two hoggs weigh 145 & 135 between 1st 7br & 1st Feby [in 152 days]. They were piggs when putt up.

☞ Elliott says they eat Heartily & grow but little in cold weather.[65]

Mr. Thomas thinks they should have the same allowance from the Sow till they are to be killed & that they rather eat less as the[y] grow bigger. Bradley says that when a hog of 140 lb. is prepared for fattg 3 bushl of pease will do it.[66]

In 1750 I had a Sow & piggs & the Sow killed herself by drinking too much new Whey & some Beef Brine being put into some whey & given to my hoggs caused them to play Antick Tricks & foam at ye mouth & 5 Piggs out of 19 Dyed.

Two piggs of ye same Litter were fed on an Equall quantity of Milk, the one had his milk mix'd wth the Equall quantity of Water & the other not, after a months feeding both were killed & the pig fed on Milk & water was much the fattest & Largest. [Cheyne's *Essay on Health & long Life* p. 204][67]

[Hoggs on Grains] Doctor Shaw an Acquaintance of mine following the Brewing business tryed whether Hoggs would feed on Grains but they would fall away tho' they had plenty—and Mr. Hartshorn[68] tryed the same & the like Effect. But Mr. Askwith[69] says that they will feed if you warm ye Grain.

[Sow Rye for Hoggs] Mr. Murfin[70] an honest Neighbour of mine tells me that he Sowed 1 & ½ bushells of Rye in the Spring & that it kept 20 hoggs in good thriving Condition till harvest—

---

[65] "Hogs will fat slowly in very cold Weather; they will eat much and fatten but little" (*op. cit.*, pp. 20-21).

[66] *General Treatise, . . . ,* I, 113.

[67] "We know from Kircher's and Dr. Woodward's Experiments, what Bulk Vegetables will thrive, by mere Element alone. Two Pigs of the same Litter, were fed upon an equal Quantity of Milk; only, to one of them, the Milk was mixt with the same Quantity of Water. After a Month's feeding, they were both killed, and that which had the Water, was found much larger and fatter than the other." (George Cheyne, *An Essay of Health and Long Life,* 7th ed., London, 1725, p. 204.)

[68] Probably Hugh Hartshorne (p. 352).

[69] Not identified. There was a Captain Samuel Askwith in charge of a company of "Battoe Men" in His Majesty's forces at New York in 1756, but there is no evidence that this was the person to whom Read referred. (*N. J. Arch.,* 1st ser., XX, 12, 14; *N. Y. Mercury,* March 22, 1756; *Penn. Gazette,* no. 1422, March 25, 1756.)

[70] Since no initial or Christian name is given here, the identification is difficult. There were several Murfins contemporary with Read, holding lands on Crosswicks Creek, in Chesterfield and Nottingham Townships, Burlington County. The plantation of William Murfin is referred to in the Northampton Township Minute Book, Dec. 31, 1781 (*op. cit.,* p. 429). This may have been near Read's Breezy Ridge.

1756 Ed. Tonkin tryed this with Success. So did Mr. Wm. Wood of Gloucester county. [Suppo[se] 15 Apr to 15 July this is 3 mo]

[Hoggs fatted on red potatoes] In the West Indies they give them Potatoe vines to Eat as I have Seen.

[Green Indn Corn] Hugh Hartshorne[71] tells me that He fed his Hogs on Green Corn, stalks and all, when ye Grains were in milk that his hoggs throve as well as they could on old Ind Corn, & he thinks it went as farr, & that they Eat and suck the Stalks.

Corn feeding of hoggs tyes up their bellies & sometimes they gett bound & wont thrive to Cure this yo may lett him run out on green Corn—[E. Hollinshead].[72]

[Fatting] Yr Rule when yo putt up ye Hoggs is to give them what they will eat clean & yt generally runs at 2 qts. a day, but if they Leave any, be more sparing. They eat less as they grow fatt, & more in clear cool weather that in moist & warm [Wm. Forster]

[Pr Letter from JB] On ye 11th 7br, 1756 Joshua Bispham Esqr[73] at my desire putt up 2 piggs of abt 70 lb. each thrifty. He

[71] Hugh Hartshorne (1719-1777) was the son of William Hartshorne, of Monmouth County, owner of a large estate at Navesink Highlands and Sandy Hook. He, with his brothers Thomas and Robert, as executors of their father's estate, in 1748 offered the property for sale. The advertisement mentioned "400 bearing Apple Trees of choice Fruit," and stated that there were "yearly winter'd on said Hook upwards of 60 Head of Neat Cattle and 20 Horses without one Lock of Hay, or any sort of Grain given them, or any Manner of Trouble to the Owner." (*N. J. Arch.*, 1st ser., XII, 466-467, 523; *N. Y. Gazette Revived in the Weekly Post Boy*, July 25, 1748, March 13, 1749.) He and Charles Read owned adjoining lots on High Street, Burlington. For several years he was assessor of Burlington and in 1757 and 1758 he served as clerk of the assembly. He was the brother-in-law of Lucy Hartshorne. (J. E. Stillwell, *op. cit.*, III, pp. 284, 288; E. M. Woodward and J. F. Hageman, *op. cit.*, p. 126.)

[72] Edmund Hollingshead lived in Chester Township, Burlington County, on a farm near Moorestown. He was at times overseer of the highway, constable, assessor, and collector of his township (E. M. Woodward and J. F. Hageman, *op. cit.*, 253-255). In 1774 he was assessed for 253 acres of land, 29 cattle and horses and 1 slave (Burl. Co. Tax Lists, 1774). At his death in 1784 his son Edmund received the home plantation and his residuary estate. To a son John he willed all his wearing apparel and "as much good Grass as will make a Load of Hay for three years, to be taken from the Plantation." (*N. J. Arch.*, 1st ser., XXXV, 198; N. J. Wills, 10955C.)

[73] Joshua Bispham came from England in 1737, and lived for a time in Philadelphia. In 1744 he purchased land in Moorestown, and there erected a home, still standing, which is said to have housed the Hessians in 1778. (G. DeCou, *op. cit.*, p. 33; also notes from N. R. Ewan.) He became a member of the New Jersey assembly in 1749, and served in several other public offices. In 1775 he offered his home plantation for sale, "containing 480 acres, situated in a high, healthy country," adding "there are on said plantation a good framed two-story dwelling-house and kitchen, a convenient wash-house, and a well of excellent water (with a pump) under the same; a good barn, stables, corn-crib, smoke-house, granary, and other convenient out-houses, three good apple-orchards, and sundry other fruit-trees; about 200 acres of cleared plow land, which is exceeding good for wheat, or other grain; the whole

gave them but half allowance viz. 1 quart of Corn raw & 1 pint in Swill pr day   at the End of 38 days viz on ye 17th or 18th of October He weighed them & they had encreased 64 lb. wch was above 3/4 lb. each pr day. He then gave them 3 quarts between them pr day   Killed them 11 January 1757 & they had encreased 1¾ lb. pr day. He weighed them alive & they weighed 292 lb. the next day after killing 230 lb. lost 62 lb. or near 1/5 of their weight. reckoning the corn @ 2s. & pork @ 3d. they cleared 12s. each. [NB He kept these too long in Cold weather & gave 2½d. pr lb. alive.]

[My own killed 1757 lost as follows: 170 lost 40 lb   110 lost 27 near ¼ of ye Weight   between the day killed & ye next day lost 2 lb. pr hogg.]

A Shoat wt alive 43 [lb.] clean 32 [lb.] lost ¼ or 25 pr ct.

As to Pasturing Hoggs it seems agreed by S. Stockes,[74] Jos. Bullock,[75] Wm. Smith & Edmund Hollingshead who are good Hogg raisers that pasture in the Spring for 2 months is very beneficial for Hoggs—That Timothy in particulr as resembling grain is as good as Clover. That a hogg does not love his Grass when old. Some give them 1 Ear Corn Morning & 1 in Evening & keep them in excellent growing order   But then remember that He must

within good fence, and properly divided into fields; 50 acres of cleared meadow . . . ; the most part of said meadow is very good, and much more may be made; the remainder . . . is woodland, the greatest part of which is well timbered" (*N. J. Arch.*, 1st ser., XXXI, 185-186; *Penn. Gazette*, no. 2435, Aug. 23, 1775). He died in 1795, leaving his residuary estate to his son Joshua. In his will he made provision for his slaves—a mulatto woman Pothena and a servant man Noah. (N. J. Wills, 11583C.)

[74] Probably Samuel Stokes, of Burlington County. When a young man he purchased 300 acres in Chester Township upon which he built his first home. In 1753 he purchased 130 acres near Moorestown where he erected a fine house which he called Harmony Hall. He was a member of the New Jersey assembly from 1757 to 1760 and held several other public offices. He died in 1781. The inventory of his estate included hogs valued at £11 and salt meat at £7-18-8. Other items in his inventory were a still listed at £10, a "Cyder trough" and a metheglin barrel, also 2 horses, a colt, sheep, several swarms of bees, a pepper mill, a "Waggon," wheat, rye, oats, and buckwheat. (G. DeCou, *op. cit.*, p. 143; E. M. Woodward and J. F. Hageman, *op. cit.*, pp. 253-255; *N. J. Arch.*, 1st ser., XXXV, 376; N. J. Wills, 10985C.)

[75] Joseph Bullock, a member of the New Jersey assembly 1769-1771, lived in New Hanover Township, Burlington County. On his death in 1792 his son Joseph inherited a plantation of 390 acres, and a son Anthony another plantation of 300 acres. His estate also included a plantation of 300 acres at Penns Neck, Salem County, and two plantations totaling 400 acres in Middlesex County. His hog-raising propensities were reflected in the inventory of his estate, which included "Pork hams & Lard" appraised at £535-0-7. The total of personal property was inventoried at £2446-14-11, including cattle, £150; horses, £154; sheep and hogs, £41-0-6; wheat, £195; rye and Indian corn, £97-2-6. (N. J. Wills, 11430C.)

354 Ploughs and Politicks

be in good case when turned in for a poor Hogg will never thrive on pasture but grow poorer [S. Stokes, Jos. Bullock.]

Mr. Bradley prefers the Black Bantam breed for a family as the best pork Bacon or Pig tho' wth the same keeping they will weigh ¼ less than the Large English Breed.[76]

[Antimony] [Ellis's *Practl Farmr* 108][77] Says he has had ye black breed of 20 stone [20 Stone 280 lbs.] that they will fatten in abt 3 Weeks—that the great Leicestershire Hoggs will sometimes weigh fifty stone or 700 lb. He gives Each hogg twice a week as much Antimony as will lye on a shilling it gives them a Stomack & keeps them from ye Gargett & Measles, gives it in Wash.

[Rooting] Ring yr Piggs as soon as they leave ye Sow or rather before & they will be kept from rooting & a hogg seldom roots from the 1st of May till ye 1st of September then they begin [Jas. Whitall].[78]

Tis better your Hoggs dig & almost burry themselves & eat Gravell than eat ye top of ye Ground   S. Stokes tells me he killed

---

[76] Bradley, in his *General Treatise* . . . (I, 68), reported the following statement as made by one W. Waller, in a letter written in 1721: "I rather preferr'd the black Bantham Breed, than the large sort common in England, though I do not believe this black sort eats less than the common large Kind, nor perhaps do they yield so much profitable Flesh for Market by one Fourth Part, as the others." Bradley also in *The Gentleman and Farmer's Guide* . . . (pp. 66, 68), writing of the Bantam breed, stated, "This is, in my Esteem, the most profitable for breeding of Pigs, for Sweetness of Flesh, and for being easily raised and fattened," adding, "These are seldom more than eight Hands high."

[77] Ellis stated that he gave his hogs antimony "in their Wash [swill] or among peas" (*op. cit.*, p. 108).

[78] James Whitall (1717-1808) lived on a plantation overlooking the Delaware at Red Bank, now National Park, Gloucester County, where he built his house in 1748. In 1777 the American forces erected a strong redoubt on the Whitall farm necessitating the destruction of his barn and his orchard. This the British attacked on October 22 in what has become known as the Battle of Red Bank. His wife, Ann Cooper Whitall (1716-1797), refused to leave her home during the battle, but, according to tradition, calmly plied her spinning wheel in an upper chamber while cannon balls and bullets rained on the house. After the battle she cared for the wounded in her home. An interesting account of how the Whitall horses, cattle, sheep, and hogs were taken by both the American and the British soldiers in the weeks following the battle, in spite of efforts to save them, is given in the diary of Job Whitall, son of James (F. H. Stewart, *op. cit.*, pp. 258-261). When in 1780 General Lafayette on a tour of the Revolutonary battlefields with the French engineer, M. deMauduit, who had supervised the construction of the fort, called on Mr. Whitall, the inhospitable host spurned the visitors and busied himself cleaning herbs in a chimney corner (F. J. de Chastellux, *Travels in North America* . . . , London, 1787, I, 258-260). In addition to the estate at Red Bank, James Whitall owned two tracts of cedar swamp, meadow land on Woodbury Creek, certain lots and a brick yard in Woodbury, and a shad fishery on the Delaware. (N. J. Wills, 2725H; F. H. Stewart, *op. cit.*, p. 315; Ralph D. Paine, "The Battle of Red Bank," 2nd ed., 1926, p. 3.)

The Husbandry of Animals

for 2 years running hoggs whose gutts were full of gravel yett they were fatt.

[Brimming] When you putt up Yong Sows not spayed to feed they are apt to be brimming wch Inclination returns at periods of abt 12 Days unless hastned by falling weather & then they will not feed. To prevent this give them weak Soap sudds. But as they will not fatt well while they are briming but that going to Boar makes them fatt soon, Give them the Boar abt 2 months before you Expect to kill them.

Bradleys *Gent. & Farmers Guide* says to give them parched oats in their food will occasion them to incline to ye boar, some say cutt a small peice of a rennet bagg into small pcs & throw them into their food   they will take boar in 3 or 4 days.[79]

Extracts from the same book from page 68 to 119 says:

The Bantam that is ye West India Shortlegged Saddleback breed best Sows.[80] They bring from 15 to 17 piggs   use a white boar of a Larger breed. They roll in Dirt to Cool themselves & will fatten better in a Cool place wth little than in a hott one wth much food.

Your boar tho' he will do at 6 mo. old should be a year old nor more than 5 & should not Serve more than 10 Sows who should take him from 9br to March.

Your sow unless yo lett her farrow too often breeds till 6 year old well. She should be a year old before she goes to boar, tho' they will @4 mo.[81]

Feed yr Sow that has piggs wth plenty of liquid food & Litter well keep her by herself & kill yr Roasters at 3 Weeks old.

dont lett them take ye boar in less than 2 mo. from pigging they often will in two weeks but tis a Chance if they have more than 3 or 4.[82]

A Hog well Littered will yeild 1½ load of Dung pr ann   tis a hott Dung.

A Hog in health Curles his Tail.

Fresh grass in too great a Quantity gives them ye Gargutt wch is known by hanging down their Heads—it kills in a few days. In a Gargutt they will eat in a fever not[83]   the Remedy in both Bleed

---

[79] *Op. cit.*, p. 92. The conclusion of the paragraph was evidently supplied by Read.

[80] "West India Short-legged Saddleback" seems to be Read's definition of the "Bantam" breed. Ellis (*The Practical Farmer* . . . , p. 108) wrote: "Swine are generally of two sorts, the small wild Black, China, or West-India Breed, and the great Leicestershire. Between these are also several sorts and mixtures."

[81] The last half of this sentence was not found in *The Gentleman and Farmer's Guide*. It was probably added by Read.

[82] The last part of this sentence apparently Read's own comment.

[83] This note seems to be in conflict with Bradley's statement (*op. cit.*, p. 113), which reads: "In the Fever, however, they will eat freely, till the very Time they drop; . . ."

under the ears & tail keep him Warm give him in his Food Madder red Oakre or Bole 1 oz.

Measled hogs known by a hoarse voice & blisters on ye Skinn. Take ½ lb. Sulphur, 3 oz Allum 2 oz Soot 1 pint Bayberries lay in his drink.[84]

A sow [goes] 4 months [16 weeks]

In Choosing a Sow do not buy a young one nor above 5 yrs old one of 3 or 4 best young sows do not bear or Suckle well—a Young boar getts small piggs.[85]

[Hams and Bacon]

To 1 Doz Hams 1¼ lb. Saltpetre 4 lb. blown Salt 6 lb. Sugar make ye whole fine & rubb yr Hams well wth this & lay a week make a Weak Lye wth ½ b[bl]. ashes putt Salt in it till it bears an [egg] boil ye pickle & skim it lett it be quite cold pour it gently on them as twill be thick at bottom keep yr hams 4 inches under ye Pickle wth a weight & if in a large Cask 6 inches lett Small hams continue 3 wks larger ones a month take them out to drain when dry rubb them wth bran hang them in a smokehouse where ye smoak passes quick.

This pickle with a small Addit of Salt & Sugr will do Tongues & Beef

[Bacon to make] [Bradley 113][86] He Chooses a hog of 6 or 9 months old & gives the followg rect for Curing bacon. Having Killed your Hog of abt 140 lb. take ½ peck common salt ¼ lb. Saltpetre ½ lb. coarse Sugar & after your Hog has lain a day or

---

Also, the amount of the remedy, which was not mentioned by Bradley, seems to have been supplied by Read.

[84] Bradley's text reads, "Bay-berries, three Quarters of a Pint" (*op. cit.*, p. 118).

[85] Apparently Read's own comment, written on the margin of Worlidge's text (p. 172). On the same page, Read underlined the following couplet from Thomas Tusser's *Five Hundred Pointes of Good Husbandry* (London, 1812 ed., p. 96) which Worlidge had quoted:

"And yet by the Year have I proved e're now,
As good to the Purse is a Sow as a Cow."

[86] These notes do not coincide in all details with the text in Bradley's *General Treatise . . .* , which reads, "The Hog must be full half a Year, or at most nine Months old. . . . about two Bushel and a half of Pease, or three Bushels at most, will bring him into good Order for killing, without making him too fat" Also ". . . we must provide half a Peck of common Salt, a Quarter of a Pound of Salt-Petre, one Pint of Petre-Salt, and a half Pound of coarse Sugar. These Quantities I use for a Hog weighing about fourteen Stone." (I, 113, 114.) Note that Read omits "One pint of Petre-Salt." Also, Bradley recommended a hog weighing 14 stone, while Read mentions "140 lbs." At 8 pounds, the standard weight of a "stone" of meat, Bradley's hog would weigh 112 pounds. However, other values of the stone were in use, varying from 4 to 18 pounds, and Read may have figured the stone as equivalent to 10 pounds. (See also *Gentleman and Farmer's Guide*, pp. 94-96.)

The Husbandry of Animals

two cutt it up & rubb your peices well therewith & sometimes shift the peices from the top to the Bottom lett them lay two weeks. ☞ I have had Excellent pork & bacon this way but I putt them abt 2 Weeks in Pickle made to bear an Egg & add to Every 5 gallon of pickle 2 quarts of a moderate weak Lye of wood ashes. I once had Bacon wch was very agreeable to the smell yett soft as butter & know not whether it was occasioned by to much Sugar or Lye.[87]

[Yr Hoggs if frozen should be putt into a Cellar to thaw for if salted when frozen twill Spoil]

The Method in wch Pork Merchants manage their Pork

A Hogg being brought into the Store house & you have a good Strong Bench 5 ft. Long & 3 Broad made of 2½ inch Oak Plank 16 or 18 High. Lay him on his Side & cut off his head as Close to the Jole as you can. Then lay him on his Back & Split down the Middle of his Back Bone or Chine then take out the Kidney, fatt and fillet & cutt your Gammon, after this is taken off, you cutt each of the Sides in half the long way when this is done drive your Broad Ax thro' the Shoulder Blade again in the Brisket Half, which will also cutt thro' 3 or 4 of the Ribs crossways, then pass your Ax thro' the foreleg before it gets to the Briskett. This is the Part of a Hog subject to Taint, and therefore must be well cutt in & the Holes well filled with Salt. Then Cut your Side in 4 Pieces in a Hog of 100 lb. or 110 lb. & 6 Pieces in a Larger, besides the Tail Pieces & the Feet cut off with a Long Hock to the Gammon; Then Throw him to the Salter who rubs them well and packs it away in a Square Bulk as close as possible taking care not to pack his Shoulder Pieces together, always observing to place the Skin downwards & that there is a small descent on the floor to carry off the Pickle. Your Gammons are packed in the Bulk as near the Middle as possible always Covered well & your whole Bulk shou'd have all the Crevices filled with Salt and a little thrown over every Layer of Pork. This Salt will make your Pickle and if you have more of it than for Pickling you may mix it in Packing tho' it won't do alone—

As to the Heads you always pack them by themselves. Laying about 4 in a Layer, Noses pointed in, this makes the Layer round, about the bigness of a Barrell, there will Come more Blood from

---

[87] Recipe noted on margin partly torn away.

a Head than from a whole Hog. Split them first and take out the brains & a Stroke Across ye Nose.

Your Pork should lay not Less than 8 days in bulk but 2 weeks is better.

If Your Pork in Damp weather keeps dripping out of the Bulk, it is not fit to pack, it should be dry & of a whitish Colour.

In Packing yr Pork lay some of yr thickest Peices at Bottom except Shoulders, Skin downwards. Mix your Shoulders well in a Barrell of which it takes 4. The Law also allows 2 Heads. These you pack near the Middle, one whole, the other Cut in two, and Put the halves next the whole head, the Nose of the whole & Butts of the other together & this fills one side of ye Barrell square—fill up every Layer with Salt and tread it down 3 times in a Barrell, minding also not to have a thick bony peice against the Bung, this would give you trouble in Pickling, fill up the last layer on the Top of ye Barrell with Belly Peices cutt to the Shape of the Cask Skin upwards. Let them stand open a Day if you have time before you head them.

If it is to go away you Pickle it off Directly.

The Law says Tread it down twice but it should be thrice & allows ½ Bushel of Salt which is too much for tis owing to too great a Quantity of Salt that it often wants Repacking. 'Twou'd be best to pickle by a Hole in the Head and drive a Plugg which is easily taken out to repickle if occasion. If You are to take out a Head mind to take out that which was last put in & generally or always has the Coopers mark on it.

It is better to sell yr Pork for less by 5s. in the Fall than to keep it till Summer, besides ye interest of yr Money, for if you keep it over you must repickle & often Repack whereby you may sometimes lose 1 Barrell in 25 or 30, as tis often packed before 'tis well shrunk.

Your Pickle should be tryed by throwing a Piece of Pork or a Gammon into it which should swim buoyant in it.

The Gammons you take out of your Bulks after 8 days or a fortnight may be packed down in a Hogshead & lay Weights upon them & fill in your Pickle. Let them lay there a fortnight, or if they lay longer either in Bulk or in Pickle it will not hurt them. They are then ready for the Smokehouse, the Longer they or your Pork lay in Bulk if it be 2 months the more effectually they shrink. In building your Smoak house Observe, to have Room enough, and that it be tight wth a Chimney at one End close stopt on the Top

The Husbandry of Animals 359

wth Holes in 2 or 3 Places in it, for the Smoak to Come out at. The making of a fire as is Customary in some Places in the middle of your House has 2 Inconveniences attending it, as it Places the fire so as to give too much Heat to the Meat, and often burns up your Smoak house as the Fat and sometimes your Gammons drop into it. [If yo have time tis safest to Smoke in moist weather & lett them only dry in windy. A Gamon must hang a mo. Green wood best to smoke & of that hickery is best]

A Gammon will be well Smoaked in a Month. But this you will know by the Colour of wch a Brown Cinnamon is better than black & dryness. Some Judge By their having their Hocks well shrunk. If you Pack them away in Cask you are to use Bran. But if you keep them over the Summer you are to repack them again the latter Part of the Summer & you are to throw out the old Bran and Repack them with fresh, So that if you have Room enough in your Smoke-house you save this Expence by keeping them there. In which you must observe to make a Small Smoak every Damp Day, otherwise there will be a Moisture upon the Surface of your Gammons, which is very Prejudicial to them.

Pickle your Pork as soon as you Pack it & before you send it off Look to it that it be full of Pickle.

Your Pork will run from 215 lb. to 230 lb. to the Barrell according to its thickness, as the thinner pork will Stow much Closer than thick.

Robert Hopkins[88] putts ye Pickle Scalding hott to his Gammons & says they are vastly ye better for it. NB I tryed it 1770 did well.

To Salt pork for a Family

After yr Hogg is cutt up & those chined wch yo intend to use as Bacon leave the Ribbs in Salt all yr bloody peices first by themselves for the earliest use. Then salt down the rest in large peices wth Lisbon Salt (fine salt it not quite Strong enough) pack down after a good rubbing wth smaller peices to make tight work Lay first a layer of Salt to cover ye bottom of the Hhd & after every 2 Layers of Pork a Layer of Salt ¼ inch thick & tread down every

---

[88] Robert Hopkins (d. 1780) had a 100-acre estate at Point-no-Point on the Delaware, five miles north of Philadelphia (*Penn. Mag. Hist. and Biog.*, XLV, 1921, p. 378). His "mansion house" when offered for sale in 1787 was described as a handsome structure 58 x 38 feet "of a beautiful hewn stone . . . elegantly finished." There were also on the premises "a coach house, a framed hay house 64 x 25 feet, a stone smoak house, and . . . a small tenement with a cellar under it, a wash house adjoining and a pump of good water before the door; an orchard of about 100 trees of the best apples, and a collection of other fruits of the best kind, with a good kitchen garden . . ." (*Penn. Packet*, Apr. 13, 1787, no. 2556, p. 3).

three layers of Pork then lay weights thereon. It will make for itself all ye pickle it requires & keeps from Rust. ¶ Good pork 29¼ wth the Salt Shaken off when boiled wt. 23½ lost 5¾ lb. or 1/5.

Calculations upon the Profits of Making & Packing Pork

The Best kind of Hogs to Barrell & for Gammons are those which Run @ abt 150 lb. or 145 lb. or abt 7 ho[gs] to the 1,000 lb. These will take there Proportion of Heads to ye Barrell wch is 2 allowed by law, if smaller you will have heads to sell, if Larger there are not heads enough to allow 2 to a Barrell, & the Gammons will be over large. These will make 3½ barrels to ye 1000 lb. Gammons taken out & the Accts will stand thus—

1000 lb. Pork @3d.	£12.10.0	
½ Bushel of Salt expended in *Salting*—½ Do. in packing & *Pickling*	0. 2.0	
3½ Barrells @4s. pr bbl.	0.14.0	
Packing 9d. freight & Com. 1s.	0. 1.9	£13.7.9
For this you have		
3½ bbls. Pork @65s.	£11. 7.6	
14 Gams. 10 lb each @6d.	3.10.0	
35 lb. of Fat at 4d.	0.11.8	£15.9.2
Profit on 3½ bbls. Pork or 1,000 lb.		£ 2.1.5

Deduct 5s. pr bbl 17s.6 leaves 23s.11 profitt[89]

If well packed 3000 [lbs.] will not fill above 10 bbls. exclusive of Gammons.

Note—That ye Pork Buyers reckon they may do to Give ¼ for each Crown pr Barrel that they sell for, but If salt should be dear & Gammons Low, tis hard work, if the afd calculatn be just, wch I had from a Friend.

Hoggs to Barrell should not fall under 110 lb. if they are, their Bones are small & flesh thin a great deal passes into a Barrel. If they Come up to 200 lb. or larger you can't make sizable Gammons.

Divide any Quantity of Pork you have Come in by 300 & it will give you the No. of Bbls, besides the fatt Gamons &c pretty nearly or Reckon 3½ bb. to ye Thousand.

The Method of trying Lard

Let your Kettle be clean & covered   cutt your Leaf Fatt in thin peices   fill the Kettle for you must not add more while this is

[89] See p. 358 for explanation of this deduction.

The Husbandry of Animals

doing when it first melts it will be thickish & of a greyish Colour you are to boil it very Slow or it will burn till it be quiat clear you should take out the Crackles when they turn very brown with a Skimmer full of holes It would be best to have as many Earthern Potts as will hold a Boiling when you take it out of your Kettle it should be strained through a seive & a thin Cloth and any dirt Crackles or Sidings Poured into another Pott as soon as it cools a Little and thickens pour into Clean Barrells for no Cask will hold it when verry hott as soon as your Cask is full head it [torn away] [In Melting down ye Lard they do not putt Salt]

[Sheep]

Mr. Ogden of Elizatown & Mr. Saltar say that if a sheep be thrifty & healthy will become fatt in Six weeks time if kept on green Corn & ½ or better 1 pint pr day of Indian Corn 1 pint pr day is abt 15*d*. pr Sheep @ 2*s*. pr bushl for 42 days. The Corn keeps from Scouring so will Turnips & Salt.

If they Creep under a hay stack the Dust & Seed etc. can never be cleaned well out of the wool.

DuHamell [120] Says that Carrots are better for Sheep than turnips give the flesh a firmness & good flavour wch Turnips do not & also feeds Cattle of any kind. [Seems to be taken from Miller][90]

[feeding at ye Stacks] If you Lett your Sheep go to the stacks of Hay the fine hay & Dust getts so into ye Wool yo can scarce ever clean [it].

[Ellis's *pr. Farmer* 109][91] Sheep fatt ye fastest when the[y] first begin to rott. The Flesh of a rotten sheep is flabby & pale red—his skin thin & no blood veins in it [Pr Mr. Skeels] The rott in Sheep is caused by moist wett feeding & so is ye red watr.

[Rot to know] To know a Sheep touched wth the rott feel ye Cod of ye weather & if there is a dry scurf or wax thereon the rott has begun but if moist & wett He is Sound, feel ye ewe between ye leggs the Gums & mouth of a rotten sheep will be white. That the Innermost part of ye white of ye Eye is Streak'd wth red is a sign of Soundness.

---

[90] *The Gardener's Dictionary,* apparently the source of Duhamel's statement. See footnote 112 of the preceding chapter. For the page reference of Duhamel's text, Read erroneously noted *120* instead of *320*.

[91] The description of the flesh and skin of the sheep affected by rot was not found in Ellis' text but apparently was the observation of Mr. Skeels.

[Bradley 7] Rubb a little of his Wool between yr Fingers & if it comes out on Gently pulling He is rotten.[92]

To prevent the rott give them dry Meat & Salt. Mr. Robins my Neighbour kept Sheep on a Deep Meadow at first He lost many but no more were lost when He gave them Salt.

Soft grass gives ye Rott to Sheep.

Moist Meadows are better to fatt off sheep than Breed them. When you want to Buy sheep take ym not off such pastures as yr own. Lett yr weathers to Stall feed be 3 or 4 years old.

Store Sheep should not be kept fatt [Mr. Skeels & Mr. Bowlsby]

To Know the Age of the Sheep remark that the Sheep Ox & all that Chew the Cudd have teeth before only on ye lower Jaw.

A Lamb is full mouthed, small teeth at a yr. old He sheds 2 teeth & getts 2 broad Teeth [Pr Mr. Skeels]

at 2 Years old He sheds 2 more & has 4 broad Teeth

at 3 yr. old He sheds 2 more & has 6 broad Teeth

@ 4 years old he sheds 2 more & has 8 broad Teeth. He never getts more but as He grows older these wear short & smooth and at length wear away.

at 3 & 4 yrs. He is at his best for feeding over Winter for at 1 & 2 he will grass feed but his teeth are too tender for hard corn.

Your Ram had best be 2 yr. old or upwards.

Cut off ye Tails of yr Yews, ye Ram getts better at them & it prevents ye fly blowing them wch they are apt to do if they are Dirty & ye Weather hott.

A Good ram will serve 20 Yews in a night & to know how many are Served paint yr Ram between ye foreleggs it will shew.

A yeo or Ewe goes 20 Weeks says Mortimer [153][93] & Bradley's *Gent. & Farmr* [122][94] & it is so.

as yr Yew goes 20 weeks they shd go to Ram abt 1st Novr & they will come abt ye Middle of March ye best time to raise 1 mo. earlier if yo desire to sell for Early Lamb a ram will serve 20 in 1 Night—paint ye breast of yr Ram will show wch he has served.

A Ram if young & fatt is eatable in June & July being farr from

---

[92] "If you open the Wool on the Side and rub it between your Fingers, and then pull a little of it gently, it will easily leave the Skin if the Sheep be rotten; but if it be sound, it will hold close to the side" (*Gentleman and Farmer's Guide* . . . , p. 7).

[93] "And therefore as an Ewe goes twenty Weeks with Lamb, you may easily calculate the time for her, to take Ram in" (*op. cit.*, I, 242).

[94] ". . . the Ewes go with Young twenty weeks, or as one may say, yeans or brings forth her Lamb in the twentieth Week. . . ." (*Gentleman and Farmer's Guide* . . . , p. 22).

rutting time. The Butchers say comon sheep 10 lb. or 12 lb. pr Quartr so to 15 lb. if large & very fatt 20 lb. They will have from 2 lb. to 10 lb. gutt fatt in them & the whole retail'd from 20 to 40 shill.

The Butchers to Judge of the Fatness of a Sheep feel of the Briskett & the Navell & if fleshy there tis a good Sign they also feel of the tail near the Rump if fatt and fleshy and across his loins. In a poor sheep you can feel the short ribs there, If fatt yo can't, they pass their fingers along his backbone to see if tis covered wth flesh—If he feels well on the ribb he is fatt tho' a yew in years may be pretty good & not feel well on the Ribb.

To those acquainted wth Sheep its apparent which are in good case on the first View, but for fatness you must Judge by feeling— The Butcher expects to clear 5s. by a Sheep 3s.6 a Lamb.

To pull the Wool wth Ease off a Skin, putt the flesh side for a small time over a kettle of boiling water & shift it so that all parts may be Steemed & it comes off easily. Some say Strew Ashes or Slack lime & fold the Skin together will do it.

[GOATS]

Of these I had 60 wch I kept at my Saw Mill to no great advantage   Many gott Destroyed by Vermin or perhaps eating the small wild Laurell, the Small sort are the best breeders & the Kids & Weathers good eating   the Larger sort are always poor breed slowly & are used to Kill brush. Scarce any fence will turn them & the[y] are Destructive to fruit trees by peeling the Bark. They breed twice a Year   the Small sort have 2 or sometimes 3 kids the large ones seldom more than one their skins make very tough Leather.

[POULTRY][95]

a Hen setts 3 weeks.
A Goose setts 4 weeks

[95] These notes on poultry were written on the margin of pages 175 and 176 of Worlidge's text. Read had underlined the following passages: ". . . the Dung of Poultry being of great use on the land, *much exceeding the Dung of any Cattle whatsoever.*"

"*If they* [poultry] *are fed with Buck, or French-wheat, or with Hemp-seed,* they will lay more Eggs than with any other sort of Grain."

"You may set them [geese] on any number of Eggs under fifteen, and above seven, giving to each Goose her own eggs; *for it's said they will not hatch a Strangers.*"

"It is observed of Geese, That in case the Waters are frozen up, . . . about their Treading time, that then the most part of their Eggs will prove Addle. The reason

1774—19 Geese well fed yeild 4¼ oz Down.
The large kind of Geese wch they have at Elizabeth town &c. are not so mischevious as ye Smaller sort—Nor are their Feathers or down so good by farr, little water serves them when fatt. D. V Vacty tells me he has had of 16½ lb.—they lay abt 60 Eggs but don't hatch well themselves.
A Duck setts 4 weeks.
In Switzerland they dip the Turkey as soon as Hatched in cold water & give them a pepper Corn   if they afterwards droop look in the tail   yo will see a bloody feather   pull it out   by this treatmt. they become as hardy as other poultry. Guinea Fowls should be left to choose their nests and hatch as they please   they will rear their Young & bring them home to you. [Wm. Logan, Esqr has proved this]

[RABBITS]

My Neighbour James Smith[96] gave a hatter 6 white Rabbitt skins out of wch he made a Hatt for his Daughter & allowed 5s. for ye furr yt was left   by this it appears that ye Skins are Valuable ye hatt—worth 40s.   the Hatter had 20s. for making out of wch the 5s. was taken. Deduct 15s. from 40s. rems 25s. for 6 Skins.

[LIVESTOCK BREEDING]
Breed to Change

It is undoubtedly of great advantage to Change ye Breed. Mr. Joseph Burr[97] who is noted for good Sheep says that the Sheep on

---

is said to be because the Goose proves more fruitful when she is trod by the Gander in the Water, *than if upon the Land.*"

"*Gravel not a little availeth,* it being usual that when Poultry are penned up, and have lost their Appetite, being set where Gravel is, they will greedily eat it."

[96] Record has been found of James Smith of Burlington City, also of James Smith of Springfield Township, either of whom might have been "neighbor" to Read. James Smith of Burlington, by will probated March 6, 1789, left his son Richard his house on the easterly side of High Street (N. J. Wills, 11312C). James Smith of Juliustown, Springfield Township, in 1776 offered his plantation of 63 acres for sale, listing "a handsome two-story dwelling-house, neatly painted, and pleasantly situated in a healthy part of the country; with good stables, a large hay-loft, waggon-house, smoak-house, &c. a good well of water, with a pump at the kitchen door; a large garden with a variety of fruit trees, neatly paled; a large orchard containing 180 apple trees; about forty five acres of cleared land, twelve acres of meadow, and six of woodland. There is also on said tract three small dwelling-houses that will rent for £12 yearly. The above place would suit a person that chooses to retire from the city" (*N. J. Arch.*, 2nd ser., I, 77; *Penn. Packet*, Apr. 1, 1776).

[97] This might refer to any one of three or four persons. Joseph Burr, Sr., (1694-1767) had a large plantation between Burlington and Mt. Holly, which he bequeathed

The Husbandry of Animals

his place have not been changed many years to his remembrance
50. That He changes his ram every two years.
Mr. Burr told me a very remarkable Story about his Fowls. That they had been kept many years on the place till at length they yeilded but few Eggs, scarce any Chickens—That some of his boys had been abroad & brot home a Couple of Cocks well grown these tho' nothing extraordinary for Size so improved his fowls that, they had Eggs & Chickens as plentyfully as ever. The same hapned to Rachel Cowgill's. Tis a Maxim among Cockers to mix the Breed to make them strike quick. The Same Rule will hold in all Animals & the Change of Grain from one Soil or if possible Country to Another is good.

[*Practical Farrier* 81] A greyhound bitch goes 6 weeks & her whelps are 12 days blind, All others go 12 weeks & whelps 7 days blind.[98]

---

to his sons, Joseph, Jr., (1732-1796) and Henry (N. J. Wills, 8135C). There was also a Joseph Burr (1726-1781), cousin of Joseph, Jr., who lived near Vincentown. He had a sawmill which he called Oak Mill, also mills and lands on Maurice River and the Mullica River. It was probably this Joseph who in 1769 purchased from Reuben Haines, one-eighth interest in the Batsto Iron Works for £700 (N. J. Deeds, Liber AC, 203). On his death his personal estate was appraised at £1900-9-10. He bequeathed the plantation where he lived to his son Joseph. (*N. J. Arch.*, 1st ser., XXXV, 65; N. J. Wills, 10761C; Charles B. Todd, *A General History of the Burr Family*, . . . 2nd ed., New York, 1891, pp. 449-454.)

[98] A note on page 81 of *The Practical Farrier* . . . published by the Society of Country Gentlemen . . . (4th ed., London, 1737), reads: "N.B. A Greyhound Bitch goes six Weeks with Whelp, and her Whelps are twelve Days blind: But all other Bitches go twelve Weeks with Whelp, and their Whelps are only seven Days blind." This reference was checked with a photostatic copy of pages from the volume in the library of the Rothamsted Experimental Station, Harpenden, England. No copy was found in the American libraries consulted.

Four: The Husbandry of Bees

HONEY was an important product of the farms of colonial Pennsylvania and New Jersey. "Bees thrive and multiply exceedingly in those Parts," wrote one of the original settlers of Philadelphia. "Honey . . . is sold in the Capital City for Five Pence per Pound. Wax is also plentiful, cheap, and a considerable Commerce."[1] In the years which followed the numbers increased. During the Revolution a British army officer, who as a prisoner had marched across New Jersey into Pennsylvania, commented upon the abundance of bees on the farms he passed. "In New England they have a very few hives of bees," he wrote, "but in this province [Pennsylvania], almost every farm house has seven or eight; it is somewhat remarkable they should be more predominant here, as all the bees upon the Continent were originally brought from England to Boston, about one hundred years ago; the bee is not natural to America, for the first planters never observed a single one in the immense tract of woods they cleared, and what I think stands forth a most indubitable proof that it is not, the Indians, as they have a word in their language for all animals, natives of the country, have no word for a bee, and therefore they call them by the name of the Englishman's Fly."[2]

Read's manuscript contains little evidence of his experience in bee husbandry. That he did actually keep bees and produce honey, however, appears from his reference to the use of honey in the making of the beverage metheglin, described in Chapter Six which follows. With this exception, and a brief observation on the gelding of combs by Mr. Ludlam, his notes are limited to abstracting the text of Bradley's *General Treatise of Husbandry and Gardening.*

[1] G. Thomas, *op. cit.* (section on Pennsylvania), p. 23.
[2] [Thomas Anburey], *Travels Through the Interior Parts of America* . . . , (London, 1791), II, 251-253.

The Husbandry of Bees

Extract from the first Vol. Bradley concerning Bees[3]

[218] The numbers of bees in a hive differ according to the different Sizes of the hives; we reckon there are eight or ten thousand Bees in a small one, & about eighteen thousand in a large one.

Three Sorts, the Bee, the drone and the queen who is the female Bee, & of these there are sometimes three in a Hive.

[232] The queen bee will produce 8 or 10 M young ones in a year. She has sometimes very little company in the hive wth her, yet in the fall the Hive shall be full of bees, she lays an egg abt the 1/24 of an inch long & 1/6 [as] thick bigger at one end than the other, of which she will lay 8 or 10 immediately in different Cells, [235] the egg in 4 days changes into a jointed maggot his two ends touch one another, it is then incompassed in a little Liquor which the Bees at that time put upon him some time after they Carry Honey for his nourishment in proportion to his Growth, [236] till the 8th day when they fill up the Cell wth food & close him up, after wch the worm remains 12 Days longer & undergoes several Changes & on the 20th day of her age endeavors to get out of her cell & Sometimes comes out of her hive and goes to work the same Day. You may know the young ones by the body being darker & the hair whiter than the old ones.

Mr. Ludlam of Cape May who keeps many bees says that if the Hive is not rich enough to preserve itself tis not worth while to feed but lett them die.

[Exsection or gelding of combs] If done at all should be done abt Wheat Harvest at wch time they will maintain themselves [Mr. Ludlam].[4]

---

[3] From Bradley's *General Treatise* . . . , I, 218, 219, 232-237. Bradley stated that the egg is "four or five Times as long as it is thick." Read writes "1/6 [as] thick."

[4] Several members of the prominent Ludlam family of Cape May County were contemporary with Charles Read, among them four sons of Anthony Ludlam (d. 1737): Providence (1726-1792), Reuben (1728-1783), Anthony (1730-1777), and Joseph (1732-1757), also his brothers Joseph (1706-1753) and Jeremiah (1708-1777). Since neither Christian name nor initial is given in the text, the person referred to cannot be identified. According to tradition, the Ludlams were probably the leading cattlemen of the county. They owned large areas of marsh and beach land, where they kept the cattle during the winter, transferring them to the woods on the upland in the summer. (Notes from H. Clifford Campion.)

Five: Farm Structures and Farm Implements

THE MORE THRIFTY FARMERS of New Jersey and of Pennsylvania in colonial times, particularly those of Dutch origin, had the reputation of giving more attention to the construction of their barns than to their dwellings. With an abundance of timber readily at hand, there was no need for skimping in size. Their great barns sheltered at once the farm animals and the produce, as well as the farm implements. Under broad sloping roofs were capacious grain lofts, and wide doors opened on the threshing floors below. These rough, substantial structures, so typical of the spirit and the character of the rural people, made a lasting impression upon Europeans who traveled the colonial highways.

Read's notes deal at length with farm structures and with other phases of farm engineering. In no branch of agriculture does he show greater genius for detail. The notes are particularly rich in descriptions of buildings and fences, and to a lesser extent of farm machines. Not content with diagrams and specifications for the construction of barracks, he gives instructions for computing their capacity, with a table for barracks of various sizes.

Plans for cowsheds are scientifically worked out. He describes the cattle barn of his "Worthy Friend Robert Ogden" of Elizabeth, which was equipped with stanchions not unlike the modern type. In similar manner he advises on the dimensions of a horse stable, allowing adequate room for manger, feed racks, and hooks for hanging up saddles and harness. Hints are given also for the construction of hanging shelves, a cheese house and an icehouse. Sir John Sinclair, who in the 1760's lived at Belleville, near Trenton, was reputed to have had the first icehouse in New Jersey,[1] and an icehouse was reported on the estate of the late Colonel Peter Schuyler, on the

[1] The Federalist and New Jersey State Gazette, Trenton, IV, no. 161, March 30, 1802, p. 3.

Farm Structures and Farm Implements 369

Passaic in 1768, but Read noted that in Canada "almost Every Farmer" had an icehouse.

As wooded areas were cleared and turned to cultivation, fencing became, in increasing degree, an indispensable part of the colonial farm. Apropos of this trend, Read gives detailed instructions for building gates, bars, and fences. Chestnut is preferred for bars and gates, locust for posts. He recommends a 5-rail fence to hold sheep and small stock. The art of splicing rails and of holing posts is fully explained. William Foster's advice on setting posts is recorded: "Sharpen ye Ends & Drive them abt 6 Inches & it make the post Stand steadier."

Wagons, harness, saddles, the plow, the roller, and the flail all come in for practical treatment. Specifications are given for materials for building "a Waggon," and a recipe for mixing paint. An entry for August, 1762, reads, "Painted the Carriage of my Chaise . . . & did it Compleatly." He paid the painter 6s. for one day's work.

The method of making bull-hide traces is described all the way from choosing the bull to fitting the cured hide for use. He recommends the use of the roller for grass and grain, and advises a roller drawn by horses "where the Land will bear them."

Following are samples of his "Miscellaneous Usefull Observations," suggestive of farmer's almanac advice:

Every Farmer should have 2 Grindstones 1 Coarse [and] 1 finer.
The best Leather for Flail thongs is a Squirrel skin raw or ½ tanned or the skin of an Eel.
Never make an ordinary or makeshift fence.

"The Levell," he writes, "is an Instrument of great Use to those who have Banks to raise or Ditches to drive." Then follows a minute description of a homemade level, and the manner of using it. Drainage and irrigation, he recognizes, are foremost problems of the farmer, and he gives special attention to Worlidge's description of a "Bucket Engine." Supplementing the text, he draws an outline of such a machine that can be propelled by windmill, water wheel, or horse, and gives a full description of its operation. It is computed to discharge 40 gallons of water a minute. Then follows the method of computing the volume of water to be removed from a piece of land.

By leaving us these notes on farm engineering, Read has made a unique contribution to our knowledge of colonial agriculture. Only in dealing with drainage and with Tull's hoe plow and drill, do Eliot's *Essays* compare with them in completeness of treatment. Except for the description of the water bucket, and a reference to Hale's description of the roller, gleanings from English books are conspicuously absent. Most of the material seems to be indigenous. Its wealth of structural detail gives an unwonted and refreshing clarity to our reconstructed picture of the American colonial farm.

[FARM STRUCTURES]

Your Grain is best Kept in barracks wth Intertices supported & the Bottom supported abt 15 or 18 inches from ye ground so that dogs & Catts may go under it & place these so near ye barn End as to pitch in at a hole on to ye thrashing floor wch should be at one End of a Small barn.

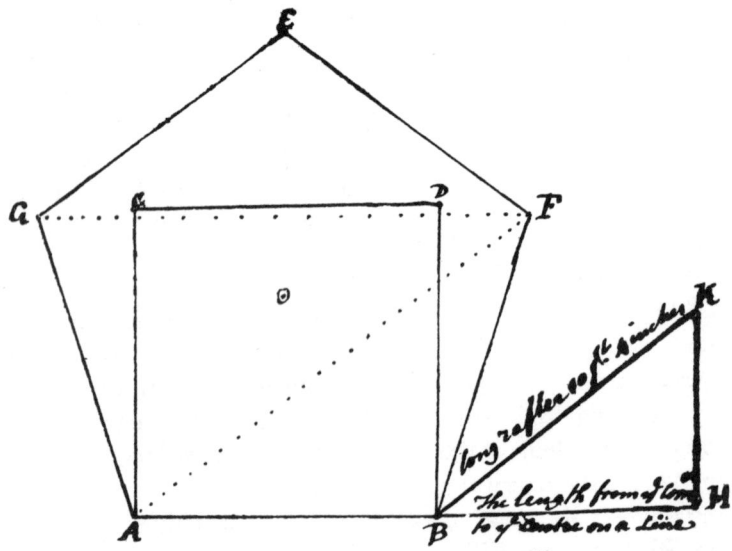

5 Post Barracks the Most Commodious

Farm Structures and Farm Implements 371

The Above Figure [see facing page] represents ye base of a Barrack of 5 posts wth a Plate & Sill 10 feet in the Clear marked A, B, F, E, G.

The Dotted Lines have no refference to ye Barrack & only divide ye outside figure to measure its contents.

The Figure wthin it is one of ye same plate & sill 4 posted whereby you may see the great Conveniency of 5 posts for allowing ye feet to be chains (in Surveying) the inner square contains 10 Acres and the outward black Lines 16½ (1/40) Acres & their proportion is as 10 to 16½ (1/40) B to H is the distance from ye Corners to ye Centre B to K ye Length of rafter of ye 5 post barrack allowing that pitch wch varies according to peoples humours. The Steeper the better it sheds the Water ye flatter ye lighter to raise or Lower. Lett yr Holes at ye Corners be pretty large that it may not pinch in lowering & yr holes not above 8 inches apart for the pins.

Rafters of 9 ft. long will do as ye corner rafters do not come down & those to ye sides are shorter the block in the middle should be of good wood trimmed to a point & tarred or painted.

Where there is much Hay to Save a plate & Sill abt 13 ft. 8 inch or 14 outside will be handy enough to raise & Lower wth help & an Engine to do it & will wth a post of 18 or 20 ft. hold a Large Quantity of Hay. Its rafter will be 6 or 8 inches longer than ye plate. I tryed May 1st 1756 & find that 2½ ft. cubical will of good Clover hay weigh 112 lb. but much of the weight must depend on the Condition tis putt up in if greener & nearer ye bottom then it will weigh the more than if well dryed & near ye top.

☞ Be sure in Building barracks that yr tennants of ye Plates go very slack into ye mortices, make yr tennants at least 6 inches long give them 1½ inch play & bind them up to that wth a wooden key behind the mortice wch is better than a pin, if they have not play Enough they will break off the Tennant in hoisting & lowering as ye Corners are of unequal higths.

The objections to Enclosed Barracks are that the Boards Intertices & Plates prevent the Hay from Settling.

Multiply the Height of the Beam into the Breadth & that Product by the Length the whole Divided By 200 & so the Quotient shews ye Load to ye Beam.

(And then ye Roof) Multiply half the Depth into the Breadth at the Beam & your Product by the Length then work as before & adding to the other shews the Things.

```
     10 feet Length              20
     20 Breadth                   5
     ───                         ───
     200                         100
      35 feet Long                35
     ────                   200)3500)  17½
     1000                        200
      600                       ────
 200)7000)  35 loads            1500
      600    17½                1400
     ────                       ────
     1000   52½ Loads            100
     1000
```

But this calculation must be when it has been awhile pressed for it will not be so heavy when 1st putt in.

1. A Barrack of 10 ft. Plate 5 posted contains 166 superficial square feet & 1 ft. & 2/10 high will hold a load of hay Settled or being 15 ft. high will hold 12 and ½ load.

2. The Surface of a 5 posted Barrack 15 ft. plate contains 380 square feet & of Settled hay 6 inches & 4/10 will make a Load or 16 ft. high will hold 30 Load.

14 ft. plate & Sill 6 7/10 of an inch 1 Load—16 ft. high 28 Load.

13 ft. Do. 8 2/10 of an inch I Load. 16 ft. high holds 23 Load.

☞ [1756] I have a Barrack 13 ft. Plate and —— ft. high it will hold but 12 Load when first halled in.

The Holes should be bored from ye o[u]t to ye inn Corner. Be sure yr Plates go loose in the mortices & that they have a key & not a pin & that when the Shoulder is up the inside of the Key be 1 inch from the morticed plate to give play in hoisting or they will break off. Holes 9 inches apart. Plates, Intertices & Brace Stuff 3 by 5 Rafter stuff 2 inches by 4 putt on yr Rafters wth a Crow foot or Bird bill.

As only one plate rests on a pin I have seen a piece shaped like ye round part of a Caskhead putt on ye head of ye pin for both plates to rest on ⌬.

Cowsheds should allow 3 ft. 2 inches to Each Cow Ox Sheds should allow 4 ft. or 4 ft. 4 inches to Each Ox. The Distance between ye posts wch Confine ye head of a Cow 6 inches an Ox 7½ or 8 inches.

These Oxe Houses would do well thatched but then yo must have a Board or two at the lower side to keep the thatching out of

Farm Structures and Farm Implements

ye way of ye oxen. Shape of the Trough Rack of ye house

[1756] My Worthy Friend Robert Ogden Esq. of Elizatown Had this Spring about thirty Cattle They were divided into Setts of Ten. One parcell fed in the Barn Yard wth Each their hay in Seperate hurdles. Ten More were under Sheds. No divisions between them open to ye South & Defended on ye other sides by being boardd wth a barrack before their Houses & long doors hung wth hinges to lift up to putt in their Hay or Grain ground. Where there was a trough upon Sleepers wch Supported that & ye floor whereon ye cattle lay the back part of this trough might be 15 or 17 inches high, the front part about 6 inches that it might not hurt ye necks of the Cattle when lying down. [The Cattle had no floor but lay on ye Ground.] There was a Plate supported by a Post here & there in wch was fixed one post fast above & below fastned to ye front of ye trough the other post wch stands at ye distance of abt 8 inches from the other had a long mortice above & moved on a pin below so as to fall off from the other atop wide enough to Lett an oxes head in & then it slipped up & an Iron long hasp came over so as to fasten in their Heads there they stand or lay down & feed unless when turned out to Water or Sun themselves wch does them good & their Stalls are easy cleaned and after a very few times using they will Each go to his own Stall.

2 Pieces above would be best it would save the trouble of morticing for the moving piece should open so as to fall back against

ye Standing one as A being the moving peice when shutt comes to (a) & falling so open the ox can't mistake his place.

☞ The length of ye 2 posts 5 ft. 10—sliding one rather longest, between ye posts 8 inches. Length for oxen from outside to ye 2 posts 8 ft. Trough Slanting 3 ft. Between ye Setts of the posts that confine their Heads 3 ft. 8 inches, height of ye lowr side of ye trough from ye ground not Exceed a foot, Heighth of ye Lowr side of ye building 6 ft.

Stable

A Stable should not be less in the Clear than Eleven ft. & if yo propose to hang up Saddles or Geers 12 ft. is better.

Your Rack should have a plank 8 or 10 inches wide at bottom a little slanting & ye Rack should stand almost upright the Distance for fine hay should be no more than 4 inches from the Center of one hole to the Center of the other. If your Hay be course then 4½ inches from ye center of one hole to ye center of the other.

Yr Rack staffs are best square & placed diamond wise ◇ with a sharp Point outwards abt 1¼ inch thro' & when yo bore ye holes leave a square ½ Inch & the barrs will not turn. Be sure you have a broad Manger made of Plank this will catch the Hay seed & what short hay they pull out & save it the Manger should be 22 inches in ye clear.

Lett yr floor be of Stone or plank a little sloping 3 inchs in 12 ft. If yr stable have not a slanting roof then lett it be —— feet High to give room for yr Rack, yr Manger bottom should be 2 ft. 8 inchs from ye Ground or floor.

In Cannada almost Every Farmer has an Ice house in order to keep his Fresh Meat in Summer as the Weather is hott [then] and the Building is Thus made—Take plank or splitt wt Oak & have dugg a hole 8 or 10 ft. Square a little narrower at bottom than at ye top & 10 ft. Deep   putt in a frame to keep yr Lining at ye proper Distance   Sett at each Corner a post to Support a plate for ye roof, 2 crotched posts [in ye] Middle for a strong ridge pole prepare ye Covering Then gett good clean [illegible] pound it to peices 2 ft. deep in ye hole havg 1st lined ye hole to ye plate [remainder illegible].

A Cheese House should have an inside door of an Open Cloth to keep out flies wch trouble yo from ye middle of July to ye Midd or End of 7br.

[Hanging shelves] Tis common to fix hanging Shelves by supporters Nail'd to ye upper Joyst but this is an exceeding bad way, for by many accidents the Supporters are Shaken Loose & if Plaistered much trouble follows   To prevent wch Nail only a ps abt 1 foot long & putt an Iron strop thereon & another on the hanging peice these will give way to any thing thrown agst it.

Barrs

They are best of Chestnutt. The best way for Pastures is to have a gate so contriv'd as to take a horse in or out & that ye gate shutt

Farm Structures and Farm Implements

against ye End of ye bars that push in & yr barrs need not Draw but when a Waggon or cart goes thro ye Gate may fasten wth a pin hung wth a piece of Leather.

Gates

In Making gates observe that they should be closest at bottom as a Fence, That your Brace shou'd foot at ye bottom of ye hanging post & go into a Jog in the outside post at or near ye top wch is better than ye reverse for yr gate cant swagg as it may when putt ye other way. I have seen the Slatts morticed thro' the middle post wch is a good way & prevents their working.

Chestnutt would be the best wood for gates. Gatepost 3 Inches square slatts 3/4 or inch.

Rivetts vastly better than nails.

As to Hanging the best way (tho' not Cheapest) is wth a hook atop drove into ye post on wch it hangs and the Eye should go thro' the gate post & have a key & a ring. At botom an arch wth small arches wch should work against two hooks drove into ye post This will work both ways. If your gate works but one it should play into a rabitt in the large post. I have seen the bottom of ye hanging post foot on ye bottom of a smooth bottle and did well, when it plays all in wood it works hard.

For small gates a small frame is best & hung wth Hooks & Eyes both thro' the posts & Gates.

I have seen large gates cheaply hung thus. On the top of the large post to wch the Gate is hung there is a Tenant about 2½ inches square then a Cap goes across it with a round hole at ye End wch the upper part of ye gate goes into, then the foot has an Iron hoop on it & a Gudgeon drove in the point of this plays on a piece of wood putt into ye ground & if after the hole was bored wherein this Gudgeon plays a piece of hard Iron or Steel was putt in ye bottom of the hole it would play Easier.

Fencing with Post & Rails

In fencing Posts & rails is to be preferred if the ground is good for posts to stand in of wch Moist ground or Clay is the best & Sand worst of all or unless you are to fence against Hoggs.

As yr rails will not all be of an Exact length Sort each pannel at ye heap where Spliced wch is done by having a block at each End & ye rails kept in place by 2 Stakes drove in ye Ground on one side & one on ye other near ye middle.

The Splicing of rails well so as to fitt yr hole wch should be 2½ inches wide & 3½ Deep is the greatest art in this fencing. They should fill the hole every way & Lap 2 inches thro' ye post yr Rail being fixed on his back yo are to mind ye twist so that in hewing the Splice at each end be plum & not all cutt away at one End so as to weaken ye rail & the sloping sides are on one side at one End & on ye opposite one at ye other End.

The End of a well spliced [rail] appears thus

Splice yr Rails only on one side & be flatt on the other & they will when drove up fill the hole better & lett them have good Lap when yo Shape them for ye hole cutt in pretty much Sloping about 4 inches from where they enter yr hole & lett ye part yt goes in be of even breadth. Take care the Lower hole is at some Distance from the ground for there is but little substance on Each Side ye Hole & it will soon Rott. The best Wood for post is 1 Locust, 2 Mulberry, 3 deepest red Cedar, 4 Wt Cedar, 5 barren oak Sassafrass & Chestnutt not good posts, ye last makes fine rails. Tis best to cutt yr Posts when the bark slips take it off & sett it green. [Jo. Burr].

Your posts should be 7 ft. at Least or 7½ is better & your Rails 11 feet Long & of good Substance a small rail Especially of Ceadar good for little, if broad they should be 3 inches on the back & 5 or 6 inches Deep.

Your Pannells will run less than 7 to a Chain or 4 Rod.

Your fence should be 5 Rails it prevents fretting on Sheep or small stock ranging about or being broke by larger Cattle by putting their Heads thro' if fewer rails. Tho' 4 or perhaps three may do on ditch as it is for bigness. In putting it up mind that your post has 4 inches if oak or other hard wood & 6 if white Cedar above ye upper hole to prevent Splitting or an Inch more is better.

A Slab of Cedar makes a fine Lower rail & tight fence.

The Gage for good Fencing is as follows always allowing 7 or 8 inchs for top. 5 holes from ye bottom of 1st hole next ye top to ye top of ye 2d 10 inches, 2d to 3d—8 inch, 3 to 4th—7 inch, 4 to 5th 6 in 4 holes from as above, 1st hole to 2d 10½ or 11 inch, 2d to 3d—9½ or 10 inches, 3d to 4th—8½ inches.

Farm Structures and Farm Implements

The holes in tourn(?) are 2-⅛ of an Inch but better bored wth 2½ Auger   In holing [with ye Ax] cut about 2/3 thro' & then turn it or it will spoile & disfigure ye underside

a holing Ax to have a good poll [should] be 1¾ or bare 2 wide, 9 inches long

He that Has such fence should have rails spliced & posts holled to repair.

Where ye Poststuff is small & ye fence wanted to be tight putt down two small hearty posts in one hole & lett yr Rail ends lay on each other & sawing ye post ends even pin a Cap on and boring hole thro' the posts about ye middle, driving a good Trunnel thro' both posts & wedging them Steadies ye Fence wch will not stand well wthout.

In Setting posts Sharpen ye Ends & Drive them abt 6 Inches & it makes the post Stand steadier [*W. Forster*] on a Ditch or loose Ground Putt yr broadest rails next ye bottom for tight fence and ye lowest should be the shortest in the pannell. The Back of all ye rails downwards. In soft ground bore a 2 inch hole in yr post near ye bottom of ye post & drive a trunnell 2 ft. long

Never make an ordinary or makeshift fence this makes yr Cattle breachy & takes near as much time as a good one.

FARM IMPLEMENTS AND OTHER EQUIPMENT

Size of Waggon Stuff from Jonathn Crispin.[2]
Bolster stuff 4½ Inchs by 5 Inchs:—5 feet long
Axeltree 5 by 8 and 4½ by 7½.
Plank for felly stuff any size from 1½ to 3 Inchs.

How to prepare Oyle for Painting

If yr Paint be of a lightish Colour use Silver Clatts & not Umber. Silver Clatts is what we call Litharge.

Grind one oz of Litharge or 2 ounces of Umber fine on a Stone. The Litharge is best & Cheapest   put this to One Gallon of Oyle stirr it constantly till it boils Then lett it Simmer one hour. White Vitriol will do the same but is Dearer.

---

[2] Jonathan Crispin lived in Evesham Township, Burlington County. In 1791 he sold to Jacob Prickett, of Northampton Township, for £100, a tract of 100 acres of "land and swamp" on Peacock's Mill stream, Northampton and Evesham townships, a tract formerly owned by Charles Read. He died in 1822, leaving a very small estate. (N. J. Deeds, Liber AR, 139; N. J. Wills, 13245C.)

[Aug. 1762] Painted the Carriage of my Chaise & used as follows & did it Compleatly.

3 quarts of Boiled oyle. @ 7s. pr gall	0. 5. 2
1 lb. red Lead	0. 8
Fess(?) wch is ½ lb. Vermilion	8. 0
Painter one day & found himself	6. 0
	0.19.10

abt 1s.9 hard varnish did ye box

Mr. Rigdens Method of Lettering is to Cutt out the Letters in a peice of large paper nothing but the Letters cut out Lay this on yr work & paint over it & it leaves the Letters of ye Colr yo want. Yo may paint it once or twice over & mend wth pencil.

Bull Hide Traces

In the Fall when you kill your Cattle, Recollect whether you want Bull hide Traces as they are usually called, Well Ropes or Leading Lines. Choose a well sized Bull (or rather a good Steer as his hide is more even). Cutt the Hide in two—Then cutt yr Traces next the Back 2½ Inches wide in general tho' it must be a Little thicker & thinner in some Places according to ye substance of the Hide of wch one good Side will make 4 pair. Throw them into a Weak Pickle, every day or two pass them thro' your hands to get them Clear of the Blood. When they are sufficiently Pickled as they will be in 6 or 8 days as the weather is Soak them 2 days in water— then one day in greazy water or Dish wash. The Boiling of Salt Beef or Pork the Best. Then Cut a Slit in the Neck End about 2 or 3 inches long. Nail the other end up to a Beam, then twist the traces Close & thro' the slitts in a Pair put a Stick like a Gambrell & hang a Sufficient Weight to the Gambrel or lay a Board one End on the Gambrell & the other on the floor or Ground & Load the board wth Stones, & in about 8 or 10 days they will be fit for use.

Thongs of 2 Inches wide wthout twisting so served make good Well Ropes.

The thin Parts of the Flank cut round like the Coils of a Snake make excellent Leading Lines served in the same manner, tho not so long Pickled as their Substance is Less & oftner greazed or soaked in Greezy Water to make them Supple.

Quere—if you can't taint off the Hair where yr Lines would be best without hair—

Farm Structures and Farm Implements

Let them Lay, a horse hide in Pickle 2 or 3 Months—the hair will come off & they make excellent strong traces, Lath Strings, or any kind of Straps—as the Hide of a horse is thin you must Cutt them 3½ Inches on ye sides & 3 on ye back.

[Backband 3 inches wide belly band 2½] Back bands should be 4 ft belly bands 3 ft. 8 besides the doubling 3 inches at Each End, buckle yr belly bands.

Miscellaneous Usefull Observations

[Coilers] June 1, 1760 I observed Geo Kemble[3] have Coilers thus. A Surcingle & a Strop to ye bottom comes between ye forelegs to ye Neck Yoke—a long Crupper to ye strops of ye Yoke & a long tongue to ye Wagon would make this Effectual & most airy.

[a Cheap Saddle] May 30, 1758 I overtook a German on the Road wth a kind of Saddle made by himself the cost but Small. It had no tree it was one foot 10½ inches long & its breadth the same nearly it was ¾ inch wider behind than before. The Outside was Leather, the inside ozenbriggs It had a Double seam all round, stitched wth Strong hemp thread, Then Stuffed wth Wool or hair. Deers hair best & had three stitchings or quiltings from the fore to the hinder part at Equal distances wch wth the Outside Stitching made two rows on each side ye horse the middle Stitching being over his back bone. There went a Strap about 2 inches broad across the Saddle abt ¼ or 1/3 from ye forepart Stitched to it before quilting when this came down to the Edge of the Saddle it was Splitt & one part served for a Girth, The other had a ring to which the Stirrup leather fastned. It had a loop for a Crupper & two thongs for tying on a great Coat, it was Easy both to Rider & horse as it Should Swell only of ye under side—the way to do this is to take up ½ or ¾ of an inch wth a loose seam about ye middle of ye Division which may be cutt out before tis stuffed.

[Grindstone] Every Farmer should have 2 Grindstones 1 Coarse 1 finer. The best Method of fixing them is in a Crotch Crosslocked wthin the Ground. They should have an Iron Axis and run on Bone wth a trough under it.

---

[3] George Kemble, of Gloucester County, apparently was a horse fancier. In 1760 he inserted the following advertisement in the *Pennsylvania Gazette* (no. 1638) for May 15: "To be sold by the subscriber, living at Great Timber Creek Bridge, near Gloucester, two very likely young horses; the one a very genteel saddle horse; the other very suitable for either the saddle or chair" (*N. J. Arch.*, 1st ser., XX, 435).

[Repairing old plows] If yr Plow or other utensils want great repair gett new, for yo will find by adding Each repair, if yo Enquire of yr workman, twill cost more than a new one, as I Experienced by a Plow. The Workman has a great Price for repairs.

[Flail thongs] The best Leather for Flail thongs is a Squirrel Skin raw or ½ tanned or the skin of an Eel therefore should be laid by for Use.

[Repair] Whenever any thing yo use getts out of order don't delay mendg till its wanted.

[Roller]

The Roller is an Excellent Instrument in farming to go over all sorts of ground that are apt to heave with the Frost to be used by Horses where the Land will bear them abt 10 ft. Long 16 or 20 inches thro'.

I saved my young grass at Springfeild & the Straw by hand rollers 3 ft. & ½ long & 7 or 9 inches over by wch a man could press down ye roots of grass & Earth so as to Cover what was raised by the Frost & do 3 acres and more in a day in The latter End of Feby & it raised no more that yr.

It is of So great use in light Soils to press ye Earth close to ye roots of yr Plants yo will not have a Crop wthout it. In the winter it prevents ye bad Effects of Frost & in the Spring if done early in the morning before the Vermin are drove into the earth by the sun it destroys Snails Sluggs & all Insects, but yo must Roll before yr Corn or grass spindles, when the leaves are well grown & it assists yr Crop to stand a Drought. Do it when the ground is dry, for tho' good in heavier Soils to break the Clods if ye Earth be wett it will press it to a Cake & hurt. [*Compl. Body of Husbandry* 311-312][4]

[The Level]

The Levell is an Instrument of great Use to those who have Banks to raise or Ditches to drive. It should be made Ten feet long (as Short levells are subject to Error). You may make it of any wood that will not spring well planed & jointed You are to have a perpendicular lett into it 15 inches long there must be a hole about as big as a dollar near the Lower end of it wth a paper pasted on ye back of it & a glass or watch crystal over it from the center

---

[4] Hale, *op. cit.*, pp. 311-312.

Farm Structures and Farm Implements

above there must be a strait Line drawn down ye perpendicular to the hole & a fine thread fixed above so that the Plummett hanging at ye End of it in the hole wch by ye paper on the back & ye Glass before is defended from the wind so that the String striking right over ye Line in the perpendicular may Shew ye Levell of ye Instrument. Yr Levell being thus prepared should be kept in a dry place on 3 Nails on a levell where yo should also have 3 sticks cutt even on the top & bottom & of exact length. Thus furnished you perform yr work thus. Drive a pin cutt flatt on ye top 1, 2, or 3 Inches above ye Ground as best suits yr work near ten feet distance drive another & when yr Levell lays on ye top of these sticks of Equal Length if the perpendicular & ye Line are even yo are right but if the plummett hangs to either side ye side it Inclines to is Lowest. When on a levell drive in a third or 4th pin & when in looking along the underside of yr Levell the top of yr 3d or moving stick is in a line wth it then the 3d or 4th pin is on a Levell wth ye 1st or 2d. Thus yo may proceed to any distance. If yo intend to give ½ inch fall in 10 feet wch is quite enough for any Ditch or 1 inch or more fall then you are to putt a ps of board or wood just the fall yo woud have in 10 feet in thickness on ye top of ye 2d pin & drive yr second pin in so low that when ye Levell lays right it strikes ye top of the third stick sett on ye top of ye 3d pin &c wth double the thickness on it or when ye two pinns are adjusted with the designed Difference move forward & then yo make what was before ye Second pin the first in the following operation.

[The Drill]

Vide ye cutt of this Instrumt page 17.[5]

[The Bucket Engine]

Of all the Engines for Draining Land Where the water is not to [be] raised above 12 or 14 foot to be thrown over a Ditch there seems none so Simple a Machine and at the same time so Expeditious as the Buckett Engine wch may be Carried either with a water Wheel, By Wind or a Horse of the Last I shall Write.

---

[5] This note is placed on the margin opposite Worlidge's description of a horse-drill for seeding grain. In the text (p. 51) Read underlined the following passage: *"One Horse and one Man may work with this Instrument, and sow Land as fast or faster than six Horses can Plough. . . ."*

Suppose you have a Horse or proper Standing Then your Work will be thus

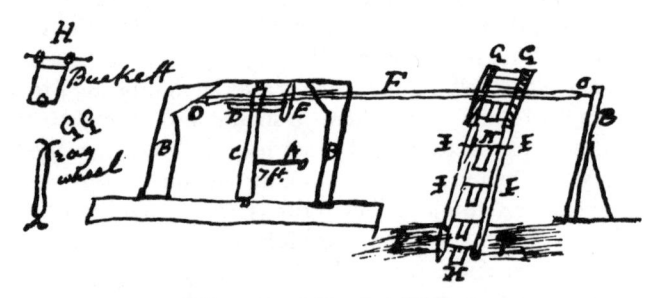

Diagram of Bucket Engine

A. The Beam at ye End of wch yr Horse goes.
B. the Frame to Support yr works.
C. an upright post standin on a steel Pivitt & working at each End in Brass.
D. A treadle Head on ye upright post.
E. A Cogg Wheel on ye shaft F.
G. two ragg wheels on ye Shaft F on each of wch goes a Chain of rings on Each and the bucketts have an Eye in each side & play on Iron bolts FF so as to turn any way HH the bucketts wch have holes in the bottom like the Clappers of a pump box wch opens as they go down into the water & assists them in filling and as soon as they are raised above the surface of the Water they shutt. There must be wooden or Iron barrs between the wheels GG or else the buckett would turn & not Empty itself. Then the Horse moves the Beam A wch move round the trundlehead on ye post C wch gives motion to ye Cogg wheel on ye Shaft F wch plays on a gudgeon at O in the Heads of the post at Each End. The Wheels GG by means of Irons of this form Y or putt into them at proper Distances hold the Chains from slipping & between them bring up the bucketts H & heaving them over the barrs between the wheels Discharge the Water into a trough placed to receive it & to Carry it where you want it   then the Empty bucketts return & again return up full. By the proportion of the wheels D & E you may Have the Shaft F go round 6, 8 or 10 times to once of the Horses round. I think it will be best to have ye Cogg wheel on ye Upright post & ye trundle Head on ye shaft F   the Larger ye wheel on ye post & ye smaller

Farm Structures and Farm Implements 383

that on ye Shaft the oftener ye Shaft will turn for one round of ye Horse 6 ft. Diameter on ye Post & 12 Inches on ye Shaft will be 6 times ye Shaft for once the Horse now Allow your Horse a Sweep of 7 ft. & that he goes round 4 times in a minute when the Shaft goes round 24 times in a Minute. Lett the wheels GG be 1 ft. Diameter (then they must have no barrs between them but ye bolts that hang the bucketts & the shaft may tip them over) & the Chains to be 26 ft. long to allow 10 ft. for ye Water to be raised then there is 6 ft. allowed for going round the wheels GG & for Dipping in ye Ditch PP so that then 3 ft. of the Chain comes up at one turn of the Shaft & 18 ft. at 6 turns, of wch there are 24 in a minute allowing ye horse to go 4 times round in a minute. Allow on these Chains 10 bucketts Each to hold 2 Gallons so that in the Horses going round thrice the whole chain comes up twice & more & thereby if only three times round in a minute for ye Horse & twice round for the chains She would Discharge 40 gallons in a Minute.

In the Fenns in Lincolnshire they have Horse Engines less lyable to be out of order they are 4, 6 or 8 Vanes wch are moved Round in a trough & whatever length they are of will raise the Water above one third of the Diameter of the Vanes or rather of the circle made by them. I have not had a perfect Description of the whole, But think the trough should be circular & it will Loose the Less Water & if the Shaft wherein the Vanes are morticed could be moved by a Water wthout ye bank as tis easy to do if a Constant Small Creek run by it vast Quantities of Water could be thrown over a Bank.

The trough the lowest part in ye bottom of a Ditch the upper above the top of ye bank so as to Convey ye Water into a trough laid across its mouth through wch it is to run over ye bank into ye Creek.

[Jany 6, 1754] I placed a receiver in a Steady but not fast rain in an hour it received ⅛ of an Inch of Water or 3 Inches for 24 hours.

A peice of Land 102 Acres is 2100 ft. Square & Suppose a ft. [of] water thereon, that is 27,022,978 galls. & An Engine is Sett to Work wch throws 50 gallons of Water in a minute it will clear it in 90 hours,[6] making no allowance for soaking, exhalation or Supply of Water to the Ground.

To try what water there is in a piece of Ground (1) Bring ye peice into a square, (2) Multiply one side by the other, both sides being in ft. (3) Multiply this by 1728 ye Cubic Inches in a foot. (4) Divide by the Cubic Inches in a gallon wch for Ale is 282 for Wine 231 & the product gives yo ye gallons of water if it be a foot Deep, if it be 18 Inches then add half the product to ye product & yo have it, if 6 inches high then ½ ye Product is the Quantity if 3 inches then a Quarter, if 4 inches then ye 3d.[7]

Weights & Measures[8]

[Duhamells preface]—An Arpent french is equal to 1 Acre & ¾ of a rood English

The French bushel 1 peck 1 quart Englis[h]

The Septier 12 french bushells

[Duhamell 359]—The English load of Hay is 18₵ weight.

A Kubell(?) of Coals 2½ bushells

[6] This appears to be an error in computation. It would require 9000 hours to clear the field.

[7] In Worlidge's section *Of the Watering of Meadows* (p. 17), Read underlined the passage from Virgil's *Georgics*, Book II, line 251 (Trans. by Fairclough, London, New York, 1922, p. 132-133): *Humida Majores herbas alit,* "A moist soil rears taller grass."

[8] Duhamel, *op. cit.,* pp. ix, x, 359.

Six: The Husbandry of the Household

THE science of home economics goes hand in hand with that of agriculture. Just as Thomas Tusser, in his *Five Hundred Pointes of Good Husbandrie*, wrote of "Housewifery" as well as of "Husbandrie," so the coupling of farm and home has come down through the years. In this dual relationship Read's notes are no exception. In this respect, however, he differs from Eliot, whose *Essays* barely touch upon matters of the colonial household.

This section is made up mainly of recipes, for the preparation of foods, beverages, and medicines. The notes suggest a rather primitive status of medical science, but give the impression of general well-being and good living, which is confirmed by the author of *American Husbandry*, who wrote:

> The inhabitants of this province [New Jersey] . . . have no town of any note . . . this circumstance keeps them very much at home and pretty free from luxury, that is from the pleasures of a capital: they live in a very plentiful manner, which indeed they could hardly fail of doing in so plentiful a country; for no where on the coast are the necessaries of life in greater plenty. Fish, flesh, fowl, and fruits, every little farmer has at his table in a degree of profusion; and the lower classes, such as servants and labourers, artisans, and mechanics in the villages are all very well cloathed and fed; better than the same people in Britain. Tea, coffee, and chocolate, among the lowest ranks, are almost as common as tea in England; they are universal articles in every farmer's house and even among the poor.[1]

Besides culinary advice, the notes include general household hints: For example, how to control vermin, how to remove spots from clothes; also directions for making objects as widely diverse as a poke bonnet, a bearskin jacket, and soap. A few of the recipes were taken from English publications; many of them were supplied by friends. Recipes for pickling oysters

[1] *Op. cit.*, I, 152-153.

and for curing sheepshead or other fish in summer, suggest the common practice of preserving seafoods. Among the beverages for which recipes are given are curant wine, metheglin, mead, cider, baum, shrubb, and imperial water. The secret of Governor Belcher's success in keeping "good Cask cider" is revealed—a small bag of hops hung with a string inside the barrel, and "bung it tight up." Mrs. Belcher's method of making soap also is given. Among the medicines described are *Elixir propietatis*, of Mrs. Logan (to which Read added a little rhubarb) and Mr. Kinsey's remedy for jaundice. Moths, rats, and moles must be dealt with, and specific advice is given for the control of each. A skunk in the cellar apparently was no uncommon experience for the colonial housewife; the procedure advised—"Seize them suddenly by ye tail."

Because of the dilapidated condition of this section of the manuscript, portions of several recipes are missing—torn away or rendered illegible. In cases where major portions are missing, so as to make questionable the meaning, the recipe is not reproduced. Titles omitted for this reason are: Recipe for making Jam of fox Grapes from Mrs. Stevens;[2] To Make Potatoe Bread; Hominy; To Cure Sheepshead or other fish in Summer; To Pickle Oysters; To Stew Catt fish or Eels; To Make Black Puddings.

[RECIPES]

As a Good farmer may afford to keep a good house & some may keep a genteel one it may not be improper to give a few good & Experienced Rects as most of the Materials are ye produce of a Farm.

Syrop for Tamarinds or punch

Putt the Sugar of 4 lb into a pott   Break 2 Eggs shells & all into

[2] Probably Elizabeth Alexander Stevens, sister of William Alexander, Earl of Stirling, and wife of John Stevens, who was intimately associated with Charles Read in public affairs. In a letter to Lord Stirling in 1753 Read made reference to Mrs. Stevens. In 1771 Stevens built an imposing residence in Lebanon Valley, Hunterdon County. Their son was Colonel John Stevens of Hoboken, famous engineer, inventor, and pioneer in steam transportation. (Charles Read to W. Alexander, Jan. 22, 1753, *op. cit.*; *N. J. Arch.*, 1st ser., IX, 335-336n.)

the Sugar   putt abt 5 Gills of water to it & beat it up till it froths or looks white wthout frothing   make a good fire at first till the scum begins to rise then draw off part of yr fire & then lett it simmer fast   there will rise a Scum wch will generally have a hole in the Middle   when it cakes take off the Cake & lett it still Simmer, & keep taking off any Scum that may rise   yo can do it to the Thickness you please. Twill take near 2 hours.

To Make Fish sauce

Take the liver of the fish after tis boiled & bruise it fine   then Draw yr butter very thick, then Season with Kyan pepper   then putt it on the fire again to keep it warm, then putt a spoonfull of Kitchup or the Liquor of Wallnutts, on ala mode to it.

To Souce Fish

Clean your fish well of the blood, cutt it in Chunks putt on a Pott of Cold water   make it pretty salt, boil it as for [word missing] but not over a feirce fire, because that will break it too much Lay it out to cool, putt it into a Liquor as much as to cover it, make of 2/3ds of the Liquor ye fish was boiled in & 1/3 vinegar & cover it close in a pott for use.

As Much Mace as yo can take up between finger & thumb & a little pepper putt into a ragg and Scalded 1/4 of a tsp[?] [illegible] in ye Liquor makes it agreeable. This is to be Done before the Vinegar is added as heat would take off the sourness & the Liquor must be putt on the fish quite cold.

To Caveach Fish

Fry it in oyl or good Butter as for eating   putt it down when cold, pour on it a Liquor made of salt[?] & water & Vinegar as above wth spices   Mind always to keep it down under the Liquor.

Perch when weather is hott may be fryed & yn next heated in a dish[?] wth some wine, Water & a few sweet herbs & stewed

To Alamode a round of beef

Take a round of Beef & let it lay in Salt 24 Hours cut Some fat Bacon & Season it with Pepper Salt & Cloves & Nutmeg   lard it well through the round pretty near together   Lay some Oak Sticks about 4 or 5 inches from the Bottom of the pott, putt the Beef on & putt abt 1 & 1/2 pint of Water & 1 pint of Wine   putt on your Pott cover and close it well atop. Let it stew over a very gentle fire, it will be done in about 8 hours. Let the Beef be once turned while Doing which you are to do by having a String around it when putt in, when Done lett your beef Cool & putt it up[?] & keep the Liquor in bottles for Gravy to yr beef & It will be savory and keep well a month in the fall of the year & gives an excellent relish to any Sauce for flesh or fish. Some stuff ye round wth herbs & Suet as for Boiling & putt it in a pan & cover it wth suett putt in Wine Water & a little Spice putt a thick Crust over it Send it to an oven keep it in 10 hours. This does well but is not so good as the first.

[Beef to pott] Take the Coarse ps of the beef. Boil them till all the bones come out, then pound fine, then Season it with pepper, Sage & Salt, then putt down a layer of beef & a Layer of marrow, wch you took out of the bones & floats on the top of ye pot. Some times (if Marrow not to be had) Butter will do Cover yr Pott wth dough or very strong brown paper may do Lett it be putt in an Oven heated for bread & stand long as the Oven has Any heat in it if yr pott be large & if Smaller till there be a Danger of burning or drying.

To Make Turtle out of a Shoulder of Veal

[Turtle] Cutt the Veal into Pieces of ¼ of a pound roll in Parsley, Thyme, Sweet Marjoram & Onions chopt fine mixed wth a little Kyan peper, wth black pepper & salt seasoned to yr Liking, grate in a small Nutmegg if yo Choose it putt a Crust in a Dish but not cover the whole bottom pack down yr Meat Close well rolled & covered in The Herbs & Seasoning, then putt ½ a pint or Less of Wine into as much Water as will cover ye Meat pretty even with the top, putt ¼ lb of butter in small peices on top of the Meat then putt in your water & wine & Send it to the Oven It stays in the oven no longer than [word missing]

To Souse a piggs head or Joule

Boil it in Salt & water & putt it in a Liquor as fish before.

Raddishes may be kept over wth yr turnips, will be crisp, may be eaten as usual or Sliced thin wth oyle Salt pepper & vinegar & give an agreable taste to other Sallads.

Bacon Fryes are made by slicing flitch bacon thin and fryed in Pancakes & are excellent eating.

The Husbandry of the Household

Sewet Pudding

One pd. of Sewet, Shred fine a pd. of flour a pd. of currants picked Clean ½ a pd. of Raisons Stoned mix all together wth Salt water very thick then boil it.

Ising glass Flummery

One pint of New Milk ½ oz. of Isinglass Shred well boiled & kept Stirring till quite dissolved then strained thro' a fine Seive sweetned to yr Taste. You may add to it a little rose or orange flower water. In cold weather less Isinglass will do in very hot weather a little more.

An Experienced Rect to make Yeast

Boil 1 Quart of wheat Bran in a gallon or 3 Quarts of Water, Strain it and Let it stand to be Cold as Milk from the Cow, then put in some Rye flour (wch is best) so as to make it as thick as Yeast usually is at the Same Time put in ½ Spoonful of Ginger & 1 Gill of Mellasses, Stir it well together. Sett it where it will be warm till it rises wch is known by its foaming and bubbling wch will be from 2 to 8 hours according to the Season, as soon as it rises set it in a Cold Place but not where it will freeze wch would spoil it—it will keep in a Cool Place 3 or 4 days use it in somewhat Larger quantity than yeast—a pint will raise 4 large Loaves.

[Mrs. Douglass][3] Take persimons & bruise them & mix them wth wheat Bran putting more bran than persimons moisten them wth a Little Water putt them in a pan & bake them dry when yo want to make Beer take abt 2 quarts of this & boil it in as much Water as will make a barrell of beer putt half the mellasses as usual. The grounds of this is good Yeast.

A Method of Preserving Barm or Brewers yeast

Take ye Yeast, beat it till thin, then wth a brush lay some on a board or tubb so that it may dry & be kept clean. When the first Coat is dry, lay on a 2d Coat, & so to any thickness by this means you may preserve it a long time.

When yo use it yo take a sufficient Quantity & dissolve in warm water. Some press out ye moisture in Canvas or woolen Baggs [*Gent Magaz*].[4]

---

[3] Probably the wife of Thomas Douglass, of Chesterfield Township.

[4] *The Gentleman's Magazine and Historical Chronicle*, London, 1746, XVI, 364-365. Published by The Dublin Society in *The Dublin Journal*. The titles in *The*

To preserve Leaven take abt 2 lb of the dough of yr last baking   keep it well covered wth flour   take ye Quantity you want & the night before yo intend to use it work it well wth yr flour or meal wth warm water Covering it up wth a linnen cloth & a blankett over it to keep it warm   in the morning twill be well risen & makes light & wholesome bread [Mrs. Burr].[5]

The 2 Last receipts were first published by the Dublin Society and afterwards incerted in ye *Gentlemans Magazine* [1746 page 365].

To make a Ferment [*from London Magazine*][6]

Take 2 quarts small beer 1 oz Issinglass boil it 5 or 6 minutes then beat it wth a whisk till it froth well & the Strings of the Issinglass disappear & it looks as thick as yeast   Lett it stand an hour & putt it to ye Wort. This is enough to Ferment a hhd of Wort.

West India way for making Leaven

Take one Egg & work up 2 oz flour wth it in water lett it stand in a bowl 24 hours & it will be sour enough. Take 3 or 4 lb flour to the leaven workd in   beat it up wth a little water blood warm then Cover it for 10 or 12 hours. In the morning when tis to be baked Add a little blood warm water to it & work it up again for yr Loaves   lett it rise in the loaf   when risen Enough Oven it.

[BEVERAGES]

To Make Currant Wine

Take 1 & ½ bushel of Currants well cleaned from the Leaves putt them into a tubb wth a plugg on one side lay a Brick or Sticks over this hole & some clean Straw   putt as much boiling Water as will cover them   lett it Lay on about 10 hours   in that time twill take most of the Colour out of them, draw this off & add three pound of Muscovado Sugar to Each Gallon immediately after drawing it off or it may turn sour putt it in a Cask wch should be full   putt the bung lightly in.   it will work in about 6 days according to the weather.

You should putt it down in a cool place & have a large bottle of

---

Gentleman's Magazine are: "A Receipt for making Bread without Barm, by the help of a Leaven," and "A Method to preserve a large stock of Barm, which will keep and be of use for several Months."

[5] Probably the wife of Joseph Burr.

[6] A study of the indices of *The London Magazine* for a period contemporary with these notes failed to reveal the source of this reference.

the Same putt by to fill it up after working then Stop it tight by ye middle of October twill be fitt to drink & will keep Several Years.

The white Currant wch is not So good a Bearer as red makes a pretty wine & if White sugar be used will be white.

Morello Cherries are bruised add ½ water & 2½ of Sugar to a gallon.

If the Weather is hott & sultry be as quick abt yr Work as possible or they will gett ye tart taste wch sours the Wine.

[*To Fine down Wine*]

Some do it & I have, by putting 3 sheets of white Paper into the Bung. Others beat up for a Quarter Cask the whites & shells of 8 Eggs beaten till they will not rope—Add to this a hand full of Gravell well washed.

The following rect (tis said) will fine a Quartr Cask in 24 hours. 1 Quart of new Milk, 1 quart of Brandy, one oz of sweet Nitre 1 oz. of allum well burnt & powdered Stirr all these well in.

To refine a hhd. of Wine take one Quart of New Milk draw off abt 2 Quarts of wine Mix this & throw it directly into the Cask it will refine well in a week *probatum est.*

To recover Wine when friet(?) or changing, for a Qr Cask ½ oz sweet Spirit of Nitre 2 lb. of loaf Sugar Cutt in peices ¼ oz. Saffron 1 Quart new milk 2 penny Weight of Cochineal well pounded putt this into ye bung & stirr it for 20 minutes.

To make a Barrill of Metheglin

By the following rect I made a barrel of Excellent Metheglin 1750 wth honey drained from ye Combs for if you wash the Combs it will give it a bad taste. It kept 6 years & was very fine.

Take 32 Gallons of Water put to it 8 Gallons of Honey, boil it over a Gentle fire, Scum as it Rises, till 8 Gallons are evaporated. Let it cool till almost Stone Cold, then put it into a New Cask. Stop it up Close, It will not be fine in Less than Nine Months.

To Make a Barrell of Mead

Wch is an Exceeding pleasant Liquor in Summer

Take 35 lb good Honey & 32 gallons of Water ¼ lb. Alspice a few cloves & a little Sweet orange peel a race or blade or two of Ginger. Boil the Honey in such Quantity of Water as may suit skimm it continually till the scumm will rise no More as the

Water consumes fill it up wth clean cold water Then throw ye whole into a clean barrell wth one head Open and fill up to your quantity wth clean water either Cold or blood Warm add one Gill & 1/2 of Yeast at the time yr Liquor is abt blood warm for if your Liquor should be too hott when the Yeast is added you Spoil the whole. Lett this Ferment for two days when tis sufficiently worked. Draw it off in Bcttles for use by a hole near the Bottom.

[Spice to be boiled by itself & that water mixd wth ye other]

As to the Spice you may use yr Discretion   some like more & some Less, some of one some another sort. Sweet orange peel very good   You are to boil ye Spice by itself to extract ye strength & mix ye Liquor. The season & place must greatly regulate ye time for Fermenting.

[I fancy these rects will not Do   look for one I approve on tryal][7]

To Make a barrell of Cyder Royall[8]

Choose sound Sweet Cider very fine   putt to it 2 lb Raisins & 2 Galln or 3 Gallns of Spirits drawn from Cider   Rum will not do so well

Cider Observations on keeping it

Caleb Haines[9] who used to have good in Cask all Summer Strained it well in from good Sound fruit filled his Hogsheads till he filled them to abt 2 inches then took half a dozen brimstone Matches & lett them when lighted into ye Cask so that the Smoke filled ye Cask then bunged them tight this checked the overworking.

Govr Belcher[10] used to have good Cask cider   his method was

---

[7] This comment by Read refers to recipes for mead given by Worlidge in his discourse *Of Bees* (p. 198). The recipe which Read approved is the one given above.

[8] "Cyder Royal," we are told, was also made by boiling regular cider down to one-fourth of its original volume.

[9] Caleb Haines (1695-1756), of Burlington, was associated with Charles Read in the ownership of woodlands in the county. Apparently he was a friend of John Woolman, who witnessed his will probated January 3, 1757. In harmony with Woolman's attitude toward slavery, Haines provided that his late negro servants, David and wife Dinah, who had already been set free, should have the possession and benefit of 10 acres of land marked for them on the plantation given to his son Josiah. The value of his personal estate was placed at £624-3-8½, including 23 head of cattle and 6 horses. (*N. J. Arch.*, 1st ser., XXXII, 137; *N. J. Wills*, 5695C-5707C; *N. J. Deeds*, Liber K, 411.)

[10] Jonathan Belcher, governor of New Jersey 1747-1757, was born at Cambridge, Mass., January 8, 1682, the son of Andrew Belcher, a wealthy merchant (*Dictionary of American Biography*, New York, 1929, II, 143-144). For facts about Belcher's career, and his interest in agriculture, see Book 1 of this work.

The Husbandry of the Household

after it came home to putt into ye bung of each barrell a small bag perhaps 1½ gill of hops hung wth a String half way down and bung it tight up.

The Method of Boiling or rather Scalding Cyder

[From Mr. Thos. Rodman][11] Take it before tis worked & putt it in a Convenient Vessell over a Gentle fire keep it Simmering not suffering it to Boil wch will cause a Scum to Rise wch must be taken off till no more will rise   then putt it Hott into yr Cask & bung it up   twill keep long.

Cider for bottling should be thus managed, in four days after it has begun to work it may be bung'd up & in 20 days after racked off into a Clean Cask & this is to be repeated twice for Cask Cyder after the first time at the space of 30 days between But if for bottling it must be rack 4 times in all.

To Make Baum beer

[an exceedg pleasant drink(?)] 14 p of Sugar to 10 Gallons of Water boiled together & [word missing] Scumed & then let it Stand till it is quite cold & Add(?) a quarter of a pound of Dry baume in the Cask, a Little Good yeast a top of the baume & then put the Liquor on, and let it Stand about 2 Weeks & then it will be fit to bottle   if you Please you may put a little peice of Lemon peel & a Clove or two in ea[ch] bottle when its bottled.

Another

7 Gallons of water & 7 Pound of Sugar & Let [boil] away till it comes to 5 Gallons   Skim it well   Let it Stand in So[me] Sweet Thing till it is Cold & then Put three qu[arts] of baume Loose in the Barrell, a Little swee[t] [marjo]ram & a Little Orange peel put in a Little bag [in] the Barrell & a Little yeast & Stir it every [Day] till it is workt Enough, & then Stop it very tight [with] Clay and Let it Lay about two weeks & twill be fit [to] Bottle.

---

[11] Thomas Rodman (1716-1796), of Burlington, was a member of the West New Jersey Board of Proprietors. From his father, John Rodman, in 1756 he inherited a residence on High Street and 200 acres of land within the city bounds, including a 5-acre orchard. His house was the first repository of the volumes composing the Burlington Library. An advertisement in 1753 refers to "the smith's shop of Thomas Rodman, Esq." which stood on High and Broad streets, Burlington. He held several public offices: justice of the peace, sheriff, judge of the court of common pleas of Burlington County, and member of the assembly. (N. J. Wills, 5754C-5768C; N. J. Arch., 1st ser., XIX, 288; Penn. Gazette, no. 1288, Aug. 30, 1753; E. J. Fisher, op. cit., p. 176.)

To Make Shrubb

To Every Pint of Limejuice add one pound of Loaf white powder Sugar, & 1 pint of Rum. This mak[es] agreeable & Cheap punch you may add rum when you make yr Punch—& fresh Lime juice. 600 Limes squeezed 11 pints of Juice took 12 lb. Sugar [Remainder illegible].

To Make Imperial Water

One oz of Cream [of] Tartar to 3 quarts of Water boiled till ye Tartar is Dissolved   pour this on the peel of a good Lemon & ½ lb of Loaf Sugar   filther this thro' a flour bagg & bottle it   twill keep a week   tis an excessive fine Cooling Draught in hott weather.

NB After once or twice making yr Taste will govern yo.

[MISCELLANEOUS]

Mr. Trenchards receipt to Make Soap

Take strong Lye that will bear an Egg sideways any Quantity yo please   putt it over ye fire   Add as much Fatt as yo can gett it to Eat in   when it will take no more boil it well (but it should be slow) To 14 gallons of this Composition add three Quarts of fine Salt & then boil it ½ an hour. Take it off the fire sett it in Tubbs to cool. When quite Cold there will be a thick hard Scum on ye top like hard Soap wch must be cutt out into Cakes or parcells in proportion to the vessell used in Boiling & ye goodness yo intend yr Soap. Eight pounds will do for a Vessell of 14 galls.   putt 8 lb of this Scum into yr Kettle of 14 gs. wth a few galls. of weak Lye.   Boil it till it be all Dissolved fill up yr Kettle wth weak Lye & boil it half an hour   take it off & sett it buy for Use.

A Kettle of 14 gallons will make enough of the first Composition to make 100 galls of Soap (if your Fatt be good & lye also) little inferior to hard Soap.   Seven Kettles full may be boild off in 12 hours, after ye Composition is prepared.

Mrs. Belcher's Method of making Soap in the Sun

Sett your Lye Tubs wth Cold water & stand a Week. Draw off the Strong Lye & to a Lye tubb of abt 90 Gallons allow 20 lb Clear fatt   putt in all at once, and well stirred, keep filling up the Lye Tubbs wth water, & try the Lye and fat in a Shell or Spoon as you do when you Boil it & put in your Small Lye by degrees as long as you find it thicken, stirr it often, which will bring it so much the Sooner. Cover it every Night, & to prevent the Rain falling into it, it will if the Sun be hot Come in 3 days if often stirred and never goes back again as it often does when Boiled. Your fatt shou'd be melted before it is put in. The Great fault when Soap is Boiled, is the forcing it too fast. It shou'd never Bubble up, but only Simmer. Then Your Soap will not go Back, or into its Liquid State.

Elixir proprietatis—of Mrs. Logan[12]

		prices
Myrrh	1 oz	1.2
Turkey rhubarb	1 oz	5.–
English Saffron	½ oz	3.–
Snake root	½ oz	0.3
Salt of Tartar	½ oz	0.3

[NB I putt 1½ oz of rhubarb]

Reduce these as fine as you can, putt them into a bottle pour on them 1½ pint of good rum lett them stand 3 or 4 days in the sun close stopped twill be fitt for use.

Mr. Kinseys[13] *Medicine for a Jaundice*

The White of an Egg beat up in a Tumbler or ½ pt Water

[Bearskin Jacket]

Lincey & bearskin ye Same Wove but ye last pulled & raised.
[3 lb. of ye Chain to ye Stripe 4 to Bearskin 7 lb. yarn to Stripe 10 lb. to Bearskin]

---

[12] Probably the wife of William Logan.

[13] Probably James Kinsey (1733-1802), lawyer, of Burlington. He was a member of the New Jersey assembly 1772-1775, where he was a leader in the opposition to Governor Franklin. After serving as a member of the committee on correspondence in 1773, he was appointed a delegate to the Continental Congress in Philadelphia in 1774. In 1789 he became chief justice of the New Jersey supreme court. The house where he lived at the northeast corner of Wood and Broad streets is still standing. In 1772 he advertised for a horse strayed or stolen from his pasture, offering 40s. for the return of the horse, and £5 for the return of the horse and capture of the thief. (*N. J. Arch.*, 1st ser., XX, 237n; XXXVIII, 216; *Penn. Gazette*, no. 2277, Aug. 12, 1772; E. J. Fisher, *op. cit.*, 313, 443; Lucius Q. C. Elmer, "The Constitution and Government of the Province and State of New Jersey . . . , *N. J. Hist. Soc. Col.*, VII, Newark, 1872, pp. 275-279; N. J. Wills, Liber XXXIX, 498.)

396 Ploughs and Politicks

17 lb. ¼ wool @20d.—Skin wool pulled £1. 8. 4
Spinning the filling @1s.6 pr lb. out of
    ye rough 1. 5.10
[Spinning came to 2.10.4]
7 lb. cotton @2s. Spinning it 3s.6 pr lb....... 1.18. 6
Weaving 20 yards plain Bearskin @8d. 13. 4
Dressing the Bearskin at the Mill @8d. 13. 4
                                                5.19.4
Weaving ye 14 Yards of Striped Stuff @10d. 11.8
                                                £6.10.0[14]

    The Bearskin was 36 inches wide & shrunk to 27 so it came to 3s.10 pr yd. (3s.10) The Stripe came to 3s.7—1 yd. 1/16 wide & 3 yds. made a doublebreasted short Jackt wth Sleeves a big man pr 4s.
    2½ of ½ thicks will do.
    Cotton loses sometimes 1/14, 1/28 1/50th.
    The Chain of Cotton run near 5 yds. to ye pound.
    Ye 14 yds. Striped Stuff made 3 Gowns & petticoats 3 yds. makes a pettycoat.
    Cotton Chain does not Shrink in length either in weaving or pulling
    March & April best to Bleach.

To Take Spots out of Cloaths

    Take two pounds of Spring Water, put in it a little potash about the Quantity of a Walnut, and a Lemon cut into small Slices; mix this well together, and let it Stand Twenty four Hours in the Sun, then Strain it through a cloath, and put the clear Liquor up for use; this water will take out all Spots, whether Pitch, Grease or Oil, as well in Hats as Cloths, Stuffs, Silk, Cottin and Linen, immediately but as Soon as Spot is taken off wash the place with Water, & when dry you will see nothing.
    If the Vestment has a Lining yo must rip it open or the first liquor getting thro' the Water does not get thro' so as to Kill the first Water & it will burn the Lining & take out its Colour.

Vermin as Moth Buggs & Fleas

    [Moths] To Preserve stockings or other woolens—putt dust of [word missing] or Sheets of paper rubbed wth Spiritt of Tur-

---

[14] Apparently Read here made an error in addition. The sum of the column is £6-11-0.

The Husbandry of the Household

[Pattern for a Cap]

One Sixth of the Cover of the Cap, ye square on ye top, to draw down a penny cover'd close down on ye top of it wth Allowance for turning in

The Poke of a Well looking Jockey Cap for a young Man wth proper allowance for turning in

½ yd. of Cotton Velvett not quite ½ yd wide covers this without a Cape

pentine among them[?] or a more cleanly way pr Mrs. Cox[15] Wash yr woolens clean then putt ym in Baggs tyed very close.

White Hellebore destroys Ratts. The Edinborough Society publish ye following rect to Kill ratts[16]—Anoint a peice of Meat with oil of aniseed drag this about till yo fix it at a place distant from ye House ye Ratts at night follow this Then have prepared some meat rubbed wth Ratsbane & Coculus India[17] they will eat of it & ye Coculus India will intoxicate them so that they cant gett

---

[15] There were so many members of the Cox (Coxe) family in New Jersey with whom Charles Read was doubtless acquainted that, without further clues, the identity of the person here referred to cannot be established.

[16] "A Receipt for destroying Baltick and other Rats, Moles, Mice, Stotes, Polecats and Weasels," (Maxwell, *op. cit.*, p. 280). The complete formula recommended is: 1 oz. oil of anniseeds, ½ lb. arsenic, 2 oz. *Nux Vomica*, grated, 1 lb. hog's lard. This note was written on the margin of Worlidge's text (p. 218), opposite the following passage: "Arsenick, or the Root of White-hellebor will destroy them . . . the last is the best, because it destroys only Rats and Mice."

[17] *Cocculus Indicus* is the trade name for berries of the *Cocculus* that are imported from the East Indies to adulterate porter. See Glossary.

back to ye house, by this ye Danger is prevented & they cant creep into holes in the House to Stink.

[S. Stokes often tryed] If a Scunk should gett into yr Cellar as they often do yo may go close to them if yo do not provoke them, Seize them suddenly by ye tail & lift them & they can not piss while held so.

putt a dead Mole into ye Holes[18]

[18] In Worlidge's text (p. 217) opposite this note, Read underlined the following: "*But the putting a dead Mole into a common Haunt, will make them absolutely forsake it.*"

Seven: Fisheries

ALTHOUGH in modern America fish culture is not commonly regarded as a branch of agriculture, the British works on agriculture in the Seventeenth and Eighteenth Centuries frequently dealt with the subject. Worlidge's *Systema Agriculturae* was one of these. The section "Of Fishing" occupies ten pages, and treats of Angling, of Taking Fish by Nets, Pots, or Engines, and of Fish-Ponds.

It is easily understood how such a treatise would be of interest to the farmers of colonial America, for fish comprised a major source of food in the New World. To the colonial agriculturist, fishing might be an important side-line to farming. It is not surprising, therefore, that Read devotes a part of his agricultural notebook to fisheries, particularly since, as we have seen, in 1763 he established a commercial fishery on the Delaware River.

Pertinent to his fishing enterprise are the description of his shad net, made in 1760, the diagram of a fishing weir, and the recipe for making glue from sturgeon. The last item reflects a colonial industry that reached beyond the scope of the household. It is of special interest in this setting because it is headed with the comment, apparently in Read's own handwriting: "In a Letter from London to B. Franklin Esqr. wch he sent me." The recipe, dated July 29, 1763, at East Smithfield, London, was from one H. Jackson. Such a subject would naturally have been of interest to Franklin. We may reasonably infer that the philosopher knew of Read's fishery and with this in view, passed the information on to him.

NOTES ON FISHING

My Nett made 1760 is knitt on a Mash Stick 2⅞ in. round & has in ye body 78 mashes in Depth. This Makes a Shad nett.

Some say that Camphire & Assafetida made into an Ointment wth sweet oyl & touch your bait wth it makes fish bite.

I have try'd this & it will not do.[1]

[Referring to special bait for carp] Take abt ¼ oz of Coculus India & pound it fine & mix it wth flour & water and break it into balls abt as big as a small pea & throw it into ye water ye fish will eat it & become drunk so that yo may take them in yr Hand in a Still water   I saw this performed by Mr. Dove[2] at Gloster June 1756 ye Coculus India is to be had at ye apothecaries at abt 6d. pr oz.

Whip an Eel on ye tail twill kill him.

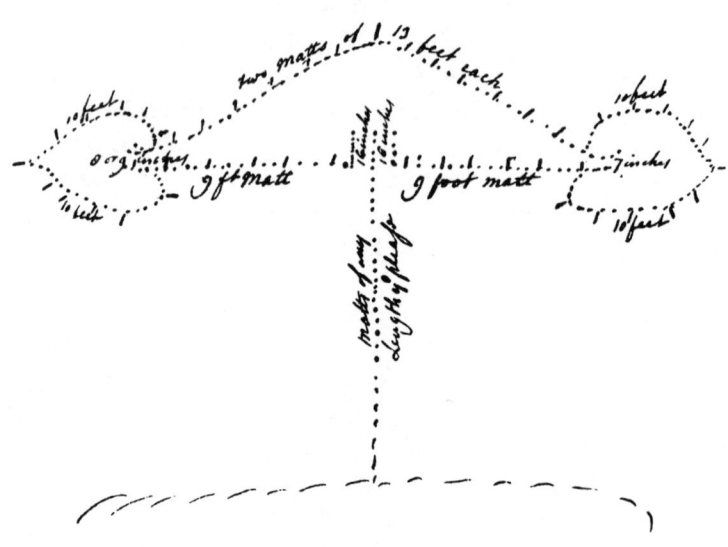

Fishing Waires

Top barrier: Two matts of 13 feet each.
End traps, each above and below: 10 feet.
Entrance to end traps: Right: 7 inches.
             Left: 8 or 9 inches.
Horizontal barrier, each side: 9 foot matt.
Entrances through horizontal barrier, each: 16 inches.
Vertical barrier: Matts of any Length yo please.

[1] This refers to the following statement in Worlidge's text (p. 259): "Gentles [maggots] anointed with Honey, and put on the Hook with a piece of Scarlet dipt in the same, is esteemed the best of all Baits for the Carp." Worlidge also mentioned a paste of bean-flower, honey and assafetida as a bait for carp.

[2] Probably David James Dove (1696-1769), English master of the Academy at

Fisheries

GLUE

A Method of Making Glue from Sturgeon &c.

[In a Letter from London to B. Franklin Esqr wch he sent me]

The Skin & Cartilages must be freed as much as possible from the other parts of the Fish & particularly from what are fat & unctuous, let them be washed clean & then boiled gently in a Sufficient Quantity of Water till they are disolved as much as possible or till little remains except some very gross Parts, the Liquor must be skimm'd pretty often during the Coction when it begins to grow thick, must be kept continually stirring to prevent burning; after wch the Decoction must be strained thro a fine wicker basket, permitted to settle an hour or 2 & then the clear Liquor must be evaporated in a proper Vessel till it is very thick, when it must be pourd into proper Moulds of Wood or Tin to stiffin, after wich it must be taken out & laid upon Netts stretchd in Frames, to dry till it becomes hard enough for Use.

Remarks

Copper Vessells for making Glues are most commodious. They must be of two kinds, one to boil the Subject in & the Other to evaporate the superfluous Moisture, The first may be constructed like a Brewers Copper fixed in Brick work with a Door, Grate, &c After the Usual Method; The other must be a double Vessel, or Water Bath, the outward Vessel of which, must be likewise set in Brick Work or the evaporating Vessel may be adapted to the Boiling Vessel which thus may answer two Purposes, first to boil the Subject in & afterwards re-

---

Philadelphia 1751-1753. He also for a time operated a private school in Philadelphia and subsequently appears to have opened a school at Gloucester, N. J. When the Germantown Academy was opened in 1761, Dove was made head of the English department, and apparently for this reason he gave up his Gloucester property which was offered as a prize in a Philadelphia lottery (*Penn. Journal*, no. 952, March 5, 1761). It was described as "a Lot of Ground, lying at Gloucester called Lilliput . . . , about two Acres of Garden-Ground well improv'd and fenc'd with Boards. One Acre is planted with fine Apple, Plum, Peach, and Cherry-Trees, which bear a great deal of choice Fruit: The other Acre is almost all planted with the largest Battersea Asparagus. There are on the Premises a Dwelling-house, and a very large Barn, fit for storing Merchant's Goods designed for Importation or Exportation, besides which there is a Cellar already dug, and walled up with good Stone, about twenty Feet square; nigh which is a Well of excellent Water. There is a small Grove of Pine-trees before the Garden, from which you are entertain'd with the most beautiful prospect of the City of Philadelphia, and of the River for four or five Miles downwards. . . . Said Place lets for £12 a Year and may be enter'd on immediately." Dove, opposing the Quaker party, produced a number of pamphlets on political questions, as well as political caricatures. (*N. J. Arch.*, 1st ser., XX, 539; *Penn. Journal*, no. 952, March 5, 1761; Joseph Jackson, *Literary Landmarks of Philadelphia*, Philadelphia, 1939, pp. 86-88; J. F. Watson, *op. cit.*, I, 561; Joseph Jackson, "A Philadelphia Schoolmaster of the Eighteenth Century," *Penn. Mag. Hist. and Biog.*, XXXV, 1911, pp. 315-332.)

ceive the other Vessel to evaporate the Liquor, but where Dispatch in Business is required it will be found to most advantage to have them seperate, that one may be boiling while the other is evaporating; the double Vessel must therefore be very large in Diameter & so shallow that altho it be large enough for its Companion to contain all the Decoction yet need not be more than 2 feet Deep or thereabouts: The two Vessels may be Set contiguous to each other with different Flews in the same Chimney; The Companion to this shallow Pan must be adapted in such manner that it must be suspended within the Other by a broad Rim of Copper rivetted to its sides 6 Inches below its Edge or upper Rim, which Rim must rest upon the Edge or Top of the fixed Copper, the inward vessel must be less in Diamiter to the other so as to leave 3 inches Space between the sides of the Two Vessels; everywhere around when the inner Vessel is suspended & 4 Inches Space from the bottom; At the Distance of every Foot in the Rim may be cut vent Holes, about Three Inches Long & two Inches Wide to let out the Steam of the boiling water in the outward Vessel; That Part of the inward Vessel above the Rim may be 6 or 8 inches that it may contain the more Matter to be evaporated, & be more readily taken out occasionally.

After the Glutinous Subjects have been dissolved [word missing] slow continued boiling with repeated Quantities of water in the boiling Vessel; The Decoctions must be strained, permitted to subside in wooden Casks or tubs drawn off pretty Clear & put into the evaporating vessel, previously suspended in the other already filled 2 thirds with Water; The Fire must then be kept up, so as to keep the water round the inner vessel in a continued boiling State which water must be supply'd Occasionally as it wastes & the Decoction stir'd frequently till all the superfluous Moisture is evaporated & the Magma becomes just thin enough to pour into the moulds very slightly greased to prevent its sticking; In this state it must dry till it is stiff enough to lay upon Netts to be further dried for use. The Tin Pans need only to be made of a Sheet of tin 6 or 8 inches Square, with the Edges turned up equal to the intended Thickness of the Glue Cake. The Nett Frames for drying resemble a little Hutt. They may be about eight Feet long as many high & four wide, fixed about 6 inches distance from each other & one over another from the Ground to the Eaves of the Roof, wich Roof is weather boarded. The Hutt with its Roof must be built first, and the Frame with the Netts fastned within it afterwards; Thus the Netts will be open on all sides to the Air except the Top wich defends them from Rain, The great Dificulty in making Glue is to prevent its burning which is effectualy done by this Method as the Decotion can receive no greater heat than [the] boiling Water; The evaporating Pan is made broad & Shallow because all evaporations is in proportion to Surface exposed to the Air; Common Glue is generally of a brown Colour

Fisheries

being generally a little burnt and full of heterogene particles, but Glue made after this Process will be the most perfect that can Possibly be made and will when finished appear transparent and of a grenish Hue, it is the strongest that can be made and grows so hard between Timber as to resist moisture incredibly. Size is only Glue left in the Consistancy between Jelly & Glue, and Jellys are only thin Size. Papins Machine will disolve Animal Sustances more readily, but is Inapplicable in a large way. Iron or Copper Ladles will be necessary to lade out the Glue; I know of nothing more worth describing but what will naturally occur to the Practitioner [Eastsmithfield, Londn July 29 1763 H Jackson].[3]

Encouragement promotes ingenuous performances. [James Logan Read, esq.]

[3] Apparently Mr. Jackson wrote the letter to Benjamin Franklin, who forwarded it to Read. Nothing to establish Mr. Jackson's identity has been found. It is possible that the letter came from Richard Jackson, Franklin's London friend and correspondent on agricultural and other matters, and the initial "R" was copied in Read's notebook as an "H," but this is pure surmise (A. H. Smyth, *op. cit.*, III, 58-60; 133-141). At the time the recipe was written, Franklin was in Philadelphia for a brief respite between London engagements.

APPENDIX A

Sketch of Charles Read

(from Aaron Leaming's Diary, November 14, 1775)

WHEN I was in Burlington Jacob Read informed me that his father The Honourable Charles Read Esqr. died the 27th of December 1774 at Martinburg on Tar River 20 miles back of Bath town in North Carolina where he had kept a small shop of goods for some time.

He was born in Philadelphia about 1713.[1] He was the son of Charles Read a merchant & sometimes mayor of Philada an active & ruling man by his wife Anne Bond.

Charles had his Education under Alexander Annard who taught him the Latin, & he was near 20 when he left that School.

About 1736[2] his father sent him to London where he was patronized by Sir Charles Wager one of the Lords of the admirality & said to be a relation. Sir Charles made him a midshipman on board the Penzance man of war of 20 guns, and his father made him remitances to support him in that rank. The Penzance saild for the West Indies; But Charles not having been bred to Sea, But used to the Philadelphia Luxuries and tasted the pleasures of London that life did not suit him. Beside there was a war approaching and Charles had not been used to that Boisterous romantic honour that characterizes the Seaman.

About 1737 or 1738 Charles Sold out, and married the daughter of a rich planter on Antiagua. She was very much of a Creole, not hansom, nor gentele but talked after the Creole accent. Charles at that time passed for a rising genteel young fellow the son of a very rich merchant & eminent grandee in Philadelphia. But its probable that Charles's father might have trusted him with the Secret of his affairs for his father died about the time of this Marriage & Its said his estate was 7000 lb. worse than even with the world & Charles had very soon the inteligence. Its suspected he knew it before he married. Charles however kept all that a Secret & soon came over with his bride to take possession of his supposed estate. When he came away his father-in-law ordered the

[1] In this date, Leaming was in error. Read was born about February 1, 1715.

[2] Although the precise date of Read's sailing is not known, there is evidence that he went abroad two or three years earlier than Leaming indicates (Library Company of Philadelphia Minutes, Dec., 1733, I, 36-37).

Sketch of Charles Read

negroes to rool him out 37 hogsheads of rum & had many more consigned to him: So that he made his appearance in Philada in quality of a rich merchant. He made the best use of all this. But determined to enter upon State for which nature seemed to have designed him.

About 1739 he bot the Clerk's office of Burlington of Peter Bard & moved to that town. Soon after the Collectors office of the Port of Burlington being vacant Sir Charles Wager gave him that with a Sallary of £60 Sterling per annum. About 1740 he got the office of Clerk of the Circuits. This was given him by Robert Morris then Chief Justice.

He now made his appearance in the world in his own proper character. He had more vices than virtues He had many of both and those of the high rank. He was intriegueing to the highest degree. No man knew so well as he how to riggle himself into office, nor keep it so long, nor make so much of it. From 1747 to about 1771 he had the almost absolute rule of Governor, Council & Assembly in New Jersey except during the short ministration of Mr. Boone who was Governor without a prime minister. I have known the Governor & Council to do things against their inclinations to please him & the Assembly have often done so. He seemed to be their leader. During that time he took the whole disposal of all offices. He little consulted the merits of the person he preferred; the sole object was whether it suited his party principles. His intrigues with women, tho' they employed a large share of his thought, were not worth naming; they were rather the foibles than the vices in so large a character yet because I know he would never have pardoned the man that should attempt his Story without making honourable mention of them I draw them into his shade. He was so vain of them that if he had penned this character they would have filled many pages.

On the death of Mr Archd Home he procured the Deputy Secretary office in about 1743. He then comenced attorney at Law and had the best run of practice of any attorney in my time. He was sometimes 3d and sometimes 2d Chief Justice of the Supreme Court. His greatest virtues were found among his vices. His offices furnished him with a constant flow of Cash. This power & flow of Cash enlarged his mind above himself. Instead of founding a fortune to his two sons as he ought to have done in those prosperous times, he ran upon schemes for the improvements of the Country, witness his Fishery at Lamberton, his Iron Works and many other schemes which tho' virtuous in a very high degree in a man of great fortune, it ought to be treated with distrust with men of little estates. He was industrious in the most unremitting degree. No man planned a scheme so well as he, nor executed them better. He loved the country better than his family. And knew no friend but the man that could serve him. He never embraced any of the sectaries; but always joined the Quakers in party except that it interfered with his politics and then he made them bow: They contributed largely towards his [rise] and he supported them, but it was with a high & prominent hand, taking to himself the mastery.

His airs and action was much after the french manner, ever on the wing & fluttering never long fixed frequently courting, frequently whispering as if to make the person believe they were in his confidence a little too severe in enmity and not grateful for good offices, high strung and selfish unwilling to forgive an injury not very faithful to his client's cause, a better judge, than Lawyer. Upright as a Judge. A fine memory, understood the Law well, spoke very well off hand but short and to the purpose, not capable of arranging and delivering a long train of Ideas, nor of replying and mending his first essay, either in speech or writings. Timerous almost to cowardice, Whimsical to the borders of insanity, which he inherited maternally, and was sometimes perceived to be of unsettled mind especially for some years before his death.

He was several years a member of the house for Burlington & Speaker and afterwards one of the Council which last however did not add much to his influence in council for he was Secretary there before, and did most of the State business and of course did it in his own way, partly because the Governor & council were ignorant and partly from his . . .

[At this point in the sketch, the pages have been torn away and lost]

APPENDIX B

Inventory of the Personal Estate of Charles Read IV[1]

His Purse & Apparel including His Watch	£ 75–10– 3
Four Team Horses	50 — —
Two Carriag Horses	40 — —
a Mule	1–10– 0
Geers for the Team	4 — —
Plows, Harrows, sundry Geers, knife Cutting Box, &c, &c	5–12 —
A Quantity of Hay, at the Stables	1–10 —
2 Coal Boxes 20s, a Log & Ox Chain 20s	2 — —
about 450 Ceeder Railes	3– 7– 6
a Dutch Fann	3–10 —
Hay & Straw in the Barn	5 — —
Rake Fork, Racks, Manger &c.	7– 6
The New Light Waggon & Swingle trees	17–10 —
Geers for 2 Horses	2 — —
a Riding Chair & Harness	20 — —
A Sulky & Harness	8 — —
A Sleigh £4 Grind Stone & Crank 45s	6– 5 —
One Good Four Horse Waggon	15 — —
One Old Do	5 — —
a Square Top'd Waggon	20 — —
Plow, Harrow chicken Coot, Garner &c	1–15 —
3 Sows & 11 Pigs at the House	6 — —
A Coulter tooth Harrow & Lie Tub	1–10 —
Cheese Press & a Hog Trough	1–18 —
A Barrell Churn & frame	1– 5 —
5 Scythes, tug chain & Cant hook	1– 4 —
7 Milch Cows	31–10 —
Green Corn	9 — —
4000 ft Pine Boards, Upper Mill	14 — —
300 ft Ceeder Do	7– 6
a Mill Saw & sundry other Tools, Do	4 — —
1 Pair of Oxen	15 — —
Grubbing Hoes, Axes, Beetle Wedges &c	3 — —
3 Sheep	1–10 —
Sundry Tools 2 Collars & old Iron	6 — —

[1] The will of Charles Read IV was probated December 6, 1783 (N. J. Wills. 10902C).

10 Geese	1- 5 —
a Brass Kettle, Cheese Vatts &c &c	2-15 —
8 Cheeses & a Hive of Honey	1- 3- 6
a Large Brass Kettle	5 — —
Two young Heiffers	7 — —
Sundrys in the Grist Mill	3 — —
Two fatt black Sows at Do	5 — —
Two Mill Stones	3 — —
A Pair of Bellows, Hammers Anvils, in Blacksmith Shop	3 — —
Wood Saw Fire Tongs, Tan Fork Jugs, Flesh Fork & Tan Yard	1 — —
2 Plows & some Corn Stalks	3- 5 —
a Broken Crank, 5 Saw Mill Saws 3 Crow Barrs, a Cross Cut Saw & other Articles, in the Lower Saw Mill	6- 5 —
the half of 15M ft. Pine Boards, at Do	25 — —
A Long Ceeder Ladder	7- 6
the half of a Pair of Timber wheels	5-10 —
a Saddle & Bridle at Wm. Coopers	1-10 —
a Wheel Barrow	10 —
Sundry old Iron at the Stables	7- 6
a Pine Desk with a Green Cover & Book Case thereon	2 — —
a Piece of Russia Sheeting	4-10 —
4 Cases of Bottles & a Meal Chest	3 — —
2 Painted Tables & a Gun	1-15 —
2 Watering Potts & 2 fire Bucketts	1 — —
a Book & Paper Case	1-10 —
a Mans Saddle & Bridle	3-10 —
3 Casks with Some flour in	1- 2- 6
Harness to the Square Top't Waggon	4-10 —
2 Brands, Hammers &c	1 — —
a Number of Books	12- 5 —
7 Books folio Volums	7 — —
a Clock	15 — —
a Maghy. Desk & Book Case	12 — —
Sundry Books & other things Contained therein	8 — —
Table & other Linen	2-18 —
a Walnut Table	1 — —
a Looking Glass	1-10 —
a Corner Cupboard, 2 Setts of Chinea & 2 of Queens Ware	2- 5 —
a Walnut Table	1 — —
6 Chairs	1- 5 —
6 Brass Candelsticks &c &c	2- 5 —
Hand Irons Fire shovel & Tongs	1-17- 6
2 Maps & sundry other Articles	5 —
a Large Looking Glass	3 — —

Inventory of the Personal Estate

6 Windsor Chairs	3-15 —
a Mahoy. Dining Table	3 — —
a Do Tea do	2 — —
2 Small Walnut Tables	2 — —
Brass Top't Hand Irons shovel and Tongs	2 — —
Beaufet & 8 Chinea Bowls	4- 5 —
2 Chinea Dishes & 12 Chinea plates	1-14 —
Chinea, Jugs, & Tea Potts	16 —
Sundry Tumblers & Glasses	12 —
a Maghy. Tea Chest Maps &c	2- 6 —
Sundry knives & forks, Earthen ware Glasses, Jugs, Bottles &c in the Pantry	5- 5 —
The Doctors Closet & the Medicines	3 — —
Sundry Earthen & Chinea Ware	3 — —
Pewter plates, fish Dish, Jarrs, Cheese Toaster Saucepan & Sundrys	4 — —
a Pine Table & a Baskitt knives & forks	15 —
Yarn & other Articles	1 — —
10 Pair Sheets & other Linen	11- 5 —
a Couch Bed Bedding & 3 Blankets	5- 5 —
a Walnut Desk & Book Case	8 — —
a Portmantua & Hand Irons	1-10 —
a Bed Bedstead & Bedding with a Red & White Coverlid on	12 — —
Bed Bedstead & Bedding with a Moss Colerd Coverlid on	8 — —
a Double Case of walnut Drawes	7-10 —
a Bedstead Bed Bedding & Curtin	15 — —
a Small Bed Bedstead & Bedding	7-10 —
a Trunnell Bed Bedstead & Bedding	2-10 —
a Walnut Bureau Table	2-10 —
a Large Looking & Shaving Glass	5 — —
a Walnut Armed Chair & 6 Rush Bottomed Chairs	2-15 —
Craddle, stand, shovel & Tongs Trunk, Hand Irons, window Curtins	2-10 —
10 Pewter Dishes & Sundrys	3-15 —
Chinea Dishes, Decanters a Caster Case & Sundrys	2-15 —
Wine Glasses, Chinea Dishes, Decanters Sauce Boats, Bottles &c	3-15 —
a Large & Small Tin Sugar Box, a Wooden Do, Bottles &c	2- 5 —
8 Jelly Glasses, a Bell Mettle Kettle a fruit Dish, a Keg, &c.	1-10 —
1 Bed Bedstead, Bedding furniture & Curtins	18 — —
a Looking Glass	3 — —
a Maghy. Dressing Table	2 — —

4 Walnut Leather Bottomed Chairs	4 — —
An Easy chair	7–10 —
2 Maghy. stands, Wash Bowl &c	2–10 —
Bed, Bedstead, Bedding and Field Curtins	14 — —
a Double Case of Walnut draws	7–10 —
4 Leather Bottomed walnut Chairs	4 — —
2 Pair sheets 26 Napkins 6 Table Cloths	
8 Large Table Cloths	14– 5 —
Looking Glass Dressing Table & Stand	3–10 —
1 Bed Bedstead & Bedding	10 — —
1 Do	8 — —
1 Do	7 — —
2 Chests Carpets & Blinds	2–12 —
Sunderies in the Lock Garratt	5 — —
Do outside Do	3–15 —
abt 30 Bushels of Rye 4s.	6 — —
16 Bushels Buckwheat 3s, 6	2–16 —
7 Baggs	10– 6
a large Meal Chest	1–10 —
Bottle Rack & Bottles	3 — —
Tubs Pails & Wooden Ware	1– 3– 6
Sundry Iron Potts & Kettles	7–10 —
Kitchin Hand Irons Fire Shovel Tongs Trammel	
Crane &c	5 — —
Earthen Ware knives & forks Coffee Mill & Sundries	3–10 —
Steelyards, Tables, Chairs Dough Trough Cupboard &c	4 — —
Sundry Plate M̶9 oz.(?) @ 8s. 6d. per oz. (?)[2]	45–18 —
Sunderies Articles, Including a	
Coverlid at B. Branins & a Hog at Is. Alloways	4– 4– 6
4 Husk, Beds, Bedsteads & Bedding	12 — —
1 Feather Do	3 — —
The Mule Cart	5 — —
Sundries in the out Cellar	1–17– 6
Meat, Soap, Cyder, Jarrs, &c.	6 — —
Punkins, Sundry old Barrells, — (?) Frames	18– 6
Suppose 50 lb. of Tallow	1–17– 6
about 300 lb. of Beef	4–10 —
Sundry Barrells of Apples &c &c	3–12 —
A Pair Stylyard Bricks, Woollen Yarn	5–15 —
Sundry Iron Backs & Plates	10 — —
70 Bushels In Corn	10–10 —
The Time of a Negro Boy named Phil	15 — —
To half of the Leather & Hides	91–10–10
20 Bushels Potatoes & 5 Bushels Turnips	2– 7– 6

[2] The weight of the plate here indicated is not clear. If the characters indicate ounces as the unit of weight, at the rate given the weight would be 108 ounces.

Inventory of the Personal Estate

Sundry Bonds & Notes, as by a Bill of particulars Signed by ye Appraisers, in the Executors Hands	1775– 1–11½
The Time of a Negro Boy Named Richard	15 — —
Total Value of Inventory	£2864–16–6½

Bibliography

I. WORKS CITED BY CHARLES READ IN HIS NOTES ON AGRICULTURE

In works of more than one edition, the edition probably used by Read is indicated by an asterisk (*). Only English editions are listed. A dagger (†) indicates an incomplete volume or set. (See key to check list, p. 428.)

BALL, JOHN

The/ Farmer's/ Compleat Guide,/ through all the/ Articles of his Profession;/ The/ Laying out, Proportioning,/ and Cropping his Ground;/ and/ The Rules for Purchasing, Managing,/ and Preserving his Stock./ In Particular,/ The Choice and Culture of/ Wheat, Barley and Oats,/ from the Seed to the Barn./ The most profitable Way of/ raising Turneps, with a/ Proposal for introducing the Northern Turnep,/ called the Naper, which/ will live on Bogs./ The Management of Mea-/dow and Pasture Ground,/ and raising of artificial/ Grasses./ The Culture of Beans, Pease,/ Tares, and Thetches./ The Raising of Hemp, Flax/ and Hops; and an Ac-/count of the New Lucerne./ The Raising of Hedge/ Shrubs, Coppice Wood,/ and Timber Trees./ The whole Doctrine of/ Soils and Manures, and/ the Ways of suiting one/ to the other in all In-/stances./ And cheap and effectual Re-/medies for all the Diseases/ of Cattle./ (London):

1760—HU, LCP, RU.

This volume, published anonymously, is attributed to John Ball, of whom no biographical record has been found (M. S. Aslin, *op. cit.*, p. 31). In the introduction, the author sets forth his qualifications for writing on agriculture with the following comment: "Scholars are too refined in their speculations [to write on agriculture]; and farmers are ignorant of the manner of instructing: the first of these want knowledge; and the latter, tho' they have a great deal, do not know how to express it. The proper person is a gentleman who has a large farm in his own hands: and so far I may be bold to tell the reader . . . I am qualified; I have not wanted a school education, I keep a good deal of ground under my own care, and I have had some experience" (*op. cit.*, p. 2). *The Farmer's Compleat Guide* was among the agricultural works consulted by George Washington (P. L. Haworth, *op. cit.*, p. 71).

BRADLEY, RICHARD

New/ Improvements/ of/ Planting and Gardening,/ both/ Philosophical and Practical./ In Three Parts./ I. Containing, A New System *of Vegetation. Explaining the Mo-/tion of the Sap, and Generation of Plants. Of Soils, and the/ Improvement of Forest-Trees. With a new Invention, whereby/ more Designs of Garden-Plats may be made in an Hour, than/ can be found in all the Books of Gardening yet extant./ II. The best Manner of Improving Flower-Gardens, or Parterres;/ of raising and propagating all Sorts of Flowers, and of the Adorn-/ing of Gardens./ III. Of Improving Fruit-Trees, Kitchen-Gardens, and Green-House-/Plants. With the Gentleman and Gardener's Kalendar./* (London):

 1717-18 (1st ed.)—BPL, HEHL†, HC, HU, MHS, MLP, USDA, YU†
 1718 (2nd ed.)—YU
 1719-20 (3rd ed., 2 vols.)—JCL, LC
 1724 (4th ed.)—PHML, USDA
 1726 (5th ed.)—NYPL, UChi, USDA
 1731 (6th ed.)—PU, UCal, UP, USDA
*1739 (7th ed.)—CorU, LC, RU.

A General/ Treatise/ of/ Husbandry and Gardening;/ Containing a New/ System of Vegetation./ Illustrated with many/ Observations and Experiments./ In Two Volumes./ Formerly publish'd Monthly, and now metho-/diz'd and digested under proper Heads, with/ Additions and great Alterations./ In Four Parts./ Part I. Concerning the Im-/provement of Land, by fer-/tilizing bad Soils. Of stock-/ing of Farms with Cattle,/ Poultry, Fish, Bees, Grasses,/ Grain, Cyder, &c./ Part II. Instructions to a Gar-/diner, wherein is demon-/strated the Circulation of/ Sap, the Generation of/ Plants, the Nature of Soil,/ Air and Situation. Of the/ Profits arising from plant-/ing and raising Timber./ Part III. Of the Management/ of Fruit Trees, with par-/ticular Observations rela-/ting to Graffing, Inarching/ and Inoculating./ Part IV. Remarks on the Dis-/position of Gardens in ge-/neral. Of the Method of/ managing Exotick Plants/ and Flowers, and natura-/lizing them to our Cli-/mate; with an Account/ of Stoves, and artificial/ Heats./ (London):

 1721 (1 vol.)—MBG, YU
 1724 (3 vol.)—HU†, NL, UI, RU†
 [1725?] (2 vol.)—JCL, USDA
*1726 (2 vol.)—AAS, LCP, MHS, NJHS, ULCH, UNC†
 1728 (4 vol.)—ColU.

A General/ Treatise/ of/ Agriculture,/ both/ Philosophical and Practical;/ Displaying the Arts of/ Husbandry and Gardening:/ In Two Parts./ Part I. Of Husbandry;/ Treats of the Nature of the Soil, Air,/ and Situation proper for the Pro-/duction of Vegetables; the dif-/ferent Methods of Improving Lands;/ the Manner of Planting and Raising/ Timber; the Stocking of Farms with/ Cattle, Poultry, Fish, Bees,/ Grass, Grain, &c. with Esti-/mates of the Profits arising thereon,/ &c./ Part II. Of Gardening;/ Treats of the Circulation of the Sap in/

Bibliography

Vegetables; the Generation of/ Plants, and their Distribution into/ Genera; the different Kinds and/ particular Management of Fruit/ and Fruit-Trees; the Methods/ of Grafting, Inarching, and Inocu-/lating; the Dispositions of Gardens/ in General; the Cultivation and Im-/provement of the Kitchen and/ Pleasure Gardens; the Man-/ner of managing Exotic Plants/ and Flowers, and naturalizing/ them to our Climate; together with/ an Account of Stoves, Artifi-/cial Heats, &c./ (London):

1757—CorU, HU, MHS, PHS, RU, UCal, UP, USDA.

The/ Gentleman and Farmer's/ Guide,/ For the/ Increase and Improvement of/ Cattle,/ viz./ Lambs, Sheep, Hogs,/ Calves, Cows, Oxen/ also/ The best Manner of Breeding, and Breaking/ Horses,/ Both For/ Sport and Burden;/ with/ An Account of their respective Distempers,/ and the most approved Medicines for the Cure/ of them./ Also some Observations/ On the many Benefits of the Woollen Ma-/nufactures of Great-Britain, and the great/ Advantages arising from Hides, Tallow, &c./ (London):

1729—CCHS, USDA.

Richard Bradley (1688-1732) was one of the most prolific of the English writers on agriculture. Between 1716 and 1730 more than twenty works were published under his name. He was a Fellow of the Royal Society, and in 1720 was elected Professor of Botany at Cambridge. He proved unfitted for the position, however, and in the years before his death fell into professional disrepute. His *New Improvements of Planting and Gardening* (1717) was among the first list of books ordered in 1732 through Peter Collinson, in London, for the public library established in Philadelphia by Benjamin Franklin. (Donald McDonald, *Agricultural Writers* . . . , London, 1908, pp. 170-176; A. K. Gray, *op. cit.*, p. 9.) Franklin listed the sixth edition of this work "with an Appendix containing *The Herefordshire Orchard*, with fine cuts, etc." for sale at his bookshop April 11, 1744 (*A Catalogue of Choice and Valuable Books* . . . , p. 10).

CHEYNE, GEORGE

An/ Essay/ of/ Health/ and/ Long Life./ (London):

1724 (1st ed.)—HU, MLP, UChi, UCin, UP, YU
1725 (2nd ed.)—CPP, PHML
1725 (3rd ed.)—LC
1725 (4th ed.)—BPL, JCL, LC, UCal, USSG
1725 (5th ed.)—ColU
1725 (6th ed.)—MLP, NYPL, USSG, UT, YU
*1725 (7th ed.)—CPP, LC, RAM, RU
1734 (8th ed.)—YU
1735 (9th ed.)—LC

1745 (10th ed.)—LCP
1813 (New York)—CorU, HEHL, LC.

George Cheyne (1671-1743), M.D., F.R.S., native of Aberdeenshire, who practiced medicine in London, wrote widely on philosophical, scientific, and medical subjects, advocating temperance and vegetarianism (*Dictionary of National Biography*, London, 1887, X, 217-219).

DUHAMEL DUMONCEAU, HENRI LOUIS

A/ Practical Treatise/ of Husbandry:/ Wherein are contained, many/ Useful and Valuable/ Experiments and Observations/ in the/ New Husbandry,/ Collected during a Series of Years, by the Celebrated/ M. Duhamel duMonceau,/ Member of the Royal Academy of Sciences at Paris, Fellow of the/ Royal Society, London, &c./ also,/ The most approved Practice of the best English Farmers,/ in the Old Method of Husbandry./ with/ Copper-Plates of several new and useful Instruments./ Compiled by John Mills (London):

*1759 (1st ed.)—APS, BU, ColU, CorU, LC, LCP, NYPL, RU, UP USDA
1762 (2nd ed.)—BPL, ColU, JCL, NYPL, RU, UCal, ULCH, USDA, YU.

Henri Louis Duhamel duMonceau (1700-1782), botanist and engineer, was a native of Paris, whose discovery of a fungus that destroyed the saffron plant led to his admission to the Academy of Sciences. His scientific studies were devoted also to vegetable physiology, the growth and strength of wood, the growth of mistletoe, corn smut, layer planting and fertilizers. He wrote a number of books, touching the fields of forestry, horticulture, agriculture, and naval architecture, several of which were published in English editions. Among the papers left by George Washington is a digest of Duhamel's *A Practical Treatise of Husbandry*. (*Encyclopaedia Britannica*, 1937 ed., VII, 719; P. L. Haworth, *op. cit.*, p. 71.)

DUPRATZ, LEPAGE

The/ History/ of/ Louisiana,/ or of/ The Western Parts/ of/ Virginia and Carolina:/ containing/ A Description of the Countries that lye/ on both Sides of the River Missisipi:/ with/ An Account of the Settlements, Inhabitants,/ Soil, Climate, and Products./ Translated from the French,/ (London):

*1763 (2 vols.)—AAS, BC, HEHL, HSP, HU, ISHS, LC, LU, MSL, NYPL, PI,UK, UP, USMA, VSL, YU
1774—AAS, APS, ColU, CorU, HML, HSP, JCBL, LC, LCP, NYPL, PU, UMich, VC.

Le Page du Pratz (d.1775), a native of the Low Countries, resided for a time in France and in 1718 migrated to North America to take up lands in the vicinity of New Orleans. After an unsuccessful attempt at colonization he spent eight years in exploring the Missouri and Arkansas river basins. On returning to France in 1743 he wrote of his observations in *Histoire de la Louisiane, avec deux voyages dans le nord de Nouveau-Mexique, dont l'un jusqu'à la mer du Sud* (Paris, 1758). (M. le Dr. Hoefer, *Nouvelle Biographie Générale* . . . publiée par MM. Firmin Didot Frères, XL, 986-987, Paris, 1862.)

ELIOT, JARED

*An/ Essay/ upon/ Field-Husbandry/ in/ New-England/ As it is or may be Ordered/ (First Essay, New London):

1748—AAS, ColU, LCP, MHiS, YU.

*A Continuation/ of the/ Essay/ upon/ Field-Husbandry/ . . . (Second Essay, New London):

1749—AAS, ColU, LCP, MHiS, YU.

*A Continuation/ of the/ Essay/ upon/ Field-Husbandry/ . . . (Third Essay, New London):

1751—ColU, HU, LCP, YU.

*A Continuation/ of the/ Essay/ upon/ Field-Husbandry/ . . . (Fourth Essay, New York):

1753—ColU, HSP, LCP, YU.

*A Continuation/ of the/ Essay/ upon/ Field-Husbandry/ . . . (Fifth Essay, New York):

1754—AAS, ColU, LC, LCP, YU.

A Continuation/ of the/ Essay/ upon/ Field-Husbandry/ . . . (Sixth Essay, New Haven):

1759—JCB, LC, YU.

Essays/ upon/ Field-Husbandry/ in/ New-England,/ As it is or may be Ordered/ (combined edition of six essays, Boston):

1760—ColU, HU, JCL, LC, NYPL UChi, USDA, YU.

Essays upon Field Husbandry in New England and Other Papers 1748-1762, By Jared Eliot, edited by Harry J. Carman and Rexford G. Tugwell, with a Biographical Sketch by Rodney H. True (Columbia University Studies in the History of American Agriculture, I, New York, 1934).

Jared Eliot (1685-1753), Congregational clergyman, physician, agriculturist, was born at Guilford, Connecticut, son of the Reverend Joseph Eliot (Harvard, 1658), and grandson of the Reverend John Eliot, "Apostle to the Indians." He was graduated from Yale College in 1706, and the following year became pastor of the church at Killingworth, now Chester, Connecticut, where he made his home the rest of his life. He served as trustee of Yale from 1730 to 1763. He was a corresponding member of the Society for the Encouragement of Arts, Manufactures, and Commerce of London. For *An Essay on the Invention, or Art of making very good, if not the best Iron, from black Sea Sand,* the society awarded him a gold medal. In 1748 he published at New London the first of his six *Essays upon Field Husbandry in New England,* the remainder following at intervals of one to five years. All six essays were published in a single volume in 1760. This was the first book on agriculture to be published in the American colonies. The first four Essays were reprinted serially in *The New York Gazette or the Weekly Post-Boy,* in the issues of May 14 to July 30, 1753 (no. 537-548). (Biographical sketch in *Essays upon the Field Husbandry in New England* . . . , 1934 ed., pp. xxv-lvi; *Dictionary of American Biography,* VI (1931), 78-79).

ELLIS, WILLIAM

The/ Modern Husbandman:/ or the/ Practice of Farming:/ As it is now carried on by the most Accurate/ Farmers in several Counties of England./ (London):

*1742-44 (complete or incomplete sets, 1 to 5 vols.)—BPL, ColU, CorU, HU, LCP, MHS, NYPL, RU, UP, USDA, YU
1750 (8 vols.)—BPL, LC, YU
1794 (5 vols.)—ULCH.

The/ Practical Farmer:/ or, the/ Hertfordshire Husbandman:/ Containing many New/ Improvements in Husbandry./ I. Of Meliorating the/ different Soils, and all/ other Branches of Business/ relating to a Farm./ II. Of the Nature of the/ several Sorts of Wheat,/ and the Soil proper for/ each./ III. Of the great Improvement/ of Barley, by Brine-/ing the Seed, after an/ entire new Method, and/ without Expence./ IV. Of increasing Crops of/ Pease and Beans by/ Horse-Houghing./ V. Of Trefoyle, Clo-/ver, Lucerne, and o-/ther Foreign Grasses./ VI. A new Method to Im-/prove Land at a small/ Expence, with Burnt/ Clay./ VII. Of the Management of/ Cows, Sheep, Suck-/ling of Calves,/ Lambs, &c. with Means/ to prevent, and Remedies to/ cure Rottenness in/ Sheep./ VIII. How to keep Pigeons/ and Tame Rabbits to/ Advantage./ IX. A new Method of Plan-/ting and Improving/ Fruit-Trees in Ploughed-/Fields./ (London):

*1732 (1st ed.)—HEHL, LCP, MLP, RU†, YU†
1732 (2nd ed.)—RU, UP
1738 (3rd ed.)—USDA, YU
1742 (4th ed.)—HSP, LCP†
1759 (5th ed.)—HEHL, USDA, YU
1732-35 (Dublin, 2 vols.)—LC, USDA, YU.

William Ellis (d. 1758), a farmer of Little Gaddesden, Herts, wrote many books on agriculture between 1730 and 1750. His earlier works are considered sound, but he did not follow up his teachings on his own farm, and, after short-lived popularity, his writings fell into disfavor. (M. S. Aslin, *op. cit.*, p. 44.)

The Gentleman's Magazine, and Historical Chronicle (London):

XVI, 1746; XX, 1750; XXIII, 1753—AAS, APS, BPL, ColU, CorU, HEHL, HSP, LC, LCP, NPL, NYPL, RU, UCal, UChi, UP, USNA, VC, YU.

HALE, THOMAS

A/ Compleat Body/ of/ Husbandry./ Containing/ Rules for performing, in the most profitable Manner,/ The whole Business of the Farmer, and Country Gentleman,/ in/ Cultivating, Planting, and Stocking of Land;/ in/ Judging of the several Kinds of Seeds, and of Manures; and in the/ Management of Arable and Pasture Grounds:/ together with/ The most approved Methods of Practice in the several Branches of/ Husbandry,/ From Sowing the Seed, to Getting in the Crop;/ and/ In Breeding and Preserving Cattle, and Curing their Diseases./ To which is annexed,/ The whole Management of the Orchard, the Brewhouse, and/ the Dairy./ Compiled from the Original Papers of the late Thomas Hale, Esq.;/ And enlarged by many New and Useful Communications on Practical Subjects,/ From the Collections of Col. Stevenson, Mr. Randolph, Mr. Hawkins, Mr. Storey,/ Mr. Osborne, the Rev. Mr. Turner, and others./ A Work founded on Experience; and calculated for general Benefit; consisting chiefly of Im-/provements made by modern Practitioners in Farming; and containing many valuable and/ useful Discoveries, never before published./ Illustrated with/ A great Number of Cuts, containing Figures of the Instruments of Husbandry; of useful and/ poisonous Plants, and various other Subjects, engraved from Original Drawings:/ Published by His Majesty's Royal License and Authority./ (London):

*1756—AAS, HU, JCL, MHS, NYPL, PHS, RU, UChi, UP, USDA
1758-59 (4 vols.)—CorU, HU, LC, LCP, MHS, MLP, RU†, UCal, USDA.

Little biographical information about Thomas Hale has been found, other than that he was a gardener. From the preface of *A Compleat Body of Husbandry* we learn that upon his death, he left a "very considerable" collection of writings on agriculture which had grown out of an active experience of more than thirty years, and which he had intended for publication. Certain men who signed themselves as "Authors and Proprietors" acquired these manuscripts, enlarged upon them and "by His Majesty's Royal License and Authority," had them published under the above title. According to the authors, Hale "actually made many Improvements in the Husbandry" of Oxfordshire and Northamptonshire. (*Op. cit.*, pp. ii, iii.) Washington ordered a copy of the book from London (John C. Fitzpatrick, editor, *The Writings of George Washington* . . . , II, 1757-1769, Washington, 1931, p. 354).

HARTLIB, SAMUEL

A/ Discours/ of/ Husbandrie/ Used in/ Brabant/ and/ Flanders;/ Shewing/ The wonderfull improvement of Land there; and/ serving as a pattern for our practice in this/ Common-wealth./ (London):

1650 (1st ed.)—HEHL, RU, USDA
1652 (2nd ed.)—HEHL
1654 (3rd ed.)—HU, LC, USDA.

The real author of this book was Sir Richard Weston.

Samuel Hartlib/ His/ Legacie:/ or/ An Enlargement of the Discourse of/ Husbandry/ used in/ Brabant and Flaunders;/ Wherein are bequeathed to the/ Common-wealth of England more/ Outlandish and Domestick Experiments and/ Secrets in reference to Universall/ Husbandry./ (London)

1651 (1st ed.)—HU, JCL, LC, NYPL, YU
1652 (2nd ed.)—APS, HU, UP
1655 (3rd ed.)—CorU, HU, JCBL, MHS, NYPL, RU, SPPL, UChi, UI, USDA, YU.

Samuel Hartlib (ca. 1600-1662) was born in Poland, and at the age of about 28 took up his residence in England. He started in business nominally as a merchant, but indulged in diverse interests and hobbies. He was an associate of Milton and assisted in establishing the embryo of the Royal Society. John Evelyn, the diarist, described him as "master of innumerable curiosities and very communicative." Several of the books on agriculture which bear his name were the work of other authors. His *Legacie* is reputed to contain the earliest reference to the growth of lucerne (alfalfa) in England. (M. S. Aslin, *op. cit.*, p. 57; D. McDonald, *op. cit.*, pp. 68-78.) Washington ordered a copy from London (J. C. Fitzpatrick, *op. cit.*, II, 354). Hartlib has been accredited with the first proposal of an agricultural college in *An Essay for Advancement of Husbandry-Learning; or Propositions for the Erecting a Colledge of Husbandry* (1651). Rowland E. Prothero (*English Farming Past and Present,* London, 1912, p. 426) questions Hartlib's authorship of this essay, and is inclined to attribute it to Gabriel Plattes, who in his *A Discovery of infinite Treasures,* etc. (1639), suggested a "Colledge for Inventions in Husbandry."

HITT, THOMAS

A/ Treatise/ of/ Fruit-Trees./ (London):

1755 (1st ed.)—AAS, CorU, HEHL
*1757 (2nd ed.)—BPL, JCL, LCP, MHS, RU, USDA
1768 (3rd ed.)—BC, ColU, RU, USDA
1758 (Dublin)—NLC.

Thomas Hitt (d. 1710) was the gardener to Lord Manners, at Bloxholme, in Lincolnshire. His *A Treatise of Fruit-Trees* was edited from manuscript and published years after his death. (M. S. Aslin, *op. cit.*, p. 62; D. McDonald, *op. cit.*, p. 211.)

HOME, FRANCIS

The/ Principles/ of/ Agriculture/ and/ Vegetation./ (Edinburgh):

1757 (1st ed.)—CorU, MHS, RU, USDA, YU
1759 (2nd ed.)—BMC, UP, LCP
*1762 (3rd ed.)—HU, JCL, PU, USDA
1763 (3rd ed.)—NYPL
1776 (3rd ed., London)—USDA, USSG.

Francis Home, M.D., (1719-1813) was appointed in 1768 the first professor of Materia Medica at Edinburgh. In *The Principles of Agriculture and Vegetation* he stressed the dependence of agriculture on "Chymistry," insisting that without a knowledge of that science, agriculture could not be reduced to principles. For this book the Edinburgh Society for the Improvement of Arts and Manufactures awarded him a gold medal. (M. S. Aslin, *op. cit.*, p. 63; R. E. Prothero, *op. cit.*, p. 216.)

HOUGHTON, JOHN

A/ Collection/ For the Improvement of/ Husbandry and Trade./ Consisting of many valuable Materials relating to/ Corn, Cattle, Coals, Hops, Wool, &c./ with/ A Compleat Catalogue of the several Sorts of/ Earths, and their proper Product; the best Sorts/ of Manure for each; with the Art of Draining/ and Flooding of Lands;/ as also/ Full and Exact Histories of Trades, as Malting, Brew-/ ing, &c. the Description and Structure of Instruments/ for Husbandry, and Carriages, with the Manner of their/ Improvement;/ An Account of the Rivers of England, &c. and how far they/ may be made Navigable; of Weights and Measures; of/ Woods, Cordage, and Metals, of Building and Stowage/ the Vegetation of Plants, &c. with many other useful Par-/ ticulars, communicated by several eminent Members of the/ Royal Society, to the Collector,/ John Houghton, . . ./ Revised . . . By Richard Bradley . . . (London):

*1727 (3 vols.)—ColU, HU†, JCL, LC, NYPL
1727-28 (4 vols.)—APS, ColU, LC, MLP, UCal, UChi, USDA, YU.

John Houghton (1640-1705) studied at Corpus Christi College, Cambridge, engaged in business in London, and was a Fellow of the Royal Society, serving on the committee that was especially concerned with agriculture. His *A Collection of Letters for the Improvement of Husbandry and Trade* (1681-83), taken from various sources on a variety of subjects, deservedly enjoyed a considerable reputation. Prothero (*op. cit.*, p. 133) describes this book as "the first attempt to found a scientific agricultural paper." (M. S. Aslin, *op. cit.*, pp. 64-65; D. McDonald, *op. cit.*, pp. 122-124.)

J., S.

The/ Vineyard:/ A/ Treatise/ Shewing/ I. The Nature and Method of Planting, Manuring/ Cultivating, and Dressing of Vines in Foreign/ Parts./ II. Proper Directions for Drawing, Pressing, Making,/ Keeping, Fining, and Curing all Defects in the/ Wine. III. An Easy and Familiar Method of Planting and/ Raising Vines in England, to the greatest Per-/fection; illustrated with several useful Examples./ IV. New Experiments in Grafting, Budding, or Ino-/culating; whereby all Sorts of Fruit may be much/ more improved than at present; particularly the/ Peach, Apricot, Nectarine, Plumb, &c./ V. The best Manner of Raising several Sorts of com-/pound Fruit, which have not yet been attempted/ in England./ Being the Observations made/ By a Gentleman in his Travels./ (London):

1727 (1st ed.)—HU, LC, MHS, UCal, USDA, YU
1732 (2nd ed.)—HU, LC, LCP, USDA.

No clue to the identity of the author has been found.

LAURENCE, JOHN

A/ New System/ of/ Agriculture./ Being a/ Complete Body/ of/ Husbandry and Gardening/ In all the Parts of them. Viz./ Husbandry in the Field, and its several Improvements./ Of Forest and Timber Trees, Great/ and Small; with Ever-Greens and/ Flow'ring Shrubs, &c./ Of the Fruit-Garden./ Of the Kitchen-Garden./ Of the Flower-Garden./ In Five Books./ Containing/ All the best and latest, as well as many new Improvements, useful to the/ Husbandman, Grazier, Planter, Gardener and Florist./ Wherein are interspersed/ Many curious Observations on Vegetation; on the Diseases of Trees, and the/ general Annoyances to Vegetables, and their probable Cures./ As also a/ Particular Account of the famous Silphium of the Antients./ (London):

*1726—ANSP, CorU, HEHL, JCL, LC, LCP, MHS, PHS, UCal, UChi, UI, UO, UP, YU
1727 (Dublin)—USDA.

John Laurence (1668-1732) was a clergyman who took up gardening as a hobby. He was born at Stamford, the son of a clergyman, and received his B.A. and his M.A. at Cambridge. For several years he lived at Yelverton, in Northamptonshire, but in 1721 he became rector at Durham. In *A New System of Agriculture*, Laurence strongly advocated enclosures as a prime factor in a more efficient and a more profitable agriculture. This book was among the trunkful of volumes which Benjamin Franklin sent to the Philadelphia Library from London in 1755. (M. S. Aslin, op. cit., p. 75; A. K. Gray, op. cit., p. 25; D. McDonald, op. cit., pp. 167-170.)

MARKHAM, GERVASE

Markham's farewell to/ Husbandry/ or,/ The enriching of all sorts of Barren and/ Sterile grounds in our Kingdome, to be as/ fruitful in all manner of Graine, Pulse, and Grasse,/ as the best grounds whatsoever:/ Together with the annoyances, and preservation of all/ Graine and Seede, from one yeare to many yeares./ As also a husbandly computation of men and cattels dayly/ labours, their expences, charges, and utmost profits,/ Now newly the third time, revised, corrected, and amended/ together with many new Additions, and/ cheape Experiments:/ For the bettering of arable pasture, and wooddy grounds. Of making good/ all grounds againe, spoyled with overflowing of salt water by/ sea-breaches, as also the inriching of the hop-garden,/ and many other things never published before./ (London):

```
1620 (1st ed.)—HEHL
1625 (2nd ed.)—HEHL, MHS, USDA
1631 (3rd ed.)—JCL, LCP, RU, USDA
1638 (4th ed.)—HEHL, RU, UP, USDA
1649 (4th ed.)—CorU, HU, YU
1653 (5th ed.)—ColU
1656 (6th ed.)—NL, RU, UChi
1660 (7th ed.)—YU
1664 (8th ed.)—LC, NYPL
*1676 (10th ed.)—CPP, HU, RU, USDA
1684 (11th ed.)—USDA, YU.
```

(The title page quoted above is of the edition of 1631.)

Gervase Markham (1568-1637) appears to have been born at Cotham, near Newark, in Nottinghamshire. He was given a good schooling, and became an excellent classical scholar. During the Civil War he held a commission in the army of Charles I. Because he was an authority on the breeding and management of horses, James I employed him to obtain for His Majesty a pure-bred Arab charger, which he imported from the East, receiving for it the handsome sum of £500. He wrote on every variety of agricultural subject, turning out so many treatises that he gained the distinction of being the "first English hackwriter," and proved that books on farming found a sale. But in spite of the popularity of his works, according to Prothero (*op. cit.*, p. 105), his reputation as an agricultural writer was doubtful. (D. McDonald, *op. cit.*, pp. 84-96.)

MAXWELL, ROBERT, editor

Select Transactions/ Of the Honourable/ The Society of Improvers/ In the Knowledge of/ Agriculture/ in Scotland./ Directing the Husbandry of the different Soils for/ the most profitable Purposes, and containing/ other Directions, Receipts and Descriptions./ Together with an Account of the Society's Endea-/vours to promote our Manufactures./ Prepared for the Press by Robert Maxwell of Arkland, a/

Member of the Society, and revised by the Presses and a Com-/mittee appointed for that End./ (Edinburgh):

*1743—ColU, JCL.

Robert Maxwell (1695-1765) was a practical agriculturist, always interested in the promotion of new methods. Peter Collinson, in a letter to Benjamin Franklin in 1751, wrote that he had sent a copy of Maxwell's "Select Transactions in Husbandry." (A. H. Smyth, *op. cit.*, III, 59.) In 1757 Maxwell published a revised edition of this work under the title *The Practical Husbandman*, which contains two of a course of public lectures, probably the first of their kind, which he gave on agriculture in Edinburgh. (M. S. Aslin, *op. cit.*, p. 92.)

MILLER, PHILIP

The/ Gardeners Dictionary:/ Containing/ The Best and Newest Methods/ of/ Cultivating and Improving/ the/ Kitchen, Fruit, Flower Garden, and Nursery;/ As also for Performing the/ Practical Parts of Agriculture:/ Including/ The Management of Vineyards,/ with the/ Methods of Making and Preserving the Wine,/ According to the present Practice of/ The most skilful Vignerons in the several Wine Countries in Europe./ Together with/ Directions for Propagating and Improving/ from real Practice and Experience,/ All Sorts of Timber Trees./ (London):

Numerous editions from 1731 to 1807—AAS, ANSP, BC, BMC, ColU, CorU, FLP, HEHL, HU, LC, LCP, MBG, MHS, NYPL, PHML, PHS, PU, RU, UCal, UChi, UP, USDA, YU.

Philip Miller (1691-1771) compiled *The Gardener's Dictionary* first published in 1731, which ran through several editions over a period of years (M. S. Aslin, *op. cit.*, p. 93). When Benjamin Franklin organized the first public library in Philadelphia in 1731-32, Peter Collinson of London expressed his interest in the project by donating a copy of *The Gardener's Dictionary*. (A. K. Gray, *op. cit.*, p. 11.) It appeared, also, in the list of books which Franklin offered for sale at his bookshop in 1744 (*op. cit.*, p. 10). John Bartram was one of Miller's correspondents, and provided him with information about American plants. (W. Darlington, *op. cit.*, pp. 382-388.)

MORTIMER, JOHN

The Whole Art of/ Husbandry:/ Or, The Way of/ Managing and Improving/ of/ Land./ Being a full Collection of what hath been/ Writ, either by Ancient or Modern Authors:/ With many Additions of New Experiments and/ Improvements not treated of by others./ As also/ An Account of the particular Sorts of/ Husbandry used in several Counties; with Proposals/ for its farther Improvement./ To which is

Bibliography

added,/ The Country-Man's Kalendar, what he is to do/ every Month in the Year./ (London):

1707 (1st ed.)—HU, USDA, YU
1708 (2nd ed.)—CorU, HEHL, HU, MHS, NYPL, YU
1712 (3rd ed.)—BMC, ColU, HU, LC, LCP, MHS, NYPL, UChi, YU
1716 (4th ed., 2 vols.)—HU, USDA, YU
**1721 (5th ed., 2 vols.)—AAS, HU, MHS, RU, USDA, YU
1761 (6th ed., 2 vols.)—AAS, CorU, HU, JCL, MLP, USDA.

The Art of Husbandry. Part II. . . . A Supplement to the whole Art of Husbandry (London):

1712—MHS

John Mortimer (ca. 1656-1736), was a merchant of Tower Hill, London, and a Fellow of the Royal Society. In 1693 he became possessed of an estate in Essex called Toppings Hall, where he carried on agricultural experiments. He studied earlier writings on agriculture, and in *The Whole Art of Husbandry* incorporated them with his own experience. This book is more systematic than many of the earlier works and represents a substantial advance in the literature of agriculture. Its translation into Swedish is evidence of its importance. (M. S. Aslin, *op. cit.*, p. 95; D. McDonald, *op. cit.*, pp. 158-160.)

PETTY, WILLIAM

Sir William Petty's/ Political Survey [Anatomy]/ of Ireland,/ with the/ Establishment of that King-/dom, when the Late Duke of Or-/mond was Lord Lieutenant;/ and also/ An exact List of the present Peers,/ Members of Parliament, and principal/ Officers of State./ To which is added,/ An Account of the Wealth and Ex-/pences of England, and the Me-/thod of raising Taxes in the most equal/ manner./ Shewing likewise that England can bear/ the Charge of Four Millions per Ann when/ the Occasions of the Government require it./ (London):

1691—HU, LCP, MLP
*1719—AAS, ColU, NYPL, PU, UCal, UChi, UMinn, USDA, YU
1769—ColU.

Sir William Petty (1623-1687), native of Hampshire and political economist, received the degree of Doctor of Physic at Oxford, was one of the founders of the Royal Society, and was knighted by Charles II. He served as surveyor-general of Ireland and purchased a large estate in County Kerry, where he developed several mining, manufacturing, and mercantile enterprises. (*Dictionary of National Biography*, 1896, XLV, 113-119; *Encyclopaedia Britannica*, 1937, XVII, 683.)

PLUCHE, NOEL ANTOINE (1688-1761)

Spectacle de la Nature:/ or,/ Nature Display'd./ On such Particulars of/ Natural History/ as were thought most proper/ To Excite the

** Used for checking, but paging different from that mentioned in notes.

Curiosity,/ and/ Form the Minds of Youth./ Translated by Mr. Humphreys (7 vols., London):

Various editions (1736-1760) in straight or mixed sets, partial or complete, at BPL, ColU, CorU, FI, HU, LC, LCP, NYPL, MLP, RU, UCal, UChi, UP, USDA, YU.

Noel Antoine Pluche (1688-1761) was born at Rheims and served for a time on the faculty of the university of that city. He was admitted to holy orders and was appointed director of the college of Clermont. Later he resided in Paris where he gave lectures upon history and geography. (Hugh James Rose, *A New General Biographical Dictionary*, XI, 163-164, London, 1853.)

ROYAL SOCIETY OF LONDON

Philosophical/ Transactions,/ giving some/ Accompt/ of the/ Present Undertakings, Studies, and Labours,/ of the/ Ingenious,/ in many/ Considerable Parts/ of the/ World.

XV, 1685. Oxford, 1686; LIII (1763), London, 1764; LXII (1772), London, 1773—BPL, ColU, CorU, ESL, HEHL, LC, LCP, NYPL, PU, UCal, UChi, UP, USSG, YU.

(Also see: *The/ Philosophical/ Transactions/ and/ Collections/ To the End of the Year MDCC./ Abridged,/ and/ Disposed under General Heads./ Vol. II./ Containing all the/ Physiological Papers./* By John Lowthorp. (The Fourth Edition, London, 1731.)

SMITH, CHARLES

The/ Antient and Present/ State/ of the/ County and City/ of/ Cork,/ in Four Books./ I. Containing, the antient Names of the Territories and In-/habitants, with the Civil and Ecclesiastical Division thereof./ II. The Topography of the County and City of Cork./ III. The Civil History of the County./ IV. The Natural History of the same./ The whole Illustrated by Remarks on the/ Baronies, Parishes, Towns, Villages, Seats, Mountains, Ri-/vers, Medicinal Waters, Fossils, Animals and Vegetables;/ together with a new Hydrographical Description of the Sea/ Coasts./ To which are added,/ Curious Notes and Observations, relating to the erecting/ and improvement of several Arts and Manufactures,/ either neglected or ill prosecuted in this County./ Embellished with new and correct Maps of the County/ and City; Perspective Views of the Chief Towns, and other/ Copper-Plates./ Published with the Approbation of the Physico-Historical/ Society./ (2 vols., Dublin):

*1750 (1st ed.)—ColU, HU, LC, NYPL, YU
1774 (2nd ed.)—BPL, HU, LCP, NYPL, PI, UChi, YU
1815 (Cork)—HEHL, HU, LC, MLP, YU
1893-94 (Cork)—HU, LC, NYPL, UCal, UChi, YU.

Charles Smith (*ca.* 1715-1762), apothecary of Dungarvan, was a pioneer in Irish topography. He was the author of several Irish county histories, and with a group of eminent physicians founded at Dublin the Medico-Philosophical Society. (*Dictionary of National Biography*, 1898, LIII, 20.)

SOCIETY OF COUNTRY GENTLEMEN, FARMERS, GRAZIERS, SPORTSMEN, &c.

The/ Practical Farrier:/ or,/ Full Instructions/ for/ Country Gentlemen, Farmers, Graziers,/ Farriers, Carriers, Sportsmen, &c./ Containing,/ Rules for Breeding and Training up of/ Colts;/ A very curious Collection of well-experienced/ Observations; and upwards of Two Hundred/ Practical Receipts, for the Cure of all com-/mon Distempers incident to/ Horses,/ Oxen,/ Cows,/ Calves,/ Sheep,/ Lambs,/ Hogs, and/ Dogs./ Digested under their proper Heads; many of which have/ been practised for many Years with great Success, and/ the rest taken from the latest and most approved Au-/thors, Noblemen, and Gentlemen, &c./ (London):

*1737 (4th ed.).

(No copy of this work was found in an American library. Photostatic copies of selected pages were procured for checking from the library of the Rothamsted Experimental Station, Harpenden, England.)

TULL, JETHRO

The/ Horse-Hoing Husbandry:/ or, an/ Essay/ On the Principles of/ Tillage and Vegetation./ Wherein is shewn /A Method of introducing a Sort of Vineyard-/Culture into the Corn-Fields,/ In order to/ Increase their Product, and diminish the common Expence;/ By the Use of/ Instruments described in Cuts./ (London):

*1733 (1st ed.)—ColU, CorU, LCP, NITC, UP, USDA, YU
1751 (3rd ed.)—AAS, BTI, CorU, HEHL, HU, LCP, NYPL, UCal, UP, USDA
1762 (4th ed.)—AC, BMC, RU, USDA
1822—HU, JCL, MHS, NYPL, PHS, UP
1829—AAS, CorU, HU, JCL, LC, NYPL, RU, SDSC, UChi, UP, USDA, YU.

Jethro Tull (1674-1740), has been called "the greatest individual improver" up to his time that British agriculture had ever known (R. E. Prothero, *op. cit.*, p. 169). Educated at Oxford, he was admitted to the bar, but illness forced him to relinquish a sedentary life. After traveling abroad, he settled on a farm "Mount Prosperous" at Shalbourn, on the borders of Berkshire and Wiltshire. In Tull's time thick broadcast sowing and the scanty use of the hoe were customary. After years of experiment and study, in the face of opposition and ridicule, he developed a new system of husbandry, based upon drill-seeding and inter-tillage. His revolutionary methods were set forth in the epoch-making work *The Horse-Hoeing Husbandry*. . . . As with many agricultural pioneers, the system of drill husbandry he advocated was not adopted until many years after his death. In this country, George

Washington was a diligent student of Tull's proposals. (D. McDonald, *op. cit.*, pp. 186-190; M. S. Aslin, *op. cit.*, p. 131; P. L. Haworth, *op. cit.*, p. 71.)

WORLIDGE, JOHN

Systema Agriculturae;/ The Mystery of/ Husbandry/ Discovered./ Treating of the several New and most Advantagious Ways/ of/ Tilling, Planting, Sowing, Manuring, Ordering, Improving/ of all sorts of/ Gardens,/ Orchards,/ Meadows,/ Pastures,/ Corn-Lands,/ Woods & Coppices./ As also of/ Fruits, Corn, Grain, Pulse, New-Hays, Cattle,/ Fowl, Beasts, Bees, Silk-Worms, Fish, &c./ With an Account of the several Instruments and/ Engines used in this Profession./ To which is added/ Kalendarium Rusticum:/ or,/ The Husbandmans Monthly Directions./ Also/ The Prognosticks of Dearth, Scarcity, Plenty, Sickness, Heat,/ Cold, Frost, Snow, Winds, Rain, Hail, Thunder, &c./ And Dictionarium Rusticum:/ or,/ The Interpretation of Rustick Terms./ The whole Work being of great Use and Advantage to/ all that delight in that most Noble Practice./ (London):

1668-69 (1st ed.)—APS, CorU, HU, LC, MHS, RU, UChi, UP, YU
1675 (2nd ed.)—BMC, HU, USDA, YU
*1681 (3rd ed.)—HU, JCL, LCP, MHS, RU, UMich, USDA
1687 (4th ed.)—ColU, CorU, HEHL, HU, JCL
1689—LCP
1697 (4th ed., enl. and rev.)—NYPL, USDA
1698 (4th ed., cor. and amend.)—HU
1716 (5th ed.)—USDA.

John Worlidge (1640-1698), resided at Petersfield, in Hampshire. His *Systema Agriculturae*, in a copy of which Charles Read kept his notes, has been described as "the first systematic treatise of husbandry on a large and comprehensive scale," bringing together the practical improvements in English agriculture that were accomplished during the first sixty years of the Seventeenth Century. (M. S. Aslin, *op. cit.*, p. 141.) It contained "much more useful and enlightened observations than any which had previously appeared" (D. McDonald, *op. cit.*, p. 116-121). Of special significance were his chapters on clovers and grasses, and on horticulture; also his treatment of farm machines, among them an improved "engine" for planting seeds, in other words, a seed-drill. On the other hand, the book was quite defective in its treatment of stock-breeding and stock-rearing. (R. E. Prothero, *op. cit.*, p. 130.)

KEY TO PRECEDING CHECK LIST[1]

AAS— American Antiquarian Society, Worcester, Mass.
AC— Amherst College

[1] The author is indebted to the librarians of the several institutions and libraries consulted, who furnished check lists of the items in their respective libraries. The list was amplified also by items from the Union Catalog of the Library of Congress, and the Union Library Catalog of the Philadelphia Metropolitan Area.

Bibliography

ANSP—	Academy of Natural Sciences of Philadelphia
APS—	American Philosophical Society, Philadelphia
BC—	Bowdoin College
BMC—	Bryn Mawr College
BPL—	Boston Public Library
BTI—	Boyce-Thompson Institute for Plant Research, Yonkers, N. Y.
BU—	Brown University
CCHS—	Chester County Historical Society, West Chester, Pa.
ColU—	Columbia University
CorU—	Cornell University
CPP—	College of Physicians of Philadelphia
ESL—	Engineering Societies' Library, New York City
FI—	Franklin Institute, Philadelphia
FLP—	Free Library of Philadelphia
HC—	Haverford College
HML—	Hayes Memorial Library, Fremont, O.
HEHL—	Henry E. Huntington Library, San Marino, Cal.
HSP—	Historical Society of Pennsylvania, Philadelphia
HU—	Harvard University
ISHS—	Illinois State Historical Society, Springfield
JCBL—	John Carter Brown Library, Providence
JCL—	John Crerar Library, Chicago
LC—	Library of Congress, Washington, D. C.
LCP—	Library Company of Philadelphia
LU—	Lehigh University
MBG—	Missouri Botanical Garden, St. Louis
MHS—	Massachusetts Horticultural Society, Boston
MHiS—	Massachusetts Historical Society, Boston
MLP—	Mercantile Library, Philadelphia
MSL—	Massachusetts State Library, Boston
NITC—	Northern Illinois State Teachers' College, DeKalb, Ill.
NL—	Newberry Library, Chicago
NLC—	Newtown Library Company, Newtown, Pa.
NJHS—	New Jersey Historical Society, Newark, N. J.
NYPL—	New York Public Library
NPL—	Newark Public Library
PHML—	Pennsylvania Hospital Medical Library, Philadelphia
PHS—	Pennsylvania Horticultural Society, Philadelphia
PI—	Peabody Institute, Baltimore
PU—	Princeton University
RAM—	Richmond Academy of Medicine
RU—	Rutgers University
SDSC—	South Dakota State College, Brookings, S. D.
SPPL—	St. Paul Public Library
UCal—	University of California
UChi—	University of Chicago
UCin—	University of Cincinnati
UI—	University of Iowa
UK—	University of Kansas
ULCH—	Union Library Company, Hatboro, Pa.
UMich—	University of Michigan
UMinn—	University of Minnesota
UNC—	University of North Carolina
UO—	University of Oregon
UP—	University of Pennsylvania
USDA—	United States Department of Agriculture, Washington, D. C.

USMA— United States Military Academy, West Point
USNA— United States Naval Academy, Annapolis
USSG— United States Surgeon General's Office, Washington, D. C.
UT— University of Texas
VC— Vassar College
VSL— Virginia State Library, Richmond
YU— Yale University

II. Sources Cited by the Author

MANUSCRIPT SOURCES

Allinson, Caroline (Yardville, N. J.)
    Allinson family papers (private collection).
American Antiquarian Society (Worcester, Mass.)
    Miscellaneous manuscripts.
American Philosophical Society (Philadelphia)
    1. Benjamin Franklin's Ledger Books AB (1728-1739) and D (1739-1747).
    2. Franklin Papers, several collections bound (79 vols.) and unbound, 1642-1810.
Berks County Historical Society (Reading, Pa.)
    Miscellaneous manuscripts.
Bucks County Historical Society (Doylestown, Pa.)
    Miscellaneous collections.
Burlington County Historical Society (Burlington, N. J.)
    Allinson Papers, 1760-1790.
Burlington County Record Office (Mt. Holly, N. J.)
    1. Burlington County Deeds.
    2. Burlington County Wills.
Campion, H. Clifford (Swarthmore, Pa.)
    Miscellaneous notes.
East New Jersey Society of Proprietors, Record Office (Perth Amboy, N. J.)
    Land grants and surveys.
Ewan, Nathaniel R. (Moorestown, N. J.)
    Miscellaneous notes and collections.
Harvard University Library (Cambridge, Mass.)
    Papers of Governor Francis Bernard, I-V, 1759-1767.
Historical Society of Pennsylvania (Philadelphia)
    1. Colonel William Bradford Papers, 27 vols., 4 boxes, 1682-1863.
    2. Cadwallader Papers, 10 boxes, 5 vols., 1630-1863.
    3. Clement Papers, numerous volumes and boxes.
    4. Ferdinand J. Dreer Collection of Letters of American Lawyers, 6 vols., 1679-1906.
    5. Elting Collections, Jurists, 1 vol., 1769-1887.

Bibliography

6. Simon Gratz Autograph Collection (American Judges), 11 boxes, 1668-1925.
7. Aaron Leaming's Diary, 4 vols., 1750-51, 1761, 1775, 1777.
8. Logan Papers, approximately 70 vols., 1664-1871.
9. New Jersey Historical Manuscripts, 2 vols., 1664-1853.
10. Pemberton Papers, 70 vols., 1641-1880.
11. Petitions and Memorials, 4 boxes, 1691-1891.
12. Shippen Papers, 33 vols., 1701-1855.
13. Charles Morton Smith Collection, 4 vols., 1685-1843.

Library Company of Philadelphia
  1. Library Company of Philadelphia Minutes, I (1731-1768).
  2. Smith Papers, 16 vols. (1678-1883).

Linnean Society, Burlington House (London)
  Peter Collinson's letter books.

Massachusetts Historical Society (Boston)
  Letter books of Jonathan Belcher, I-IX, 1731-1755.

New Jersey Adjutant-General's Office (Trenton)
  1. Records of the Colonial Wars.
  2. Records of the Revolutionary War.

New Jersey Historical Society (Newark)
  1. Belcher Papers, copies of portions of the letter books of Jonathan Belcher dealing with his administration as governor of New Jersey (see above), 4 folders, nos. 1-580 (1731-1755).
  2. Aaron Leaming's diaries, 1750-51, 1760-61.
  3. Miscellaneous manuscripts.
  4. Paris Papers, correspondence of Ferdinand John Paris, London agent of the New Jersey Proprietors, and other papers, 1681-1784.
  5. Jacob Spicer's Diary, 1755-56.
  6. Spicer Papers, 1745-1780.

New Jersey Public Record Office (Trenton)
  1. Burlington County tax lists, 1773, 1774, 1779, 1780.
  2. Miscellaneous manuscripts.
  3. Minutes of Northampton Township, Burlington County, 1697-1803.

New Jersey Office of the Secretary of State (Trenton)
  1. Books of Commissions.
  2. Books of Deeds.
  3. Books of Wills, and files of original wills.
  4. Letters of Administration.
  5. Marriage Records.
  6. [Jeremiah] Basse's Surveys of New Jersey.
  7. Other surveys.

New Jersey Office of the Clerk of the Supreme Court (Trenton)
  Minutes [Docket] of the Supreme Court, 1749-1775.

Bibliography

New York Historical Society (New York City)
    Alexander Papers, 1717-1798.
    Miscellaneous manuscripts.
New York Public Library (New York City)
    1. Correspondence of Lord Stirling, 1765-1767.
    2. Work-Book of the Printing House of Benjamin Franklin and David Hall, 1759-1766.
New York Department of State (Albany)
    Book of Patents, XIV, 1770.
Nolan, J. Bennett (Reading, Pa.)
    Private collection of manuscripts.
Philadelphia County Record Office (Philadelphia)
    1. Books of Administration.
    2. Books of Deeds.
    3. Books of Wills.
Princeton University (Princeton, N. J.)
    Miscellaneous manuscripts.
Read, James Charles (Greenwich, Conn.)
    Read family papers.
Rutgers University (New Brunswick, N. J.)
    1. Boggs Collection (uncatalogued).
    2. Allinson Papers, 1763-1792.
    3. Miscellaneous manuscripts.
St. Mary's Church (Burlington, N. J.)
    Parish Register, I (1703-1836).
Stevens Institute of Technology (Hoboken, N. J.)
    Theodosius Stevens Collection, 1737-1830.
West New Jersey Society of Proprietors, Record Office (Burlington, N. J.)
    Land grants and surveys.
Wright, Sydney L. (Philadelphia)
    Private collection of manuscripts.

ARCHIVES, LAWS, COURT RECORDS, GOVERNMENT REPORTS, LEGISLATIVE PROCEEDINGS

ALLINSON, SAMUEL, *Acts of the General Assembly of the Province of New-Jersey, from the Surrender of the Government to Queen Anne, on the 17th day of April, in the Year of our Lord 1702, to the 14th Day of January, 1776, to which is annexed, The Ordinance for regulating and establishing the Fees of the Court of Chancery of the said Province . . .* (Burlington, 1776).

An Answer to a Bill in the Chancery of New-Jersey, at the suit of John Earl of Stair, and others, commonly called Proprietors of the Eastern Division of New-Jersey, against Benjamin Bond, and

Bibliography

others, claiming under the original Proprietors, and Associates, of Elizabeth-Town . . . (New York, 1752).

Archives of the State of New Jersey: First Series, I-XXXV, *Colonial Documents* (1631-1776), *Journal of Governor and Council* (1682-1775), *Newspaper Extracts* (1704-1775), *Marriage Records* (1665-1800), and *Abstracts of Wills* (1670-1785), (Newark, Paterson, Somerville, and Trenton, 1880-1939); *Second Series*, I-IV, *Newspaper Extracts* (1776-1782), (Trenton, 1901-1917).

LEAMING, AARON, AND SPICER, JACOB, *The Grants, Concessions, and Original Constitutions of the Province of New-Jersey; The Acts Passed during the Proprietary Governments, and other material Transactions* . . . (Philadelphia, 1755).

Minutes of the Provincial Congress and the Council of Safety of the State of New Jersey, [1774-1776] (Trenton, 1879).

MUNRO, JAMES, editor, *Acts of the Privy Council of England, Colonial Series, IV, 1745-1766* (London, 1911).

NEVILL, SAMUEL, *The Acts of the General Assembly of the Province of New-Jersey* [I], *From the Time of the Surrender of the Government in the Second Year of the Reign of Queen Anne to this present Time* . . . (Philadelphia, 1752); II, *From the Year 1753,* . . . *to the Year 1761* . . . (Woodbridge, 1761).

Pennsylvania Archives, First Series, Selected and Arranged from Original Documents in the Office of the Secretary of the Commonwealth . . . by Samuel Hazard, I-XII (Philadelphia, 1852-1856); *Third Series*, edited by William H. Engle, I-XXX (Harrisburg, 1894-1899).

"Rutgers University, Federal and State Relations," *Rutgers University Bulletin, Fifth Series*, no. 1 (New Brunswick, 1928).

Session Laws of the Province of New Jersey, published under the title "Acts of the General Assembly of the Province of New Jersey," between 1702 and 1776.

The Votes and Proceedings of the General Assembly of the Province of New Jersey (Assembly Minutes), 1702-1776.

DIRECTORIES, GAZETTEERS, MEMOIRS, CATALOGS, LETTERS, TRAVELS, EARLY AGRICULTURAL TEXTS, SOCIETY PROCEEDINGS, COLLECTIONS, AND MAPS

ACRELIUS, ISRAEL, *A History of New Sweden; or, the Settlements on the River Delaware. Memoirs of the Historical Society of Pennsylvania*, XI (Philadelphia, 1874).

ALLINSON, WILLIAM J., *Memorials of Rebecca Jones* (2nd ed., Philadelphia, 1849).

American Husbandry, containing an Account of the Soil, Climate, Production and Agriculture of the British Colonies in North

Bibliography

America and the West Indies . . . by an American (2 vols., London, 1775; also New York, 1939, edited by Harry J. Carman).

AMERICAN PHILOSOPHICAL SOCIETY, *Transactions, of the American Philosophical Society, held at Philadelphia, for promoting Useful Knowledge,* I (1769-1771), (Philadelphia, 1771).

———, *Early Proceedings of the American Philosophical Society for the Promotion of Useful Knowledge, compiled by one of the Secretaries, from the Manuscript Minutes of its Meetings from 1744 to 1838* (Philadelphia, 1884).

[ANBUREY, THOMAS], *Travels through the Interior Parts of America; in a series of letters. By an Officer* (2 vols., London, 1791).

ASLIN, MARY S., *Catalogue of the Printed Books on Agriculture published between 1471 and 1840, with notes on the authors* (Rothamsted Experimental Station Library, Harpenden, England, 1926).

BARBER, JOHN W., and HOWE, HENRY, *Historical Collections of the State of New Jersey:* . . . (Newark, 1861).

BASSUS, CASSIANUS, *Geoponica* (Venice, 1538; also Cambridge, 1704; and Leipzig, 1895, H. Beckh, editor).

BOYD, JULIAN P., editor, *Indian Treaties printed by Benjamin Franklin, 1736-1762, With an Introduction by Carl VanDoren and Historical & Bibliographical Notes* . . . (Historical Society of Pennsylvania, Philadelphia, 1938).

BURNABY, ANDREW, *Travels through the Middle Settlements in North-America, in the Years 1759 and 1760, with Observations upon the State of the Colonies* (London, 1775).

Catalogue of Books belonging to the Burlington Library (pamphlet, Burlington, N. J., 1876).

A Catalogue of Books belonging to the Library Company of Philadelphia (pamphlet, Philadelphia, 1741).

The Charter, Laws and Catalogue of Books of the Library Company of Burlington (pamphlet, Philadelphia, 1758).

The Charter, Laws and Catalogue of Books of the Library Company of Philadelphia (pamphlet, Philadelphia, 1764).

CHASTELLUX, [FRANCOIS J.], MARQUIS DE, *Travels in North-America, in the Years 1780, 1781, and 1782.* Translated from the French by an English Gentleman (2 vols., London, 1787).

[CLUNY, ALEXANDER], *The American Traveller: containing Observations on the Present State, Culture and Commerce of the British Colonies in America, and the further Improvements of which they are capable,* . . . *In a series of letters by an old and Experienced Trader* (Philadelphia, 1770).

DARLINGTON, WILLIAM, *Memorials of John Bartram and Humphry Marshall, with Notices of their Botanical Contemporaries* (Philadelphia, 1849).

DOUGLASS, WILLIAM, *A Summary, Historical and Political, of the First*

Bibliography 435

Planting, Progressive Improvements, and Present State of the British Settlements in North-America (2 vols., Boston, 1755).

[DUANE, WILLIAM], *Letters to Benjamin Franklin, from His Family and Friends, 1751-1790* (New York, 1859).

DUER, WILLIAM ALEXANDER, *The Life of William Alexander, Earl of Stirling; Major General in the Army of the United States, during the Revolution: with selections from his correspondence.* New Jersey Historical Society Collections, II (New York, 1847).

DULLARD, JOHN P., compiler, *Manual of the Legislature of New Jersey, One Hundred and Fifty-fifth Session* (Trenton, 1931).

ELIOT, JARED, *An Essay on the Invention, or Art of making very good, if not the best Iron, from black Sea Sand* (New York, 1762).

ELMER, LUCIUS Q. C., *The Constitution and Government of the Province and State of New Jersey, with biographical sketches of the Governors from 1776 to 1845, and reminiscences of the bench and bar, during more than half a century.* New Jersey Historical Society Collections, VII (Newark, 1872).

EVANS, LEWIS E., *A Map of Pensilvania, New-Jersey, New-York and the Three Delaware Counties . . .* Published by Lewis Evans, March 25, 1749, according to Act of Parliament (Philadelphia, 1749).

FADEN, WILLIAM, [*Map of*] *The Province of New Jersey, Divided into East and West. Commonly called The Jerseys.* Engraved & Published by Wm. Faden, Charing Cross, December 1st, 1777. (Reprinted by the Geological Survey of New Jersey, Trenton, 1877.)

FIELD, RICHARD S., *The Provincial Courts of New Jersey, with Sketches of the Bench and Bar.* New Jersey Historical Society Collections, III (New York, 1849).

FITZPATRICK, JOHN C., editor, *The Diaries of George Washington, 1748-1799* (4 vols., Boston, 1925).

FITZPATRICK, JOHN C., editor, *The Writings of George Washington from the Original Manuscript Sources, 1745-1799,* II (1757-1769), 354 (Washington, 1931).

FRANKLIN, BENJAMIN, *A Catalogue Of Choice and Valuable Books, Consisting Of Near 600 Volumes, . . . To Be Sold for Ready Money only, by Benj. Franklin, at the Post-Office in Philadelphia, on Wednesday, the 11th of April 1744 . . .* (pamphlet, Philadelphia, 1744).

FRONTO, MARCUS CORNELIUS, *The Correspondence of, with Marcus Aurelius Antoninus, Lucius Verus, Antoninus Pius, and various friends,* edited and translated by C. R. HAINES (2 vols., Loeb Classical Library, London, New York, 1919).

HILDEBURN, CHARLES R., *Records of Christ Church, Philadelphia, Baptisms, 1709-1760* (Philadelphia, 1893).

KALM, PETER, *Travels into North America; containing Its Natural History, and A circumstantial Account of its Plantations and*

Bibliography

Agriculture in general, with the civil, ecclesiastical and commercial state of the country, The manners of the inhabitants, and several curious and important remarks on various subjects. Translated into English by John Reinhold Forster (2nd ed., 2 vols., London, 1772).

Kemble Papers, New York Historical Society Collections, 1883, 1884 (2 vols., New York, 1884, 1885).

LEFROY, J. HENRY, editor, *The Historye of the Bermudaes or Summer Islands* (Works issued by the Hakluyt Society, LXV, London, 1882).

The Letters and Papers of Cadwallader Colden, III, *New York Historical Society Collections,* 1919 (New York, 1920).

MYERS, ALBERT COOK, editor, *Hannah Logan's Courtship; A True Narrative: The wooing of the daughter of James Logan, colonial Governor of Pennsylvania, and divers other matters, as related in the diary of her lover, the Honorable John Smith, assemblyman of Pennsylvania and King's Councillor of New Jersey, 1736-1752* (Philadelphia, 1904).

NELSON, WILLIAM, *New Jersey Biographical and Genealogical Notes, From the Volumes of the New Jersey Archives, with Additions and Supplements. New Jersey Historical Society Collections,* IX (Newark, 1916).

New Jersey Historical Society Proceedings, Second Series, I-XIII (1867-1895), *New Series,* I-XIV (1916-1929), (Newark).

The Papers of Lewis Morris, Governor of the Province of New Jersey, from 1738 to 1746. New Jersey Historical Society Collections, IV (Newark, 1852).

P., W., *A Letter from a Gentleman at Elizabeth-Town to his Friend in New-York* (pamphlet, Philadelphia, 1764).

READ, CHARLES, *Copy of a Letter . . . to The Hon. John Ladd, Esq.; and his Associates Justices of the Peace for the County of Gloucester* (pamphlet, Philadelphia, 1764).

ROGERS, ROBERT, *A Concise Account of North America: Containing a Description of the several British Colonies on that Continent, including the Islands of Newfoundland, Cape Breton, &c. . . .* (London, 1765).

SAUNDERS, WILLIAM C., editor, *The Colonial Records of North Carolina, . . . I-X,* 1662-1776 (Raleigh, 1886-1890).

SMITH, JOHN JAY, *Letters of Dr. Richard Hill and His Children; or, The History of a Family, as Told by Themselves* (Philadelphia, 1854).

SMITH, SAMUEL, *The History of the Colony of Nova-Caesaria, or New-Jersey; containing an account of its first settlement, progressive improvements, the original and present constitution, and other events, to the year 1721. With some particulars since; and a short view of its present state* (Burlington, 1765; also 2nd ed., Trenton, 1877).

Bibliography

SMYTH, ALBERT HENRY, editor, *The Writings of Benjamin Franklin* (10 vols., New York, 1905-07).

SPARKS, JARED, *The Works of Benjamin Franklin; containing several political and historical tracts not included in any former edition, and many letters official and private not hitherto published; with notes and a life of the author* (10 vols., Boston, 1840).

STATE HISTORIAN OF THE STATE OF NEW YORK, *Second Annual Report, Colonial Series*, I (Albany, 1897).

STEVENS, HENRY, compiler, *An Analytical Index to the Colonial Documents of New Jersey, in the State Paper Offices of England. New Jersey Historical Society Collections*, V (Newark, 1858).

STEVENS, HENRY N., *Lewis Evans His Map of the Middle British Colonies in America . . .* (3rd ed., London, 1924).

SUETTER, MATTHEW, *Pennsylvania, Nova Jersey et Nova York cum Regionibus ad Fluvium Delaware in America sitis, Nova Delineatione ob oculos posita* (map, ca. 1748).

THOMAS, GABRIEL, *An Historical and Geographical Account of the Province and Country of Pensilvania; and of West-New-Jersey in America . . .* (London, 1698).

TUSSER, THOMAS, *Five Hundred Points of Good Husbandry, as well for the champion, or open country, as for The Woodland or Several; together with A Book of Huswifery* (London, 1812 ed.).

VIRGIL, with an English translation by H. Rushton Fairclough, in two volumes, I, *Eclogues, Georgics, Aeneid I-VI* (Loeb Classical Library, London, New York, 1922).

W., J., *An Address to the Freeholders of New-Jersey; On the Subject of Public Salaries* (pamphlet, Philadelphia, 1763).

WHITEHEAD, WILLIAM A., *East Jersey under the Proprietary Governments: a narrative of events connected with the settlement and progress of the province, until the surrender of the government to the Crown in 1702. New Jersey Historical Society Collections*, I (Newark, 1846).

WIGGINS, FRANCIS S., *The American Farmer's Instructor, or Practical Agriculturist; comprehending the cultivation of plants, the husbandry of the domestic animals, and the economy of the farm; together with a variety of information which will be found important to the farmer* (Philadelphia, 1840).

NEWSPAPERS AND PERIODICALS

(Dates indicate the volumes cited)

Burlington, New Jersey: *The Burlington Advertiser, or Agricultural and Political Intelligencer*, I-II, 1790-1791.

Camden, New Jersey: *West Jersey Press*, 1888.

Cape May Court House, New Jersey: *The Cape May County (New*

Jersey) *Magazine of History and Genealogy*, I, nos. 1-6, 1931-1936, Cape May County Historical and Genealogical Society.

Charleston, South Carolina: *South Carolina Historical and Genealogical Magazine*, XI, 1910.

Mt. Holly, New Jersey: *Mt. Holly Mirror*, 1892.

New York City: *New York Gazette*, 1763; *New York Gazette and Weekly Mercury*, 1772-1774; *New York Gazette, or the Weekly Post Boy*, 1747, 1753, 1766, 1768-1770; *New York Gazette revived in the Weekly Post Boy*, 1747-1749, 1752; *New York Mercury*, 1756; *New York Weekly Post Boy*, 1745, 1746; *Rivington's New-York Gazetteer*, 1773.

Newark, New Jersey: *Centinel of Freedom*, 1801; *Newark Evening News*, 1925; *Newark Gazette*, 1801.

Philadelphia, Pennsylvania: *American Weekly Mercury*, 1720, 1737, 1739, 1740; *The Bulletin*, 1886; *Claypoole's Daily Advertiser*, 1800; *Dunlap's Pennsylvania Packet*, 1774; *North American*, 1908, 1909; *Pennsylvania Gazette*, 1737, 1751, 1753, 1756-1760, 1762, 1764, 1766-1775; *Pennsylvania Journal*, 1749, 1751, 1758, 1761, 1764, 1765, 1768-1771, 1774; *Pennsylvania Packet*, 1772, 1776, 1787.

Pittsburgh, Pennsylvania: *Pittsburgh Press*, 1935.

Trenton, New Jersey: *The Federalist and New Jersey State Gazette*, IV, 1802.

Woodbridge, New Jersey: *New American Magazine*, I-III, 1758-1760.

SECONDARY AUTHORITIES—BOOKS, ARTICLES IN PERIODICALS, ENCYCLOPEDIAS

ALLINSON, SAMUEL, "Fragmentary History of the New Jersey Indians," in *New Jersey Historical Society Proceedings*, 2nd series, IV (1875), 31-50.

AMERICAN COUNCIL OF LEARNED SOCIETIES, *Dictionary of American Biography* (20 vols., New York, 1928-1937).

BAILEY, L. H., *The Standard Cyclopedia of Horticulture* . . . (6 vols., New York, 1914-1916).

BERRY, HENRY FITZ-PATRICK, *A History of the Royal Dublin Society* (London, 1915).

BIRDSALL, KATHARINE H., "The Historical VanVeghten House," in *Somerset County Historical Quarterly*, I, pp. 92-97 (Somerville, N. J., 1912).

BOYER, CHARLES S., *Early Forges & Furnaces in New Jersey* (Philadelphia, 1931).

———, "Pomona Hall, the Home of Joseph Cooper, Jr.," in *Camden History*, I, no. 7 (reprinted from *West Jersey Press*, May 2, 9, and 16, 1935, Camden).

BRAINERD, THOMAS, *The Life of John Brainerd, the brother of David*

Bibliography

Brainerd, and his successor as missionary to the Indians of New Jersey (Philadelphia, 1865).
BRIDENBAUGH, CARL, *Cities in the Wilderness, The First Century of Urban Life in America, 1625-1742* (New York, 1938).
BRUSHFIELD, T. N., "Raleghana, Part II, The Introduction of the Potato and of Tobacco into England and Ireland," in *Report and Transactions of the Devonshire Association for the Advancement of Science, Literature, and Art*, XXX (1898), pp. 158-197, Plymouth.
CABEEN, FRANCIS VON A., "The Society of the Sons of Saint Tammany of Philadelphia," in *Pennsylvania Magazine of History and Biography*, XXVI (1902), (installment) 335-347.
CLOKIE, HUGH MACD., "New Jersey as a Separate Royal Province, 1738-62," in *New Jersey—a History*, Irving S. Kull, editor (New York, 1930), I, 207-227.
COLLINS, JOHN, *Views of the City of Burlington, New Jersey, taken from Original Sketches* (Burlington, 1847).
CONDIT, MRS. BENJAMIN SMITH, "The Story of Beverwyck," in *New Jersey Historical Society Proceedings*, new series, IV (1919), 128-141.
CUSHING, THOMAS, and SHEPPARD, CHARLES E., *History of the Counties of Gloucester, Salem, and Cumberland, New Jersey, with Biographical Sketches of their Prominent Citizens* (Philadelphia, 1883).
DECOU, GEORGE, *Moorestown and Her Neighbors, Historical Sketches* (Philadelphia, 1929).
Dictionary of National Biography (63 vols., London, New York, 1885-1900).
EDDY, GEORGE S., *A Work-Book of the Printing House of Benjamin Franklin & David Hall, 1759-1766* (pamphlet, New York, 1930).
Encyclopaedia Britannica (14th ed., London, 1937).
ELZAS, BARNETT A., *The Jews of South Carolina from the Earliest Times to the Present Day* (Philadelphia, 1905).
FISHER, EDGAR J., *New Jersey as a Royal Province, 1738-1776* (Columbia University Studies in History, Economics and Public Law, XLI, New York, 1911).
FOLSOM, JOSEPH, F., "Colonel Peter Schuyler at Albany," in *New Jersey Historical Society Proceedings*, new series, I (1916), 160-163.
GRAY, AUSTIN K., *Benjamin Franklin's Library (Printed, 1936, as "The First American Library"), A Short account of the Library Company of Philadelphia, 1731-1931* (New York, 1937).
GREEN, CHARLES F., *Pleasant Mills, New Jersey, Lake Nescochague, a Place of Olden Days, An Historical Sketch* (pamphlet, 3rd ed.).
GUMMERE, AMELIA MOTT, *Friends in Burlington*. Reprinted from the *Pennsylvania Magazine of History and Biography*, VII-VIII, 1883-1884 (Philadelphia, 1884).
HART, CHARLES HENRY, "Who Was the Mother of Franklin's Son?" in

Pennsylvania Magazine of History and Biography, XXXV (1911), 308-314.
HAWORTH, PAUL LELAND, *George Washington: Farmer; Being an account of his home life and agricultural activities* (Indianapolis, 1915).
HESTON, ALFRED M., *Absegami: Annals of Eyren Haven and Atlantic City*, 1609 to 1904 . . . (2 vols., Atlantic City, 1904).
HOEFER, M., *Nouvelle Biographie Générale* . . . publiée par MM. Firmin Didot Frères (46 vols., Paris, 1852-1866).
HORNOR, WILLIAM M., JR., *Blue Book [of] Philadelphia Furniture, William Penn to George Washington* . . . (Philadelphia, 1935).
JACKSON, JOSEPH, *Literary Landmarks of Philadelphia* (Philadelphia, 1939).
JOHNSON, ALLEN and MALONE, DUMAS, editors, *Dictionary of American Biography*, under the auspices of the American Council of Learned Societies (20 vols., New York, 1928-1937).
KEITH, CHARLES P., *The Provincial Councillors of Pennsylvania who held office between 1733 and 1776, and Those Earlier Councillors who were some time Chief Magistrates of the Province, and their Descendants* (Philadelphia, 1883).
KEMMERER, DONALD L., *Path to Freedom, The Struggle for Self-Government in Colonial New Jersey, 1703-1776* (Princeton, 1940).
KULL, IRVING S., editor, *New Jersey—A History* (4 vols., New York, 1930).
LANE, WHEATON J., *From Indian Trail to Iron Horse; Travel and Transportation in New Jersey, 1620-1860* (Princeton, 1939).
LAUFER, BERTHOLD, *The American Plant Migration, Part I: The Potato* (Field Museum of Natural History, Anthropological Series, XXVIII, no. 1, Publication 418, Chicago, 1938).
LEACH, J. GRANVILLE, "Colonel Charles Read," in *Pennsylvania Magazine of History and Biography*, XVII (1893), 190-194.
LIPPINCOTT, BERTRAM, *An Historical Sketch of Batsto, New Jersey* (pamphlet, 1933).
MCDONALD, DONALD, *Agricultural Writers, From Sir Walter of Henley to Arthur Young, 1200-1800* . . . (London, 1908).
MILLER, GEORGE J., *The Courts of Chancery in New Jersey, 1684-1696* (Perth Amboy, 1934).
MOTT, GEORGE S., "The First Century of Hunterdon County, State of New Jersey," in *New Jersey Historical Society Proceedings*, 2nd series, V (1877-79), 59-111.
NELSON, WILLIAM, *The Indians of New Jersey: Their Origin and Development; Manners and Customs; Language, Religion and Government, with Notices of Some Indian Place Names* (Paterson, 1894).

Bibliography

New Jersey, A Guide to Its Present and Past (American Guide Series, New York, 1939).
NOLAN, J. BENNETT, *The Foundation of the Town of Reading in Pennsylvania* (Reading, 1929).
———, *Neddie Burd's Reading Letters, an epic of the early Berks bar* (Reading, 1927).
———, *Printer Strahan's Book Account: A Colonial Controversy* (Reading, 1939).
OLIVER, VERE LANGFORD, *The History of the Island of Antigua, one of the Leeward Caribees in the West Indies, from the first settlement in 1635 to the present time* (3 vols., London, 1899).
PAINE, RALPH D., *The Battle of Red Bank* (compiled by the Ann Whitall Chapter, Daughters of the American Revolution, pamphlet, 2nd ed., Woodbury, N. J., 1926).
PARKER, R. WAYNE, "New Jersey in the Colonial Wars," in *New Jersey Historical Society Proceedings*, new series, VI (1921), 193-217.
PETERSON, CHARLES J., *Kate Aylesford, a story of the refugees* (Philadelphia, 1855).
PHILHOWER, CHARLES A., "The Aboriginal Inhabitants of New Jersey," in *New Jersey—A History*, Irving S. Kull, editor (New York, 1930), I, 14-53.
PRINCETON UNIVERSITY, *Princeton University Library, American Library Association Visit, June 29, 1916* (Princeton, 1916).
PROTHERO, ROWLAND E., *English Farming Past and Present* (London, 1912).
PROUD, ROBERT, *The History of Pennsylvania in North America . . .* (2 vols., Philadelphia, 1797-98).
RAWLE, WILLIAM BROOKE, "Laurel Hill and Some Colonial Dames Who Once Lived There," in *Pennsylvania Magazine of History and Biography*, XXXV (1911), 385-414.
READ, JOHN MEREDITH, "Charles Read," in *Pennsylvania Magazine of History and Biography*, IX (1885), 339-343.
ROSE, HUGH JAMES, *A New General Biographical Dictionary* (12 vols., London, 1853).
SALEM COUNTY HISTORICAL SOCIETY, *Colonial Roof Trees and Candle Ends* (pamphlet, Salem, N. J., 1934).
SCHARF, J. THOMAS, and WESTCOTT, THOMPSON, *History of Philadelphia, 1609-1884* (3 vols., Philadelphia, 1884).
SMITH, R. MORRIS, *The Burlington Smiths: A Family History* (Philadelphia, 1877).
STARR, SARAH LOGAN WISTAR, *History of Stenton* (pamphlet, Philadelphia, 1938).
STEWART, FRANK H., compiler and editor, *Notes on Old Gloucester County, New Jersey. Records published by the New Jersey Society of Pennsylvania*, I (Philadelphia, 1917).

STILLWELL, JOHN E., *Historical and Genealogical Miscellany* . . . (5 vols., New York, 1903-1932).

STRYKER, WILLIAM S., *The Reed Controversy: Further Facts with reference to the character of Joseph Reed, Adjutant General on the Staff of General Washington* (pamphlet, Trenton, 1885).

———, *The Old Barracks at Trenton, N. J.* (New Jersey Historical Society, Jan. 20, 1881, pamphlet, Trenton, 1885).

SYPHER, J. R., and APGAR, E. A., *History of New Jersey, from the earliest settlements to the present time* . . . (Philadelphia, 1871).

TRENTON HISTORICAL SOCIETY, *A History of Trenton, 1679-1929, Two Hundred and Fifty Years of a Notable Town with Links in Four Centuries* (2 vols., Princeton, 1929).

TODD, CHARLES BURR, *A General History of the Burr Family, with a Genealogical Record from 1193 to 1891* (2nd ed., New York, 1891).

WALKER, EDWIN R., "The Old Barracks, Trenton, N. J.," in *Pennsylvania Magazine of History and Biography*, XXXVI (1912), 187-208.

WATSON, JOHN F., *Annals of Philadelphia and Pennsylvania in the Olden Time; being a collection of memoirs, anecdotes, and incidents of the city and its inhabitants, and of the earliest settlements of the inland part of Pennsylvania . . . enlarged, with many Revisions and Additions*, by Willis P. Hazard (3 vols., Philadelphia, 1898).

WHEELER, WILLIAM OGDEN, *The Ogden Family in America, Elizabethtown Branch, and Their English Ancestry*, . . . (Philadelphia, 1907).

WHITE, BARCLAY, "Early Settlements in Springfield Township, Burlington County, N. J.," in *Proceedings, Constitution, By-Laws, List of Members, &c., of the Surveyors' Association of West New Jersey* . . . (Camden, 1880), pp. 62-68.

WHITEHEAD, WILLIAM A., *Contributions to the Early History of Perth Amboy and Adjoining Country with sketches of men and events in New Jersey during the provincial era* (New York, 1856).

WINFIELD, CHARLES H., *History of the County of Hudson, New Jersey, from its earliest settlement to the present time* (New York, 1874).

WOODWARD, CARL R., *The Development of Agriculture in New Jersey, 1640-1880* . . . (New Jersey Agricultural Experiment Station, Bulletin 451, New Brunswick, 1927).

———, *Agriculture in New Jersey* (reprinted from *New Jersey—A History*, Irving S. Kull, editor, New York, 1930).

WOODWARD, E. MORRISON, and HAGEMAN, JOHN F., *History of Burlington and Mercer Counties, New Jersey, with Biographical Sketches of many of their Pioneers and Prominent Men* (Philadelphia, 1883).

WRIGHT, ELIAS, *A Short History of the several Tracts of Land once Containing Iron Furnaces, &c., Lying within the boundary lines of Joseph Wharton's Lands* (pamphlet, Atlantic City, 1898).

Glossary

(First spelling as in Read's manuscript)

ADDER'S TONGUE: A fern of the genus *Ophioglossum*, so called from the shape of its fruiting spike.
ALLSPICE: The berry of the pimento or allspice tree, of the West Indies, also a mildly pungent spice prepared from it, supposed to combine the flavor of cinnamon, nutmeg, and cloves.
ALLUM (ALUM): Aluminum sulfate, either alone or in combination.
ANISEED (OIL OF): Anise oil, or oil made from the seed of anise, an herb (*Pimpinella anisum*) growing naturally in Egypt, and cultivated in many lands for its aromatic seeds.
ARPENT: An old French measure of land varying from .84 to 1.26 acres, here given as 1 3/16 acre.
ASSAFETIDA (ASAFETIDA): The fetid gum resin of various Persian and East Indian plants, used in medicine, having a strong odor and taste of garlic.
AXLETREE: The spindle or axle of a wheel.
BANDY WICKETT: A curved wicket.
BARM: Yeast formed on brewing liquors.
BATT (BAT): A crooked piece of wood (attached to the horns of oxen; see "bandy wickett").
BAUM (BALM) BEER: Beer flavored with balm, an aromatic herb.
BAY SALT: A coarse-grained variety of common salt, originally obtained from sea water.
BILGE: The protuberant part near the middle, as of a cask.
"BLACK" CATTLE: Bovine cattle in general, of any color.
BLAIN: An inflammatory swelling or sore; a blister.
BOILERS: Peas especially suitable for boiling.
BOLE: A fine, compact, soft clay, usually colored yellow, brown or black by iron oxide, formerly used as a pigment, and in medicine.
BOLSTER: A transverse bar above the axle of a wagon on which the bed or body rests.
BONNY CLABBER: Coagulated sour milk; clabber.
BREAK: An implement for the crushing or breaking of flax and hemp plants in the process of recovering the fiber.
BRISKET: In domestic animals, the breast or lower part of the chest in front of and between the forelegs.
BUDDING: The inserting of a bud into an opening in the bark of a different stock for propagating desired varieties of plants.
BURNFIRE: A bonfire.

Glossary

BUSHED IN: Harrowed in with brush harrow or by dragging brush across the field.

CALAVANCE PEAS: Probably cowpeas.

CAVEACH: To cut up and pickle fish and fry in oil.

CHINCE (CINCH): To tighten up; to girth tightly.

CHINE: A piece of the back of an animal; to cut into chines.

CINKFOIL (CINQUE-FOIL): Any of several plants of the genus *Potentilla*; also called "five-finger."

COCULUS INDIA: *Cocculus Indicus* is the trade name for berries of the cocculus that are imported from the East Indies to adulterate porter. They are used by the Chinese in catching fish. The berries contain an acrid poison which intoxicates or stuns the fish until they can be caught.

COILERS: Part of the harness of a cart-horse; the breeching.

CONVOLVULUS: A genus of erect or trailing or twining herbs and shrubs, including common species popularly called bindweed, related to the morning glory and the sweet potato.

COSTESE (COSTIVE): State of constipation; presumably medicine to relieve this condition.

CREAM OF TARTAR: Potassium tartrate or bitartrate, a white crystalline substance with a gritty, acid taste, used as an ingredient of baking powder.

CRIBBS (CRIBS): Mangers for feeding animals.

CROW GARLICK: The wild onion, *Allium vineale*.

CRUPPER: A leather loop passing under a horse's tail and fastened to the saddle to keep it from slipping.

CUT: To castrate.

CUVE: A tub of wine.

DEW-ROTT (DEW RET OR DEW ROT): To ret, as flax or hemp, by exposure to rain, dew, and sun.

DOCK: Coarse weedy herbs, with crowded panicled racemes of small, inconspicuous, mostly green flowers, and leafstalks somewhat sheathing the stem.

ELIXIR PROPRIETATIS: Tincture of aloes and myrrh, commonly abbreviated *elixir pro.*

FELLY: Segment of the rim of a wooden wheel; sometimes, in familiar speech, the entire rim.

FENNY: Boggy, marshy; pertaining to, inhabiting, or grown in a fen.

FESS (FESSE): Divided horizontally through the middle, usually with one tincture above the line and a different one below, as in heraldry.

FILLY: A female foal or colt, a young mare.

FIRKIN: A small wooden vessel or cask of indeterminate size used for butter, lard, etc.

FLAIL: An implement used for threshing or beating grain from the ear by hand, consisting of a wooden handle, at the end of which a stouter and shorter stick called a swingle is so hung as to swing freely.

FLUMMERY: A soft, jellylike food made of flour or meal; a kind of custard.

FODDER: Leaves and tops, dry-cured, for use as stock feed.

FOGG: Long grass remaining in pasture until winter.

FOWL MEADOW GRASS: A grass similar to redtop which thrives on wet meadow lands.

FREESTONE: Light chocolate in hue; colored like Connecticut freestone.

FRENCH MALLOW: The tree mallow, *Lavatera albia*.

FRIET (FRIT): To decompose and partly melt.

FUNDAMENT: The buttocks; also the anus.

GAMBRELL (GAMBREL): A stick or iron, crooked like a horse's hind leg, used by butchers in suspending slaughtered animals.

GAMMON: A ham or flitch of bacon salted and smoked or dried.

GARGUTT (GARGET): A disease in swine and cattle marked by inflammation of the head or throat. A diseased condition of the udders of cows, arising from the inflammation of the mammary glands.

GRAFT: To insert a cion (or bud) into a stem, root, or branch of another plant, so that a permanent union is effected, especially for purposes of propagation.

GUDGEON: An iron or steel pivot fixed in the end of a wooden bar or shaft.

GUM: To exhude a colloidal substance commonly called gum.

HARLE (HARL): The filaments of flax or hemp.

HASSOCKS: Rank tufts of bog grass or sedge; tussocks.

HATCHELL (HATCHEL): To draw through a hatchel, an instrument with long iron teeth set in a board for cleansing flax or hemp from the coarse and refuse parts.

HAULM: Collectively the stems or stalks of cultivated plants, especially after the crop has been gathered; straw or litter.

HELLEBORE: The powdered root of the American white, or swamp hellebore (*Veratrum viride*), used for destroying vermin.

HERD GRASS (HERD'S GRASS): Also called redtop, an upright perennial grass (*Agrostis alba*) distributed widely, used especially in lawn mixtures.

HOCK: Tarsal joint or its region in the hind limb of horses and other quadrupeds, corresponding with the ankle of a man.

HOVES (OF CATTLE): Swelling of the stomach due to excess gas caused by overeating of green clover or grass.

HURDLE: A screen, or sieve.

IMPERIAL WATER (also IMPERIAL): A beverage made of hot water, sugar, cream of tartar, and lemon peel.

INOCULATING: Grafting by inserting buds; budding.

ISINGLASS: A semitransparent, whitish, and very pure form of gelatin,

Glossary

originally prepared from the air bladders of certain fish, used for making jellies, and glue, as a clarifier, etc.

JOULE (JOWL): Under-jaw, or cheek.

LAY DOWN: To plant or sow a field with a certain crop.

LAZY-BEDS: Beds in which potatoes are grown, formed by placing the potatoes on the surface of the ground, and covering them with earth from trenches dug at the sides of the bed.

LEA (LEY): Meadow, pasture.

LIMEKILL (LIMEKILN): A kiln or furnace in which limestone or shells are burned and reduced to lime.

LINSEY: Linsey-woolsey: Coarse cloth made of linen and wool, or cotton and wool.

LITHARGE: Lead monoxide.

LIVRES: $13\frac{1}{2}$ livres = £1.

LOCK: A handful, armful, or small bundle, as of hay.

LOOMY (LOAMY): Of the nature of loam, a soil consisting of a friable mixture of clay, sand, and organic matter.

LUCERNE: Alfalfa (*Medicago sativa*), a deep-rooted long-lived perennial leguminous forage plant, a native of southwestern Asia, in use centuries before the Christian era.

MACE: A kind of spice consisting of the dried arillode or external fibrous covering, of the nutmeg.

MADDER: A herbaceous perennial (*Rubia tinctorum*), with long fleshy roots which are ground to powder; formerly important source of turkey red dye.

MAGMA: A suspension of a large amount of precipitated material in a small volume of liquid.

MANGER: A trough or open box in which fodder is placed for horses or cattle to eat.

MARL: An earthy crumbling deposit consisting chiefly of clay mixed with calcium carbonate in varying proportions; greensand marl.

MEAD: A fermented drink made of water and honey with malt, yeast, etc.; metheglin.

METHEGLIN: A beverage made of fermented honey and water; mead.

MILLET: Several varieties of small-seeded cereal and forage grasses, belonging to the genus *Panicum* or to closely allied genera; among the most ancient of food grains, generally grown in this country as a supplementary or catch-crop.

MORTICE: A cavity, usually rectangular, into or through which some other part or any arrangement of parts fits or passes; a cavity cut into a piece of timber to receive a tenon.

MOULD: To cover with "mould" or soil; to ridge.

MUSCOVADO: The dark, moist, impure sugar obtained by evaporating cane juice and drawing off the molasses; unrefined sugar.

MUZZLE: A plow clevis.

Glossary 447

NECK-YOKE: A bar by which the end of the tongue of a wagon or carriage is suspended from the collars of the harness.

N[EW] S[TYLE]: The "new style" calendar, adopted about 1750.

NITRE: Saltpetre.

O[LD] S[TYLE]: The "old style" calendar, under which the year began in March.

OSNABRIGGS (OSNABURG, OZENBRIGGS): A species of coarse linen, a stout coarse cotton fabric used for overalls, sacking, etc.

PENNYROYAL: A perennial mint plant yielding the commercial oil of pennyroyal.

PETRE (PETER, PETRE SALT): See saltpetre.

PIPKIN: A small earthen pot, usually one having a horizontal handle.

PLAISTER: Plaster.

PLATE: A horizontal member at top or bottom of a stud partition between which the studs are placed.

PLIGHT: Condition (as "in good plight").

POKE: Pokeweed, a coarse American perennial herb (*Phytolacca Americana*) with white flowers and dark purple juicy berries.

POLL: The blunt or round end or butt of a hammer or axe.

POMPIONS: Pumpkins.

POT-BELLIED: Having a protuberant belly.

QUERE (QUAERE): An inquiry to be resolved; a question.

RABITT (RABBET): A recess or rectangular groove made in a piece of wood to receive the edge of another cut to fit it.

RACE (OF GINGER): A root or a sprig of ginger.

RAGG (RAG) WHEEL: A wheel with tooth-like projections for engaging with the links of a chain; a sprocket wheel.

RATSBANE: Rat poison, specifically white arsenic (arsenic trioxide).

RAY GRASS (RYE GRASS): A perennial grass (*Lolium perenne*) with spikelets borne in zigzag spikes, useful in meadows and pastures; also called "English meadow grass."

RED OCHER: A red, earthy variety of hematite (iron oxide) used as a pigment.

RENNETT (RENNET): An extract prepared from the stomach of calves used to curdle milk.

ROTT (ROT): Any of a number of parasitic diseases which chiefly attack sheep and are characterized by rotting or necrosis of the tissues.

ROWEN: A stubble field left unplowed until late in the autumn, to be grazed by cattle; a second growth crop, or aftermath.

RUE: A strong scented perennial woody herb (*Ruta graveolens*) having yellow flowers, with a bitter taste, used in medicine.

RUNNET: Rennet.

RUSHES: Any of various plants of the genus *Juncus*, having cylindrical, often hollow stems.

Glossary

RUTTING TIME: Reproductive period in animals; time of being "in heat."

SADDEN: To render firm, solid or thick, as soil by treading.

SAFFRON: A species of crocus (*Crocus sativus*) with purple flowers, cultivated for the drug and dyestuff it yields; a deep orange-colored substance consisting of the stigmas of the plant, used as a dyestuff, to flavor foods, and in medicine.

ST. FOIN (SAINFOIN): A leguminous perennial forage plant (*Onobrychis sativa*—Holy Clover), with rose colored flowers and deep penetrating roots, introduced into England in the Seventeenth Century.

SALLAD (SALAD): An early term commonly applied to lettuce.

SALTPETRE: Potassium nitrate, nitre; when in its native state, sometimes called petre salt; when refined, saltpetre.

SEDGE: Plants of the genus *Carex*, grasslike herbs, often growing in dense tufts in marshy places.

SHRUBB (SHRUB): A liquor composed of fruit acid, especially lemon juice and sugar, usually with spirit to preserve it.

SILK (OF CORN): The silky styles on an ear of Indian corn.

SILL: A horizontal timber which forms the base of a frame.

SKIPPERS: The larvae of the cheese fly, a black dipterous insect (*Piophila casei*) which in the larval state lives in cheese, ham, and smoked beef; it can jump several inches.

SLACKING (SLAKING): To cause (lime) to heat and crumble by treatment with water; to hydrate; also to alter (lime) by exposure to air.

SLINK (FOAL): To give birth to (a foal) prematurely.

SMUTT (SMUT): Diseases of plants caused by parasitic fungi characterized by black, often dusty, masses of spores.

SOUCE (SOUSE): Certain cuts of meat (e.g., fish, or pig's feet or ears) steeped in pickle.

SPEARGRASS: Kentucky bluegrass (*Poa pratensis*).

SPITT (SPIT): The depth of a spade thrust.

SQUILLS: A bulbous herb (*Urginea scilla*), having long racemes of small white flowers; the cut and dried fleshy inner scales; the red variety used chiefly as rat poison.

STOCK: The trunk or main stem of a tree or other woody plant; a stem from which slips are made.

STOOL: To send up shoots or suckers.

STROUD: A blanket manufactured for barter or sale in trading with the North American Indian, or the material of which these blankets were made.

SURCINGLE: A belt, band or girth passing over a saddle or over anything on a horse's back to bind it fast.

SWEET MARJORAM: An aromatic European herb (*Majorana hortensis*) closely related to the wild marjoram.

SWILL: A semi-liquid food for animals, especially for swine, composed

Glossary

of animal or vegetable refuse from kitchens, markets, etc., mixed with water or milk.

SWINGLE: A wooden instrument like a large knife about 2 feet long with one thin edge used for beating and cleaning flax.

TAILINGS: Chaff and residue separated in the first grinding.

TAMARINDS: The fruit of a tropical tree, a brown shelled pod 3 to 6 inches long, used in making a cooling beverage and in cookery; when pressed in sirup or sugar, the "preserved tamarind" of commerce.

TANNER'S BARK MILL: A mill in which bark is ground for tanning.

TARES: Any of several vetches; tare vetch.

TENNANT (TENON): A projection, properly of rectangular cross section, at the end of a piece of timber, to be inserted in a socket or mortice in another timber to make a joint.

TOP (CORN): To cut off the upper part of the corn stalk.

TOSLE: Tassel.

TREACLE: Originally a medical compound of various ingredients deemed an antidote for poisons; the sirup obtained in refining sugar; a saccharine fluid consisting of juices of certain plants.

TREFOIL: Bird's foot clover (*Lotus corniculatus*), a perennial clover-like plant with a long taproot, valuable for pastures on light dry soils.

TRUNNEL (TREENAIL): A slender piece of hard wood used in fastening together timbers; a wooden nail.

UMBER: A brown hydrated ferric oxide, containing manganese oxide and clay, used as a pigment.

VIGNERON: A wine grower; one who cultivates vines.

WAIRE (WEIR): A large fixed fish trap consisting of fences of stakes or matted brush, sometimes with netting, forming successive enclosures, to prevent escape.

WEN: An indolent encysted tumor, occurring commonly on the scalp.

WHEY: Clear straw-colored liquid, consisting of water and milk-sugar, that remains when casein and other ingredients in milk are coagulated as in making cheese.

WHITE VITRIOL: Zinc sulfate, a crystalline compound used as an astringent and an emetic.

WORT: The unfermented infusion of malt that when fermented becomes beer.

Index

Academy of Sciences at Paris, 416
Admiralty courts, 195, 196
Affane cherry, 295
Agricultural literature, 46-48, 413-430
Agriculture, xi-xxiii, 14-16, 70-85, 104-106, 112, 116, 151, 157, 159, 183, 184, 221, 224, 229-384
Aitkin, Charles S., xxiv
Aix-la-Chapelle, Peace of, 167
Albany, N. Y., 69, 165, 169
Albany Convention, 167
Ale, 110
Alexander, James, 51, 100, 102, 103, 138
Alexander, William, see Stirling, Lord
Alfalfa, 254, 274-276
Allen, Justice John, 136, 197, 198
Allinson, Caroline, xxv
Allinson, Samuel, 69, 143, 144, 199, 200, 214, 221
Almonds, 263, 313
Alum, 285
American Antiquarian Society, xxv
American Husbandry, 12, 232, 233, 235, 258, 259-262, 266, 322, 385
American Philosophical Society, 21, 49, 224
American Weekly Mercury, 33
Anderson, Enoch, 128
Animal breeds, 328
Animal husbandry, 322-365
Annard, Alexander, 27, 30, 404
Anne, Queen, 9
Antigua, xiv, 32, 40, 216, 218, 298
Antill, Edward, 110
Antimony for hogs, 354
Apple brandy, 56
Apple trees, 75, 76, 184, 265, 320
Apples, 106, 230, 231, 262, 263, 319
Apricots, 263, 312
Arthur, Captain, 41
Arthur, John, 70
Ash, 320
Ashes, 79, 244, 245, 249, 250, 300, 315
Ashfield, Lewis M., 148
Askwith, Capt. Samuel, 351
Aslin, Mary S., xxv
Asparagus, 261, 293, 304, 305
Assembly of New Jersey, 9-11, 17-20, 98, 99, 102, 116-144, 147, 150, 152, 158, 161-178, 185, 207, 208, 223

Atsion, 15, 88, 95, 96; River, 67, 88
Austin, Amos, 281
Austin, Caleb, 281

Bache, Franklin, xxiv
Bacon, 80, 231, 326, 354, 356, 357, 359, 388
Bainbridge, John, 287
Bakehouses, 59
Ball, John, *The Farmers' Compleat Guide*, 254, 274, 275, 413
Bancker, Gerard, 158
Bancroft, George, 220
Banking meadows, 79, 161
Bantam breed of hogs, 355
Baptists, 5
Barbadoes, 57
Bard, Peter, 405
Barley, 213, 229, 231, 241, 252, 268, 272, 283, 284, 287, 303, 324
Barm, 389, 390
Barnett, Claribel R., xxiv
Barns, 368
Barracks, 18, 78, 175-177, 368, 370-372; supplies, 162, 163
Bars, 374-377
Bartram, John, 26, 31, 49, 83, 107, 242, 252, 253, 303, 350, 424
Basking Ridge, 157, 158
Bath, N. C., 217
Batsto, 15, 88, 93-96, 337, 365; Creek, 88, 93, 94
Battle of Trenton, 176
Baum beer, 386, 393
Beach, 319
Beans, 232, 259, 287, 293, 294
Bear Key Inn, 79, 287
Bear Swamp, 68
Bearskin (jacket), 385, 395, 396
Beaver Pond, 67, 70
Beck, Henry C., xxv
Becker, E. Marie, xxv
Bedford, Duke of, 103, 104
Beef, 11, 56, 57, 60, 170, 232, 324, 337, 338, 387, 388; curing, 340
Beer, 110, 390
Bees, 366, 367; wax, 366, 367
Beets, 302, 317
Belcher, Andrew, 392

Index

Belcher, Jonathan, xv, 10, 14, 17, 19, 40, 62, 80, 87, 100, 101, 103-137, 146, 151, 161, 167-170, 173, 180, 197-200, 231, 334, 386, 392-395
Belleville, 272, 368
Bergen, 6
Bergen County, 4, 5, 57, 86, 196, 205; militia, 280
Berkeley, Lord, 7, 8
Berks County, Pa., 43
Bermuda, 295, 303; potatoes, 259, 296
Bernard, Francis, 48, 137-139, 143, 148, 154, 155, 184, 185, 187-189, 201
Berrien, John, 202, 203, 210
Beverages, 385, 386, 390-394
Beverwyck, 230
Bibliography, 413-442
Biddle, Joseph, 325, 344, 345
Billeting of troops, 174, 175
Bills of credit, see Currency
Birds, protecting corn against, 280
Bispham, Joshua, 69, 326, 352
Bispham, Thomas, 79
Black, Harry, 343, 344
Black, Samuel, 70
Black, Thomas, 325, 343
Black, William, 343
Black Bantam hogs, 354
Black Point, 346
Black pudding, 386
Blacksmiths, 55
Blackwood, Mr., 199, 200
Blight, 314, 315
Blissard, Dorothy, 32
Blockhouses, 169
Blood as a manure, 250
Bluegrass, 81
Board of Trade, 13, 104, 145, 149
Boarding School, Burlington, 48
Bogs in Ireland, 244
Bond, Anne, 23, 24, 404
Books, 29, 30, 45
Boone, Thomas, 15, 137, 138, 405
Borden, Joseph, Jr., 60, 61
Bordentown, 6
Borton, William, 307
Boston, 28, 105; fire, 142
Bound Brook, 141; bridge, 142
Bounties, 135, 157-159; on hemp and flax, 158, 159, 258; on silk, 158
Bow Hill, 278, 279
Bowlsby, Thomas, 245, 260, 300, 362
Boycott, 151
Boyd, Julian, xxiv
Brabant, 267

Braddock's Bridge, 67
Braddock's defeat, 168
Bradford, Andrew, 25, 33
Bradford, William, 135
Bradley, Richard, *A General Treatise of Husbandry and Gardening*, &c., 254, 263, 265, 268-271, 312, 313, 315, 319, 342, 343, 347, 348, 351, 354, 356, 357, 366, 367, 414; *New Improvements of Planting and Gardening*, 46, 261, 284, 285, 304, 305, 308, 309, 310, 413, 414; *The Gentleman and Farmers' Guide*, 324, 335, 341, 342, 347, 354-356, 362, 415; biographical note, 415
Brainerd, David, 180, 181
Brainerd, John, 180-183
Bramham, 345
Branin, Anne, 212
Branin, Elizabeth, 212
Branin, Michael, 212
Bread, 56, 57, 59, 287
Bread and Cheese Run, 184
Breeding chickens, 365; dogs, 365; hogs, 354, 355; horses, 328, 329; sheep, 362, 364, 365; principles of, 328
Breezy Ridge, 14, 76-79, 84, 351
Breweries, 56, 59, 110
Brewers' yeast, 389, 390
Brick, John, 49
Brickyards, 56
Bridges, 140, 160, 161
Brimstone, 286
Bristol, Pa., 49
British Board of Agriculture, 233
British troops, 165, 167, 168, 174-177, 219, 220, 223
Brotherton, 19, 95, 182-186
Brushfield, T. N., 295
Buck-bean, 26, 27
Bucket engine, 369, 370, 381-384
Bucks County, Pa., 26, 34, 61, 86
Buckwheat, 229, 249, 255, 257, 273, 283, 286, 363
Budding, 261, 263, 311, 314, 315, 319
Bullock, Anthony, 353
Bullock, Joseph, 326, 353, 354
Burlington, xix, 3, 4, 6, 7, 10, 11, 20, 28, 37-39, 46, 48, 49, 71, 78, 90, 99, 102, 111, 131, 140, 146, 210, 212, 216-219, 404; port of, xv, 54, 57-62
Burlington barracks, 175, 176; fair, 82, 83; library, 21, 224; militia, 169, 219; residences, 42
Burlington County, 4, 14, 15, 57, 64, 65, 68, 84, 87, 99, 196, 200, 212

Index

Burlington Society for the Promotion of Agriculture, etc., 238
Burnaby, Andrew, 143, 318
Burnet, William, 10
Burning clay, 243, 249; grass, 244
Burr, Aaron, 115
Burr, Joseph, 65, 88, 364, 365, 376, 390
Burr, Mrs. Joseph, 390
Butchering pork, 357; sheep, 362, 363
Butter, 57, 59, 61, 80, 109, 231, 261, 277, 322, 325, 326, 342-346, 348-350
Buttermilk, 347
Buttonwood, 265, 319

Cabbage, 84, 231, 260, 262, 302-304, 310, 314, 350; roots, 317
Cadwalader, Dr., 99
Cadwallader, James, 220
Calavance peas, 261, 305-308, 332
Calendar of gardening, 293, 294
Calves, 324, 334-340, 343-345
Calvin, Stephen, 185
Cambridge University, 415, 421, 422
Campion, H. Clifford, xxv, 367
Canada, 164, 171
Canadian expedition, 100, 164, 169
Canary Islands, 54, 295
Candles, 122
Cap, 397
Cape May, 4, 50, 52, 54, 135; militia, 166, 300
Cape May County Historical and Genealogical Society, xxv
Carman, Harry J., xxv, 233
Carnegie Library, Pittsburgh, xxv
Carolina cattle, 331
Carolina Gazette, 291
Carp, 400
Carrots, 250, 262, 301, 346
Carteret, Philip, 8
Carteret, Sir George, 7, 8
Cassianus Bassus, 317
Catalpa, 265, 320
Caterpillars, 319
Catfish, 386
Catholics, 5
Cato's Letters, 45
Cattle, 49, 73, 74, 78, 80, 136, 188, 221, 230-232, 247, 260, 269-272, 275, 277, 283, 284, 313, 314, 320, 322-325, 331-350, 367, 372, 378; diseases, 341, 342; feed, 271-273, 299, 301, 303, 304, 306, 324, 325, 332-340; fodder, 333, 334; wild, 324
Cauliflower, 84, 231

Cedar, 265, 320, 376; swamps, 64-66, 87, 88, 221
Celery, 302
Chaise, 80, 108, 369, 377
Chambers, Alice T., xvi
Chambers, Charles, xvi
Chambers, Elizabeth, xvi
Chambers, Jane, xvi
Chambers, Lucy Ann, xvi
Chancellor, 208-210
Chancery court, 196, 208-210
Chancery, master of, 199
Charity School, 43
Charles I (of England), 423
Charles II (of England), 31
Charleston, S. C., 32
Check list of libraries, 428-430
Cheddar cheese, 347
Cheese, 57, 59, 80, 82, 231, 322, 325, 326, 340, 342-348
Cheese house, 374
Cherries, 76, 230, 231, 262, 263, 311-314, 319
Chesnick, Barney, xxiv
Chestnuts, 107, 265, 320, 374, 375
Chew, Benjamin, 188
Cheyne, George, *An Essay of Health and Long Life*, 351, 415, 416; biographical note, 416
Child, Amy, 22
Chocolate, 385
Christ Church, Philadelphia, 24, 292
Church Lane, Burlington, 42
Church lotteries, 141
Church of England, 22, 23
Cider, 55, 56, 106, 109, 110, 230, 314, 386, 392, 393
Cider royal, 392
Citrus fruits, 318
Clams, 184
Claret, 110
Clay, xvii, 71, 242, 243, 245, 249-252, 273, 283, 310, 315
Clayton, David, 348
Clinton, Governor, of New York, 165, 166
Clothing, 385, 395-397
Clover, xvii-xxii, 71-73, 76, 79, 80, 107, 237, 241, 244, 249, 254-257, 266-273, 277, 330, 342, 350
Clover seed, 267-271, 278
Coates, Christopher, 101, 153
Cocculus Indicus, 397, 398, 400
Cockerels, 365
Coffee, 59, 385
Cohansey, 166

454 Index

Colden, Mr., 51
Coles, Mary, 345
Coles, Samuel, 242, 265, 320
College of New Jersey, 113-116, 142, 161, 224, 280, 290
Collinson, Peter, 26, 31, 44, 83, 97, 242, 415, 424
Colonial colleges, 161, 162
Colonies, union of, 130, 131
Colts, 322, 323, 328, 329
Columbia University, xxv
Commerce, xv, 6, 11, 49, 54-63, 112, 149, 150, 159, 160, 192, 195, 196, 258, 259
Commissions, 123, 155; for trying pirates, 195, 196, 201; on military supplies, 165, 166, 169, 170; to treat with Indians, 181-184, 187
Committees in council, 161; of correspondence, 149, 173, 174; on observation for Burlington, 219
Compost, 248
Conarro, Andrew, 76
Conarro, Isaac, 105
Conferences at Burlington, 183; Crosswicks, 181, 182; Easton, 186-189
Congregationalists, 5, 6
Conscription, 173
Conservatory, 318
Continental army, 220; Congress, 217, 395
Convolvulus, 293, 294
Cooke, M., *The Manner of raising, ordering and improving Forrest Trees*, 47
Cooke, William, 166, 236, 239
Cooper, Henry, 332
Cooper, Isaac, 333
Cooper, John, 94
Cooper-Hewitt Works, 26
Cooperage, 55
Copper mines, 86, 278
Corn, 71, 107, 159, 184, 245, 247, 248, 250, 260, 266, 268, 276, 322, 326, 350, 380; cultivation of, 278-280; soaking for feed, 350; topping, 279, 280; see also Indian corn
Cornbury, Lord, 10
Cornwall, 3
Cosby, William, 10
Cotoxing Creek, 66; Pond, 67
Cotton, 395, 396
Council, New Jersey, 9, 17-20, 51, 97, 99, 100, 102, 103, 116, 118, 138, 145-163, 170, 177, 178, 189, 197, 199, 208, 210, 213, 215, 223
Counterfeiting, 196
County seats, 205, 206

Courts, 195-211; clerk, 10, 39, 102; fees, 196
Cowgill, Rachel, 325, 345, 348, 365
Cowpeas, 261, 305-308
Cowperthwaite, Hugh, 75
Cows, 106, 206
Cowsheds, 372
Cox, John, 96
Cox, Mr., 197
Cox, Mrs., 397
Coxe, Daniel, 148-152
Craftsmen, 56
Cranberries, 57, 181, 263, 266
Cranbury, 181
Crane, Stephen, 182
Cress, 293, 294
Crime, 206
Crispin, Jonathan, 377
Crooked Billet wharf, 49
Crop rotations, 238
Crops, 143, 144, 254-321
Cross Roads, 220
Crosswicks, 19, 181, 182
Cucumbers, 259, 261, 293, 294, 309, 310
Cumberland County, 49, 128, 196, 219; militia, 166
Currants, 263, 312; wine, 390, 391
Currency, 18, 116, 131-133, 139, 143, 144, 148, 149, 151, 153, 159, 164, 165, 169, 171-173, 223
Customs collector, 10-12, 37, 54-63; fees, 60, 151

Dairy, 231, 322; cattle, 331-350; products, 322, 324, 325; see also Butter, Cattle, Cheese, and Milk
Davis, David, 242
Davis, Martha, 242
Davis, Samuel, 239, 240
Dealy, Joseph, 78
Death penalty, 206, 207
Debts, 215
Declaration of Independence, 215
DeCou, George, xiii, xxiv, xvi
Deer, 106, 318; dung, 319; trapping, 140, 180, 181, 206
Delaware Bay, 4, 166; River, 4, 6, 13, 15, 26, 59, 64, 170, 177
Delaware Indians, 179, 187
deMauduit, M., 354
DeNormandie, John A., 49, 65, 238
Deputy secretary, 97-120
Dey, Derrick, 280
Dey (Dye) Tunis, 81, 159, 280
Dey mansion, 280

Index

Diancourt, Captain, 275
Diaries of George Washington, 233
Dickinson, Jonathan, 113
Dikes, 318
Distilleries, 55
Ditches, 72, 79, 237, 252, 380-384
Dock, 253
Dougharty, William, 290
Doughty, Daniel, 46, 75, 324, 325, 334, 342, 345
Douglass, John, 83, 84
Douglass, Thomas, 389
Douglass, Mrs. Thomas, 389
Dove, David J., 400
Drainage, 161, 380-384
Drill, 370-381; plow, 81
Drum, Hugh, 78
Dublin Society, &c., 390
Ducks, 232, 328, 364
Duhamel Dumonceau, H. L., *A Practical Treatise of Husbandry*, 254, 260, 273-277, 301, 361, 384, 416; biographical note, 416
Dung, 245-248, 251, 252
Dunker, Henry, 96
DuPratz, LePage, *The History of Louisiana*, 298, 416; biographical note, 416, 417
Durham boats, 61; ironworks, 26
Dutch Reformed Church, 5, 162
Dutch settlers, 5-7, 18
Dye plants, 257-259, 290-293

Earlham, 239
Earmarks of cattle, 78
Easton, 179, 186-189; conference, 19; treaty, 19
Eddy, George Simpson, xxiv
Edge Pillock, 67, 183
Edgon, Captain, 52
Edinburgh Society, etc., 421
Edmunds, John, 243, 244
Eels, 386, 400
Egg Harbor, 88; cattle, 324, 331
Egg production, 365
Elder, 252, 314
Eliot, Jared, xi, xvi-xxi, 70-73, 81, 86, 238, 255, 263, 268, 270, 303, 308, 309, 318; biographical note, 417, 418; *Essays upon Field Husbandry . . .* , xvi, xviii, 70, 232, 237, 246, 251, 254, 255, 257, 259, 260, 262, 265-271, 277, 283, 299, 301, 323, 324, 330, 335, 350, 351, 370, 385
Eliot, John, 417
Eliot, Joseph, 417

Elixir proprietatis, 386, 395
Elizabethtown, 6, 109, 111, 126, 129, 132, 134, 146, 171; barracks, 175, 176
Ellis, Daniel, 69, 215, 216
Ellis, William, *The Modern Husbandman . . .* , 301, 418; *The Practical Farmer*, 46, 241, 243, 244, 247-251, 254, 255, 258, 260, 268-270, 275-277, 284, 285, 290, 324, 334, 340, 341, 354, 355, 361, 418; biographical note, 418, 419
Elm, 265, 320
Elton, William, 308
Embargo, 62
English law, 209, 210; settlers, 5, 6
Epidemic, 111
Episcopalians, 5
Equity, 210
Erwin, John, 77
Essex County, 4, 57, 205
Etna, 15, 88, 92, 96, 212, 214, 220
Evans, Dr. Cadwallader, 213
Evans, Issac, 96
Evans, Lewis, *Map of Pennsylvania . . .* , 75
Evelyn, John, 419
Everlasting grass, 269
Evesham Township, 64, 90, 92, 183
Ewan, Nathaniel R., xxv
Exports, 11, 12, 56, 57, 87, 158, 167, 231

Faden, William, Map of New Jersey, 75, 90
Fairs, 82, 83
Farm demonstration, 263, 264; engineering, 368-384; equipment, 368-370, 377-384, 407, 408, 410; implements, 80, 368-370, 377-384, 407, 408, 410; inventories (estates), 238-240, 242, 245, 246, 253, 267, 278, 279, 281, 296, 300, 302, 306, 307, 331, 332, 336, 338-340, 343-346, 353, 365, 391, 407-411; machinery, 368-370; 377-384; management, 265, 266; property assessed for taxes (tax lists), 240, 253, 290, 296, 297, 352; structures, 368-377; utensils, 221
Farms and houses for sale, descriptions of, 239, 240, 242, 249, 253, 271, 272, 279, 287, 291, 297, 302, 306, 308, 318, 331-333, 344, 345, 352, 353, 359, 364, 400
Farms of Charles Read, xi-xxii, xxiv, 64-85
Feather grass, 73
Feeding cattle, 271-273, 324, 325; 332-334; colts, 329; hogs, 326, 350-355; sheep, 361, 362

Index

Fees, court, 196; legal, 152; of secretary, 102
Fencing, 74, 252, 369, 374-377; posts, 74, 375-377; splicing rails, 376, 377
Fenimore, Joshua, 238
Ferment, 390
Ferries, 160
Fertilizers, see Manures
Fever, 109
Fibre crops, 257-259, 287-290
Financial depression, 143, 144, 146, 151, 152, 212
Finnish settlers, 4
Fire Insurance Company, 303
Firewood, 107, 109, 115, 122
Fish, 59, 83, 106, 386, 387; cooking, 387; curing, 386; nets, 399; sauce, 387
Fisher, Hendrick, 132, 162
Fisher, Thomas, 215
Fisheries, xvi, 15, 53, 84, 85, 184, 399-403
Fishing weirs, 84, 399, 400
Flail, 369; thongs, 380
Flanders, 267-330
Flax, 55, 146, 158, 159, 230, 246, 257-259, 268, 273, 290
Flaxseed, 283
Fleas, 396, 397
Flour, 55, 57, 59-61, 93, 136, 169
Flummery, 389
Flying Camp, 219
Fodder, 333, 334
Foods, 385-390
Forage crops, 254, 255, 266-278
Forests, 13, 14; conservation, 139, 140; trees, 265, 319-321
Fork, 270
Fort, Marmaduke, 259, 260, 296
Fortifications, 169, 170
Foster, William, 46, 69, 181, 325, 339, 340, 344, 369, 377
Fowl meadow grass, xvi, 81, 254, 266
Fowls, 365
Franklin, Benjamin, xi-xiii, xv-xxii, 20, 28-30, 33, 36, 39, 41, 43, 44, 49, 52, 70, 84, 111, 112, 146, 149, 151, 153, 156, 167, 190, 193, 213, 223, 224, 252, 263, 268, 303, 308, 309, 318, 337, 399, 400, 415, 422, 424
Franklin, Sarah, 52
Franklin, William, xi, 10, 11, 19, 20, 94, 137, 145-147, 150-152, 156-162, 177, 178, 190, 200, 202, 203, 209, 210, 212, 213, 215, 395
Franklin & Hall, 190
Franklin Institute, xxv

Franklin's printshop, 29, 30, 45; purchases from, 39, 40
Freeland, Rebecca, 23
Frelinghuysen, Theodore, 6
French and Indian War, 16, 18, 116, 129, 130, 139, 143, 144, 147, 150, 164-178, 180, 186, 223
French naval forces, 166
Friendly Association, etc., 187
Frontier, 168-171; defense, 173, 177, 186-189, 223
Fronto, Marcus Cornelius, 317
Frontoniani, 317
Fruit, 59, 230, 231, 385; culture, 262-264, 311-319; trees, 78, 106, 107, 230
Fulling mills, 55
Fur hat, 364
Furs, 57, 192

Gallows, 206
Gambling, 141
Gambold, J., 47
Gammons, 358-360
Gardening, 84; calendar, 293, 294; see also Vegetables
Gardiner, Thomas, 64
Garlic, 350
Gates, 369, 374, 375
Gates, Horatio, 203
Geese, 221, 232, 327, 363, 364
General Loan Office of Pennsylvania, 34
Gentleman's Magazine and Historical Chronicle, 297, 298, 335, 389, 390, 419
Geoponica, 317
George II, 145, 207
George III, 145, 207
German settlers, 25
Germantown, 27; Academy, 400
Gibbon, Mr., 66
Glass, 55
Glassworks, 135
Globe mill, 308
Glossary, 443-449
Gloucester, N. J., 6, 79, 236, 401
Gloucester County, 4, 5, 57, 64, 65, 67, 68, 82, 87, 161, 178, 200, 205, 219; militia, 169
Glue, 399-403
Goats, 80, 327, 363
Goatskins, 363
Gooseberries, 263
Goshen Neck, 67
Grafting, 78, 263, 312-315; grapes, 316

Index

Grain, 57, 61, 75, 146, 158, 184, 229, 230, 255, 257, 278-287, 326, 329; drill, 81; prices, 287; processing, 286; steeping, 257, 284-286; 330
Grapes, 106, 315-317, 386; culture, 264
Grass, 72, 73, 81, 246, 332-334; seed, 76, 79, 81, 254, 278
Gravel, 244
Gray, Austin, xxiv
Gray, Charles, 114, 231
Great Timber Creek, 67
Green Bank, 291
Green Spring, 308
Greene, Belle de Costa, xxiv
Greene, General, 283
Greene, Maud H., xxiv
Greenhouses, 264, 318
Greyhound, 365
Griffin, Samuel, 220
Griffith, John, 296
Grindstones, 369, 379
Gristmills, 55, 56, 93, 135, 184, 221, 286
Grouse, 232
Grover, James, 86
Guinea fowls, 81, 82, 327, 328, 364
Gull, The, 216

Hackensack River, 4, 57
Haight, Joseph, 216
Haines, Caleb, 245, 392
Haines, Caroline, xxv
Haines, George, 245, 344
Haines, Margaret, xxv
Haines, Reuben, 94, 245, 337, 365
Haines, Thomas, 344
Haines Creek, 90
Hainesport, xiii, xvi
Hale, Thomas, *A Compleat Body of Husbandry*, 47, 241, 247-250, 254, 260, 269, 270, 272, 273, 276, 285, 286, 295, 300, 301, 324, 330, 337, 348-350, 370, 380, 419; biographical note, 419
Halifax, Lord, 201
Hamilton, John, 100, 164
Hampton, Jonathan, 332, 334
Hampton, Va., 49
Hams, 231, 326, 327, 356-360
Hancock, William, 81, 166, 236, 253
Hancock House, 253
Hancock's Bridge, 253
Hanson, John, 328, 329
Hardy, Josiah, 62, 101, 137, 145
Harmony Hall, 353
Harness, 55, 368, 369, 378, 379

Hartlib, Samuel, *A Discours of Husbandry . . .* , 250, 254, 267, 420; *His Legacie;* 250, 254, 267, 269, 420; biographical note, 420
Hartshorne, Hugh, 46, 176, 326, 346, 351, 352
Hartshorne, John, 346
Hartshorne, Lucy, 346, 352
Hartshorne, Robert, 302
Hartshorne, William, 352
Harvard University, 104
Harwood, Joseph, 24
Harwood, Sarah, 24
Hassocks, 237-239
Hat making, 56, 327, 385
Hay, 78, 107, 146, 183, 255, 256, 266-278, 322, 333-336, 361, 371, 372, 374; tea, 324
Hedges, 237, 252
Hellebore, 397
Hemp, 146, 157-159, 230, 257-259, 287-290; culture, 287-290; harvesting, 288, 289; seed, 290; soil, 287, 288; treatment 289, 290
Hens, 328, 363
Herbs, 232
Herd grass, xix, 71, 72, 107, 254, 266
Hessians, 176
Hewes, Joseph, 217
Hewlings, Abraham, 246
Hickory, 265, 320
Hickory Grove, 338
Hicks, Austin, 302
Hides, 324, 338
High Street, Burlington, 42
High (Market) Street, Philadelphia, 24, 25
Hillhouse, James, 298
Hillsdown, 307
Historic American Buildings Survey, xxv
Historical Society of Pennsylvania, xiv, xxiv
Hitt, Thomas, *A Treatise of Fruit Trees*, 47, 246, 263, 311-313, 315, 319, 420; biographical note, 420, 421
Hogs, 73, 74, 78, 221, 230, 231, 238, 247, 260, 262, 269, 272, 273, 280, 281, 296, 301, 302, 322, 326, 327, 343-345, 350-363; diseases, 355, 356; farrowing, 355, 356; feeding, 326, 350-355; rooting, 354
Hollingshead, Edmund, 326, 352, 353
Hollingshead, John, 352
Homan, Andrew, 221
Home, Archibald, 97, 405

458 Index

Home, Francis, *The Principles of Agriculture and Vegetation*, 251, 252, 421; biographical note, 421
Home, James, 97
Home economics, 385-398
Hominy, 386
Honey, 366, 367, 391, 400
Honey comb, 367
Hopkins, Robert, 359
Hopkinson, Francis, 215
Hops, 261, 304, 386, 393
Horner, Isaac, 207
Horse-radish, 263, 312
Horses, 56, 74, 83, 105, 108, 169, 221, 230, 247, 269, 272, 276, 322, 323, 328-330, 381-383; dung for hotbeds, 308, 309; racing, 82, 83, 141; teeth, 329, 330; trading, 83
Horticulture, 259-264, 293-319
Hotbeds, 261, 308-310
Hough, Jonathan, xii, xiii, xv, 70, 267
Houghton, John, *A Collection for the Improvement of Husbandry and Trade*, 421; biographical note, 421
Housatonic, 193
House of Commons, 17, 134
House of Lords, 17
Household goods, 408-410; husbandry, 385-398; industries, 55, 56
Huckleberries, 293, 294
Hudson River, 4, 6, 57
Hugg, William, 236, 239
Hugg's Inn, Gloucester, 239
Hunt, Mr., 101
Hunter, Henry, 197
Hunter, Robert, 10, 208
Hunterdon County, 4, 5, 57, 61, 65, 75, 117, 128; agriculture, 229, 230
Husband-man's Guide, The, 47
Husbandry, of animals, 322-365; of bees, 366-367; of the household, 385-398; of plants, 254-321; of the soil, 235-253
Hutchinson, William T., xxvi

Icehouses, 368, 369, 374
Illness, 111, 148, 212-214
Imperial water, 386, 394
Imports, 11
Imprisonment for debt, 140, 151, 152, 206
Independence Hall, 236, 242, 303
Indian corn, 146, 230, 231, 238, 255, 257, 261, 278-280, 283, 286, 293, 294, 296, 304, 337, 361; for hogs, 352
Indian Mills, 184
Indian peas, 305

Indian Treaties Printed by Benjamin Franklin, 51
Indians, 18, 19, 87, 95, 116, 141, 164, 168, 170, 171, 177, 179-194, 223, 366; claims, 141; conferences, 19; policy toward, 18, 19, 189-192; reservation, 19; treaty, 183
Indigo, 283, 284, 290-292
Ingoldsby, Richard, 10
Inoculation, 262, 263, 311, 312, 319
Insects, 314
Inskeep, John, 340
Inter-colonial wars, 164-178
Inventory, 407-411
Ireland, 57, 295
Irish potatoes, 298
Irish settlers, 5
Iron, 57, 67, 417; forges, 86-96; furnaces, 15, 16, 86-96; industry, 49, 87; manor, 15, 90-92; manufacturing, 86-96, 157, 222, 242; mining, 86; ore, 86-96; ware, 42
Ironworks, 15, 16, 25, 53, 86-96, 135
Ising-glass flummery, 389

Jacks, 323, 330
Jackson, H., 399, 403
Jackson, Richard, 403
Jail breaking, 126
Jam, 386
Jamaica ballast, 245
James, Abel, 96
Jaundice, 386, 395
Jersey Magazine, 47
Jockey cap, 397
John of Salisbury, 317
Johnson, James, 328
Johnson, John, 65
Johnston, Andrew, 114, 181
Johnston, David, 302
Johnston, John, 172
Johnston, Lewis, 339
Johnston, Samuel, 217
Jones, Hezekiah, 344
Jones, Nathaniel, 145
Juniata Valley, 179
Junto Club, 29, 303
Justice of peace, 195, 196

Kalm, Peter, 235, 322
Kate Aylesford, 94
Keith, George, 22
Kellar, Herbert A., xxvi
Kelly, William, 230
Kelsey, Jonathan H., xxv
Kemble, George, 379

Index

Kemble, Peter, 80, 148, 199, 201, 236, 244, 245
King George's War, 100
King of Spain, 323
King, Thomas, 187, 188
King, William, 244
Kinsey, James, 69, 200, 395
Kinsey, John, 114
Knollenberg, Bernhard, xvii, xxiv
Knott, David, 331
Knott, Peter, 331

Ladd, John, 132, 148, 189-191
Lafayette, General, 354
Lamberton, 53, 84
Lambs, 362
Lancaster, Pa., 190
Lands, 64-69, 111, 159; claims, 141, 161, 180-183, 186-189, 206; disputes, 8, 9, 17; grants, 12, 13; riots, 98, 116, 126, 127; speculation, 12, 13, 64-69; taxes, 122-126; tenure, 229
Lard, 360, 361
Laufer, Berthold, 295
Laurel, 363
Laurel Hill, 283, 284
Laurence, John, *A New System of Agriculture . . .*, 347, 422; biographical note, 422
Lawrence, John, 69, 148, 216
Lawrence, Robert, 131
Laws, colonial, 134, 135, 195-211
Leach, M. Atherton, xxv
Leaming, Aaron, xxiii, 16, 19, 20, 33, 52, 75, 81, 84, 116, 131, 138, 199, 217, 222, 404; diary, 52, 53, 404-406
Leather, 55, 363, 378, 379
Leaven, 390
Lee, Linwood, xvii, xxiv
Legislation, colonial, 121-163, 185, 195, 223
Leicestershire hogs, 354, 355
Lenni Lenape Indians, 179, 185
Leonard, Thomas, 138
Letter from a Gentleman at Elizabethtown, 192, 193
Letter to Jared Eliot, xi, xvi-xxii
Letter to John Ladd, 189-192
Level, 369, 380, 381
Levi, Isaac, 292
Library, 20, 21, 27, 34, 35, 45-48, 116
Library Company of Burlington, xiv, 46-48, 393
Library Company of Philadelphia, xiv, xxiv, 20, 21, 29, 45, 46, 224, 286
Library of Congress, xxv

Lime, 245, 246, 248, 249, 251, 252, 284, 286, 300
Limes, 394
Limonite, 86-96
Lincoln, Abraham, 278
Lincoln, Mordecai, 278
Lindo, Moses, 291
Linen, 55-57
Linsey, 395
Lippincott, Bertram, xxv
Liquor, 92, 110, 179-181, 184, 187, 188
Lisbon, 57, 59; salt, 246
Little Egg Harbor River, 68
Little Timber Creek, 68
Livestock, 59, 73, 74, 80, 105, 106, 221, 230, 322-365, 407, 408; breeding, 364, 365
Lloyd, Sally, 51
Loan offices, 161, 162
Locust, 265, 320, 376
Locust Hall, 343, 344
Log College, 113
Log stable, 66
Logan, George, 286
Logan, Hannah, 28, 50, 345
Logan, James, 22, 26, 27, 37
Logan, Sarah, 27, 28
Logan, William, xii, xiii, xvi, 27, 46, 49, 81-84, 89, 167, 180, 187, 212, 213, 218, 257, 262, 286, 314, 328, 364
Logan, Mrs. William, 386, 395
Loganian library, 46
Loller, Michael, 334-336
London, 27, 31, 40, 79, 80, 84, 97, 101, 104, 107, 108, 153, 154, 223
London Coffee House, 22-24, 44, 45
London Gazette, 287
Long-a-Coming Tavern, 340
Long Branch, 249
Lords of Trade, 10, 61, 62, 87, 131, 138, 158, 177, 203, 216, 258
Lords of Treasury, 174
Lotteries, 114, 139-143, 160, 400
Loudon, Lord, 172
Lovelace, Lord, 10
Lucerne, 254, 274-277
Lucius Verus, 317
Ludlam, Anthony, 366, 367
Ludlam, Jeremiah, 366, 367
Ludlam, Joseph, 366, 367
Ludlam, Providence, 366, 367
Ludlam, Reuben, 366, 367
Lumber, 49, 115, 116; business, 39
Lumberton, xiii, 66
Lutherans, 5

Index

Lyde, Byfield, 106
Lye, 394, 395

Mackibbinn, John, 75
Madder, 292, 293
Mail service, 7
Makefield Township, Pa., 34
Malaria, 109
Mallow, 259, 293, 294
Malt, 287, 337
Mangers, 374
Manufactures, 11, 56, 112, 116, 151, 224
Manures, 76, 79, 84, 235-252, 261, 266, 267, 304, 307, 310, 315, 319
Maples, 265, 320
Marcus Aurelius, 317
Market days, 58, 59, 82, 83
Market gardening, 259, 266
Markets, 82, 83, 265, 266; in Philadelphia, 35, 36; regulations, 60, 61
Markham, Gervase, *Farewell to Husbandry*, 248, 423; biographical note, 423
Marl, xvii, 71, 237, 250, 252
Marpole, George, 74
Martinburg, N. C., 216, 218
Maryland, 49, 280; cattle, 331
Mason, Samuel, 292
Massachusetts, 104, 137
Massachusetts Historical Society, xxv
Masters, William, 47, 261, 308, 309
Matthers, James, 240, 241
Maxwell, Robert, *Select Transactions*, 252, 295, 298, 299, 331, 397, 423, 424; biographical note, 424
May, Robert, 64
McCormick Historical Association, xxvi
Mead, 386, 391, 392
Meadows, xvii, xix, 79, 161, 237, 239, 266, 267, 318, 362
Meager, Leonard, *The New Art of Gardening*, 47
Meat, 169, 170, 322-365, 385-388
Medford, 90, 200; Friends' cemetery, 220, 221
Medford Lakes, 92
Medicines, 385, 386, 394, 395
Medico-Philosophical Society of Dublin, 427
Melons, 261, 262, 309, 310
Mercer County, 229, 230
Merlet, John, 315
Metheglin, 80, 386, 391
Middle Colonies, 4
Middle Temple, 145, 154

Middlesex County, 4, 57, 117, 133, 161, 196; courthouse, 161
Military operations, 131
Militia, 18, 19, 164-178, 219, 223, 280
Milk, 80, 231, 324, 325, 337, 342-346, 351
Miller, George, xxvi
Miller, Philip, *The Gardener's Dictionary*, 46-48, 252, 301, 350, 361, 424; biographical note, 424
Millet, 78, 255, 257, 283
Milling grain, 257, 286
Milton, 105-107
Minisinks, 186-188, 193
Minsis, 179
Missionary, Swedish, 51; to Indians, 180-183
Mistletoe, 50, 51
Mob assembly, 120
Moffett, Herbert N., xxv
Mohawk River, 69
Mohawks, 188
Mohegans, 185
Molasses, 57, 62
Moles, 386, 398
Monmouth County, 4, 5, 8, 57, 67, 98, 99, 117, 152, 205, 219
Monmouth County Historical Society, xxv
Montgomerie, John, 10
Montgomery, Charles, xxv
Montreal, 176
Moore, Benjamin, 66
Moore, Joseph, xii, xiii, xvi, 76, 261, 307
Moorestown, 220, 337, 353
Moravians, 5
Morgan, Alexander, 240
Morgan, Benjamin, 239, 240
Morgan, Maurice, 153, 154, 156, 157
Morello cherries, 391
Morris, J., 249, 283
Morris, Lewis, 10, 13, 14, 62, 97-100, 164, 196, 258
Morris, Robert Hunter, 20, 86, 97-99, 108, 127, 133, 134, 136, 145, 148, 166, 196-198, 201-203, 249, 404
Morris County, 4, 5, 57, 65, 75, 86, 117, 135, 196, 205
Morrisania, 98
Morristown, 244
Mortimer, John, *The Whole Art of Husbandry*, 241-243, 245, 247-250, 254, 263, 277, 285, 312, 313, 329, 330, 362, 424, 425; biographical note, 425
Moss, 249, 251, 252
Moths, 386, 396, 397
Motto, Read family, 41

Index

Mt. Holly, 14, 65, 76, 219, 220, 237; sawmill, 244
Mt. Holly Herald, xiii
Mt. Kemble, 244
Mt. Laurel Friends' Cemetery, 220; Meeting House, 214
Mud, as a fertilizer, 237-241, 248, 250, 251, 310
Mulberries, 158, 262, 263, 312, 313, 376
Mules, 230, 323, 328-330; breeding, 323
Mullica River, 94
Munitions, 16, 19, 165, 169
Murfin, Mr., 326, 351
Murfin, William, 351
Muscadine grapes, 316
Musconetcong River, 75
Muscovado sugar, 218, 390
Muskmelons, 262
Mutton, 338, 362, 363

Nassau Hall, 21, 115, 116
Naval forces, 164, 166, 167
Naval stores, 54, 57
Navigation, 139, 161
Navy, British, 31
Negroes, 280, 283, 284
Nescochague, 94
Neshaminy, Pa., 113
Nevill, Samuel, 133, 134, 137, 189, 199, 201, 204, 205, 208
New American Magazine, 48, 189
New Brunswick, 6, 86, 115, 146, 162; barracks, 175, 176
New Castle, Del., 166
New England, 4, 94
New Jersey, settlement of, 4-8
New Jersey Archives, xiii
New Jersey Association for Helping the Indians, 182
New Jersey Historical Society, xxiv
New Jersey State Library, xxiv
New Sweden, 7
New York, province of, 10, 168; agriculture, 229
New York City, 6, 7, 11, 51, 55, 57, 58, 94, 95, 105, 165, 171, 310
New York Historical Society, xxv
New York Public Library, xxiv, 190
Newark, N. J., 6, 80, 146, 171
Newbold, Gertrude, xxv
Newbold, Horace, xxv
Newcastle, Duke of, 103
Nichas, 188
Nicolls, Governor, 236

Nicolls patent, 8
Nitrogen-fixation, 261
Noble, Mrs., 50
Noble, Samuel, 42
Nolan, J. Bennett, xxv
Norris, Isaac, 28
North Carolina, 21, 216-218, 404
North Carolina State Library, xxv
Northampton Township, 78, 84, 92
Nova Scotia, 224

Oak, 265, 320
Oats, 71, 229, 231, 268, 276, 287, 303, 310
Odell, Jonathan, 69
Ogden, David, 95, 145, 148, 209
Ogden, Robert, xv, 81, 150, 158, 170, 176, 182, 262, 311, 320, 324, 327, 334, 339, 361, 368, 373
Ogdensburg, 339
Ohio, 222, 224
Okie, John M., xxv
Oliphant, David, 281
Oliphant's Mill, xii, xiii, xiv, 281
Olive oil, 57
Oliver, Mr., 106
Olives, 313
Oneida Indians, 187, 193
Oneida Lake, 185
Onions, 260, 302, 303, 305
Oranges, 264, 318, 319, 346
Orchards, see Fruit Trees
Orleans, 32
Osborn, George A., xxiv
Oswego, 171
Otsego County, N. Y., 69; tract, 69
Oxen, 74, 80, 81, 221, 272, 280
Oxford University, 137, 145, 425, 427
Oyer and terminer, court of, 197, 205
Oysters, 184, 385, 386

Pahaquarry, 332
Painting, 377, 378
Paltsits, Victor Hugo, xxiv
Pancakes, 388
Paoqualin, 332
Paper mills, 55
Paris, Ferdinand, 103, 104
Paris, Peace of, 177
Parker, James, 48, 135, 148, 153
Parliament, 87, 109, 150, 174
Parmesan cheese, 347
Parr, William, 310, 311
Parsley, 277
Parsnips, 260, 301

Partridge, Richard, 97, 101, 103, 104, 136, 149
Partridges, 232
Passaic County Park Commission, 280
Passaic River, 4, 57, 161
Pastures, 254, 255, 266-278, 332-334
Paxton volunteers, 190
Pea worms, 306
Peace, 147
Peaches, 76, 80, 230, 231, 262, 293, 294, 311, 314; borer, 262, 263, 311
Pears, 230, 231, 262, 263, 313, 319
Peas, 84, 232, 252, 259, 261, 287, 293, 294, 305-308, 351, 356
Pemberton, Israel, 22, 28, 44, 49, 183, 187, 188, 190, 216
Pemberton, Israel, Jr., 28
Pemberton, James, 28, 40-42, 64, 101, 189, 198, 214, 218, 219
Pemberton, John, 28, 214
Pemberton, Rachel Read, 28
Pemberton Papers, xiv
Penn, John, 252
Penn, Laetitia, 22
Penn, Richard, 332
Penn, Thomas, 332
Penn, William, 8, 22
Pennsylvania, 14, 19, 69, 95, 145, 168; agriculture, 229; supreme court, 199
Pennsylvania Gazette, 30, 33, 34, 48, 112
Pensauken Creek, 239, 240, 306
Penzance, 31, 32
Penzance Mill, 67
Perkins, Abraham, 306
Perth Amboy, 6, 7, 11, 20, 50, 54, 58, 65, 102, 125, 126, 129, 134, 140, 143, 146, 175, 176, 201, 210
Petersborough, 80, 108, 318
Peterson, Charles J., 94
Pettit, Charles, 157
Petty, William, *Political Survey of Ireland*, 342, 343, 425; biographical note, 425
Philadelphia, 3, 6, 7, 11, 13, 22-24, 29, 31, 32, 38 55, 57, 58, 64, 83, 87, 91, 94, 95, 105, 218, 220, 221
Philadelphia in Eighteenth Century, 35-38, 45; commerce, 35-37; hospital, 303; library, 415, 422, 424; mercantile life, 25; shipping, 36
Philadelphia Academy, 43, 112, 224
Philadelphia Society for Promoting Agriculture, 286
Philhower, Charles A., xxvi
Philips, Abner, 287

Pickersgill, Harold E., xxv
Pierpont Morgan Library, xxiv
Pig-sties, 350
Pigeons, 232
Pine Lane Farm, xxv, 75
Pine land, 65-67
Piracy, 195, 196, 201
Pitt County, N. C., 217
Pittsburgh, 22
Pius, Antoninus, 317
Plant husbandry, 254-321
Plaster of Paris, 238
Plattes, Gabriel, 420
Pleasant Mills, 94
Plows, 80, 251, 369, 380
Pluche, Noel Antoine, *Spectacle de la Nature*, 309, 425, 426; biographical note, 426
Plums, 83, 84, 230, 231, 263, 312-314, 319
Poke, 293, 294; berries, 291
Political parties, 7, 8, 19
Pomona Hall, 333
Pompton Indians, 186-188
Pontiac's War, 189
Population, 4, 146
Pork, 11, 56, 57, 59-61, 80, 231, 296, 326; curing, 327, 356, 357; marketing, 327, 360; packing, 327, 358-360; pickling, 327, 356-360; profits, 360; salt, 359, 360; smoking, 356-359
Portugal, 54
Post, Edward, xxv
Posts, 375-377
Potato bread, 386
Potatoes, 106, 259, 293-300, 326; origin, 295; fed to hogs, 352
Poultry, 59, 106, 230, 232, 247, 248, 327, 328, 363; crossing breeds, 365
Powell, Robert, 67, 68
Powell, Thomas, 48, 137
Prerogative court, 196
Presbyterian Church, Trenton, 290
Presbyterians, 5, 113-116, 161, 184
Price, Robert F., 159
Prices, 143, 144, 159; of cattle, 337, 338
Prickett, Abraham, 67, 68
Princeton, N. J., 146, 171
Princeton (College) University, xxiv, xxv, 21, 49, 113-116, 142, 161
Printing, 48, 134, 135
Prisoners, 171
Proprietors, 7-9, 65, 117, 118
Protecting trees, 313, 314
Prothero, Roland E., 420, 423
Pruning, 262, 313; orange trees, 318

Index

Public debt, 139; funds, 140; offices, 123; records, 140
Pudding beans, 307
Pumpkins, 259, 280, 293, 294, 311
Punch, 386, 387
Puritans, 5, 7, 9

Quails, 232
Quakers, 5-7, 9, 18, 19, 22, 25, 34, 37, 40, 45, 82, 83, 114, 117, 148, 165, 166, 173, 180, 182, 185, 200, 214, 217, 220, 406
Quebec, 171, 176
Queen bee, 367
Queen's College, 21, 161, 162, 280
Quinces, 263, 313
Quit rents, 8, 9, 17, 69, 117, 131

Rabbits, 232, 364; skins, 327, 364
Racoon (New Sweden), 51
Radishes, 79, 259, 293, 294, 301, 388
Rails, 375-377
Rainfall, 80, 383, 384
Rakestraw, Thomas, 79
Raleigh, Sir Walter, 295, 298
Rancocas Creek, xiii, xvi, 13, 66, 76-78, 92, 157, 238, 307
Range cattle, 231
Raritan Bay, 4; River, 4, 6, 15, 57, 183
Raspberries, 83
Rats, 386, 397, 398
Ratsbane, 397
Rawhide, 378, 379
Rawle, Frances, 284
Ray grass, 273
Razors, 41
Read, Alice (daughter of Charles IV), 220
Read, Alice Thibou, 32-34, 39, 40, 111, 212-214, 220; funeral, 213; will, 214
Read, Amy Bond, 23
Read, Andrew, 24, 42
Read, Ann, 220
Read, Charles I, 3, 22; death, 23; will, 23
Read, Charles II, 3, 22-24, 29-35, 404; public official, 26; death, 33
Read, Charles III—Ancestry, 3, 22-26; summary of his life, 3-21
    Youth, 22-38; boyhood in Philadelphia, 23; schooling, 27, 30; relatives, 27, 28; association with Benjamin Franklin, 29; sent to England, 30, 31; in British Navy, 31; visit to Antigua, 31, 32; married Alice Thibou, 32; return to Philadelphia, 33, 34; merchant, 34-36; court clerk, 37; removal to Burlington, 37
    New Jerseyman, 39-53; home life, 39-42; relations with brother James, 43, 44; founder of Burlington library, 45-49; literary interest, 48; member of American Philosophical Society, 49; mercantile ventures, 49, 50; friends, 50-53; characteristics, 53
    Customs collector, 54-63
    Land speculator, 64-69; land purchases and sales, 64, 65; sawmills and timber trade, 66-69; interest in Otsego Tract, 69
    Countryman, 70-85; purchase of Sharon, 70; letter to Jared Eliot, 70-73; farm implements, 73, 74; Sharon sold, 75; farm in Morris County, 75; purchase of Breezy Ridge, 76; experimental farming, 76, 78-80; draining meadows, 77-79; Breezy Ridge sold, 79; farmer friends and correspondents, 80-84; Burlington fairs, 82, 83; horse-trading, 83; fishery, 84, 85
    Ironmaster, 86-96; acquiring lands, 88, 89; Taunton, 90-92; Etna, 92, 93; Batsto, 93, 94; Atsion, 95; disposal of ironworks, 96
    Secretary, 97-120; commissioned, 97, 98; justice of the peace, 98; association with Governor Morris, 99; functioning without a governor, 100; duties as secretary, 101, 102; clerk of council, 102; surrogate, 102, 103; Jonathan Belcher, governor, 103-112; association with Belcher, 111; College of New Jersey at Princeton, 113-115; building of Nassau Hall, 115, 116; land and tax disputes, 116-120
    Legislator, 121-144; elected to assembly, 121; speaker, 121; sponsoring legislation, 122-128; controversy with governor, 129-131; reelected to assembly, 131; currency legislation, 131-133; death of Belcher, 137; commissions renewed by Bernard and Hardy, 138; appointed to council, 138; continued in assembly, 138-144
    Councillor, 145-163; seated in council, 145; counsel for R. H. Morris, 145; association with Governor Franklin, 146, 147; associates in council, 148; member of finance committees, 148, 149; on committee of correspondents, 149; communications on Stamp Act,

Index

149, 150; legislation on legal fees, 152; on committee to investigate robbery of provincial treasury, 152; advocate of union of colonies, 153; loss of secretary's office, 153-156; appeal to Governor Bernard, 154-156; custodian of royal seals, 157; agricultural legislation, 157-159; internal improvements, 159, 161; chartering of Queen's College, 161, 162; influence on legislation, 162, 163

Colonel, 164-178; major in third intercolonial war, 164; commissioner of military supplies, 165, 166; legislative leadership in French and Indian War, 167-169, 171-177; colonel of militia, 169, 170; colonial barracks, 174-176

Indian commissioner, 179-194; first Crosswicks conference, 181; second Crosswicks Conference, 181, 182; Burlington conference, 183; Indian reservation, 182, 183; visit to Brotherton, 184; Easton conference, 186-189; letter on Indian policy, 189-193; advocate of justice for Indians, 193

Jurist, 195-211; court clerk, 196; surrogate, 196; justice of peace, 196; advocate of court reform, 196, 197; offered place on supreme court, 197; acceptance, 198; reappointed, 198; resigned, 199; attorney, 199; partner with Samuel Allinson, 199, 200; commissioner for trying pirates, 201; again appointed to supreme court, 201, 202; named chief justice, 202, 203; associate justice, 204; traveling the circuits, 205, 206; opinion on chancery, 208-210

Exile, 212-225; financial reverses, 212; illness, 212-214; death of wife, 213; death of grandchildren, 214; assignment of estate, 215; departure for West Indies, 215; death, in North Carolina, 216, 217; descendants, 217-222

Biographical sketch, 404-406

Read, Charles IV, xxii, 66, 91-94, 96, 212, 214-217, 219, 220, 281; estate, 221; inventory, 407-411; will, 220, 221
Read, Charles V, 214, 220, 221
Read, Charles (infant son of Charles Read IV), 220
Read, Charles H., 222
Read, Collinson, 28, 29, 43
Read, Deborah, xv, 28, 146
Read, Elizabeth, 220
Read, Honore, 95
Read, Jacob, 39, 212-214, 217-219, 404
Read, James, 24, 26, 28, 34, 43, 44, 83, 84, 89, 220
Read, James (son of Charles III), 39, 213
Read, James Charles, xxv-xxvi, 222
Read, James Logan, xii, xvi, 222, 403
Read, John, 28
Read, Mary, 24
Read, Rachel, 22, 23
Read, Robert, 24
Read, Samuel, 220
Read, Sarah, 22, 23,
Read, Sarah (II), 43, 280
Read, Sarah Harwood, 24, 33, 44
Read, Thomas, 23
Read, William Logan, 220
Read family arms, 41
Read residence in Philadelphia (London Coffee House), 22-24, 44, 45
Reading, John, 100, 137, 138, 176
Reading, Pa., 84, 89
Read's Mill, 66, 90
Read's Mill Creek, 90, 242
Real estate, 64-69
Recipes, 385-395
Red Bank, Battle of, 354
Red House, 345
Reed, Joseph, 153-157, 220
Reed, William, 207, 221, 222
Reeve (Reeves), Captain Peter (?), 41, 254, 278
Religion in colonial New Jersey, 4-7
Repeal of Stamp Act, 150, 151
Reservation, 182-186
Revolutionary War, 16, 51, 96, 164, 219, 220, 224
Rice, 57, 255, 257, 283, 284
Richards, Burnet, 87
Richards, William, 95
Rigden, William, 297, 378
Riots, 7, 117, 126, 127, 130, 131, 152
Roads, 6, 7, 116, 140, 160
Roberts, Hugh, 303, 313, 314
Robins, Aaron, 325, 327, 343, 362
Robinson, P., 83
Rocky Hill, 86
Rodman, John, 100, 393
Rodman, Thomas, 46, 69, 393
Rogers, Katharine E., xxiv
Roller, 73, 369, 370, 380
Rothamsted Experimental Station, xxv, 427
Royal Dublin Society, 335

Index

Royal Society of London, 237, 415, 419, 421, 425; *Philosophical Transactions*, 244, 291, 314, 426
Rum, 34, 62, 392-394
Runaway servants, 74, 78, 240
Rushes, 253
Rutgers University, xii, 21, 161, 162, 234
Rye, 184, 229, 231, 255, 257, 281, 282, 286, 326; feed, 342; for hogs, 351; flour, 287
Rye grass, 73, 254, 268-270, 272, 273, 278, 325

Saddles, 80, 368, 369, 374, 379
Sainfoin, 81, 254, 276, 277
Salad, 259, 293, 294
Salaries of public officials, 204, 205; of the governor, 119, 120, 151
Salem, N. J., 6, 11, 54, 55, 58, 166
Salem County, 4, 57, 196
Salem grass, 73
Salt, 57, 96, 237, 246, 248, 285, 286, 312, 349, 350, 356, 358, 360, 361; industry, 49, 50
Saltar, Hannah Lawrence, 303
Saltar, Lawrence, 95
Saltar, Richard, 81, 86, 95, 104, 136, 148, 176, 181, 199, 201, 202, 204, 260, 278, 300, 303, 320, 327, 346, 361
Saltpeter, 284-286, 356
Sand, 243-246, 250-252, 283
Sandin, Mr., 51
Sassafras, 376
Savoy cabbage, 303
Sawmills, 13, 54, 55, 66, 88, 91, 93, 115, 116
Saws, 67, 68
Schuyler, Adoniah, 302
Schuyler, John, 302
Schuyler, Col. Peter, 80, 86, 108, 138, 165, 166, 168-171, 173, 176, 262, 264, 318, 368
Scotch settlers, 5
Seashore, 249
Secretary of New Jersey, 10, 97-120, 153-155, 198, 201; fees, 102
Semi-tropical fruits, 264
Seneca Indians, 187
Sergeant, Jonathan, 116
Servants, 74, 78, 91, 93, 109, 229, 411
Sewet pudding, 389
Shad, 399
Shade trees, 265, 319-321
Sharon, xv, xvii, xix, 14, 70-75, 237, 267, 279, 331, 334, 338, 343, 345

Sharp Isaac, 166
Shaw, Dr., 99, 326, 351
Shaw, Samuel, 304, 330
Shaw, Thomas, 165, 330
Sheep, 57, 105, 221, 230, 231, 247, 248, 250, 272, 273, 275, 277, 278, 301, 319, 322, 327, 341, 361-365; feeding, 361, 362; rot, 327, 361, 362
Sheepshead, 386
Shelbourne, Earl of, 157, 264
Shelves, 374
Sheppard, William, 47
Sheriff Act, 127, 128
Sherwood, Joseph, 149, 153, 156
Shield, The, 3
Shinn, Solomon, 253
Shipbuilding, 25, 61
Shipping, xv, 11, 24, 49, 59, 60, 216, 266
Shirley, General, 167
Shoemaker, Rebecca, 43, 280
Shoemaker, Thomas, 43
Shrewsbury, 67, 86
Shrub, 386, 394
Shute, Joseph, 283, 284
Silk, 57, 158; culture, 213, 263; filature, 240
Silliman, Eben, 270
Silver plate, 40, 41
Silviculture, 265, 319-321
Sinclair (St. Clair), Sir John, 168, 272, 368
Skeeles, William, 47, 176, 327, 335, 336, 350, 361, 362
Skinner, Stephen, 148, 152
Skins, 57, 61
Skippers, 346
Skunks, 386, 398
Slaves, 4, 15, 41, 45, 74, 91, 93, 220, 221, 229, 230
Sleeper, George M., xiii, xxiv
Smallpox, 100
Smith, Charles, *The Ancient and Present State of the County and City of Cork*, 295, 426; biographical note, 426, 427
Smith, Daniel, 47, 69, 166, 182, 338
Smith, Daniel Doughty, 75
Smith, James, 47, 364
Smith, John, 28, 40, 42, 48-50, 69, 81, 106, 138, 148, 157, 182, 201, 210, 216, 322, 345
Smith, Joseph, 74
Smith, Leonard R., xii, xxiv
Smith, Mary, 334
Smith, Richard, 42, 105, 114, 115, 166, 345

466   Index

Smith, Robert, 116
Smith, Samuel, 3, 47-49, 69, 81, 109, 148, 149, 157, 168, 181, 182, 200, 208-210, 324, 338, 339, 345; *The History of the Colony of Nova-Caesarea, or New Jersey*, 48
Smith, William, 47, 326, 353
Smith, William Lovett, 334, 345, 346
Smith, Mrs. William Lovett, 334, 346
Smuggling, 61, 62
Smyth, Frederick, 148, 204, 209, 210
Smythe, William, 302
Soap, 385, 386, 394, 395; suds, 355
Society for Arts, Manufactures and Commerce, 158
Society for Promoting Christian Knowledge, 181, 182, 185
Society for Promoting Husbandry and Manufactures in Ireland, 335
Society for the Encouragement of Arts, Manufactures and Commerce of England, 417, 418
Society of Country Gentlemen, &c. *The Practical Farrier*, 365, 427
Soiling, 255, 271, 272
Soils, 235-253; for alfalfa, 275; conservation, 237, 238, 283; improvement, 237
Somers, Richard, 297
Somerset County, 4, 5, 57, 205
Soot, 248, 249, 276, 300, 315
Souse (of fish), 387; (of pig's head), 388
South Carolina, xv, 32, 43, 137
South Carolina Gazette, 291
Spain, 54, 164
Spanish naval forces, 166
Spanish potatoes, 296, 298
Sparks, Jared, *The Works of Benjamin Franklin*, xi
Sparta, N. J., 339
Spear grass, 71, 266
Speltz, 255, 281
Spicer, Jacob, 50, 81, 131, 132, 149, 153, 165, 169, 170, 182, 183, 300
Spices, 57
Spinach, 259, 293, 294
Spirits, 392-394
Springfield meadows, 239
Springfield Township, xii, xiii, xv, xvii, 14, 64, 70, 74
St. Croix, xv, 21, 215, 216, 218, 264, 317
St. Mary's Church, Burlington, 39, 142, 279, 304; cemetery, 213
St. Peter's Church, Perth Amboy, 142
Stables, 66, 368, 372-374
Stage coaches, 160

Stamp Act, 19, 45, 149-151, 153; Congress, 150
Stamping mill, 93
Steel industry, 222
Stenton, 27, 81
Steuart, Andrew, 190
Stevens, John, 51, 80, 86, 138, 148, 152, 167, 181, 386
Stevens, Mrs. John, 386
Stevens Institute of Technology, xxv
Stiles, Nicholas, 306
Stilton, 347; cheese, 347
Stirling, Lord, 20, 49, 51, 148, 152, 157, 158, 201, 202, 264, 386
Stockton, Richard, 49, 148, 209
Stokes, Samuel, 326, 353, 354, 398
Strahan, William, 43, 44
Straw, 310
Strawberries, 263
Stray cattle, 78
Sturgeon, 399-403
Success Mills, 67
Suetter, Matthew, *Map of Pennsylvania, New Jersey, &c*, 75
Sugar, 57, 62, 218
Summerill, Thomas C., xxv
Supreme court, 11, 133, 134, 154, 155, 189, 193, 195, 197-211, 223
Surrey, 31
Surrogate, 10, 102, 103, 154-156, 196
Surveys of land grants, 64-69
Susquehanna River, 69
Sussex County, 57, 61, 86, 128, 186, 193, 205, 229, 230
Swamps, 161; see also Cedar swamps
Swedish settlements, 4
Sweet-briar, 252
Sweet potatoes, 79, 259, 260, 295-298
Sweetwater, 94
Swine, see Hogs
Switzer, Mr., 299

Tallow, 57, 324, 337, 338
Tallow candles, 55
Tamarinds, 386, 387
Tanbark, 261, 308
Tannenbaum, Samuel, xvii, xxiv
Tanneries, 55, 56
Tanning leather, 230, 231
Tar River, N. C., 217
Taunton, 15, 88, 90-93, 96
Taverns, 92, 160
Taxes, 116, 118, 122-126, 135, 139, 149, 151, 159, 178, 223, 287; on tea, 151
Taylor, Israel, 307

Index

Tea, 59, 151, 385
Tedyuscung, 182, 187
Tennent, Gilbert, 6, 36, 105, 106
Tennent, William, 113
Thibou, Alice, see Read, Alice Thibou
Thibou, Gabriel, 32
Thibou, Jacob, 31, 32, 34
Thomas, Jonathan, 67
Thomas, Nathaniel, 326, 350, 351
Thompson, Charles A., xi, xii, xxiv
Thompson, Robert T., xxvi
Thorn, Joseph, 297
Thorn, Thomas, 297
Three Brothers, The, 218
Threshing, 136, 266; clover seed, 270, 271; peas, 305-308; wheat, 281
Tibout, Captain, 266, 310
Ticonderoga, 304
Tiebout, Albartus, 310
Tilghman, Hilary, xxv
Timber, 11, 13, 14, 54, 57, 59, 61, 76, 88, 139, 140, 184, 188, 265, 320; lands, 64-69
Timber Creek, 65
Timothy, 73, 76, 79, 254, 255, 266, 273, 274; seed, 274
Tinton Manor, 67, 86, 98
Tobacco, 266, 295
Tonkin, Bathsheba, 348
Tonkin, Edward, xii, xiii, xv, 67, 265, 279, 320, 326, 335, 336, 345, 348, 351
Tonkin, Mrs. Edward, 80
Tonkin, John, 279
Tonkin, Samuel, 279
Tools, 407, 408, 410
Tories, 219, 220
Townsend, Mary, xiv, xxiv
Townshend Acts, 151
Townshend, Heywood, 47
Trade, see Commerce
Travel, 6, 7, 160
Treasury of New Jersey, 121, 125, 136, 148; robbed, 151, 152
Treaties, Indian, 51
Treaty at Easton, 186-189
Trees, 232; felling, 320; transplanting, 313, 319
Trefoil, 254, 210, 217, 218; marsh, 26, 27
Trenchard, George, 80, 272
Trenchard, Mr., 394
Trenton, xvi, 6, 15, 38, 100, 146, 154, 174, 197; Barracks, 18, 175, 176
Trenton State Teachers' College, xxv
Trevascon, 3
Tropical fruits, 318

Troth, Jane, xvi
Troughs, 373
Truck crops, 259, 266
Trying lard, 360, 361
Tull, Jethro, *Horse Hoeing Husbandry*, 276, 299, 370, 427; biographical note, 427, 428
Turkeys, 232, 328, 363
Turner, James, 41
Turnips, 250, 259, 260, 262, 293, 294, 299-301, 310, 314, 337, 350, 361, 388; fly, 300; seed, 300
Turtle, 388
Tusser, Thomas, *Five Hundred Pointes of Good Husbandry*, 356, 385

Unalachtigo, 179
Unami, 179
U. S. Department of Agriculture, xxiv
U. S. Soil Conservation Service, xvii
University of Chicago, xxvi
University of Edinburgh, 421, 424
University of Pennsylvania, xxv, 43
University of Pittsburgh, xxv
Unknown Swamp, 67
Upper Freehold Township, 67
Urine, 249, 284, 285, 300

Van Beverhoudt, Lucas, 230
Vanrow, Doctor, 264, 317
VanSciver family, 79
VanSciver, Joseph B., xxv
VanVeghten, Derrick, 283, 364
Veal, 232, 335, 338, 339, 388
Vegetables, 59, 79, 84, 230, 259-262, 293-311; transplanting, 309, 310
Venison, 108, 232
Vermin, 396, 397
Verree, James, 69, 291, 298
Vetch, 81
Villages, 55, 56
Vinegar, 317
Vines, 315-317
Vineyard, The, 264, 315-317, 422
Vineyards, 158, 315-317
Virginia, 295, 298
Viticulture, 264, 315-317

Waddingham, Samuel, xv, 43, 280
Wager Park, 31
Wager, Sir Charles, 30, 31, 37, 43, 404, 405
Wagons, 221, 369, 377, 378, 407, 410
Waller, W., 354
Walnuts, 107, 311, 319

Index

Wapings, 187
Ward, John Talbot, xxv
Warrell, Joseph, 81, 204, 271
Warren County, 229, 230
Warren, Sir Peter, 119
Washington, George, 154, 220, 233, 323, 413, 416, 419, 427, 428; diaries, 233
Water engine, 79
Water wheels, 271, 369, 381-384
Watermelons, 259, 261, 293, 294
Wax, 57
Wayne, Anthony, 244
Weather, 80, 143, 144
Weaving mills, 55
Weeds, 71, 79, 235, 237, 238, 247, 248, 252, 253, 278, 279
Weights and measures, 384
Weirs, see Fishing weirs
Wens in cattle, 333
Wesley, Charles, 43
West, Charles, 245
West, John, 74
West, Roscoe L., xxv
West India breed of hogs, 355
West Indies, xiv, 21, 31, 32, 36, 57, 59, 62, 164, 214, 215, 222, 323
West New Jersey Society of Proprietors, 113, 114, 118, 300
Westcott, Richard, 88
Weston, Richard, 250, 420
Wetherill, John, 132, 204
Wheat, 136, 158, 159, 229, 231, 238, 255, 257, 281, 285-287, 310; flour, 324; smut, 285
Wheelwrights, 55
Whipping post, 206
Whip-poor-will peas, 305
Whitall, Ann Cooper, 354
Whitall, James, 354
Whitall, Job, 354
White, Barclay, xxv, 74, 75
White, Elizabeth, xii
White, Joseph J., xii, 75
White, Walter R., xxv

White, William, 306
Whitefield, George, 6, 36, 43
Whiteoak leaf, 321
Wickoff, Simon, 125
William I, 116
Willingboro Township, 64
Wilson, John, 94
Windpuff in cattle, 324, 325, 332, 333, 341, 342
Wine, 41, 57, 59, 109, 110, 264, 316, 386, 390, 391
Wood Street, Burlington, 42
Wood, William, 326, 331, 352
Woodbridge, 6, 48
Woodbury, 190
Woodrow, Henry, 68, 169
Woodruff, Mr., 110
Woodruff, Samuel, 148
Woods, Mr., 272
Wool, 55, 231, 361-363, 396, 397
Woolley, Edmund, 236, 242
Woolman, John, 45, 307, 337
Worlidge, John, *Systema Agriculturae*, xii, xv, 244, 246, 247, 250, 255, 260, 263, 265, 269, 270, 273, 276, 281-283, 286, 290, 293, 301-303, 309, 310, 312, 314, 320, 330, 363, 364, 369, 381, 384, 392, 397-400, 428; biographical note, 428
Wright, Joseph, 331
Wright, Sydney L., xxv
Writing Master's Assistant, 48
Wyalusing, Pa., 179
Wyoming Valley, Pa., 179

Yager, Henry, 197
Yale University, xi, xvi, 417
Yard, Joseph, 169, 174-176, 182, 289
Yeast, 389
Yellow fever, 44
Yelverton, Mr., 285
York, Duke of, 7
Young, Arthur, 233
Young, Col. Henry, 166, 200, 300